Waste Recovery and Management

Sustainable development approaches cannot be met unless waste management is addressed as a priority. *Waste Recovery and Management: An Approach Toward Sustainable Development Goals* presents a comprehensive examination of environmental pollution and health hazards caused by differing types of waste, its recycling and other e-waste management strategies, and potential political and legal interventions. It also presents the available carbon-recycling methods and investigates how these might be applied to reinforce waste management in industrialized countries as well as developing and emerging economies. Each chapter includes valuable data and case studies that serve as practical guidance for academicians, researchers, and stakeholders for quantifying the impacts of waste, and for planning integrated solid waste collection and treatment systems, thereby working toward sustainability at a global level.

Features:

- Covers both traditional and new technologies for identifying and categorizing the sources and nature of various types of waste.
- Provides methods for the safe disposal of municipal solid wastes, plastic waste, bio-medical wastes, hazardous wastes, and e-wastes.
- Explains practical measures to cover the broad spectrum of everyday applications of waste management for environmental sustainability.
- Contains a focused discussion of the current scenario and future research directions for different types of waste in each chapter.

Waste Recovery and Management
An Approach Toward Sustainable Development Goals

Edited by
Ajay, Parveen, Ashwini Kumar,
Ravi Kant Mittal, and Rajesh Goel

CRC Press
Taylor & Francis Group
Boca Raton London New York

CRC Press is an imprint of the
Taylor & Francis Group, an **informa** business

Designed cover images: © Shutterstock

First edition published 2023
by CRC Press
6000 Broken Sound Parkway NW, Suite 300, Boca Raton, FL 33487-2742

and by CRC Press
4 Park Square, Milton Park, Abingdon, Oxon, OX14 4RN

CRC Press is an imprint of Taylor & Francis Group, LLC

© 2023 selection and editorial matter, Ajay, Parveen, Ashwini Kumar, Ravi Kant Mittal, and Rajesh Goel; individual chapters, the contributors

Library of Congress Cataloging-in-Publication Data
Names: Ajay, editor. | Kumar, Parveen (Professor of mechanical engineering), editor. |
Kumar, Ashwini (Professor of mechanical engineering), editor. |
Mittal, Ravi Kant, editor. | Goel, Rajesh, editor.
Title: Waste recovery and management : an approach toward sustainable development goals /
edited by Ajay, Parveen, Ashwini Kumar, Ravi Kant
Mittal, and Rajesh Goel.
Description: First edition. | Boca Raton : CRC Press, 2023. |
Includes bibliographical references and index. | Identifiers: LCCN 2022055108 (print) |
LCCN 2022055109 (ebook) | ISBN 9781032281933 (hardback) | ISBN 9781032418032 (paperback) |
ISBN 9781003359784 (ebook)
Subjects: LCSH: Refuse and refuse disposal. | Pollution. | Sustainable engineering.
Classification: LCC TD793 .W385 2023 (print) | LCC TD793 (ebook) |
DDC 628.4/40286–dc23/eng/20230123
LC record available at https://lccn.loc.gov/2022055108
LC ebook record available at https://lccn.loc.gov/2022055109

ISBN: 9781032281933 (hbk)
ISBN: 9781032418032 (pbk)
ISBN: 9781003359784 (ebk)

DOI: 10.1201/9781003359784

Typeset in Times
by codeMantra

Contents

Preface...ix

Acknowledgements xiii

Editors.. xv

Contributors xix

Chapter 1   An Introduction to E-Waste Management: Indian Perspective............ 1

Anubhav Kumar Prasad and Raviranjan Kumar Singh

Chapter 2   Re!turn It Back to Nature: Composting at Home Just Got Easier...... 13

Achintya Lal and Ganesh S. Jadhav

Chapter 3   Sustainable Recent Development Practices for the Plastic
Waste Management: A Review 27

*Madhavi Konni, Bhavya Kavitha Dwarapureddi, and
Manoj Kumar Karnena*

Chapter 4   Review on Sustainable Technologies for Liquid Waste
Treatment and Management...47

Gaydaa Al Zohbi

Chapter 5   A Review: Medical Wastes Management and Disposals: Cases
of MENA Countries 69

*Ghita Amine Benabdallah, Najoua Labjar, Mohamed Dalimi,
and Souad El hajaji*

Chapter 6   A Comprehensive Review on Performance Evaluation Methods
of Biomass Cook Stoves..95

*Chandrika Samal, Ramesh Chandra Nayak,
Manmatha Kumar Roul, Diptikanta Das, and Ramanuj Kumar*

Chapter 7   Performance Evaluation of an Improved
Biomass-Fired Cook Stove: A Comparative Analysis 107

*Chandrika Samal, Ramesh Chandra Nayak,
Manmatha Kumar Roul, Diptikanta Das, and Ramanuj Kumar*

Chapter 8   The Glimmers of Hope: Transforming COVID-19 Medical
Wastes into Value-Added Products as an Immediate Step to
Encounter Medical Waste 121

R. Sharmila, K. Akila, S. Danushri, and M. Zion Mercy

Chapter 9   Control of Transfrontier Movement of Hazardous Waste:
Africa the Final Destination? 145

M. B. Gasu, G. N. Gasu, and E. J. Mbeng

Chapter 10 E-waste Management Strategies and its Opportunities:
Toxic but Beneficial 163

Falguni Shinde, Priyanka Vibhandik, and Khalid Alfatmi

Chapter 11 Performance Evaluation of Concrete Incorporating Admixtures
and Stone Waste Powder 189

Kiran Devi, Babita Saini, and Paratibha Aggarwal

Chapter 12 Feasibility Study on Utilization of Waste Materials in Concrete..... 205

R. Padmapriya, J. S. Sudarsan, and N. Sunmathi

Chapter 13 Decontamination of Pollutants Present in the Total
Environment Using Microorganisms 225

*M. Supreeth, Siddesh V. Siddalingegowda, H. G. Lingaraju, and
Shankramma Kalikeri*

Chapter 14 State of Agro-wastes Management in Nigeria:
Status, Implications, and Way Forward 241

*Toyese Oyegoke, Ayandunmola Folake Oyegoke, Opeoluwa
Olusola Fasanya, and Abdul-Alim Gambo Ibrahim*

Chapter 15 Experimental Investigation of MPFI and GDI
Engines using Gasoline Fuel 273

Ufaith Qadiri

Chapter 16 Utilization of Waste Ceramic Tiles as Coarse
Aggregates in Concrete 291

*Priyanka Dhurvey, Harsh Panthi, Parth Verma, and
Chandra Prakash Gaur*

Chapter 17 E-Waste Generation, Flow, and Management in Eastern
Region of Sri Lanka 303

A. K. Hasith Priyashantha, N. Pratheesh, and P. Pretheeba

Chapter 18 Sustainable Municipal Solid Waste Management through
Membrane Science and Technology...323

R. K. Prajapati, Mohd. Ayub Ansari, and Haider Iqbal

Chapter 19 Study of Ultra-acoustic Behavior of Aspartic Acid in
Water and Aqueous Potassium Sorbate: An Insight into
Interactional Features..343

*Kshirabdhitanaya Dhala, Sulochana Singh, and
Malabika Talukdar*

Chapter 20 Proposal of an Improved Waste Collection System for
Urban Environments 361

J. D. C. da Costa and R. D. S. G. Campilho

Index..387

Preface

Waste disposal ends up in direct and indirect environmental impacts, like land occupation, resource depletion, amplification of worldwide warming thanks to alkane series and different greenhouse emission emissions, water intoxication thanks to landfills, in addition to action and cytotoxic effects from emissions to air within the case of combustion. Direct impacts of waste represent a major, however relatively little, share of global climate change, whereas resource depletion among similar effects is coupled to indirect environmental impacts. Indirect results of waste are coupled with the extraction and process of various resources to provide differing kinds of merchandise whereas specializing in the output instead of the input in several industries. Waste management is outlined because the totally different approaches and procedures designed and enforced to spot and handle the various sorts of waste from generation till disposal. Full implementation of waste management processes, together with waste interference and apply, and employment where doable, has and may additionally facilitates to avoid goodish environmental impacts once assessed from a life-cycle perspective – considering direct effects like emissions and indirect effects like resource depletion.

This book is an outcome of the extensive research accomplished by various researchers, academicians, scientists, and industrialists in waste recovery and management. The area is under-explored, and the outcomes are worth the research effort. Experimentation, modeling and simulation, and management techniques are the powerful tools for developing new concepts, approaches, and solutions to devise valuable information on the process which led to need of compiling this work. Since the information related to waste recovery and management is scattered into patents and research publications and not at one place in a systematic form, editors recognize their ethical responsibility to compile, share, and spread the knowledge accumulated and technology developed, with the students, researchers, and industry people, to draw the benefits of this work in the form of a book and gain technical competence in the frontal area.

This book systematically covers a comprehensive coverage of environmental pollution and health hazards caused by differing types of waste, its recycling and other e-waste management strategies, and required political and legal interventions. It also provides carbon-recycling routes available and investigates how within the long-term they might be applied to reinforce waste management in industrial countries as well as developing and emerging economies. This book consists of 20 chapters that describe "Waste Recovery and its Management in different aspects." Chapter 1, "An Introduction to E-Waste Management: Indian Perspective," summarizes the insight into an introduction to electronic waste and its prevention. Chapter 2, "Re!turn It Back to Nature: Composting at Home Just Got Easier," focuses on the re-utilization of the kitchen waste/wet waste generated in daily households and to find a way of reusing the kitchen/wet waste leveraging it to a cleaner and greener environment. Chapter 3, "Sustainable Recent Development Practices for the Plastic Waste Management: A Review," reports the sustainable plastic waste treatment technologies from the literature available and analyzes the mechanisms. Chapter 4, "Review on Sustainable Technologies for Liquid Waste Treatment and Management," presents the different types of waste water, and the different technologies and methods used in their treatment and management and focuses on the sustainable technologies of liquid waste treatments. Chapter 5, "A Review - Medical Waste Management and Disposals: Cases of MENA Countries," focuses on an analysis of the current practices in medical waste management worldwide, particularly in MENA region and to comparative study of the most waste treatment methods and waste management. Chapter 6, "A Comprehensive Review on Performance Evaluation Methods of Biomass Cook Stoves," emphasizes on a systematic and comprehensive review of internationally developed standardized biomass cook stove evaluation methods. Chapter 7, "Performance Evaluation of an Improved Biomass-Fired Cook Stove: A Comparative Analysis," emphasizes on a systematic and comprehensive review of internationally developed standardized biomass cook stove evaluation methods. Chapter 8, "The Glimmers of Hope: Transforming COVID-19 Medical Wastes into Value-Added Products as an Immediate Step to Encounter Medical Waste," aims at assessing the ability to transform COVID-19 medical waste into beneficial value-added products such as fuel, energy, fabrics, road construction materials, bricks, and other functional outputs. Chapter 9, "Control of Transfrontier Movement of Hazardous Waste: Africa the Final Destination?," critically examines global and national regimes for controlling the trans-border transportation of hazardous waste and its effects on the human environment by assessing the impact of the illegal trade in hazardous waste on Africa. Chapter 10, "E-waste Management Strategies and Its Opportunities: Toxic but Beneficial," focuses on some reviewed work for understanding and analyzing the rate of e-waste generation and recovery of the rare metals from this e-waste. Chapter 11, "Performance Evaluation of Concrete Incorporating Admixtures and Stone Waste Powder," is based on the utilization of stone waste in concreting to improve the performance by minimizing the negative impact on the environment by reducing cement content and cost of construction. Chapter 12, "Feasibility Study on Utilization of Waste Materials in Concrete," focuses on different types of wastes, which have been tried to find alternatives for cement, fine aggregate, and coarse aggregate as a part of the concrete to fetch a better outcome to achieve sustainability, and help in achieving the 3R (Reduce, Reuse, Recycle) principle. Chapter 13, "Decontamination of Pollutants Present in the Total Environment using Microorganisms," reviews emerging pollutants in the environment and microorganisms capable of successfully removing those. Chapter 14, "State of Agro-wastes Management in Nigeria: Status, Implications and Way Forward," attempts to present the trend of annual agricultural production and corresponding annual wastes generated; to review the current practice deployed and its implication on the health, safety, and environment of Nigerian residents; and offer better approach to managing these agro-wastes in an environmentally sustainable manner. Chapter 15, "Experimental Investigation of MPFI and GDI Engines using Gasoline Fuel," is related to the MPFI and GDI spark ignition engines using conventional gasoline. Chapter 16, "Utilization of Waste Ceramic as Coarse Aggregate in Concrete," intends to recycle these tile wastes efficiently as a partial replacement of nominal coarse aggregates with tiles aggregates. Chapter 17, "E-waste Generation, Flow, and Management in Eastern Region of Sri Lanka," attempted to understand the e-waste scenario in Batticaloa District, eastern region of Sri Lanka, via a qualitative research approach. Chapter 18, "Sustainable Municipal Solid Waste Management through Membrane Science and Technology," presents the various membrane technologies,

including separation by pressure-assisted membrane filtration processes, e.g., microfiltration, ultrafiltration (UF), reverse osmosis (RO), and nanofiltration (NF), with a focus
on RO applied at the primary level and the merits and demerits of the different existing
leachate treatments for sustainable MSW management by using these membrane separation technology. Chapter 19, "Study of Ultra-acoustic Behavior of Aspartic Acid in
Water and Aqueous Potassium Sorbate: An Insight into Interactional Features," focuses
on aspartic acid (Asp) and potassium sorbate (PS) that were taken for their research.
Solutions of Asp in water and aqueous PS media with different compositions were
studied at 298.15 K for acoustic investigation. Chapter 20, "Proposal of an Improved
Waste Collection System for Urban Environments," proposes improved equipment for
underground waste disposal, with emphasis to safety devices in accordance with safety
regulations, and automated mechanisms for platform opening and closing.

This book is intended for both academia and industry. The postgraduate students,
Ph.D. students, and researchers in universities and institutions, who are involved in
the areas of sustainable waste recovery and management, will find this compilation
useful.

The editors acknowledge the professional support received from CRC Press and
express their gratitude for this opportunity.

Readers' observations, suggestions, and queries are welcome.

Editors

Dr. Ajay
Mr. Parveen
Dr. Ashwini Kumar
Prof. (Dr.) Ravi Kant Mittal
Mr. Rajesh Goel

Acknowledgements

The editors are grateful to the CRC Press for showing their interest to publish this book in the buzz area of waste recovery and management techniques. The editors express their personal adulation and gratitude to Mr. Joe Clements, (editor), CRC Press, for giving consent to publish our work. He undoubtedly imparted the great and adept experience in terms of systematic and methodical staff who have helped the editors to compile and finalize the manuscript.

The editors wish to thank all the chapter authors to contribute their valuable research and experience to compile this volume. The chapter authors, corresponding author in particular, deserve special acknowledgements for bearing with the editors, who persistently kept bothering them for deadlines, and with their remarks.

The editors, Dr. Ajay, Mr. Parveen, Dr. Ashwani Kumar and Mr. Rajesh Goel, wish to thank Prof. R. K. Mittal for his unreserved guidance, valuable suggestion, and encouragement in nurturing this work. Prof. Mittal is a wonderful person and the epitome of simplicity, forthrightness, and strength and is a role model for editors.

Finally, the editors obligate this work to the divine creator and express their indebtedness to the "ALMIGHTY" for gifting them power to yield their ideas and concepts into substantial manifestation. The editors believe that this book would enlighten the readers about each feature and characteristics of Waste Recovery and Management techniques.

Dr. Ajay
Mr. Parveen
Dr. Ashwini Kumar
Professor (Dr.) Ravi Kant Mittal
Mr. Rajesh Goel

Editors

Dr. Ajay is currently serving as an Associate Professor in the Mechanical Engineering Department, School of Engineering and Technology, JECRC University, Jaipur, Rajasthan, India. He received his Ph.D. in the field of Advanced Manufacturing from Guru Jambheshwar University of Science & Technology, Hisar, India, after B.Tech. (Hons.) and M.Tech. (Distinction) from Maharshi Dayanand University, Rohtak, India. His areas of research include Artificial Intelligence, Materials, Incremental Sheet Forming, Additive Manufacturing, Advanced Manufacturing, Industry 4.0, Waste Management, and Optimization Techniques. He has over 60 publications in international journals of repute, including SCOPUS, Web of Science, and SCI-indexed database and refereed international conferences. He has also co-authored the textbook *Incremental Sheet Forming Technologies: Principles, Merits, Limitations, and Applications* (CRC Press, Taylor and Francis). He has recently edited a book, entitled *Advancements in Additive Manufacturing: Artificial Intelligence, Nature Inspired and Bio-Manufacturing* (Elsevier). He has organized various national and international events, including an international conference on Mechatronics and Artificial Intelligence (ICMAI-2021), as conference chair. He has more than 15 national and international patents in his credit. He has supervised more than eight M.Tech, Ph.D. scholars, and numerous undergraduate projects/thesis. He has a total of 13 years of experience in teaching and research. He is a Guest Editor and Review Editor of reputed journals, including *Frontiers in Sustainability*. He has contributed many international conferences/symposiums as a session chair, expert speaker, and member of editorial board. He has won several proficiency awards during the course of his career, including merit awards, best teacher awards, and so on.

He has been adviser of Association of Engineers and Technocrats (AET) and has also authored many in-house course notes, lab manuals, monographs, and invited chapters in books. He has organized a series of faculty development programs, international conferences, workshops, and seminar for researchers, Ph.D.-, UG- and PG-level students. He teaches the following courses at the graduate and postgraduate level: Additive Manufacturing, Manufacturing Technology, Smart Manufacturing, Advanced Manufacturing Processes, Material Science, CAM, Operations Research, Optimization Techniques, Engineering Mechanics, Computer Graphics, Design of Experiments and Research Methodology, and so on. He is associated with many research, academic, and professional societies in various capacities.

Key links:
Google Scholar: https://scholar.google.co.in/citations?user=TmZS4JIAAAAJ&hl=en
Publons Profile: https://publons.com/researcher/1596469/ajay-kumar/
Research Gate: https://www.researchgate.net/profile/Ajay_Kumar349
ResearcherID: D-5813-2019
ORCID ID: https://orcid.org/0000-0001-7306-1902

Mr. Parveen is currently serving as an Assistant Professor in the Department of Mechanical Engineering, Rawal Institute of Engineering and Technology, Faridabad, Haryana, India. Currently, he is pursuing Ph.D. from National Institute of Technology, Kurukshetra, Haryana, India. He completed his B.Tech. (Hons.) from Kurukshetra University, Kurukshetra, India, and M.Tech. (Distinction) in Manufacturing and Automation from Maharshi Dayanand University, Rohtak, India. His areas of research include Materials, Die-less Forming, Additive Manufacturing, CAD/CAM, and Optimization Techniques. He has over 20 publications in international journals of repute, including SCOPUS, Web of Science, and SCI-indexed database and refereed international conferences. He has five national and international patents in his credit. He has supervised four M.Tech. scholars and numerous undergraduate projects/thesis. He has a total of 12 years of experience in teaching and research. He has organized a series of faculty development programs, workshops, and seminar for researcher and UG-level students. He is associated with many research, academic, and professional societies in various capacities.

Key links:
Google Scholar: https://scholar.google.com/citations?hl=en&authuser=1&user=
    ESQ-RnYAAAAJ
ORCID ID: https://orcid.org/0000-0002-2922-6228

Dr. Ashwini Kumar is currently working as an Associate Professor at the Department of Mechanical Engineering, SGT University, Gurugram, India. Prior to joining the university, Dr. Kumar gained 11 plus years of experience holding different academic and research responsibilities in different organizations. He has completed his research in Optimal Thermohydraulic Performance of Three Sides Artificially Roughened Solar Air Heaters at the National Institute of Technology, Jamshedpur. He received the Doctor of Philosophy from the Department of Mechanical Engineering at the National Institute of Technology, Jamshedpur. He has authored several scientific articles in high-impact journals. His research contribution includes articles on Artificial Roughness, Heat Transfer, Composites, Solar Air Heaters, Coatings, CFD Analysis, Fins and Flow Analysis, and Thermal Optimization Techniques. He has received the Best Session Paper Award at international conferences, 2016 in NIT Patna and "Academic Excellence Award" for the session 2018–2019 by Jayoti Vidyapeeth Women's University, Jaipur, during the international conference ETCDWP-II, 2019, February 12–14, 2019, Jaipur, India. Dr. Kumar is associated with many universities; along with that he is an editorial board member of reputed journals and also a reviewer of international journals (SCI and Scopus Index). He has published a total of 93 research items to date, which include 20 papers in SCI/Scopus, 18 international/national conferences, books, and book chapters with a total Google Scholar citation of 1,150. He has 13 patents in his credit to date. He also looks forward to guiding many research scholars often developing their own interest in the field for which he is an expediter. He believes that Science, Engineering, and Technology are advancing at a fast pace and obsolescence of physical infrastructure, skills, and competence take place rapidly. He is keen on taking steps to improve the existing infrastructure, investment, and intellectual strength wherever they exist and network them so as to utilize them effectively and optimally for meeting changing needs. Not only does he profess the values of Indian culture but he also practices many of the fundamental principles of humanity and society. While dedicating himself to the cause of creating a world-class infrastructure, he has also undertaken much community service in various parts of India out of his quest to contribute more to society.

Key links:
ORCID ID: https://orcid.org/0000-0003-3120-0294
Google Scholar: https://scholar.google.co.in/citations?user=brGZQk0AAAAJ&h
    l=en&authuser=0
Research Gate: https://www.researchgate.net/profile/Dr-Kumar-68
Scopus Author ID: https://www.scopus.com/authid/detail.uri?authorId=
    56420717600

Professor (Dr.) R.K. Mittal began his academic career at the Birla Institute of Technology and Science, Pilani (BITS Pilani), India, one of the top universities of India, in 1975 and retired from there as a Senior Professor in 2018. He headed the offshore campus of BITS Pilani at Dubai, UAE, as Director, and was the Vice-Chancellor of a private university, K.R. Mangalam University, Gurugram, India. In his academic endeavor, he has introduced, developed, and taught a wide spectrum of courses in Mechanical Engineering, Computer Science, and in emerging and inter-disciplinary areas, at the graduate and postgraduate levels, including product design, reliability engineering, control systems, systems modelling, robotics and intelligent systems, Mechatronics, MEMS, and nanotechnology. He is a life member of international professional societies such as IEEE and ACM.

Prof. Mittal is author/co-author of over 85 technical papers and has co-authored popular textbooks, including *Elements of Manufacturing Processes* (PHI Learning); *Robotics & Control* (McGraw-Hill); *Incremental Sheet Forming Technologies: Principles, Merits, Limitations, and Applications* (CRC Press, Taylor and Francis); and edited the book *Advancements in Additive Manufacturing: Artificial Intelligence, Nature Inspired and Bio-Manufacturing* (Elsevier). He edited two international conference proceedings and has also authored many in-house course notes, lab manuals, monographs, and invited chapters in books. Seven students have completed Ph.D.'s under his supervision, and his research interests include Robust Robot Design, Robot-Path Planning, Micro-electromechanical Systems (MEMS), Nanotechnology, Software Engineering, Software Testing, and e-waste. He has guided more than 60 M.E. and M.S. dissertations and numerous undergraduate projects/theses.

At BITS Pilani, he held a series of leadership positions, including Director (Special Projects); Director BITS-Pilani Dubai Campus, Deputy Director Administration; Dean, Academic Registration and Counselling Division, and Founder Chief, Computer Assisted Housekeeping Unit, IEEE Student Branch Counsellor, among others. He was instrumental in establishing the BITS Alumni Association (BITSAA) and was its Founder President. He received his B.E. (Hons.) and M.E. in Mechanical Engineering, and Ph.D. degrees from BITS Pilani, India. He obtained the highest rank in order of merit in M.E. and was awarded the Institute's Gold Medal.

Key links:
ORCID ID: https://orcid.org/0000-0001-6761-0345
Google Scholar: https://scholar.google.com/citations?hl=en&user=rwWOPY8A
    AAAJ
Research Gate: https://www.researchgate.net/profile/R_Mittal3
Scopus Author ID: https://www.scopus.com/authid/detail.uri?authorId= 36869748000

Mr. Rajesh Goel completed Mechanical Engineering at Regional Engineering College (NIT Kurukshetra) in 1998. He started his career as design engineer. He completed his M.S. in Quality Management from Birla Institute of Technology (BITS), Pilani. He continued to grow his industrial career and worked for 14 years in different roles in quality and environmental functions. He worked in business excellence for year and on a project of Business Excellence Sustainability Task (BEST), where the integration of all management systems has been done (QMS + EMS + OHSAS + Social Accountability). The project has been awarded the Siemens Excellence Award.

For industrial technology transfer, he traveled and stayed in Japan for six months in Osaka and Toyama and learned the Japanese language. For continued learning he joined the Senior Management Program at IIM Calcutta in 2012. He has participated in and presented research papers on Integrated Management, Energy Efficiency, Sustainable Development, Sustainable Manufacturing, Behavior-based Safety at numerous international conferences. Two research papers have been published in research journals, and around six papers have been published in conference proceedings.

He has worked in different industries and had leadership roles in different functions in manufacturing. He continues to learn and has joined the PhD program at the Swiss School of Business Management (SSBM) Geneva, completing research on Sustainable Manufacturing for MSME in Indian industries. He is currently working at Sage Metals Ltd as Vice President and heading all Indian operations, handling a revenue of $65MN. He is managing three manufacturing facilities in India along side his continued learning approach.

Contributors

Partibha Aggarwal
National Institute of Technology,
Kurukshetra
Kurukshetra, India

K. Akila
Bishop Heber College
Tiruchirappalli, India

Khalid Alfatmi
SVKM's Institute of Technology
Dhule, India

Mohd. Ayub Ansari
Bipin Bihari College
Jhansi, India

Ghita Amine Benabdallah
Mohammed V University
Rabat, Morocco

R. D. S. G. Campilho
Pólo FEUP
Porto, Portugal

J. D. C. da Costa
Polytechnic of Porto
Porto, Portugal

Mohamed Dalimi
Mohammed V University
Rabat, Morocco

S. Danushri
Bishop Heber College
Tiruchirappalli, India

Diptikanta Das
KIIT Deemed to be University
Bhubaneswar, India

Kiran Devi
Shree Guru Gobind Singh Tricentenary
University
Gurugram, India
National Institute of Technology,
Kurukshetra
Kurukshetra, India

Kshirabdhitanaya Dhala
Siksha 'O' Anusandhan Deemed to be
University
Bhubaneswar, India

Priyanka Dhurvey
Maulana Azad National Institute of
Technology
Bhopal, India

Bhavya Kavitha Dwarapureddi
GITAM Institute of Science, GITAM
(Deemed to be) University
Visakhapatnam, India

Souad El hajaji
Mohammed V University
Rabat, Morocco

Opeoluwa Olusola Fasanya
National Research Institute for
Chemical Technology
Zaria, Nigeria

G. N. Gasu
Redeemer's University
Ede, Nigeria

M. B. Gasu
Osun State University
Osogbo, Nigeria

Chandra Prakash Gaur
Maulana Azad National Institute of
    Technology
Bhopal, India

Abdul-Alim Gambo Ibrahim
Federal University Wukari
Wukari, Nigeria

Haider Iqbal
Bipin Bihari College
Jhansi, India

Ganesh S. Jadhav
Dr. Vishwanath Karad MIT World
    Peace University
Pune, India

Shankramma Kalikeri
JSS Academy of Higher Education &
    Research Mysuru
Mysore, India

Manoj Kumar Karnena
GITAM Institute of Science, GITAM
    (Deemed to be) University
Visakhapatnam, India

Madhavi Konni
GITAM Institute of Science, GITAM
    (Deemed to be) University
Visakhapatnam, India

Ramanuj Kumar
KIIT Deemed to be University
Bhubaneswar, India

Najoua Labjar
Mohammed V University
Rabat, Morocco

Achintya Lal
Dr. Vishwanath Karad MIT World
    Peace University
Pune, India

H. G. Lingaraju
JSS Academy of Higher Education &
    Research Mysuru
Mysore, India

E. J. Mbeng
Osun State University
Osogbo, Nigeria

M. Zion Mercy
Bishop Heber College
Tiruchirappalli, India

Ramesh Chandra Nayak
Synergy Institute of Technology
Bhubaneswar, India

Ayandunmola Folake Oyegoke
Salama Infirmary Hospital & Maternity
Zaria, Nigeria

Toyese Oyegoke
Ahmadu Bello University
Zaria, Nigeria
l'Universite de Lyon
Lyon, France

R. Padmapriya
Sathyabama Institute of Science and
    Technology
Chennai, India

Harsh Panthi
Maulana Azad National Institute of
   Technology
Bhopal, India

R. K. Prajapati
D. J. College Baraut
Baraut Baghpat, India

Anubhav Kumar Prasad
United Institute of Technology
Prayagraj, India

N. Pratheesh
Eastern University Sri Lanka
Chenkalady, Sri Lanka

P. Pretheeba
Eastern University Sri Lanka
Chenkalady, Sri Lanka

A. K. Hasith Priyashantha
Eastern University Sri Lanka
Chenkalady, Sri Lanka

Ufaith Qadiri
Malla Reddy Engineering College
Hyderabad, India

Manmatha Kumar Roul
GITA Autonomous College
Bhubaneswar, India

Babita Saini
National Institute of Technology,
   Kurukshetra
Kurukshetra, India

Chandrika Samal
GITA Autonomous College
Bhubaneswar, India

R. Sharmila
Bishop Heber College
Tiruchirappalli, India

Falguni Shinde
SVKM's Institute of Technology
Dhule, India

Siddesh V. Siddalingegowda
JSS Academy of Higher Education &
   Research Mysuru
Mysore, India

Raviranjan Kumar Singh
Kashi Institute of Technology
Varanasi, India

Sulochana Singh
Siksha 'O' Anusandhan Deemed to be
   University
Bhubaneswar, India

J. S. Sudarsan
National Institute of Construction
   Management and Research
   (NICMAR)
Pune, India

N. Sunmathi
Sathyabama Institute of Science and
   Technology
Chennai, India

M. Supreeth
JSS Academy of Higher Education &
   Research Mysuru
Mysore, India

Malabika Talukdar
Siksha 'O' Anusandhan Deemed to be
  University
Bhubaneswar, India

Parth Verma
Maulana Azad National Institute of
  Technology
Bhopal, India

Priyanka Vibhandik
SVKM's Institute of Technology
Dhule, India

Gaydaa Al Zohbi
Prince Mohammad Bin Fahd University
Al Khobar, Saudi Arabia

1 An Introduction to E-Waste Management
Indian Perspective

Anubhav Kumar Prasad and
Raviranjan Kumar Singh

CONTENTS

Methodology 1
1.1   Introduction 1
     1.1.1   Some of the Common E-Waste Items.. 2
1.2   History 3
1.3   Impact on Society 4
1.4   Reforms... 5
     1.4.1   Measures Taken within India 5
     1.4.2   Measures Taken in Other Countries 6
     1.4.3   Uniform Law 8
1.5   Additional Measures... 9
     1.5.1   Social Awareness.. 9
1.6   Conclusion 10
References.. 10

METHODOLOGY

From around 28 research papers and survey papers, websites have been taken into consideration with government reports.

1.1   INTRODUCTION

The term "e-waste management" or "electronic waste management" can be understood as disposable issue that arises due to electronic items that are discarded as being useful anymore. E-waste does not mean that whatever e-products we do not use creates such problem. It means that either one's rejected product is not meant to be used by some other or that product is simply thrown away as waste or rubbish item. In other words, the quantity of e-waste that is recycled is significantly small as compared to e-waste being generated, and this is what causing this issue to need worldwide concern.

The next point needs to be addressed is that it is the toxic elements [1–3] inside that rubbish product that comes into picture when such product is buried inside ground or

DOI: 10.1201/9781003359784-1

1

TABLE 1.1

Electronic Components and its Effect on Health [4]

Components	Constituents	Affected Body Parts
Printed circuit boards	Lead and cadmium	Nervous system and kidney
Mother boards	Beryllium	Lung and skin
CRT cathode ray tubes	Lead oxide, barium, and cadmium	Heart, liver, and muscles
Switches and flat-screen monitors	Mercury	Brain and skin
Computer	Cadmium	Kidney, liver
Cable insulating	Polyvinyl chloride (PVC)	Immune system
Plastic housing	Bromine	Endocrine system

that toxic element comes in contact with air or water. The scenario's complexity can be understood by the following facts (Table 1.1).

The above data are itself sufficient to prove danger of e-waste, and therefore, reforms began to handle such situation (discussed later). The severity of problem is already at its alarming situation due to rapidly changing technology, which makes even a month-old product looks outdated. The second reason is that many countries are trying to dispose such waste in other countries. This increases the dump size and related health hazards for such countries even more.

1.1.1 SOME OF THE COMMON E-WASTE ITEMS

    a. Home appliances
- Microwaves
- Home entertainment devices
- Heaters
- Fans

    b. Home entertainment devices
- DVDs
- Blu-ray Players
- Televisions
- Video game systems
- Printers

    c. Communications and IT devices
- Cell phones
- Smartphones
- Desktop computers
- Computer monitors
- Laptops
- Circuit boards
- Hard drives

    d. Electronic utilities
- Massage chairs
- Heating pads

- Remote controls
- Television remotes
- Electrical cords
- Lamps
- Smart lights
- Night lights
- Treadmills
- Fitbits
- Smart watches
- Heart monitors
- Diabetic testing equipment

e. Office and medical equipment
- Copiers/printers
- IT server racks
- IT servers
- Cords and cables
- WiFi dongles
- Dialysis machines
- Imaging equipment
- Phone and PBX systems
- Audio and video equipment
- Network hardware (i.e. servers, switches, hubs, etc.)
- Power strips and power supplies

f. Uninterrupted power supplies (UPS systems)
- Power distribution systems (PDUs)
- Autoclave
- Defibrillator

1.2   HISTORY

The Technical or Second Industry Revolution (1870–1920) [3] started the e-waste issue, as this was the period when new technologies started taking place and made older technological products look obsolete and waste. E-waste was then initially dumped into landfills or was incinerated. It took a long time when governments started realising the severity of it. The first step taken was Resource Conservation and Recovery Act (RCRA) [5], taken by US government in 1976. However, historic incidents shook the world and many reforms and laws were made. Khian Sea waste disposal incident (August 31, 1986): The ship was supposed to deliver the e-waste ash of about 14,000 tons from Philadelphia to New Jersey, but at the end, e-waste was disposed to various parts of sea as New Jersey refused to accept the e-waste ash [6]. Over the next 16 months, Khian Sea searched whole Atlantic for a dump site, but could not succeed, as Honduras, Bermuda, Dutch Antilles, Dominican Republic, Honduras, Panama, Bermuda, Guinea Bissau, and the Dutch Antilles refused, and return to Philadelphia was also not possible. At the end, it dumped 4,000 tons of the waste near Gonaïves in Haiti as "topsoil fertiliser". Later, the captain admitted that he dumped the remaining 10,000 tons of waste into the Indian and Atlantic oceans. The Haitian government subsequently banned all waste imports. Local clean-up crews later buried some of the waste in a bunker inland, but the rest remained on the beach [7–9].

Koko, a small fishing village, was in news during 1988, when world come to know that two Italian firms were dumping their toxic wastes into Sunday Nana's vacant yard for $100 per month [10]. The issue was severe as the firm in the name of fertilisers were dumping toxic wastes, which eventually started causing health issues to locals of that village. Later in May 1988, a government-run Nigerian newspaper named "The Daily Times" investigated and found over 2,000 drums, sacks, and containers with some of them having letter R (international symbol indicating toxic and harmful industrial waste), which was the reason for villagers' sickness. After Nigerian government protest and independent survey report of British environmental group that proved the presence of polychlorinated biphenyl (PCB), the Italy government agreed to remove the waste. In mid-July, a barge called Karin B docked in Koko to collect the dented and leaky drums. After it attempted to dock at four different European ports, crew members started complaining of chest pain.

Such events led the way for the Basel Convention in 1989 [11]. This is an international treaty responsible for restricting the movement of hazardous waste between countries, especially between developed and less developed countries.

The cause for the movement of e-wastes is that no country is able to recycle or dispose the e-wastes in an efficient manner. It is estimated that 23% of e-waste generated in developed countries is exported to seven developing countries. Today, about 20% of the e-waste is recycled officially. In India, 95% of the e-waste is recycled by informal sector and the rest by formal sector. Formal and informal sectors are discussed latter in this chapter. US recycled only 15% of the e-waste in 2019 [12]. The rough estimate is of 70%–80% of global e-waste that is actually shipped to landfill sites in developing nations, where it is sorted and sold or burned to extract materials.

In 2019, developed countries shipped around 9.7 million tons of e-waste to developing countries. Agbogbloshie, the world's largest e-waste dump site, annually receives 225,000 tons of e-waste [12].

1.3   IMPACT ON SOCIETY

a. Impact on India

To understand the impact on India, it is important to understand why this problem is arising. India is the third largest e-waste market in the world next to USA and China. The reason for this is unawareness and loopholes in the laws. India in the Hazardous and Other Wastes (Management and Transboundary Movement) Rules, 2016, has allowed the import of refurbishment and re-exportation of second-hand goods. However, a clear-cut distinction between e-waste and second-hand goods is not emphasised. This increases the e-waste even further. The effect of globalisation in itself has created the problem in India with countries like China continuously exporting electronic items. This has affected India in terms of landfills and human health issues. These issues vary from state to state with Delhi at top followed by Greater Mumbai. There are 10 states that contribute to 70% of the total e-waste generated in the country, with 65 cities generating more than 60% of the total e-waste in India [13].

The following human health issues are observed in India [14]:

E-Waste Sources	Constituents	Health Effects
Solder in printed circuit boards, glass panels, and gaskets in computer monitors	Lead	• Damage to central and peripheral nervous systems, blood systems, and kidney damage
		• Adverse effects on brain development of children; causes damage to the circulatory system and kidney
Chip resistors and semi-conductors	Cadmium	• Toxic irreversible effects on human health
		• Accumulates in the kidney and liver
		• Causes neural damage
Relays and switches, and printed circuit boards	Mercury	• Chronic damage to the brain
		• Respiratory and skin disorders due to bioaccumulation in fishes
Galvanised steel plates and decorator or hardener for steel housing	Chromium	• Causes bronchitis
Cabling and computer housing	Plastics and PVC	• Burning produces dioxin that causes reproductive and developmental problems
Electronic equipment and circuit boards	Brominated flame retardants	• Disrupt endocrine system functions
Front panels of CRTs	Barium, phosphorus, and heavy metals	• Cause muscle weakness and damage to the heart, liver, and spleen
Copper wires, printed circuit board tracks	Copper	• Stomach cramps, nausea, liver damage, or Wilson's disease
Nickel–cadmium rechargeable batteries	Nickel	• Allergy of the skin to nickel results in dermatitis, while allergy of the lung to nickel results in asthma
Lithium-ion battery	Lithium	• Lithium can pass into breast milk and may harm a nursing baby, and inhalation of the substance may cause lung oedema
Motherboard	Beryllium	• Carcinogenic (lung cancer) and inhalation of fumes and dust cause chronic beryllium disease or berylliosis

The landfill issue is also of great concern; this is because improper dumping of e-waste is making lands barren.

1.4   REFORMS

1.4.1   MEASURES TAKEN WITHIN INDIA

The first rule called "Management and Handling Rule" was proposed on 14 May 2010 [15]. However, it was introduced in the year 2011; although very late, it provided some rules for producer, consumer, or bulk consumer regarding electronic items. It was basically the reform on Environmental Protection Act 1986. The rule extended its purview to consumables or components or spares or parts of electrical and electronic equipment (EEE), along with their products. The rule worked on the extended producer responsibility (EPR), which is the global best practice to ensure the take-back of the end-of-life products.

To further strengthen the EPR, a new organisation named "Producer Responsibility Organisation" (PRO) was formed in the year 2018, with the following objective: "A professional organisation authorised or financed collectively or individually by producers, which can take the responsibility for collection and channelisation of e-waste generated from the 'end of life' of their products" [16]. The PRO will have consumer or consumers to assist them in meeting the legal obligations. The target for collection and channelisation was set to 20%, and this is increasing by 10% for the next 5 years. The producers have to meet targets, which should be 20% of the waste generated by their sales. This will increase by 10% annually for the next 5 years. The PRO was initially started in E-Waste (Management) Rules, 2016, and later with amendments, it was finalised in the year 2018. The PRO license was supposed to get cancelled if it does not satisfy the criteria mentioned. The PRO also emphasises on the monitoring of the e-waste, such that it should reach the authorised recycler/dismantler.

Another amendment was proposed with the named "Hazardous Wastes" (Management and Handling) in the year 2002, but was later modified and was finally implemented in 2003 [17].

The amendment focused on three prime wastes: bio-medical waste covered under the Bio-Medical Wastes (Management and Handling) Rules, 1998 [18]; wastes covered under Municipal Solid Wastes (Management and Handling) Rules, 2000 [19]; and lead acid batteries covered under the Batteries (Management and Handling), Rules, 2001 [20].

The above amendment emphasised on proper monitoring of handling or recycling of hazardous waste by an occupier. In other words, this amendment is for any person who intends to be an operator for the collection, reception, treatment, transport, storage, and disposable of hazardous wastes.

Apart from this, many other reforms/amendments were introduced. But the main point is that all such rules and amendments are still not strictly followed. The informal sector is still handling about 80% of the e-waste.

As per the reports of ASSOCHAM (Associated Chambers of Commerce and Industry of India), in India, the e-waste is increasing at a compound rate of 30% each year.

1.4.2 Measures Taken in Other Countries

As a measure to tackle this situation, developed countries like America, Australia, and others have introduced many laws and amendments for e-waste management. Let's start with America first.

USA brought the first law on 20 October, 1965, named "Solid Waste Disposal Act" (SWDA) [21]. The United States Environmental Protection Agency described it as the first effort to improve waste disposal technology [22].

This was to provide a framework for states to make effective control over solid waste disposal, and to define minimum safety requirements for landfills. However, during 1976, American Congress figured out that the provisions made were not capable enough to properly manage the nation's waste, and as a measure to it, Resource Conservation and Recovery Act (RCRA) in 1976 [23] was introduced. Resource Recovery Act in 1970 was passed in American Congress to emphasis on recycling and energy recovery rather than disposal of e-wastes.

The US EPA was formed in the interim, which issued the final report to Congress: Disposal of Hazardous Wastes.

The Hazardous and Solid Waste Amendment [24] was regulated in the year 1984. It directed EPA to revise the criteria for landfills receiving hazardous household waste. It focused on the treatment of contaminated water running off the landfills. Likewise, other suggestions were implemented in this amendment. Let's see the works of Australia. The National Waste Policy, 2009 [25], of Australia provides a framework for waste and resource recovery. It highlighted on bringing together businesses, governments, communities, and individuals for waste and resource recovery. It highlighted the five key points:

- Avoid waste
- Improve resource recovery
- Increase use of recycled material and build demand and markets for recycled products
- Better manage material flows to benefit human health, the environment, and the economy
- Improve information to support innovation, guide investment, and enable informed consumer decisions

An update was made on 2018. In 2019, National Waste Action Plan pointed out seven targets:

- Regulate waste exports
- Reduce total waste generated by 10% per person by 2030
- Recover 80% of all waste by 2030
- Significantly increase the use of recycled content by governments and industry
- Phase out problematic and unnecessary plastics by 2025
- Halve the amount of organic waste sent to landfill by 2030
- Provide data to support better decisions

Let's move to European Union.

The European Union's policy [26] treats waste as a resource making of a European recycling society in which the Member States can develop autonomous waste elimination systems.

Waste Framework Directive, 2018, provides definitions to key concepts like waste, recycle, and prevention. It defined prevention in two aspects: prevention of waste and prevention of harmful effects of wastes.

Polluter-pays principle aims for prevention and remedy environmental damage.

Extended producer responsibility strengthens the re-use and the prevention, recycling, and other recovery of waste.

The **Waste Shipment Regulation** supervises and make control of such shipments which is used for the movement of e-waste with Union and outside Union to help protect both human health and the environment. It is basically the inclusion of the Basel Convention.

Besides this, there are **Industrial Emissions Directive (Waste Incineration)** and **Landfill Directive** to achieve high level of human health and environmental protection.

Including all country's reforms and laws is not possible, and the authors have tried to provide some of the developed countries scenario. The reason for not emphasising on developing countries is that they import e-wastes also, and this disrupts their laws and reforms.

1.4.3 UNIFORM LAW

The need for Uniform Law is to handle the movement of e-waste between countries, especially between developed and developing countries. The reason for uniform law dated back to when Khian Sea, 1986, and Koko (Nigerian Village), 1988, incidents took place, leading to Basel Convention. Till now, 78 countries of the world have agreed on global legislation. This is significant improvement from 44% of the population in 2014. The International E-Waste Day is observed on 14 October every year since 2018. The main purposes of global legislation towards e-waste management are the following:

- To ensure the establishment of a clear legal framework to manage e-waste collection and recycling.
- To facilitate EPR for ensuring that the producers are financing the collection and recycling of e-waste.
- To help countries for experienced recyclers for bringing the required technical expertise.
- To have international standards enforced for licensing system collection and recycling.
- The use of incentives to facilitate informal collections to be sent to licensed recyclers.
- Internationally licensed treatment facilities must be used in case a country is not able to set up such facility within country.
- Ensure that costs to run the system are transparent, and stimulate competition in the collection and recycling system to drive cost-effectiveness.
- To make sure stakeholders know the impact of e-waste on environment and also the necessity of e-waste collection and recycling.
- To make consumers aware of the environmental benefits of recycling.

Apart from that other organisation such as the International Solid Waste Association (ISWA), the International Telecommunication Union (ITU) and the United Nations University – SCYCLE programme (UNU-SCYLE) joined forces to create the

Global E-waste Statistics Partnership in close collaboration with the United Nations Environment Programme (UNEP) in order to create the Global E-waste Statistics Partnership as a way of addressing the challenges associated with managing e-waste in 2017.

1.5   ADDITIONAL MEASURES

1.5.1   SOCIAL AWARENESS

The reasons for social awareness as a key attribute towards e-waste management are as follows:

- Importance of recycling
- Importance to minimum e-waste generation (frequently changing of electronic items)
- Human health and environment issues
- Training of informal sector

Importance of Recycling: It is important for the following reasons:

- It is critical to keep electronic waste out of landfills.
- Reclaiming of precious metals like gold, silver, and platinum for e-waste.
- Reclaiming of valuable materials will result in decreased demand of raw materials.
- Using recycled material will decrease greenhouse gas emission.
- Reuse, recycle, or refurbished e-waste will eventually decrease landfills.

Importance to minimum e-waste generation: Minimum e-waste generation is directly related to checking of frequently changing of electronic items. If necessary, the previous one should be donated, recycled, or used for refurbishing purpose. The impact of globalisation is making many companies to use India as their prime market. This has led to products with frequent technological changes, tempting people to change their products frequently.

Human health and environment issues: Damaging environment will eventually harm us, and this needs to be communicated to each and every individual. The health issues are critical and must be known by every citizen; their ignorance towards e-waste generation and management can result in drastic changes to human race.

Training of informal sector: India e-waste collection and management is mainly done by informal sector. About 90% of e-waste was managed by this sector. It is true that this sector is providing livelihood for many and contributes to high collection rates; the problem is with their substandard working style and lack for proper expertise persons. This can pose, and is posing, major human health and the environment issues, as well as the loss of valuable and scarce materials. So, it is of prime importance that formal and informal sector should work together.

1.6 CONCLUSION

The biggest reason for e-waste generation can be said to technological evolution (Second Industrial Revolution). The second point is globalisation, which is providing many options for the same product; this indeed tempts person to change their electronic items at a faster rate. The changing of items frequently needs to be properly managed by recycling, reuse, or refurbishment of such items. Recycling not only decreases e-waste size, but also helps in providing valuable materials like gold, silver, and platinum. This also helps in recovering valuable materials, which in return reduces demand for raw materials.

Landfills used for e-waste dumping is another challenge, and this should be done when proper recycling and reuse still lefts e-waste. To make people aware of the cons of e-waste in terms of health issue and environment issue is equally important.

In India, Gujarat is ranked number 1 in e-waste recycling, and many recycling plants in India and other countries, especially developed countries, are being getting set up with expertise guidance. The governments are working on both fronts: recycling and proper storage of e-waste and awareness among citizens.

The Basel Convention pointed out on the movement of e-waste among countries, but still e-wastes are getting stored in developing countries either legally or illegally. In India also, this is the issue, and this is mainly because of the loopholes (differentiation between waste and second-hand goods) and ignorance.

The challenges related to it can be presented in the following points [27]:

- The electronic devices are thrown away and not given for recycling nor donating
- Involvement of child labour
- Ineffective legislation
- Health hazards
- Lack of incentive schemes
- E-waste imports
- Reluctance of authorities involved
- Security implications

The responses and management measures to be followed are given below:

- E-waste recycling practices
- Adopting of a circular economy
- Urban mining
- Research and data gathering
- Policies and regulations

REFERENCES

1. Dahl R. Who pays for e-junk? *Environ Health Perspect.* 2002;110:A196–A19C.

2. CPCB. *Guidelines for Environmentally Sound Management of E-Waste* (As approved vide MoEF letter No. 23-23/2007-HSMD). Delhi: Ministry of Environment and Forests, Central Pollution Control Board, March 2008. Available from: http://www.cpcb.nic.in.

3. Baud I, Grafakos S, Hordjik M, Post J. Quality of life and alliances in solid waste management. *Cities*. 2001;18:3–12.

4. http://www.legalservicesindia.com/article/2249/E-Waste-Management-IssuesChallenges-and-Proposed-Solutions.html.

5. https://en.wikipedia.org/wiki/Second_Industrial_Revolution#:~:text=The%20 Second%20 Industrial%20Revolution%2C%20also, into%20the%20early%2020th%20 century.

6. https://www.epa.gov/rcra.

7. https://www.linkedin.com/pulse/history-e-waste-karan-thakkar#:~:text=The%20 first%20step%20taken%20towards, waste%20in%20less%20developed%20countries.

8. Cunningham WP, Mary A. *Principles of Environmental Science*. Boston, MA: McGraw-Hill Further Education. 2004. ISBN 0072919833.

9. Knight D. *U.S. Toxic Waste to Be Returned to Sender*. Rome: Inter-Press Agency. 1998. Retrieved 6 April 2013.

10. Leonard A. *The Story of Stuff*. New York: Simon & Schuster. 2010, pp. 224–226. ISBN 9781-4391-2566-3.

11. https://timeline.com/koko-nigeria-italy-toxic-waste-159a6487b5aa.

12. http://www.basel.int/TheConvention/Overview/tabid/1271/Default.aspx.

13. https://theprint.in/india/pm-calls-for-week-long-garbage-free-country-but-india-is-the-worlds-highest-waste-generator/478889/.

14. Puckett J, Byster L, Westervelt S, Gutierrez R, Davis S, Hussain A, et al. *Exporting Harm: The High-Tech Trashing of Asia*. Seattle: Basal Action Network. Available from: http://www.ban.org.

15. https://www.meity.gov.in/writereaddata/files/1035e_eng.pdf.

16. https://cpcb.nic.in/uploads/Projects/E-Waste/Guidelines_for_PRO_23.05.2018.pdf.

17. http://extwprlegs1.fao.org/docs/pdf/ind40674.pdf.

18. https://dhr.gov.in/sites/default/files/Biomedical_Waste_Management_Rules_2016.pdf.

19. https://cpcb.nic.in/municipal-solid-waste-rules/.

20. https://www.mpcb.gov.in/sites/default/files/batteries/BatteriesAnnualreport201920 MPCB24052021.pdf

21. https://www.govinfo.gov/content/pkg/COMPS-893/pdf/COMPS-893.pdf.

22. EPA. Solid Waste Management on Tribal Lands - The Solid Waste Disposal Act (SWDA) of 1965. United States Environmental Protection Agency. Retrieved 24 January 2014, https://19january2017snapshot.epa.gov/www3/region9/waste/tribal/reg.html.

23. https://www.epa.gov/laws-regulations/summary-resource-conservation-and-recovery-act.

24. https://www.congress.gov/bill/98th-congress/house-bill/2867.

25. https://www.awe.gov.au/environment/protection/waste/how-we-manage-waste/national-waste-policy.

26. https://www.etui.org/topics/health-safety-working-conditions/hesamag/waste-and-recycling-workers-at-risk/eu-waste-legislation-current-situation-and-future-developments.

27. https://www.drishtiias.com/daily-updates/daily-news-analysis/e-waste-generation.

2 Re!turn It Back to Nature
Composting at Home Just Got Easier

Achintya Lal and Ganesh S. Jadhav

CONTENTS

2.1 Introduction 13
2.2 Methodology and Results 18
    2.2.1 Field Study 18
    2.2.2 Problems 18
    2.2.3 Concept Generation 18
    2.2.4 Screening of Concepts 18
    2.2.5 Design Development... 19
    2.2.6 Validation of the Proposed Design Intervention
        with the Stakeholders.. 23
2.3 Discussion... 23
2.4 Conclusion 24
References... 25

2.1 INTRODUCTION

Waste management using the right methods is becoming the biggest modern-day problem throughout the globe [1]. With the beginning of the disposable material culture, mankind is just using resources and disposing of them without giving them a second thought and the adverse effects it could have on the planet. With the increase in urbanization, there is a need to pay more attention to ideal waste management. Figure 2.1 represents a majority of the waste being disposed of in open areas.

Presently, in households, people tend to just collect waste. It is handed over to the municipality trucks. In some upper-end societies, people are asked to segregate the waste into wet and dry waste also known as "biodegradable" and "non-biodegradable waste", respectively. In some metropolitan cities, it was observed during the field survey, the majority of the citizens are mixing dry and wet waste. But at the end, even the segregated waste is handed over to the municipality trucks. It is often noticed that the waste is mixed and, later, is used as a landfill. With rapid urbanization, there are all sorts of waste being generated and the consistency has multiplied significantly [2].

Examples of waste are municipal solid waste, industrial waste, mining hazardous waste, agriculture waste, medical waste, packaging waste, e-waste, construction

FIGURE 2.1   Waste being disposed of in open areas.

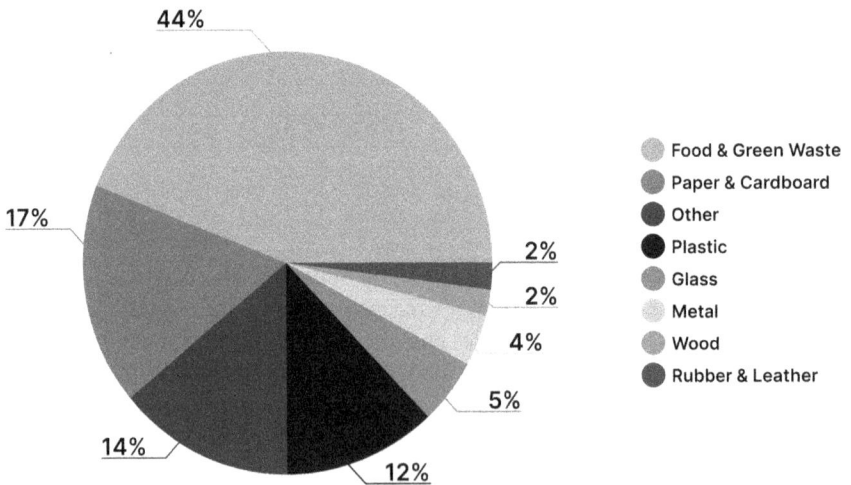

FIGURE 2.2   Quantity and types of waste being generated.

and demolition waste, etc. [2]. They are all used as landfills. Waste like plastic takes approximately 1,000 years to decompose naturally [3].

Amongst all, this waste food and green waste form the majority. 44% of the world waste is food and green waste, and 17% is paper and cardboard, which is biodegradable. Hence, 61% of the world waste is fit to be recycled and reused [4] (Figure 2.2).

At the current rate, the increase in the total world waste is estimated to grow by 13% by 2030 and by 17% by 2050 (Figure 2.3).

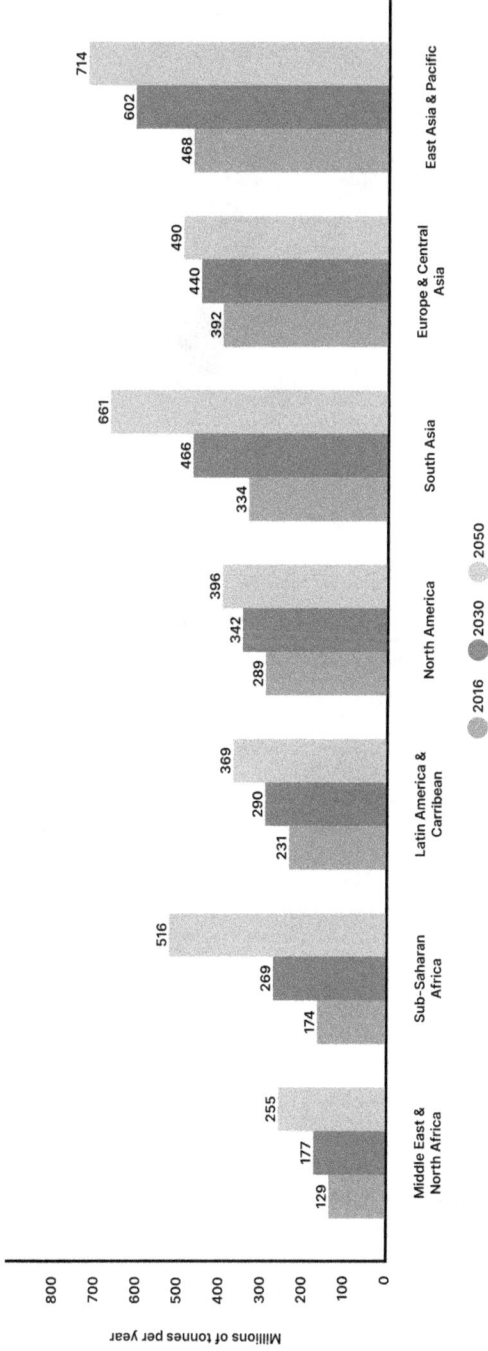

FIGURE 2.3 Estimation of waste generated by 2050.

FIGURE 2.4   Kitchen waste.

FIGURE 2.5   Insects attracted to the kitchen.

Along with all these drawbacks, kitchen waste tends to smell pungent. This smell is not welcoming for anyone into a place where food is prepared (Figure 2.4). At the same time, it is observed that flies and other insects are also attracted to the kitchen (Figure 2.5). This may lead to the contamination of food. People often put air fresheners and insect poison to avoid such instances. But to an extent, all this will also affect food. Air fresheners and insect poison are forms of chemicals that are fatal. Keeping these substances in the kitchen would also affect the quality of food.

Composting was initially termed as a "basic farming activity" but with the awareness in the twenty-first century [1]. Composting cannot be forgotten to be considered as a recourse to a sustainable future. Home gardeners, nowadays, tend to buy fertilizers from the market but those need not be solely organic. Ergo composting can be used as a common treatment for kitchen waste, which is biodegradable, which means it can break down biologically to create fertilizers to boost the quality of soil significantly [1].

Kitchen waste is all organic. It consists of fruit and vegetable peels, seeds, tea leaves, tissue papers, rotten/stale food, eggshells, plant and animal remains, etc. Concisely, they have an animal or plant origin. In consequence, they are potential sources of organic matter and plant nutrients. If these are utilized well, they are great sources for soil fertility [5].

Many kitchen gardeners have an empty flower pot to store the kitchen waste to create compost. This process would naturally take approximately 2 weeks for the waste to break down and be ready as a fertilizer.

Composting allows us to recycle our general kitchen waste along with the potential to contribute to an ecological and sustainable environment [5]. It is a solution readily available to all without a levy promising an enthralling outcome. Creating compost at home is an easy solution.

This study focuses on the disposal of kitchen waste sustainably. The solution provided in this study comprises the complete disposal of kitchen waste with the help of compost bins. This creates a good opportunity of creating organic waste at an individual's home (Figure 2.6).

FIGURE 2.6 Disposal of kitchen waste.

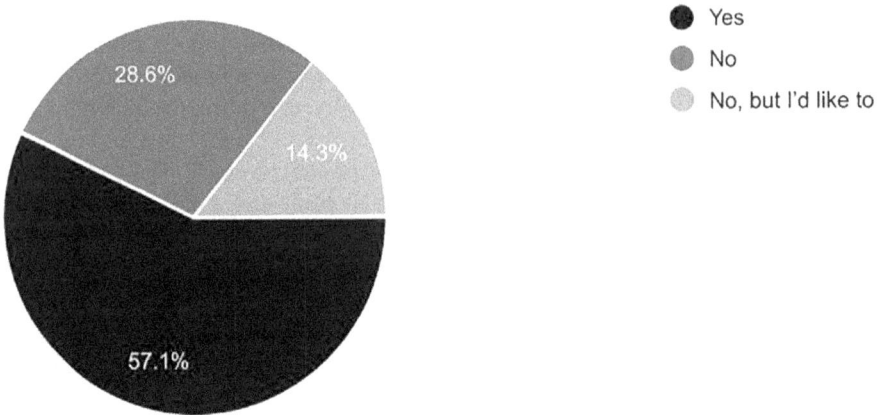

FIGURE 2.7   Chart of number of people creating compost at home.

2.2   METHODOLOGY AND RESULTS

2.2.1   FIELD STUDY

A short survey was carried out amongst a community. This was to get an idea of the people creating compost at home or the people interested in doing so (Figure 2.7).

Amongst 100 people, 57 create compost at home. Out of these 57 people, 34 are from the age group of 18–25.

Inclusive Criteria: Wet waste generated in households.

2.2.2   PROBLEMS

During the field survey, it was observed that the problem faced by the majority of the people was that the container used would smell pungent. This smell would attract flies and insects that are an annoyance at home. The container would take up a lot of space at home. Additionally, the smell is not welcoming to be kept in indoor conditions at home. A balcony or an open space is required to do so. It was also a complaint that it is a long and time-consuming process taking an average of 2 weeks.

2.2.3   CONCEPT GENERATION

After the survey, some concepts were created, which tackled some problems. A concept was finalized, which solved most of the problems and complaints (Figure 2.8).

2.2.4   SCREENING OF CONCEPTS

All these concepts were compared using a Pugh Chart and are presented in Table 2.1. The most optimized concept was selected.

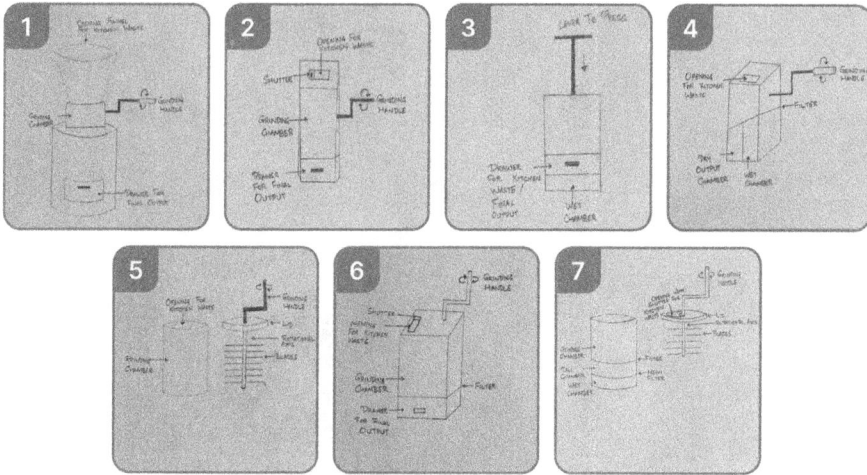

FIGURE 2.8 Concepts of compost bins.

TABLE 2.1
Pugh Chart

Criteria	Space Required (Out of 10)	Smell Prevention (Out of 10)	Final Output (Out of 10)	Total Score (Out of 30)
Concept 1	2	1	4	7
Concept 2	5	7	4	16
Concept 3	4	8	6	18
Concept 4	7	7	6	20
Concept 5	8	3	2	13
Concept 6	5	7	5	17
Concept 7	8	9	8	**23**

2.2.5 Design Development

Working: Re!turn comprises three chambers, namely the grinding chamber, the drying chamber, and the wet chamber (Figures 2.9–2.14).

The first chamber is the grinding chamber where the waste is put in via the opening on top. It has blades set-up on a rotating axis working by rotating the lever on top, which will break the waste down into smaller bits. There is a mesh separating this chamber from the second chamber that is the drying chamber to maintain the consistency of the particles. The second chamber is where the final product is collected but is meant to be left there for a while for the excess water to percolate to the third chamber, which is the wet chamber leading to a dry compost output in the second chamber.

FIGURE 2.9   My design: Re!turn.

FIGURE 2.10   Re!turn front view.

LEFT RIGHT

FIGURE 2.11 Re!turn side views.

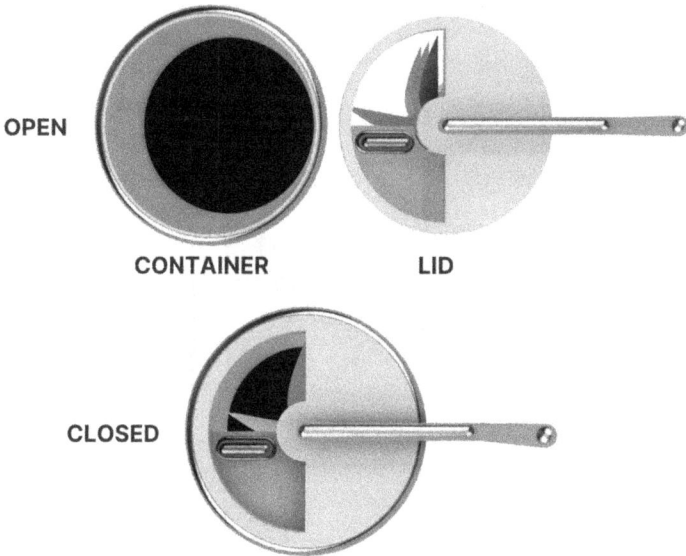

OPEN

CONTAINER LID

CLOSED

FIGURE 2.12 Re!turn top view.

FIGURE 2.13   Re!turn bottom view.

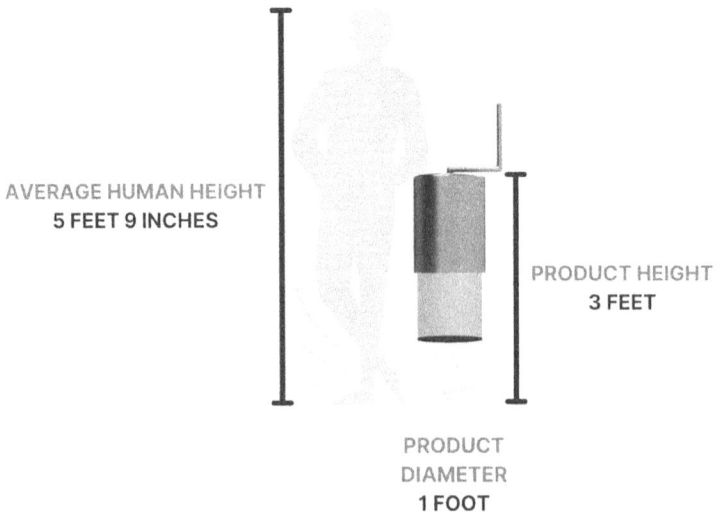

AVERAGE HUMAN HEIGHT
5 FEET 9 INCHES

PRODUCT HEIGHT
3 FEET

PRODUCT
DIAMETER
1 FOOT

FIGURE 2.14   Concept along with human figure with dimensions.

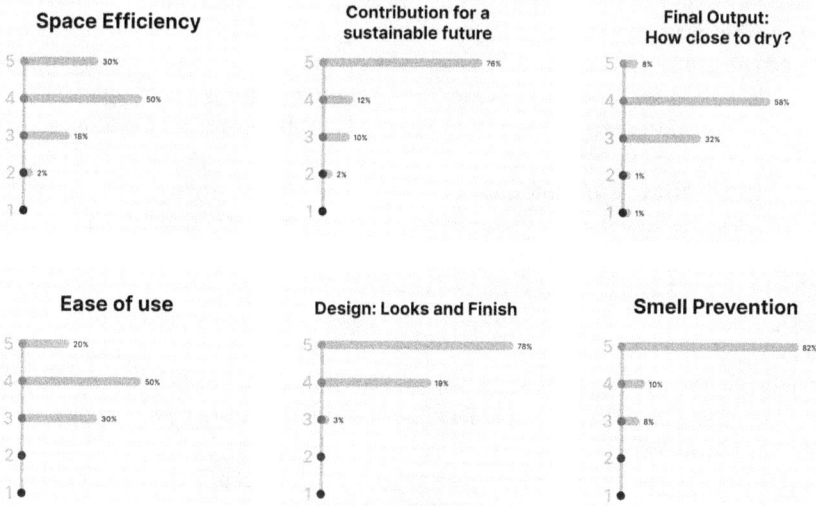

FIGURE 2.15 People's rating for the product in the parameters.

Output: The output is dry compost, which is collected in the second chamber of the bin after the straining of water. The second chamber is lined with a bag made of muslin cloth. This prevents one from touching the compost and is ready for use.

Modelling software: The CAD model is modelled and rendered on Autodesk Fusion 360.

2.2.6 VALIDATION OF THE PROPOSED DESIGN INTERVENTION WITH THE STAKEHOLDERS

A survey was conducted for stakeholders to rate the product on the given parameters, which were space efficiency, a contribution for a sustainable future, final output: how close to dry was the output, ease of use, design: looks and finish and smell prevention (Figure 2.15). One questionnaire was formed and based on which responses were recorded. The majority of the stakeholders have shown their wide acceptance of the proposed design intervention.

2.3 DISCUSSION

It was observed that with time, there was a significant increase in the quantity of waste being generated and most of it is biodegradable, which can be a huge reason for promoting composting in households.

As suggested, prevention and reduction of waste are important. But recovery and reutilization of food waste/kitchen waste and organics contribute to a sustainable future. Reusing kitchen waste forms the majority of the waste being discarded or being used as landfill in the municipality waste. This serves as an opportunity for recycling this waste into compost, animal feed, or biofuel [6]. Many places have developed organics collection programs rather than putting a ban on landfilling. They collect food waste from specific parts of a town and is later separated, collected, and transported to a centralized facility [7]. This creates a low-cost source-separated feedstock [8].

The biological cycle of nutrients is irreplaceable for all living organisms. The kitchen/wet waste found in our households is biodegradable. Biodegradation is the segmentation of nutrients into simpler parts. This may not be a simple fragmentation but also be termed as the creation of new compounds by other atoms [9]. Composting seems to complete this cycle in a simpler way. Compost contains most of the essential plant nutrients in an organic form [10]. During the creation of compost, nutrients take time to break down to reach the stage where they can readily be absorbed by the plant roots in comparison with inorganic fertilizers, hence increasing the composition of organic matter in the soil.

People have started to upcycle the waste as in creating chairs from the traditional furniture making joinery techniques [11].

During the field survey, it was observed that the average time to create compost at home out of kitchen waste took around 2 weeks. Re!turn is a compost bin that speeds up this process efficiently in approximately 3 days. It has a grinder that breaks down the particles of our daily kitchen, which makes it easier to be incorporated by the soil. This process being carried out at homes can reduce the generation of waste by putting the waste to use, promoting a greener environment. A greener environment inculcates a cleaner atmosphere as there will be lesser waste to be put up to be disposed of in the open, by the municipality trucks, or as a landfill.

Creating compost at home is inexpensive as it can be the utilization of the waste being generated at home. Composting reduces the manure volume/inorganic fertilizers volume by 30%–50%, making it more affordable and providing a much healthier soil output [8].

2.4 CONCLUSION

The kitchen waste, which is generated in cities, is often handed over to the municipality trucks, which is either burned in small cities or is used as a landfill in metropolitan cities. In some of the higher-end societies, citizens are often asked to keep the waste segregated but only some follow. This creates unorganized waste, which is all dumped into the soil. Many socially aware individuals have also started to make their own compost at home by collecting the kitchen waste in empty flower pots and letting it biologically degrade over a span of weeks. Handing over waste to the municipality trucks and letting them treat it and later buying fertilizers for the home gardens may not be totally organic. It may contain some traces of inorganic fertilizers, which may not be a good decision in a long term. Creating compost at home can be termed as the most convenient and sustainable alternative to reusing waste, contributing to a greener future.

With the implementation of the proposed design, every household will be able to create their compost at home promoting home gardening without any additional fertilizers. This would, hence, reduce the amount of kitchen waste being generated and dumped as municipal landfills, leading to the reutilization of kitchen waste in daily households.

Limitations in the study: A CAD model was made as a physical prototype could not be made. The long-term behaviour of the product could not be analysed due to time constraints in the building of a physical prototype.

REFERENCES

1. Akinbile, C. O., & Yusoff, M. S. (2012). Solid waste generation and decomposition using compost bin technique in Pulau Pinang, Malaysia. *Waste Management & Research*, *30*(5), 498–505.
2. Zuberi, M. J. S., & Ali, S. F. (2015). Greenhouse effect reduction by recovering energy from waste landfills in Pakistan. *Renewable and Sustainable Energy Reviews*, *44*, 117–131.
3. Kale, S. K., Deshmukh, A. G., Dudhare, M. S., & Patil, V. B. (2015). Microbial degradation of plastic: A review. *Journal of Biochemical Technology*, *6*(2), 952–961.
4. https://datatopics.worldbank.org/what-a-waste/trends_in_solid_waste_management.html.
5. Adegunloye, D. V., Adetuyi, F. C., Akinyosoye, F. A., & Doyeni, M. O. (2007). Microbial analysis of compost using cowdung as booster. *Pakistan Journal of Nutrition*, *6*(5), 506–510.
6. Dusoruth, V. (2018). *Household Food Waste Generation and Organics Recycling: Too Time Consuming or for the Better [Public] Good?*, Washington, DC: Agricultural & Applied Economics Association.
7. Badgett, A., & Milbrandt, A. (2020). A summary of standards and practices for wet waste streams used in waste-to-energy technologies in the United States. *Renewable and Sustainable Energy Reviews*, *117*, 109425.
8. Chen, L., de Haro Marti, M., Moore, A., & Falen, C. (2011). The composting process. *Dairy Manure Compost Production and Use in Idaho*, *2*, 513–532.
9. Riddech, N., Klammer, S., & Insam, H. (2002). Characterisation of microbial communities during composting of organic wastes. In: Insam, H., Riddech, N., Klammer, S. (Eds) *Microbiology of Composting* (pp. 43–51). Springer, Berlin, Heidelberg.
10. Pace, M. G., Miller, B. E., & Farrell-Poe, K. L. (1995). *The Composting Process*. Logan: Utah State University.
11. Galdon, F., Bertelsen, S., Hulse, J., & Hall, A. (2021). Object-oriented upcycling: An object-based approach to the circular economy. In: Sung, K., Singh, J., Bridgens, B. (eds) *State-of-the-Art Upcycling Research and Practice* (pp. 9–13). Cham: Springer.

3 Sustainable Recent Development Practices for the Plastic Waste Management
A Review

Madhavi Konni, Bhavya Kavitha Dwarapureddi, and Manoj Kumar Karnena

CONTENTS

3.1   Introduction 28
3.2   Types of Plastic 29
     3.2.1   Thermoplastics..29
     3.2.2   Thermosetting Plastics..29
3.3   Recycling 30
3.4   Pyrolysis...31
3.5   Hydrocracking 32
3.6   Gasification 33
3.7   Degradation Technologies 34
     3.7.1   Types of Oxo-Biodegradations 34
           3.7.1.1   Photo-oxidative Degradation 34
           3.7.1.2   Thermal Degradation..35
           3.7.1.3   Ozone Degradation 35
           3.7.1.4   Mechanochemical Degradation 35
           3.7.1.5   Catalytic Degradation 35
     3.7.2   Biodegradation...35
           3.7.2.1   Oxic Biodegradation 35
           3.7.2.2   Anoxic Biodegradation 36
3.8   Composting of Plastics 36
3.9   Future Scope of Research...37
3.10 Conclusions..40
References...40

DOI: 10.1201/9781003359784-3

3.1 INTRODUCTION

Industrial development increased the manufacturing of products for human activities, leading to the generation of waste and causing environmental degradation as humans discharge these products (Zhang et al. 2021). Plastic is a versatile waste generally used for packing in industries for various applications, and these products have become an indispensable part of people's daily life (Sharuddin et al. 2016). These materials have gained much prominence due to their wide range of applications, i.e., affordability, durability, and relative accessibility compared to other technologies (Jiang et al. 2022). The industries' general types of plastics are polyethene, high-density polyethene, low-density polystyrenes, polyvinyl alcohols, etc. Ritchie and Roser (2018) stated that since 1950, plastic usage in industries has increased by 200-folds, with an annual rate of increase of up to 8.4%. Geyer et al. 2017 stated that improper management and longer decay of biodegradability of these products lead to significant accumulation and generation of waste. Neo et al. (2021) stated that by 2025, the plastic production rate might reach 500 million tonnes per year due to consumer demand and an increase in the global population. Plastic production negatively impacts the environment, forming debris in the oceans and terrestrial habitats (Walker and Xanthos 2018; Huang et al. 2022). According to the International Union for Conservation of Nature reports, it was estimated that approximately 14 million tonnes of plastic debris ended up in the deep-sea sediments; due to their chemical nature and non-biodegradability, the marine species ingest them causing severe injuries and death. Furthermore, microplastics cause severe health effects as they accumulate over time in the food chains (Campanale et al. 2020). The lack of adequate and affordable alternatives for plastic manufacturing was still unknown (Vara and Karnena 2019). Even though several technologies like incineration, landfill, etc., are available worldwide to reduce the harmful effects of the chemicals from plastics that pose a threat to the environment, they are ineffective in controlling them (Aryan et al. 2019; Kerdlap et al. 2022). According to the estimates, the plastic contribution to the total waste generation was too high, i.e., 89% in Myanmar, and too low, i.e., 2% in the United States (Jambeck et al. 2015). Management of plastic waste and associated pollution pose a severe threat to the environment and become a challenge to policymakers in developing countries like India. Plastic waste management is becoming a problem in developing countries due to the adaptation of the effective waste management system as they need more high capital for initial setups. Neo et al. (2021) stated that India will be ranked 12th and 5th highest contributor to plastic waste mismanagement by 2025. India is an emerging economic country in Asia and the most imperative producer and consumer of plastic during various industrial manufacturing processes (Zhu et al. 2022). The management strategies in India are ineffective and, at a gross level, during the segregation, recovery, and proportioning of the waste in the dumping yards (Lahiry 2019). Jambeck et al. (2015) stated that India contributed nearly 1.64% to the total global waste and generated 85% of the country's plastic waste in 2010. India annually generates 9.64 million tonnes of plastic, in which 40% of the waste is caused only by the inefficiency in managing during the collection to the end life (BusinessLine 2019; de Mello Soares et al. 2022). According to the Central Pollution Control Board (2018), municipal solid waste consists of nearly 6.9% of plastic waste of the total waste generated (CPCB 2018). As the country with the highest population and land-scarce, the waste dumping capacities and landfill sites are inadequate. The Indian government flagships like the Swachh Bharat Mission have somehow pitched some initiatives to reduce and manage plastic waste. However, these flagships are localized and shorter and do not wholly reduce plastic waste (Dhanshyam and Srivastava 2021). Integrating policies and assessments might be productive in the management of waste. Appropriate integrations and proper policy implementations might be imperatives for plastic waste management. Conventional techniques like incineration and landfill still operate in India, causing environmental degradation (Vara and Karnena 2019). The incineration process for plastic reduction requires a lot of energy and releases hazardous chemicals as by-products like carbon dioxides, sulfur dioxide, dioxin, heavy metals, etc. (Franchini et al. 2004). This leads to global warming during the operation and poses a severe threat to human health (Ashworth et al. 2014). Pramila and Vijaya (2011) stated that landfills are regularly used to treat plastic waste with the highest proportions in India. Due to the plastic's non-biodegradable nature, these substances will degrade slower and occupy larger space by piling up. Further, these wastes release leachate, interact with groundwater, cause toxicity, and deteriorate the adjacent lands (Teuten et al. 2009). Thus, many researchers consider these conventional techniques unsuitable for plastic waste management (Vara and Karnena 2019; Zhang et al. 2021; Okan et al. 2019). To overcome the problems with conventional techniques, several researchers have developed various technologies worldwide to replace these technologies. The current book chapter reviews the modern methods that have been adopted and utilized for waste management.

3.2 TYPES OF PLASTIC

Polymerization of the smaller molecules can synthesize the plastic. Based on the thermal properties, the plastics are divided into the following.

3.2.1 THERMOPLASTICS

These plastics can be moulded many times, and the chemical composition and nature of the plastic materials cannot be changed upon heating. The atomic units of these plastics range from 2×10^{-4} to 5×10^{-5}. The molecules are arranged linearly and form chains; these plastic atoms are attached to the ends of the carbon atoms. The free radical techniques process the introduction of double bonds responsible for creating the macromolecules. This type of polymerization is also called addition polymerization.

3.2.2 THERMOSETTING PLASTICS

These plastics can be moulded only for one time; the chemical compositions of these plastics are altered upon heating; no modifications are done once moulded. Ghosh et al. (2013) stated that irreversible chemical alterations couldn't be considered examples of these polymers. Under favourable conditions, the phenol–formaldehyde and polyurethanes are formed by using the growth polymerization process. The water vapours condense the bifunctional molecules formed during the reactions during each response step. The monomers of these plastics, upon heating, are converted into infusible substances and change their chemical structures (Zhongming et al. 2017).

The recycling methods for plastics have received widespread attention from researchers worldwide due to environmental problems (Howard 2002). These methods recover valuable products from waste plastics and have become a consistent theme for sustainable development. Garcia and Robertson (2017) stated that recycling one ton of waste plastic can save nearly 130 million kilojoules of energy. Further, in their review, Geyer et al. (2017) mentioned that 3.5 billion barrels of oil annually could be saved by recycling waste plastics globally. Thus, the new studies on recycling technologies have tremendous significance for impending development.

3.3 RECYCLING

These methods generally consist of primary/secondary recycling systems; the primary process gained much prominence worldwide due to its low cost and easy operation procedures. Al-salem et al. (2010) stated that primary recycling requires higher grade and uncontaminated plastics for recycling to exhibit similar properties to virgin plastics. Secondary recycling involves a mechanical process; the reprocessed plastic products from the plastic waste exhibit lower performance than virgin plastic (Kumar et al. 2011). Plastic treatment might involve recycling techniques like reducing the size by shredders, cleaning, and drying. The disadvantage of these processes is the weakening quality of recycled products. During the process, the molecular weight of the products is reduced by chain scissions after a series of recycling (Jagtap et al. 2015; Jacobsen et al. 2022). The secondary recycling process for plastic recovery is limited to some thermoplastics sensitive to temperatures (Okan et al. 2019; Suzuki et al. 2022). The physical recycling methods are eco-friendly and economically viable. Due to these constraints, plastic waste's physical recovery process is narrowly applied. Direct application of this technology is garbage classification, mainly for the segregation of plastic waste. Thus, these technologies require further government policy attachment to classify plastic waste.

Energy recovery from plastic waste means obtaining energy from non-recyclable plastics through incineration. The calorific values of some plastic materials are similar to the crude oil derivatives. The primary by-products of incineration are water and carbon dioxide; thus, this process can replace traditional fuels. The energy value of polyvinyl chloride is much lower than other materials. It is unsuitable for incineration as these plastics generate highly acidic toxic gases, which can corrode the equipment (Thanh et al. 2011). Further to the recovery of energy, obtaining monomers and petrochemicals from plastic waste gained much prominence among researchers. Proper plastic waste treatment can generate various types of hydrocarbons essential for fuel with different mixtures. At present, resource recovery is made by thermolysis using plastic waste. In this process, plastic polymers undergo chain scission and generate lower molar mass monomers (Okan et al. 2019). Thermolysis can be divided into three categories (pyrolysis, gasification, and hydrocracking) using plastic materials during fuel recovery (Brems et al. 2013; Fakirov 2021). The pyrolysis process uses catalysts in inert atmospheres, and the gasification process is done under stoichiometric atmospheres. The hydrocracking process is done under the hydrogen gas; the main aim is to convert the longer polymer chains to smaller chains to develop substances like char and oil at higher temperatures. The significant difference between these processes is operating procedures and products. The coming sections will review some of the processes currently adopted by many researchers to recover valuable products from plastics.

3.4   PYROLYSIS

The long-chain polymers are converted into smaller chains with the help of thermal degradation at 300–900°C in inert atmospheres (Chen et al. 2014). Valuable products like oil, gas, char, etc., have been recovered and produced during the operation. The pyrolysis process is suitable for all types of plastics except polyvinyl chloride, as the heating of this plastic releases toxic fumes (Wong et al. 2015). The key factors that affect the operation of pyrolysis are operating temperatures, the type of reactor, the catalyst used, etc. Thus, more research is needed on the pyrolysis mechanisms, which might help control the process parameters; it will effectively recover valuable resources. The graphic representation of the pyrolysis process is shown in Figure 3.1. Scheirs and Kaminsky (2006) stated that the pyrolysis mechanism couldn't be explained in a single operation; it does not have a systematic tool and involves a series of complex reactions. However, the following mechanisms are proposed for pyrolysis (Cullis and Hirschler, 1981): depolymerization, fragmentation, chain stripping, and cross-linking. Nevertheless, decomposition of plastics mainly depends on the structures of the polymers.

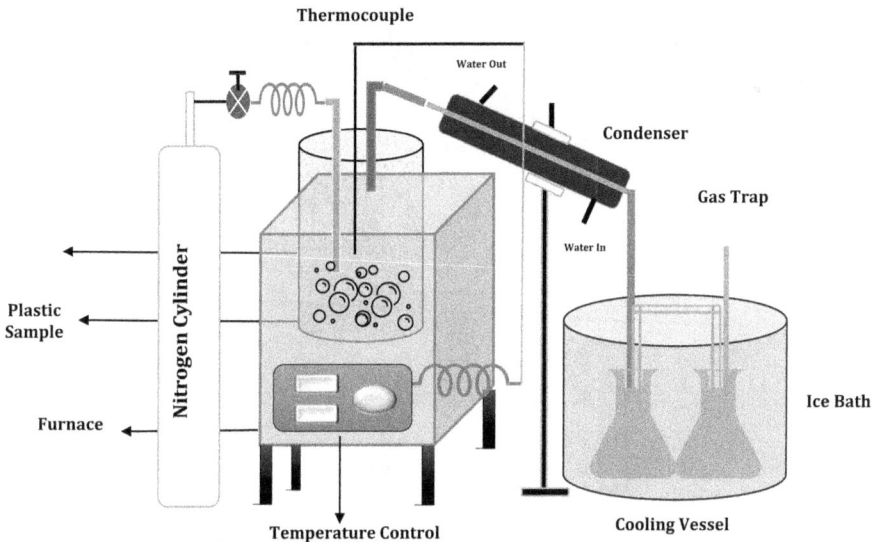

FIGURE 3.1   Diagrammatic representation of pyrolysis.

3.5 HYDROCRACKING

This process is also called hydrogenation and involves breaking the longer-chain hydrocarbons into smaller chains with catalysts under high hydrogen pressure. The broken chains' small structures are gasoline analogies (Al-Salem et al. 2010). The diagrammatic representation of the process is shown in Figure 3.2. Compared to the conservative techniques for decomposing plastics, these methods have more significant advantages. The plastic waste is converted into a high-grade fuel with high calorific values (Kasar et al. 2020). This process is more selective than pyrolysis and produces liquid gas from C_6 to C_{12} (Munir et al. 2020). The hydrogenation process improves the gasoline quality by reducing the aromatic compounds (Fuentes-Ordóñez et al. 2013; Schwarz et al. 2021) and further enhances the heat and mass transfer during the reactions.

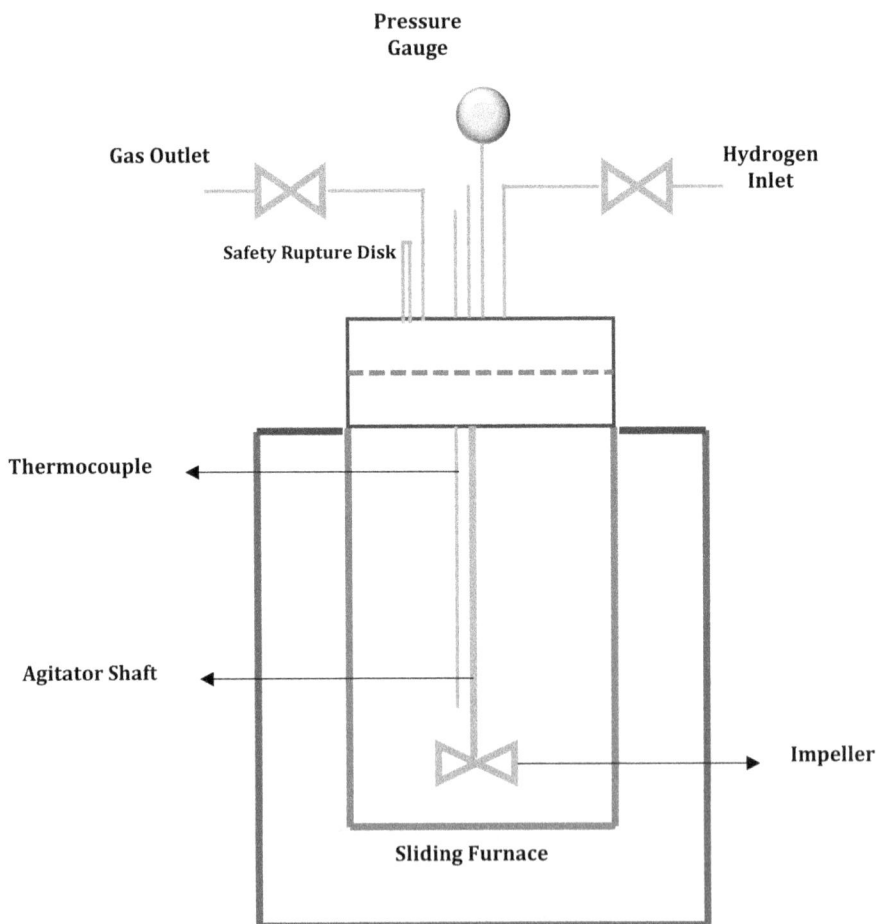

FIGURE 3.2 Diagrammatic representation of hydrocracking.

In contrast, in the pyrolysis process, the transfer of energy in heat has limitations as the larger polymers melt faster into liquid (Miranda et al. 2013). In the liquid phases, the heat transfer is not faster and has disadvantages (Gupte and Madras 2004). Continuous pressure by applying hydrogen is compulsory for the reactions, but this process is expensive compared to traditional pyrolysis. Further, this process is performed under high pressures, which might require sophisticated instrumentation. Thus, this process gained lesser attention than the traditional process, even though this process is eco-friendly compared to the other conventional techniques. Therefore, we have not discussed the current technologies in detail in this chapter as they do not attract the industrial sectors. There is a need to conduct more research on the current technologies to move them for industrial application by reducing the associated costs.

3.6 GASIFICATION

This process is also a thrombolysis technique compared to other conservative plastic recycling methods. The waste plastic is converted to syngas during gasification at a high temperature in the presence of oxygen (Robinson 2009). The industries used this technique as Akzo, Battelle, and Texaco gasification. The Texaco methods are the most common technique adopted widely and used by the recycling industries worldwide. Figure 3.3 shows the process of gasification using plastic waste as feed. Liquefaction and entrained bed processes are the standard processes involved in gasification. Liquefaction consists of depolymerizing the plastic waste into several valuable products like oils. The non-condensable fumes generated during the operation are used as fuel. The oil obtained is further filtered to eliminate the contaminants; these oils are injected into the gasifier. This process is done in the presence of oxygen and steam and operated at 1200–1500°C. This process has even not gained attention from researchers as they require high temperatures and are prone to release toxic gases.

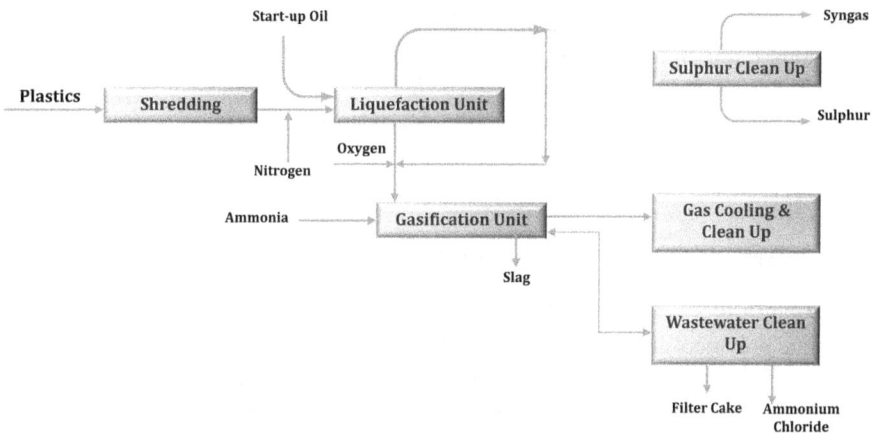

FIGURE 3.3 Diagrammatic representation of gasification.

Chemolysis is another process used for plastic recycling waste. Chemolysis is also called depolymerization (Ragaert et al. 2017). The plastics are depolymerized into monomers at 80–280°C (Payne et al. 2019). Even though several plastics are available, polyurethane and polyethene terephthalate are mostly recycled using these methods. The parameters used for the operation of chemolysis mainly depend on the type of plastics, which leads to the difficulty in recycling the monomer polymers from the mixture of the plastics. Thus, the application of these techniques to plastic recycling becomes narrow. Only a few countries have adopted these techniques for recycling plastic waste as the current technologies are not linked directly to mainstream recycling technologies. Even though the above technologies are proven to be efficient in recycling plastics, they are not sustainable both economically and environmentally friendly. Thus, there is a need for the development and continuous research on the sustainable recycling of plastic waste. The coming sections discuss the sustainable plastic waste recycling technologies currently practising the researchers worldwide.

3.7 DEGRADATION TECHNOLOGIES

As previously discussed, recovering plastic resources from waste is advantageous in the capital. Recycled plastic using the physical recycling methods resembles pristine plastics; however, after a few recycling processes, the mechanical and physical characteristics of the recycled products are lost. Further recycling is impossible and eventually requires degradation to reduce the volumes of waste in the environment. In addition, the recovery of plastics by using recycling procedures requires high capital and energy. Further, toxic by-products are released during the recycling process, which causes a severe threat to the environment. Due to these problems, various technologies for plastic degradation have their advantages. In addition, degradation of these polymers releases only carbon dioxide and water vapour. The degradation methods do not require high pressures and temperatures compared to plastic recycling technologies. Thus, degradation of the polymers of the plastics requires eco-friendly and non-hazardous methods (Moharir and Kumar 2019; Volk et al. 2021). Therefore, research on these technologies to generate other valuable chemicals like C2 fuels during degradation is much needed. Thus, based on the available literature, the current sections show findings on the degradation of the polymer treatment for sustainable entertainment (Jiao et al. 2020).

3.7.1 TYPES OF OXO-BIODEGRADATIONS

3.7.1.1 Photo-oxidative Degradation

Synthetic polymers are degraded by the ultraviolet radiation released by the sun. The lifetime of the polymers is determined by the ultraviolet radiation extending from 300 to 400 nm; solar light is the leading radiation source (Elahi et al. 2021). Elahi et al. (2021) stated that ultraviolet radiations could easily break carbon-to-carbon bonds in the polymers.

3.7.1.2 Thermal Degradation

Thermal degradations are classified under oxidative degradation. The depolymerization reaction does this process, and temperatures and ultraviolet rays are required to initiate the response (Teare et al. 2000). Depolymerization generally starts at the imperfections of bonds present in the polymer chains. Further, it is evident that at more significant temperatures, polymers are depolymerized.

3.7.1.3 Ozone Degradation

In the atmosphere, ozone is generally present and causes degradation of polymers. Due to the inactiveness of the oxidative process, polymers stay longer in the environment (Teare et al. 2000). Ozone in the atmosphere degrades the molar masses of polymers by forming reactive oxygen species (Andrady et al. 1998).

3.7.1.4 Mechanochemical Degradation

This process generally involves breakdown of the polymer chains using mechanical stress and ultra-sonic methods; radical reactions enhance the long-chain branches and decrease the molecules' distribution (Li et al. 2005). In this method, the oxide molecule of the nitrogen terminates the longer chains and breaks down the macromolecules of the polymers.

3.7.1.5 Catalytic Degradation

In these methods, the plastic's depolymerization occurs when heated above 38°C, and free radicals are released during these reactions and break down the plastic polymers (Elahi et al. 2021). Catalytic waste transforms the polymers into hydrocarbons. The most commonly used catalysts are platinum, cobalt, silicon dioxide, molybdenum, etc. Even though these methods are available for the degradation of plastics and have gained much prominence from many researchers worldwide, they have disadvantages in terms of capital, skilled labour, and the duration required. Thus, there is a need for sustainable methods to degrade plastics. Bioremediation was proven to be one such method in recent years, gaining the attention of many researchers.

3.7.2 BIODEGRADATION

Characteristic changes of polymers, i.e., physical or chemical caused by the microorganisms, are called biodegradation. Microbes degrade natural and artificial plastics faster in real-time applications (Alshehrei 2017). Biodegradation occurs in both oxic and anoxic conditions.

3.7.2.1 Oxic Biodegradation

In oxic degradation, the microbes break down the organic polymers into smaller substances in oxygen. Figure 3.4 shows the schematic representation of the microbial degradation accepting electrons. Müller (2005) reported that the by-products released during this process are carbon dioxide and water.

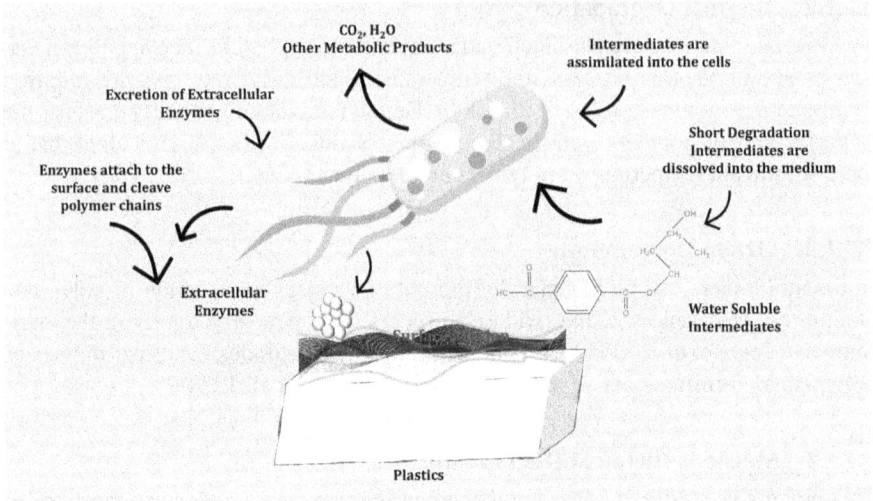

FIGURE 3.4   Microbial degradation of plastics.

3.7.2.2   Anoxic Biodegradation

In anoxic degradation, the breakdown of the polymers was done by using microbes without oxygen. For this method, oxygen is not the main compound for attenuating the pollutants at the harmful waste sites. The microbes (anaerobic) use sulphates, manganese, and iron and accept carbon dioxide as an electron to break down the complex polymers into more minor complexes. These polymers cannot be absorbed directly into the microbial cells as they are larger and not water-soluble. These polymers are the energy source for the microbes and secrete extracellular enzymes to degrade the complex polymers. Enzymes secreted by microbes act as essential compounds for the degradation of polymers. Biological degradation of the microbes involves mineralization and depolymerization processes. The microbial extracellular enzymes convert the larger molecules of the polymer into smaller molecules and make them soluble in water. The smaller molecules are semipermeable to the microbial membrane and used as energy sources (Gu 2003). Figure 3.5 illustrates polyethene breakdown and the components involved in the process (Shahnawaz et al. 2016) (Table 3.1).

3.8   COMPOSTING OF PLASTICS

These methods transform waste plastics into humus-like substances and control biodegradation (Shah et al. 2008). During composting, the dry plastic was mixed with mature compost and treated with the microbes at 40°C, and the water content was maintained up to 60–80% (Semitela et al. 2019; Unmar and Mohee 2008). Figure 3.6 shows microbial compositing of plastic materials. The by-products of composting are minerals, methane, water vapour, etc. (Unmar and Mohee 2008). In addition, immense heat is generated in composting, which might destroy the valuable microbes

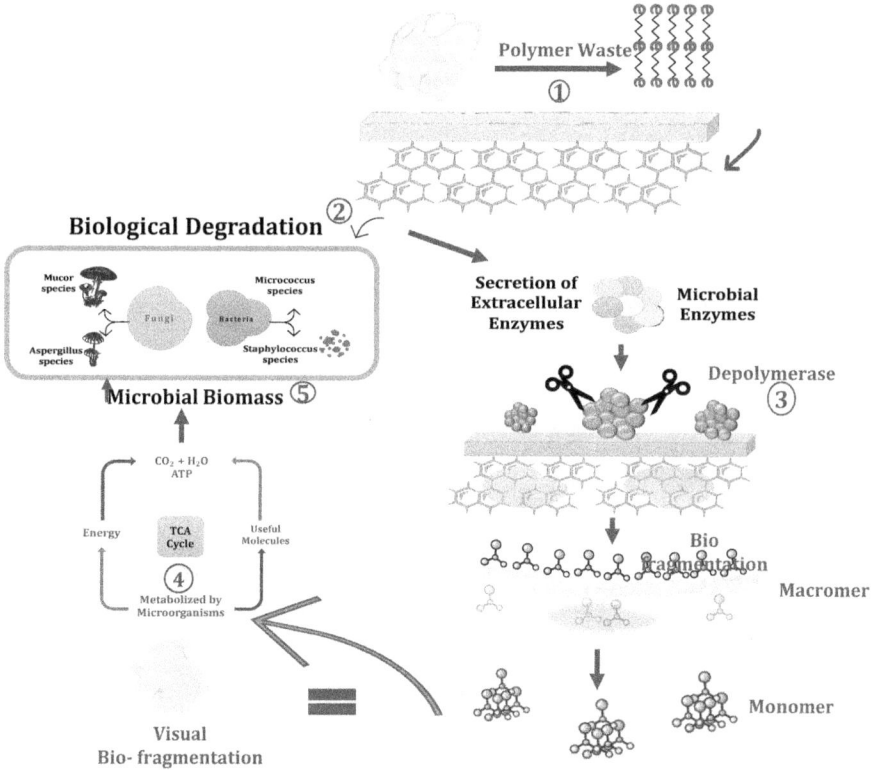

FIGURE 3.5 Diagrammatic representation of polyethene breakdown.

responsible for composting (Shah et al. 2008). However, these technologies are mainly used for biodegradable plastics. Bench-scale studies showed practical results in degrading polyethene substances (Körner et al. 2005). These technologies can't be adopted in countries with land scarcity and lower temperatures. Further, these techniques must be integrated with nanotechnology like nano-bioremediation to treat plastic waste more effectively and achieve good results. Researchers have to focus on the biological methods of treating plastic waste for a sustainable environment.

3.9 FUTURE SCOPE OF RESEARCH

Integration technologies should be adopted to degrade waste plastics; in recent years, nanotechnology has gained much prominence in allied applications; integrating nanotechnology with bioremediations will sustainably enhance the removal of waste plastics. Even though several researchers were researching these technologies individually and stated that they are efficient in pollutant degradation, the integration of these technologies for the degradation of plastics is not studied yet. Priyadarsini and Biswal (2021) reviewed these integration technologies to degrade the pollutants having similar structures relevant to the plastics and showed how efficient are these

TABLE 3.1
Widely Used Microbes for the Degradation of Plastics

S.no	Bacteria Name	Family	Order	Plastic Degraded	References
1	*Aspergillus niger*	Trichocomaceae	Eurotiales	Films	Konduri et al. 2010
2	*A. nidulans*	Trichocomaceae	Eurotiales	Plastic cups	Priyanka and Archana 2011
3	*Arthrobacter*	Micrococcaceae	Actinomycetales	High-density polyethene	Balasubramanian et al. 2010
4	*Aspergillus flavus*	Trichocomaceae	Eurotiales	Plastic films	El-Shafei et al. 1998
5	*Aspergillus niger*	Trichocomaceae	Eurotiales	Carry bags	Aswale and Ade, 2008
6	*Aspergillus oryzae*	Trichocomaceae	Eurotiales	Low-density polythene	Konduri et al. 2011
7	*Brevibacillus borstelensis*	Paenibacillaceae	Bacillales	Low-density polyethylene	Hadad et al. 2005
8	*Bacillus*	Bacillaceae	Bacillales	Carry bags	Vijaya and Mallikarjuna 2008
9	*Bacillus cereus*	Bacillus	Bacillus	Polythene bags	Aswale and Ade 2008
10	*Bacillus cereus*	Bacillaceae	Bacillales	Low-density polythene	Suresh et al. 2011
11	*Bacillus megaterium*	Bacillaceae	Bacillales	Low-density polythene	Abrusci et al. 2011
12	*Bacillus mycoides*	Bacillaceae	Bacillales	Polyethene	Seneviratne et al. 2006
13	*Cladosporium cladosporioides*	Cladosporiaceae	Capnodiales	Polythene	Bonhomme et al. 2003
14	*Listeria monocytogenes*	Listeriaceae	Bacillales	Polyethene	Kumar et al. 2007
15	*Micrococcus*	Micrococcaceae	Actinomycetales	Plastic cups	Kathiresan, 2003
16	*Pseudomonas aeruginosa*	Pseudomonadaceae	Pseudomonadales	Low-density polythene films	Kyaw et al. 2012
17	*Penicillium funiculosum*	Trichocomaceae	Eurotiales	Low-density polyethene	Gilan et al. 2004
18	*Penicillium pinophilum*	Trichocomaceae	Eurotiales	Powdered polythene	Volke-Sepulveda et al. 2002
19	*Phanerochaete Chrysosporium*	Phanerochaetaceae	Polyporales	High-density polyethylene	Iiyoshi et al. 1998
20	*Pseudomonas putida*	Pseudomonadaceae	Pseudomonadales	Polythene bag waste	Nwachukwu et al. 2010
21	*Pseudomonas spp.*	Pseudomonadaceae	Pseudomonadales	Artificial polyethene	Nanda et al. 2010

(Continued)

TABLE 3.1 (*Continued*)
Widely Used Microbes for the Degradation of Plastics

S.no	Bacteria Name	Family	Order	Plastic Degraded	References
22	*Rhodococcus rhodochrous*	Nocardiaceae	Mycobacteriales	Polyethene	Fontanella et al. 2009
23	*Rhodococcus ruber*	Rhodococcus	Mycobacteriales	Low-density polyethylene	Chandra and Rustgi, 1997
24	*Serratia marcescens*	Enterobacteriaceae	Enterobacterales	PE carry bags	Aswale and Ade 2008
25	*S. marcescens*	Enterobacteriaceae	Enterobacterales	Carry bags	Aswale and Ade 2008
26	*Staphylococcus epidermidis*	Staphylococcaceae	Bacillales	Low-density polyethene	Chatterjee et al. 2010
27	*Streptomyces viridosporus*	Streptomycetaceae	Actinomycetales	Polyethene pro-oxidant plastics	Pometto et al. 1992
28	*Streptomyces*	Streptomycetaceae	Actinomycetales	Polyethene powder	Usha et al. 2011
29	*Trichocomaceae*	Trichocomaceae	Eurotiales	Low-density polyethylene	Pramila and Vijaya Ramesh 2011

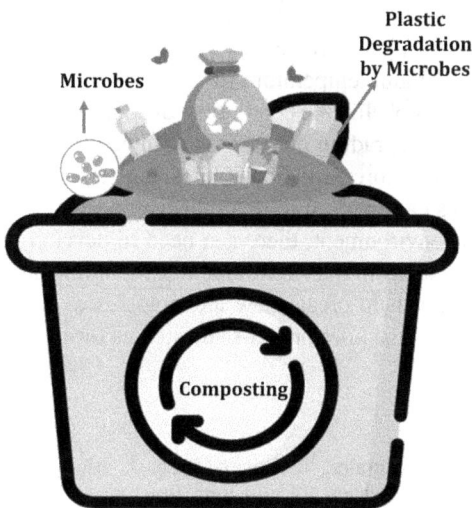

FIGURE 3.6 Microbial composting of plastic waste.

technologies. Since no full-scale/bench studies are conducted on integrating efficient technologies for removing plastic materials, research is needed. Academic institutes should collaborate with industries that promote new novel technologies. Developing countries like India have to adopt and promote these integrations by funding the research as these technologies are more advantageous and eco-friendlier than conventional techniques in India. In addition, toxicological studies have to be conducted simultaneously to check the effects of bioremediation and nanotechnology in the degradation of plastics. Nano-bioremediation is the current new novel technology gaining prominence among researchers worldwide. Thus, this technology has to be tested to degrade plastic waste; still, there is a need for continuous research to develop novel technologies as new contaminants have been evolving every day.

3.10 CONCLUSIONS

The current chapter provided insight into the technologies currently adopted for treating plastic waste. Conservative technologies have advantages and disadvantages as they produce secondary pollutants in the environment. Many countries abandoned these technologies for the treatment of plastics. Even though physical recycling methods for plastic treatment are proven to be efficient in treating plastic waste, continuous recycling alters the structures of the plastics and can't be processed further. Thermolysis and chemolysis processes are widely used in developing countries as they are easy to operate. These processes help recover gasoline and valuable products from waste plastics and reduce the stress on natural resources. However, these technologies can't be employed in the lower-developed countries and land-scarce countries as they require enormous capital and land for establishment. Currently, degradation technologies are efficient in sustainably treating plastics. Bioremediation is a technique that has gained prominence among many researchers worldwide as it doesn't release secondary pollutants into the environment. However, there is a need to conduct more studies on these technologies as, to date, no toxicological studies have been undertaken during their operation. Compositing techniques are currently confined only to tropical countries as they require optimum temperature for microbial activation, and they can't be adopted in temperate or cool climatic conditions. In addition, there is a need to explore the microbes capable of degrading plastic waste. Integration of bioremediation with nanotechnology, i.e. nano-bioremediation studies, has to be conducted to understand how these technologies can be used to treat plastics. As new pollutants are continuously emerging in the environment, there is a need to develop novel technologies to degrade the contaminants from the environment. Combining academic research with industrial research might help create new technologies as lab-scale testing can be implemented directly into the pilot or full-scale studies through these collaborations.

REFERENCES

Abrusci, C., Pablos, J. L., Corrales, T., López-Marín, J., Marín, I., & Catalina, F. 2011. Biodegradation of photo-degraded mulching films based on polyethylenes and stearates of calcium and iron as pro-oxidant additives. *International Biodeterioration & Biodegradation*, 65(3): 451–459.

Al-Salem, S. M., Lettieri, P., & Baeyens, J. 2010. The valorization of plastic solid waste (PSW) by primary to quaternary routes: From re-use to energy and chemicals. *Progress in Energy and Combustion Science*, 36(1): 103–129.

Alshehrei, F. 2017. Biodegradation of synthetic and natural plastic by microorganisms. *Journal of Applied & Environmental Microbiology*, 5(1): 8–19.

Andrady, A. L., Hamid, S. H., Hu, X., & Torikai, A. 1998. Effects of increased solar ultraviolet radiation on materials. *Journal of Photochemistry and Photobiology B: Biology*, 46(1–3): 96–103.

Aryan, Y., Yadav, P., & Samadder, S. R. 2019. Life cycle assessment of the existing and proposed plastic waste management options in India: A case study. *Journal of Cleaner Production*, 211: 1268–1283.

Ashworth, D. C., Elliott, P., & Toledano, M. B. 2014. Waste incineration and adverse birth and neonatal outcomes: A systematic review. *Environment International*, 69: 120–132.

Aswale, P., & Ade, A. 2008. Assessment of the biodegradation of polythene. *Bioinfolet*, 5: 239.

Balasubramanian, V., Natarajan, K., Hemambika, B., Ramesh, N., Sumathi, C. S., Kottaimuthu, R., & Rajesh Kannan, V. 2010. High-density polyethylene (HDPE)-degrading potential bacteria from marine ecosystem of Gulf of Mannar, India. *Letters in Applied Microbiology*, 51(2): 205–211.

Bonhomme, S., Cuer, A., Delort, A. M., Lemaire, J., Sancelme, M., & Scott, G. 2003. Environmental biodegradation of polyethylene. *Polymer Degradation and Stability*, 81(3): 441–452.

Brems, A., Dewil, R., Baeyens, J., & Zhang, R. 2013. Gasification of plastic waste as waste-to-energy or waste-to-syngas recovery route. *Natural Science*, 5(6): 695–704.

BusinessLine. 2019. India generates 9.46 mn tonnes of plastic waste annually. https://www.thehindubusinessline.com/news/science/india-generates-946-mn-tonnes-of-plastic-waste-annually/article29299108.ece (accessed 18 March 2022).

Campanale, C., Massarelli, C., Savino, I., Locaputo, V., & Uricchio, V. F. 2020. A detailed review study on potential effects of microplastics and additives of concern on human health. *International Journal of Environmental Research and Public Health*, 17(4): 1212.

Central Pollution Control Board (CPCB). 2018. Life Cycle Assessment (LCA) study of plastics packaging products. https://cpcb.nic.in/uploads/plasticwaste/LCA_Re port_15.05.2018.pdf (accessed 18 March 2022).

Chandra, R., & Rustgi, R. 1997. Biodegradation of maleated linear low-density polyethylene and starch blends. *Polymer Degradation and Stability*, 56(2): 185–202.

Chatterjee, S., Roy, B., Roy, D., & Banerjee, R. 2010. Enzyme-mediated biodegradation of heat treated commercial polyethylene by *Staphylococcal* species. *Polymer Degradation and Stability*, 95(2): 195–200.

Chen, D., Yin, L., Wang, H., & He, P. 2014. Pyrolysis technologies for municipal solid waste: A review. *Waste Management*, 34(12): 2466–2486.

Cullis, C. F., & Hirschler, M. M. 1981. *The Combustion of Organic Polymers* (Vol. 5). Oxford University Press, USA.

de Mello Soares, C. T., Ek, M., Östmark, E., Gällstedt, M., & Karlsson, S. 2022. Recycling of multi-material multilayer plastic packaging: Current trends and future scenarios. *Resources, Conservation and Recycling*, 176: 105905.

Dhanshyam, M., & Srivastava, S. K. 2021. Effective policy mix for plastic waste mitigation in India using system dynamics. *Resources, Conservation and Recycling*, 168: 105455.

Elahi, A., Bukhari, D. A., Shamim, S., & Rehman, A. 2021. Plastics degradation by microbes: A sustainable approach. *Journal of King Saud University-Science*, 33(6): 101538.

El-Shafei, H. A., Abd El-Nasser, N. H., Kansoh, A. L., & Ali, A. M. 1998. Biodegradation of disposable polyethylene by fungi and *Streptomyces* species. *Polymer Degradation and Stability*, 62(2): 361–365.

Fakirov, S. 2021. A new approach to plastic recycling via the concept of microfibrillar composites. *Advanced Industrial and Engineering Polymer Research*, 4(3): 187–198.

Fontanella, S., Bonhomme, S., Koutny, M., Husarova, L., Brusson, J. M., Courdavault, J. P., ... Delort, A. M. 2010. Comparison of the biodegradability of various polyethylene films containing pro-oxidant additives. *Polymer Degradation and Stability*, 95(6): 1011–1021.

Franchini, M., Rial, M., Buiatti, E., & Bianchi, F. 2004. Health effects of exposure to waste incinerator emissions: A review of epidemiological studies. *Annali dell'Istituto superiore di sanità*, 40(1): 101–115.

Fuentes-Ordóñez, E. G., Salbidegoitia, J. A., Gonzalez-Marcos, M. P., & Gonzalez-Velasco, J. R. 2013. Transport phenomena in catalytic hydrocracking of polystyrene in solution. *Industrial & Engineering Chemistry Research*, 52(42): 14798–14807.

Garcia, J. M., & Robertson, M. L. 2017. The future of plastics recycling. *Science*, 358(6365): 870–872.

Geyer, R., Jambeck, J. R., & Law, K. L. 2017. Production, use, and fate of all plastics ever made. *Science Advances* 2(7): 82.

Geyer, R., Jambeck, J. R., & Law, K. L. 2017. Production, use, and fate of all plastics ever made. *Science Advances*, 3(7): e1700782.

Ghosh, S. K., Pal, S., & Ray, S. 2013. Study of microbes having potentiality for biodegradation of plastics. *Environmental Science and Pollution Research*, 20(7): 4339–4355.

Gu, J. D. 2003. Microbiological deterioration and degradation of synthetic polymeric materials: Recent research advances. *International Biodeterioration & Biodegradation*, 52(2): 69–91.

Gupte, S. L., & Madras, G. 2004. Catalytic degradation of polybutadiene. *Polymer Degradation and Stability*, 86(3): 529–533.

Hadad, D., Geresh, S., & Sivan, A. 2005. Biodegradation of polyethylene by the thermophilic bacterium *Brevibacillus borstelensis*. *Journal of Applied Microbiology*, 98(5): 1093–1100.

Howard, G. T. 2002. Biodegradation of polyurethane: A review. *International Biodeterioration & Biodegradation*, 49(4): 245–252.

Huang, J., Veksha, A., Chan, W. P., Giannis, A., & Lisak, G. 2022. Chemical recycling of plastic waste for sustainable material management: A prospective review on catalysts and processes. *Renewable and Sustainable Energy Reviews*, 154: 111866.

Iiyoshi, Y., Tsutsumi, Y., & Nishida, T. 1998. Polyethylene degradation by lignin-degrading fungi and manganese peroxidase. *Journal of Wood Science*, 44(3): 222–229.

Jacobsen, L. F., Pedersen, S., & Thøgersen, J. 2022. Drivers of and barriers to consumers' plastic packaging waste avoidance and recycling–A systematic literature review. *Waste Management*, 141: 63–78.

Jagtap, M. M. D., Khatavkar, M. S. S., & Quazi, T. 2015. Methods for waste plastic recycling. *International Journal on Recent Technologies in Mechanical and Electrical Engineering*, 2: 120–122.

Jambeck, J. R., Geyer, R., Wilcox, C., Siegler, T. R., Perryman, M., Andrady, A., Narayan, R., & Law, K. L. 2015. Plastic waste inputs from land into the ocean. *Science*, 347(6223): 768–771.

Jiang, J., Shi, K., Zhang, X., Yu, K., Zhang, H., He, J., ... Liu, J. 2022. From plastic waste to wealth using chemical recycling: A review. *Journal of Environmental Chemical Engineering*, 10(1): 106867.

Jiao, X., Zheng, K., Chen, Q., Li, X., Li, Y., Shao, W., Xu, J., Zhu, J., Pan, Y., Sun, Y. & Xie, Y. 2020. Photocatalytic conversion of waste plastics into C2 fuels under simulated natural environment conditions. *Angewandte Chemie International Edition*, 59(36): 15497–15501.

Kasar, P., Sharma, D. K., & Ahmaruzzaman, M. 2020. Thermal and catalytic decomposition of waste plastics and its co-processing with petroleum residue through pyrolysis process. *Journal of Cleaner Production*, 265: 12.

Kathiresan, K. 2003. Polythene and plastics-degrading microbes from the mangrove soil. *Revista de Biologia Tropical*, 51(3–4): 629–633.

Kerdlap, P., Purnama, A. R., Low, J. S. C., Tan, D. Z. L., Barlow, C. Y., & Ramakrishna, S. 2022. Comparing the environmental performance of distributed versus centralized plastic recycling systems: Applying hybrid simulation modeling to life cycle assessment. *Journal of Industrial Ecology*, 26(1): 252–271.

Konduri, M. K., Anupam, K. S., Vivek, J. S., DB, R. K., & Narasu, M. L. 2010. Synergistic effect of chemical and photo treatment on the rate of biodegradation of high density polyethylene by indigenous fungal isolates. *International Journal of Biotechnology & Biochemistry*, 6(2): 157–175.

Konduri, M. K., Koteswarareddy, G., Rohini Kumar, D. B., Venkata Reddy, B., & Lakshmi Narasu, M. 2011. Effect of pro-oxidants on biodegradation of polyethylene (LDPE) by indigenous fungal isolate, *Aspergillus oryzae*. *Journal of Applied Polymer Science*, 120(6): 3536–3545.

Körner, I., Redemann, K., & Stegmann, R. 2005. Behaviour of biodegradable plastics in composting facilities. Waste Management, 25(4): 409–415.

Kumar, S., Hatha, A. A. M., & Christi, K. S. 2007. Diversity and effectiveness of tropical mangrove soil microflora on the degradation of polythene carry bags. *Revista de Biología Tropical*, 55(3–4): 777–786.

Kumar, S., Panda, A. K., & Singh, R. K. 2011. A review on tertiary recycling of high-density polyethylene to fuel. *Resources, Conservation and Recycling*, 55(11): 893–910.

Kyaw, B. M., Champakalakshmi, R., Sakharkar, M. K., Lim, C. S., & Sakharkar, K. R. 2012. Biodegradation of low density polythene (LDPE) by *Pseudomonas* species. *Indian Journal of Microbiology*, 52(3): 411–419.

Lahiri, S. 2019. India's challenges in waste management: the key to efficient waste management is to ensure segregation source and resource recovery. Down To Earth, 08 May 2019. https://www.downtoearth.org.in/blog/waste/india-s-challenges-in-waste-management-56753.

Li, J., Guo, S., & Li, X. 2005. Degradation kinetics of polystyrene and EPDM melts under ultrasonic irradiation. *Polymer Degradation and Stability*, 89(1): 6–14.

Miranda, M., Cabrita, I., Pinto, F., & Gulyurtlu, I. 2013. Mixtures of rubber tyre and plastic wastes pyrolysis: A kinetic study. *Energy*, 58: 270–282.

Moharir, R. V., & Kumar, S. 2019. Challenges associated with plastic waste disposal and allied microbial routes for its effective degradation: A comprehensive review. *Journal of Cleaner Production*, 208: 65–76.

Müller, R. J. 2005. Biodegradability of polymers: Regulations and methods for testing. Biopolymers Online. doi: 10.1002/3527600035.bpola012.

Munir, D., Amer, H., Aslam, R., Bououdina, M., & Usman, M. R. 2020. Composite zeolite beta catalysts for catalytic hydrocracking of plastic waste to liquid fuels. *Materials for Renewable and Sustainable Energy*, 9(2): 1–13.

Nanda, S., Sahu, S., & Abraham, J. 2010. Studies on the biodegradation of natural and synthetic polyethylene by *Pseudomonas* spp. *Journal of Applied Sciences and Environmental Management*, 14(2): 57–60.

Neo, E. R. K., Soo, G. C. Y., Tan, D. Z. L., Cady, K., Tong, K. T., & Low, J. S. C. 2021. Life cycle assessment of plastic waste end-of-life for India and Indonesia. *Resources, Conservation and Recycling*, 174: 105774.

Nwachukwu, S., Obidi, O., & Odocha, C. 2010. Occurrence and recalcitrance of polyethylene bag waste in Nigerian soils. *African Journal of Biotechnology*, 9(37): 6096–6104.

Okan, M., Aydin, H. M., & Barsbay, M. 2019. Current approaches to waste polymer utilization and minimization: A review. *Journal of Chemical Technology & Biotechnology*, 94(1): 8–21.

Payne, J., McKeown, P., & Jones, M. D. 2019. A circular economy approach to plastic waste. *Polymer Degradation and Stability*, 165: 170–181.

Pometto, A. L., Lee, B. T., & Johnson, K. E. 1992. Production of an extracellular polyethylene-degrading enzyme(s) by *Streptomyces* species. *Applied and Environmental Microbiology*, 58(2): 731–733.

Pramila, R., & Ramesh, K. V. 2011. Biodegradation of low density polyethylene (LDPE) by fungi isolated from marine water a SEM analysis. *African Journal of Microbiology Research*, 5(28): 5013–5018.

Priyadarsini, M., & Biswal, T. 2021. Bioremediation of plastic material by using nanotechnology. In: Acharya, S.K., Mishra, D.P. (Eds.) *Current Advances in Mechanical Engineering* (pp. 27–38). Springer, Singapore.

Priyanka, N., & Archana, T. 2011. Biodegradability of polythene and plastic by the help of microorganism: A way for brighter future. *Journal of Environmental and Analytical Toxicology*, 1(4): 1000111.

Ragaert, K., Delva, L., & Van Geem, K. 2017. Mechanical and chemical recycling of solid plastic waste. *Waste Management*, 69, 24–58.

Ritchie, H., & Roser, M. 2018. Plastic pollution. Our World in Data, https://ourworldindata.org/plastic-pollution?utm_source=newsletter

Robinson, G. 2009. Recovering value from mixed plastics waste. In: G. Robinson (Ed.) *Proceedings of the Institution of Civil Engineers-Waste and Resource Management* (Vol. 162, No. 4, pp. 207–213). Thomas Telford Ltd, United Kingdom.

Scheirs, J., & Kaminsky, W. (Eds.). 2006. *Feedstock Recycling and Pyrolysis of Waste Plastics: Converting Waste Plastics into Diesel and Other Fuels*. John Wiley & Sons Incorporated.

Schwarz, A. E., Ligthart, T. N., Bizarro, D. G., De Wild, P., Vreugdenhil, B., & Van Harmelen, T. 2021. Plastic recycling in a circular economy; determining environmental performance through an LCA matrix model approach. *Waste Management*, 121: 331–342.

Semitela, S., Pirra, A., & Braga, F. G. 2019. Impact of mesophilic co-composting conditions on the quality of substrates produced from winery waste activated sludge and grape stalks: Lab-scale and pilot-scale studies. *Bioresource Technology*, 289: 121622.

Seneviratne, G., Tennakoon, N. S., Weerasekara, M. L. M. A. W., & Nandasena, K. A. 2006. Polyethylene biodegradation by a developed Penicillium–Bacillus biofilm. *Current Science*, 90(1): 20–21.

Shah, A. A., Hasan, F., Hameed, A., & Ahmed, S. 2008. Biological degradation of plastics: A comprehensive review. Biotechnology Advances, 26(3): 246–265.

Shahnawaz, M., Sangale, M. K., & Ade, A. B. 2016. Bacteria-based polythene degradation products: GC-MS analysis and toxicity testing. *Environmental Science and Pollution Research*, 23(11): 10733–10741.

Sharuddin, S. D. A., Abnisa, F., Daud, W. M. A. W., & Aroua, M. K. 2016. A review on pyrolysis of plastic wastes. *Energy Conversion and Management*, 115: 308–326.

Suresh, B., Maruthamuthu, S., Palanisamy, N., Ragunathan, R., Pandiyaraj, K. N., & Muralidharan, V. S. 2011. Investigation on biodegradability of polyethylene by *Bacillus cereus* strain Ma–Su isolated from compost soil. *International Research Journal of Microbiology*, 2(2): 292–302.

Suzuki, G., Uchida, N., Tanaka, K., Matsukami, H., Kunisue, T., Takahashi, S., S., Viet, P.H., Kuramochi, H., & Osako, M. 2022. Mechanical recycling of plastic waste as a point source of microplastic pollution. *Environmental Pollution*, 303, 119114.

Teare, D. O. H., Emmison, N., Ton-That, C., & Bradley, R. H. 2000. Cellular attachment to ultraviolet ozone modified polystyrene surfaces. *Langmuir*, 16(6): 2818–2824.

Teuten, E. L., Saquing, J. M., Knappe, D. R., Barlaz, M. A., Jonsson, S., Björn, A., ... Takada, H. 2009. Transport and release of chemicals from plastics to the environment and to wildlife. *Philosophical Transactions of the Royal Society B: Biological Sciences*, 364(1526): 2027–2045.

Thanh, N. P., Matsui, Y., & Fujiwara, T. 2011. Assessment of plastic waste generation and its potential recycling of household solid waste in Can Tho City, Vietnam. *Environmental Monitoring and Assessment*, 175(1): 23–35.

Unmar, G., & Mohee, R. 2008. Assessing the effect of biodegradable and degradable plastics on the composting of green wastes and compost quality. *Bioresource Technology*, 99(15): 6738–6744.

Usha, R., Sangeetha, T., & Palaniswamy, M. 2011. Screening of polyethylene degrading microorganisms from garbage soil. *Libyan Agriculture Research Center Journal International*, 2(4): 200–4.

Vara, S., Karnena, M. K., Dwarapureddi, B. K., & Chintalapudi, B. 2019. Will single use products lead to sustainability? *International Journal of Social Ecology and Sustainable Development (IJSESD)*, 10(2): 37–52.

Vijaya, C. & Mallikarjuna Reddy, M. 2008. Impact of soil composting using municipal solid waste on biodegradation of plastics. *Indian Journal of Biotechnology*, 7: 235–239

Volk, R., Stallkamp, C., Steins, J. J., Yogish, S. P., Müller, R. C., Stapf, D., & Schultmann, F. 2021. Techno-economic assessment and comparison of different plastic recycling pathways: A German case study. *Journal of Industrial Ecology*, 25(5): 1318–1337.

Volke-Sepúlveda, T., Saucedo-Castañeda, G., Gutiérrez-Rojas, M., Manzur, A., & Favela-Torres, E. 2002. Thermally treated low density polyethylene biodegradation by *Penicillium pinophilum* and *Aspergillus niger*. *Journal of Applied Polymer Science*, 83(2): 305–314.

Walker, T. R., & Xanthos, D. 2018. A call for Canada to move toward zero plastic waste by reducing and recycling single-use plastics. *Resources, Conservation & Recycling*, 133: 99–100.

Wong, S. L., Ngadi, N., Abdullah, T. A. T., & Inuwa, I. M. 2015. Current state and future prospects of plastic waste as source of fuel: A review. *Renewable and Sustainable Energy Reviews*, 50: 1167–1180.

Zhang, F., Zhao, Y., Wang, D., Yan, M., Zhang, J., Zhang, P., ... Chen, C. 2021. Current technologies for plastic waste treatment: A review. *Journal of Cleaner Production*, 282: 124523.

Zhongming, Z., Linong, L., Xiaona, Y., Wangqiang, Z., & Wei, L. 2017. IUCN Director General's statement on World Oceans Day. http://119.78.100.173/C666/handle/2XK7JSWQ/103244.

Zhu, B., Wang, D., & Wei, N. 2022. Enzyme discovery and engineering for sustainable plastic recycling. *Trends in Biotechnology*, 40(1): 22–37.

4 Review on Sustainable Technologies for Liquid Waste Treatment and Management

Gaydaa Al Zohbi

CONTENTS

4.1 Introduction 47
4.2 Types of Liquid Waste 49
4.3 Liquid Waste Treatment...49
4.4 Sustainable Technologies for Wastewater Treatment 50
    4.4.1 Phycoremediation 50
    4.4.2 Phytoremediation...53
    4.4.3 Membrane Distillation Bioreactor 58
    4.4.4 Nanotechnology...58
    4.4.5 Eco-friendly Adsorption Technology 60
    4.4.6 Advanced Oxidation 61
4.5 Conclusions..63
References..64

4.1 INTRODUCTION

In recent years, industrial activities, population growth, and urbanization have led to environmental degradation due to the rise of waste generation. The waste disposal has negative and dangerous ramifications on the environment and on human health. Liquid can be contaminated by many ways, such as by natural catastrophes like hurricanes, floods, and tornados or by human-made calamities and industrial defilement. Inadequate management of waste can lead to serious health effects as a consequence of combustion, detonation, and pollution of soil, water, and air. Furthermore, inappropriate waste treatment and elimination affect those living near such communities, resulting in expensive cleanups [1]. The disposal of untreated waste into water bodies causes dangerous effects on marina life, and it can conclusively influence human health since poisoned compounds can be transmitted by the food chain by means of bioaccumulation. Moreover, leaking of waste may poison soils and water flow and can cause air pollution due to the emission of pollutants like persistent organic pollutants. Many other issues could be caused by mishandling of waste dumping, for

DOI: 10.1201/9781003359784-4

instance, pollution of water and air, and scenery depravation. The release of organic matter into water, originating from land, causes a supersaturation of rivers with dissolved greenhouse gases [2]. The rise of organic matter and nutrient in rivers may boost riverine metabolic processes like nitrification, denitrification, and aerobic and anaerobic bio-degeneration [3]. According to Ref [4], the discharge of treated and untreated wastewater into rivers leads to an increase in the concentration of dissolved nutrients and dissolved organic carbon. Enriched bioactivity of organic matter and nutrients causes an increase in GHG emissions from down rivers [5].

Fresh water represents mere 2.5% of all of the available earth's water, and only 1% of fresh water is promptly ready for human and ecosystem's use. According to Ref [6], industries, heating up and cooling, consumed 20% of the existing fresh water in 2010. Wastewater is considered a valuable resource, mainly due to frequent droughts and water scarcity in many places. Thus, treating wastewater offers double benefits, iterating the water supply and preserving the planet from pollutants. Urban, agricultural, and industrial wastewater consist of various kinds of contaminants like nutrients, suspended solids, organic matter, heavy metals, and pathogens. The evacuation of wastewater without treatment causes deleterious impacts to the human health and environment. These impacts include unfavorable alterations to ecosystems, human health hazards, lessening of the economic worth of resources, and aesthetic detriment [7]. The contaminants present in wastewater involve pathogens, suspended solids, nutrients, heavy metals, oil, biological oxygen demand, and grease. Pathogen affects human health. Nutrients increase algal growth, resulting in deterioration of the coral reef environment, lowering the quantity of light reaching the coral and causing blooms of unfavorable toxic phytoplankton. High BOD lowers the dissolved oxygen amount in receiving waters and leads to reduction in the survivorship of several organisms. Oil suffocates benthic organisms and breathing roots and contaminates commercial species [8]. Coral reefs and sea grass are affected by suspended solids by lowering light permeation and suffocating benthic organisms. Heavy metals, which contaminate water, for example, arsenic, lead, cadmium, mercury, and chromium, can kill or minimize survivorship of individuals and bio-magnify in the food chain, resulting in neuronal deterioration and restraining the normal metabolic process and embryonic development in humans [8]. By recognizing the dangerous effects of wastewater, we can realize that it is considered a global issue that should be handled.

Water remediation techniques comprise various technologies of various nature, such as physical techniques, chemical techniques, and biological techniques. The remediation methods ought to be appropriate for the nature of contaminant to maximize their effects. Hence, the remediation method of wastewater should be sustainable to be a vital pillar in achieving sustainability goals. Sustainable liquid waste treatment is defined as a green technology used to treat liquid waste without harming the environment and humans. It requires hindering of diseases, saving energy, using environmentally friendly methods, lowering generation of excess sludge, and effectively controlling their scent issues, as well as ensuring healthy living by retaining high environmental fineness and reducing effluents. Many advantages may be gained by using sustainable technologies for liquid waste treatment. It involves lowering water scarcity by reusing treated water, minimizing the dependency on surface and ground water sources, improving human health by decreasing pathogens in surface water and ground water, and avoiding the eutrophication of waterways.

4.2   TYPES OF LIQUID WASTE

Liquid waste is classified into three categories: sewage, trade liquid waste, and hazardous liquid waste [9].

Sewage is a waste liquid derived from domestic activities and consists of urine and excrement. It comprises house, restaurant, hospital, school, hotel, and public toilet wastes. Sewage water involves gray water and black water. Gray water is defined as the wastewater from bathrooms, sinks, washing machines, and showers [10]. Its treatment process is characterized by low contamination levels. Recycled gray water can be used for irrigation, flushing toilets, and washing. Black water, also called brown water, refers to the wastewater that comes from toilets and bathrooms, which includes urine, fecal matter, soap residues, and sanitary products [11]. It is contaminated by pathogens and greases and is harmful to human. Flood water derived from flooding of water bodies a result of heavy rain and hurricanes is also considered black water since it merges with sewage water and is contaminated with bacteria. Black water can be used as fertilizer after treating, recycling, and composting.

Trade liquid waste is all non-human liquid waste produced from industrial and commercial premises, commercial properties, trade, and business [12]. It involves liquid waste like oils, chemicals, and greases but does not involve wastewater from showers, laundry, and toilets. Storing trade waste without treatment that includes banned substances like grease, burnable liquids, acids, and solids may encumber the sludge system, deteriorate components, and impact the treatment procedure.

Hazardous liquid wastes are liquid wastes that are dangerous or probably harmful to the environment and human health [13]. They are released from industrial activities, hospital, laboratories, and chemical production. Inappropriate storage, treatment, transportation, and elimination processes of hazardous waste may lead to contaminated surface water and ground water and cause land pollution. Hazardous wastes are categorized based on their chemical, physical, and biological features, resulting in materials that are reactive, toxic, ignitable, infectious, radioactive, and corrosive. These wastes might be in different forms, such as gas, liquid, solid, and paste. Much income can be generated by recycling hazardous waste. Regarding the environmental benefits, a reduction of pollution, energy use, volume of waste, and consumption of raw materials can be achieved. Also, an increase in production performance and a decrease in costs related to buying raw materials and waste management can be achieved.

4.3   LIQUID WASTE TREATMENT

A treatment process generally involves four consecutive stages: (i) pretreatment (mechanical and physical), (ii) primary treatment (chemical and physicochemical), (iii) secondary treatment (biological and chemical), and (iv) tertiary (chemical and physical). The liquid waste treatment can be principally categorized into biological, physical, and chemical techniques. These techniques can be used individually or collectively, depending on the pollutant's kind, and the size and volume of the contaminant. In the first treatment, water is stored in a tank, where massive materials sink at the base, and the sprightly material prime on the top. In the secondary treatment, microbes are separated from treated water. A series of chemical and physical techniques are used in the tertiary treatment to remove harmful microbiological pollutants from polluted water. In addition, the tertiary treatment comprises filtration and disinfecting remediation.

In the biological treatment, microorganisms are used to decay organic wastes into carbon dioxides, water, and simple inorganic substances or simpler organic substances. The biological treatment aims to boost the development and activity of microorganisms by controlling their environment and preserving a physical contact of high concentration of microorganisms with the liquid waste. The chemical and physical treatments are used to remove inorganic materials before the biological process since the inorganic materials cannot be decomposed by biological treatment, and the high intensity can robustly restrain the decay activity. Chemical materials are used in the chemical treatment to decompose pollutants and remove many contaminants such as dyes, antifreeze, metal residues, and paints and solvents. Chemical and physical treatments are costlier than biological treatment.

4.4   SUSTAINABLE TECHNOLOGIES FOR WASTEWATER TREATMENT

4.4.1   PHYCOREMEDIATION

Micro- or macro-algae are used in this technique to eliminate contaminants from liquid waste. Phycoremediation is considered a sustainable technology for wastewater treatment as algae are a stellar/premium carbon capture, resulting in minimization of the impacts of carbon footprint [14]. The advantages of this technology can be derived from its excellent nutrient elimination capability, its low operational costs and the acclimation of P and N intro algal biomass, and obviating the necessity for sludge treating and oxygenation of pollutant before its evacuation into water body. In addition, this method is environmentally safe, given that the algal biomass could be used as fertilizer after nutrient abstraction without producing any secondary pollutants [15]. Many treatment mechanisms to eliminate hazardous substances from wastewater can be used in phycoremediation. This technique has the potential to exclude nitrogen and phosphorous from wastewater. Micro-algae synthetize nucleic acids, proteins, and phospholipids, which are used to remove several pollutants, like $NO(3-)$, $PO(3-, 4)$, and N-oxide, and to digest/predigest significant amount of N and P nutrients [16]. Also, microalgae have the ability to reduce the organic matter in pollutants and diminish BOD, total organic carbon, and chemical oxygen demand (COD) levels in contaminants. Moreover, the pH of wastewater can be controlled by algae. Hence, an increase in the level of hydroxide in water caused by the assimilation of CO_2 by algae through photosynthesis leads to increase in the pH of wastewater, resulting in reduction in the acidity of wastewater. In addition, many algal types have the ability to absorb heavy metals. Microalgal remediation generates less sludge than bacterial remediation since microalgae grow at a slow pace and can eliminate more nutrients [17]. Moreover, microalgae play an essential role in improving the quality of water. Many studies reported the efficiency of various microalgae in reducing the amount of BOD, PH, TDS, COD, NO(-3), PO(3-, 4), ammonia, and heavy metals in wastewater. The efficiency of removing contaminants depends on the type of microalgae. For instance, *Chlorella vulgaris* is more efficient in reducing COD and BOD than *C. salina* [18]. In addition to its vital role in treating wastewater, microalgae might be a crucial source of biomass for energy production. Biomass is considered a renewable energy resource and plays an important role in reducing the carbon footprint and in achieving sustainable economy. Microalgae produce biomass by using P and N, and CO_2 and sunlight from the atmosphere. By using processing technologies, the generated biomass might be used to synthesize green biofuel such as biodiesel, bioethanol, and biomethane. The generated biomethane can be used to produce carbon neutral energy, and the generated biogas can be used to generate energy, liquid fuels, and fuel cells.

Many treatment methods use microalgae to remove dangerous compounds from wastewater including both organic and inorganic waste, and heavy metals. These methods are discussed below.

Constructed wetlands (CWs): These are systems that use the natural processes including soil, vegetation, and the linked microbial assemblage to facilitate the remediation of wastewaters. Based on the dominant macrophyte, CWs can be categorized into four types, namely, floating leaved, submerged, free-floating, and rooted emergent macrophytes [19]. Also, this method may be classified into free water surface and subsurface systems relative to the wetland hydrology [20]. The subsurface flow may be categorized into horizontal and vertical flow based on the flow direction (Figure 4.1).

All kinds of CWs are highly effective in eliminating organic compounds by aerobic, anoxic, and anaerobic microbial degradation. The vertical flow CW shows the highest treatment performance attributable to its high inflow concentrations. The horizontal flow constructed wetland is used in the remediation of wastewater diluted with stormwater runoff [20]. The vertical flow CW is used in the primary and secondary stages of remediation. However, the free water surface CW is used in the tertiary stage of treatment [22]. For suspended solids, all kinds of CWs are used for their removal. CWs are not effective in eliminating nutrients. The soil/peat accretion is used for long period phosphorus sewers. The free water surface CW has the highest potential in removing phosphorus due to the friction of water with soil particles that is bounded, and phosphorus is absorbed and/or precipitated by soil particles. However, the phosphorous retention capacity is very small compared to the phosphorus components present in wastewaters [23]. Recently, some filtration materials like lightweight clay aggregates and by- and waste products like steel slags have been developed and shown to enhance phosphorous removal [24,25]. Free water surface CWs coupled with the emergent macrophyte system is a series of tanks with a water profundity of 20–40 cm, comprising 20–30 cm of implanting soil. This technique is effective in eliminating organic compounds via microbial decay/disintegration and sediment of colloidal particles. Besides, subsequent denitrification, ammonia

FIGURE 4.1   Schematic of CW [21]. CW, constructed wetland.

volatilization in the circumstances of high photovoltaic (PV) values attributable to algal photosynthesis, and nitrification in the water column are used to remove nitrogen. Also, it can efficiently eliminate suspended solids through sediment and filtration via bushy vegetation. Since the friction of water with soil particles is bounded and the phosphorus is absorbed and/or precipitated by soil particles, the detention of phosphorus is low. Moreover, many kinds of wastewaters have been treated using this technique, including animal wastes, paper and pulp waste, agricultural drainage, dairy pasture runoff, and landfill leachate.

Elimination of nitrogen by using the CW technology is low or zero based on the method used, owing to weak nitrification in saturated water. It is low in horizontal-flow CWs and FWs, and it is zero in vertical flow CWs [20,23]. Ammonia volatilization could be an appropriate solution/correct approach to solve the low nitrogen removal in CWs with open water surface. Hence, subsequent denitrification, ammonia volatilization in the circumstances of high PV values attributable to algal photosynthesis, and nitrification in the water column are used to remove nitrogen [26]. In the case of VF CWs, a combination of VF CWs with HF CWs could be an appropriate solution by supplying/making suited environment to minimize the nitrogen generated throughout/through nitrification in VF beds [23,27].

- **Waste stabilization ponds (WSPs)**: These are used for the treatment process of industrial and municipal wastewater. This technology is appropriate for developing and low-income countries. WSP systems offers many advantages such as a simple design and building, low capital, maintenance, and operational costs, significant efficiency in pathogen elimination if properly designed, and a low generation of biological sludge. In addition, WSP systems are sturdy, credible, and less delicate to shock loading. They comprise single chains of aerobic, facultative, and maturation lagoons, or many chains in parallel. Facultative and anaerobic lagoons are modeled to remove BOD, while maturation lagoons are designed for pathogen removal [28].

The anaerobic lagoons are the smallest series and are utilized in the primary remediation process. These are more appropriate in the treatment of wastewater with high concentrations of BOD. Sedimentation and anaerobic digestion are used to reduce BOD and solid concentration (IETC-UNEP, 2002). The efficiency of sedimentation and anaerobic digestion depends on the temperature. However, the required/desired volumetric BOD loading determines the time of hydraulic retention, and it can reach 20 days [29]. At 20°C, 60% of BOD could be removed by using a duly designed anaerobic lagoon. In the case of temperatures more than 20°C and a BOD concentration up to 300 mg/L, 1-day hydraulic retention is enough [30]. The decrease in temperature has a negative effect on the anaerobic digestion process, and the sedimentation becomes the dominant process in this case [28]. On the contrary, anaerobic ponds have many disadvantages and issues related to smell (odor) and the increased concentration of sulfide and ammonia, owing to the anaerobic processes [28,31]. The increase in sulfide concentrations has the potential to remove *Vibrio cholera* [32] (Figure 4.2).

4.4.2 PHYTOREMEDIATION

Phytoremediation is defined as green and ecofriendly mechanisms based on the use of plants to treat organic and inorganic contaminants and other pollutants that could be difficult or arduous to metabolize. Also, phytoremediation is low-cost treatment and has the ability to help in the removal of heavy materials, which are considered as dangerous pollutants and deposit as sediments [34]. Phytoremediation is classified into two categories: continuous or natural, and chemically enhanced [35]. The continuous phytoremediation uses natural hyperaccumulator plants, whose shoots have an extraordinary metal accumulation and root have an excellent endurability to metal poisonousness [36]. However, continuous phytoremediation has many drawbacks such as low biomass, sluggish extensions and low relocation of hyperaccumulator plants, immobility of metallic element, and low capturing element by roots [37]. Chemically enhanced phytoremediation is developed to mitigate the restrictions of the continuous/natural phytoremediation by improving and establishing many features of phytoremediation founded on pollution types, decontamination technique, influenced agent, and treatment environment [38]. Phytoremediation is a technology involves mainly five methods, namely, phytoextraction, phytovolatilization, phytodegradation, rhizofiltration, floating treatment wetland (FTW), and phytostabilization [39].

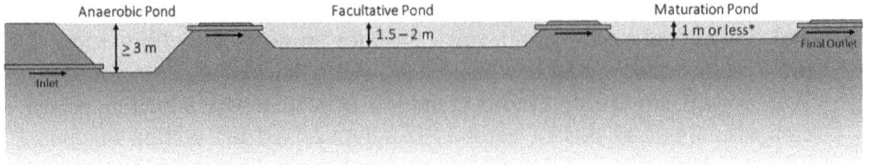

FIGURE 4.2   Schematic of a waste pond system [33].

- **Phytoextraction**: Phytoextraction or phytoaccumulation is a technology based on the absorption and relocation of metal pollutant in the soil through plant roots into the aboveground parts of the plants. Hyperaccumulators are plants that have the ability to absorb uncommonly larger quantities of metals than other plants. One or more mixture of hyperaccumulators are chosen and grew at a place derived from the metal's kind. After sundry weeks or months of growing, the plants should be cropped and then burned or composted to recycle the metals. This process could be occurring several times to minimize the contaminant levels to permissible limits. Table 4.1 shows the names of plants used in the removal process of different contaminants.

- **Phytovolatilization**: In this method, contaminants are removed and fractionated into the plant air area, to be then spread within the air surrounding [47]. So, specific metals could be volatilized by plants, like volatile and semivolatile organic compounds [48]. In this process, contaminants are

TABLE 4.1

Names of Plants Used to Remove Different Kinds of Contaminants

Contaminants	Plants Used in the Removal Process	References
Cadmium	Water spinach (Ipomoea aquatic)	[40]
Zinc	Duck wood (*Lemna gibba*)	[41]
Chromium	Small pondweed (*Potamogeton pusillus*)	[42]
Pb	*Ceratophyllum demersum* and *Myriophyllum spicatum*	[43]
Heavy metals from coastal water	Water hyacinth (*Eichhornia crassipes*)	[44]
Crude oil from artificial wastewater	Urea fertilizer and palm oil mill effluent treatment	[45]
Heavy metals from industrial wastewater	Vetiver (*Chrysopogon zizanioides*)	[46]

absorbed and concentrated in roots. Simultaneously, some ions, predominantly metals, start to displace to minimize the generated tension in plant tissues. The remaining pollutants are displaced to the photosynthetic site and lead to two operations: extraction of pollutants through evapotranspiration in the presence of water to transfer them into the air space. Meantime, the volatilization of the contaminants is induced by temperature and ultraviolet (UV) rays near the leaves stomata, resulting in deposition of a great amount of the contaminants into the air space and converting them into less ecotoxic components. Phytovolatilization is viewed as valuable since it results in photochemical disintegration and an important reduction of contaminants. Contrariwise, phytovolatilization may be regarded as a threat to urban areas owing to the decline in the air quality.

- **Phytodegradation**: It is the ability of plants to generate enzymes that accelerate/stimulate the degeneration reactions of xenobiotics [49]. The process of degradation starts with the generation of enzymes by plants that are secreted in the root area. Phytodegradation can take place inside and outside of the plants and is useful for the treatment of river, surface and ground water, and soil. The secreted enzymes mineralize the contaminants into inorganic compounds or degenerate the contaminants to less harmful compounds that is pulled by the cell wall or isolated by the vacuoles [50]. The plant exudates can enhance the microbial metabolic activity and development, resulting in boosting the degradation of pollutants [51,52].
- **Phytostabilization**: The plants used in this method prevents the motion of the pollutants to ground water and their decampment to the surface. The plants used in this process should have some specific features: an advanced root system that enhance absorption, adsorption, and aggregation of pollutants in the tissues and their transforming to less solvable compounds in the rhizosphere [53], as well as low accretion ability for contaminants in their aboveground section, and high tolerance to pH changes, saltiness, and soil humidity [54]. The main advantages of this method is the zero secondary waste, low cost, less required materials, and improved soil fertility.

- **Rhizofiltration**: It is used to treat surface and ground water, polluted by contaminants evacuated from agricultural drainage, industries, as well as acid mine runoff. This technique can competently eliminate heavy metals like cobalt, copper, arsenic, lead, and cadmium and is considered sustainable and cost-efficient. The roots of hydroponic plants are used to eliminate metal pollutants by imbibing and precipitating the heavy material from the marine environment (Figure 4.1). Either ground-dwelling and marine algae could be used in this technology. Water hyacinth and duckweed are examples of marine plants used in rhizofiltration, while tobacco, spinach, sunflower (has the greatest potential to remove lead from contaminants), and Indian mustard are examples of terrestrial plants [55]. The ground-dwelling plants are more auspicious due to their string and prolonged roots along with their rapid growth and greater effective area [56]. Moreover, this technology is used to remediate high volumes with a low amount of toxic chemical, to decrease the volume of ancillary waste and the potential of recycling [57]. In addition, this method can be used on-site and off-site (Figure 4.3).
- **FTWs**: Plants are able to absorb nutrients and desorb poisoned composites. In this technology, plants are implanted on a buoyant carpeting, while their roots are expended down to the polluted water. The roots act as biological filters, hence taking up toxic metals and elements, and nutrients from the wastewater. However, microorganisms are used to deteriorate organic matter and compose biofilms on the carpeting surface and the roots. Low-density plants or plants with aerenchymatous capabilities are used to maintain floating [59]. The choice of the plant depends of many factors such as kind of the contaminator, workability to the water component and environmental conditions, capability of plants to synthesize a thick root frame, and local accessibility [60] (Figure 4.4).

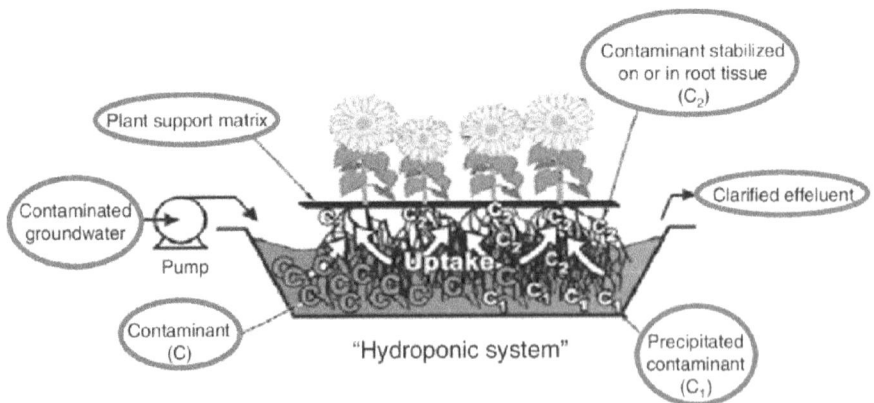

FIGURE 4.3 Schematic of rhizofiltration [58].

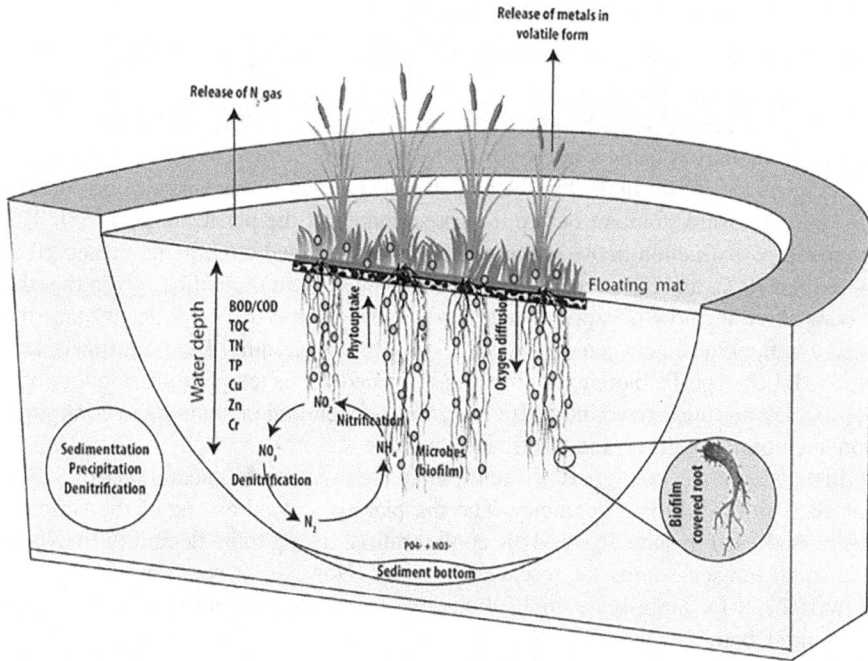

Release of metals in volatile form

Release of N₂ gas

Floating mat

Water depth

BOD/COD
TOC
TN
TP
Cu
Zn
Cr

Phytouptake

Oxygen diffusion

Sedimenttation
Precipitation
Denitrification

NO₃
Nitrification
NH₄⁺
Microbes
(biofilm)

NO₃
Denitrification

N₂

PO₄ + NO₃

Sediment bottom

Biofilm covered root

FIGURE 4.4 Visual representation of FTW [61]. FTW, floating treatment wetland.

The design of FTWs depends on many factors like water depth. The kind of wastewater, stream change, and treatment aim determine the water depth. Actually, low water depth is convenient to strip/slough suspended solid and fine particles, while high water depth is more appropriate to rectify rough suspended solids. The effectiveness of FTWs relies on air temperature and solar radiations. This is due to the deterioration procedure of contaminants from wastewater by integrated biogeochemical processes that are commanded/piloted by the weather parameters. Hence, the enhancement of plant growth and microbial proliferation during spring promote the contaminant degradation processes, while the reduction of microbial and plant proliferation leads to lessen the extraction of pollutants. Aeration of wastewater can maximize the extraction of nutrients from wastewater. This is due to creation of aerobic micro-zones by aeration, which catalyzes the generation of biofilms in the substrate. FTWs can be used to strip organic and inorganic contaminants from wastewater through different processes. The extraction of organic matter from waste is performed by emplacing/fixing to the roots and substrate. However, other mechanisms could be used in the removal process of organic matter such as oxygenation, filtration, and nutrient assimilation. The elimination process of nutrients is conducted by many biogeochemical mechanisms. Nitrogen, ammonia, and NO-2 are taken up by microalgae and plants through their roots [62]. The elimination of phosphorus is performed by many processes such as precipitation, complexation, sorption, and incorporation into plant biomass [63].

4.4.3 MEMBRANE DISTILLATION BIOREACTOR

The membrane distillation bioreactor is a sustainable technology used to treat the waste heat generated in industries and generate high goodness water. It is a hybrid system of the thermophilic bioprocess and the MD process. The operating principle of this technology is using a microporous hydrophobic membrane to create a vapor–liquid interface. Water in the hot feed gasifies on the membrane surface and spreads the vapor pressure gradient before it is precipitated at the permeate part [64]. The temperature distinction between the permeate and the feed ought to be preserved at more than 30°C, and the permeate must be ideally at more than 50°C, given that the driving force for flow is vapor pressure gradient. Specific thermophiles (organisms need a temperature between 45 and 80°C to grow) are required to be utilized, and the used thermophilic bioreactor must be performed at this temperature domain [65]. MBRs may be categorized into different groups depending on membrane configuration, membrane position, and bio-treatment process.

In the membrane configuration factor, there are two kinds of membranes of MBR: flat plate and cylindrical geometry. The flat plate is viewed as one of the simplest design and used in submerged MBR configuration, owing to its flexibility to evolve into small lab-scale units for research activities. However, it could not be used or is overlooked for large-scale applications due to its low packing density and large footprint. Contrarily, the cylindrical shape form has high packing density and space economy, resulting in making this technology more competitive [66]. The membrane of cylindrical geometry can be HG or multitube based on the flow direction and lumen diameter. In HF, water streams from the shell side to the lumen side through tight lines with a diameter fluctuate from 0.4 to 2.6 mm. However, water streams from the lumen side to the shell side and have a comparatively larger diameter (5–15 mm) [67].

The bio-treatment is classified into three categories: anoxic, aerobic, and anaerobic [68]. In the aerobic process, organic carbon and ammonia, present in the waste flow, are transformed, using dissolved oxygen, into carbon dioxide and nitrate, consecutively. In the anoxic process, the oxidation of carbon dioxide occurs due to the lack of dissolved oxygen in the presence of another oxygen source. In the case of absence of dissolved and oxygen sources for biochemical transformation of organic carbon, the process is called anaerobic, and it is used usually to support phosphorus disposal [69].

In membrane position, MBR is designed in two shapes: membrane located/implanted outgoing the bioreactor (side-stream-MBR), and membrane inglorious/obscure inward the bioreactor (submerged-MBR). The extraction of remediated water from the membrane is occurred by suction pressure actionable on the permeate part, and/or status head on the feed side in the case of submerged MBR. However, water needs high trans-membrane pressure to perforate out of the membrane [67] (Figure 4.5).

4.4.4 NANOTECHNOLOGY

Nanotechnology is viewed as a sustainable technology for wastewater treatment. This technology uses special materials that are highly different from the conventional one, in the matter of electrical, optical, magnetic, and mechanical features,

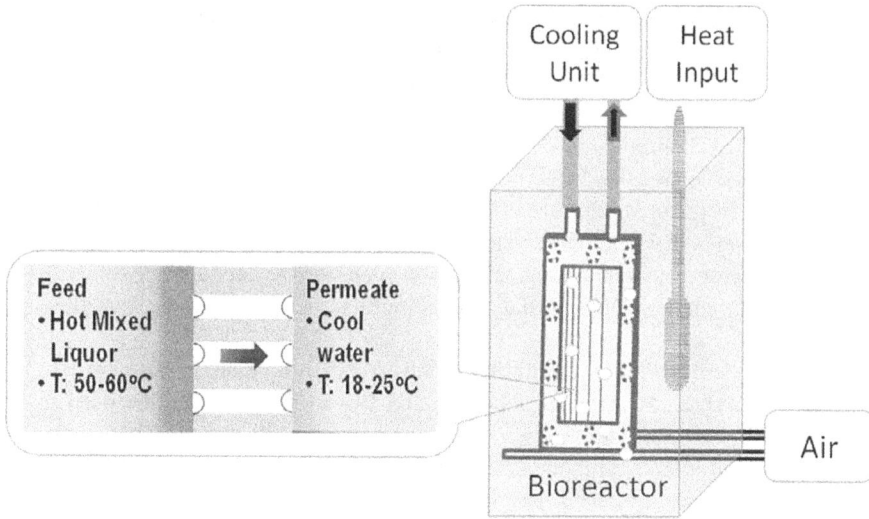

FIGURE 4.5 Schematic representation of an MDBR method [67].

ascribed to their nanoscale dimensions. These materials are characterized by small dimension and large surface, resulting in high absorbency and reactivity. Many studies demonstrated the ability of nanomaterials in removing organic contaminants, inorganic anions, heavy metals, and bacteria. The nanomaterials involve metal oxide nanoparticles, nanocomposites, zero valent metal, and carbon nanotubes.

- **Metal oxide nanoparticles**: Photocatalytic degradation technology demonstrated its ability in wastewater remediation by oxidizing pollutants into low-molecular weight products. These products convert into H_2O, CO_2, and anions like PO_3^- and Cl^- for reuse. The materials used as photocatalysts are metal oxides and sulfide semiconductors. Titanium dioxide (TiO_2) is regarded as one of the greatest photocatalyst subsisted, assignable/owing to its high photocatalytic activity, low cost, harmless feature, easy obtainability, and chemical and biological stability [70,71]. ZnO nanoparticles have appeared as a beneficial photocatalyst for wastewater remediation. Their advantages derive from their high oxidation ability, and eco-friendly and good photocatalytic characteristic [72]. ZnO nanoparticles are considered an ideal candidate for sewage remediation due to its compatibility with organisms. The metallic doping of ZnO nanoparticles by metal dopants such as cationic, codopants, and anions plays an important role in enhancing their photodegradation [73]. Also, boosting photodegradation performance can be achieved by coupling ZnO nanoparticles with other semiconductors like cadmium oxide, graphene oxide, and stannic dioxide [74–76]. In addition, ZnO is more cost-effective than TiO_2 [77].
- **Carbon nanotubes**: Carbon nanomaterials have demonstrated a great potential in wastewater treatment due to their high strength in absorbing

a large range of pollutants, large specific surfaces, and selectivity toward aromatics. Carbon nanotubes are one kind of carbon nanomaterials, have attracted more attention, and have developed rapidly, ascribed to their high absorptivity for a wide range of pollutants and high specific surface areas. The combination of carbon nanotubes with metals can enhance their surface areas and electrical, mechanical, and optical features [78,79]. Moreover, nanoparticles are effective against various kinds of bacteria and fungi. Nanoparticles emancipate metal ions that react with cellular elements via different ways involving reactive oxygen species production, cell wall deterioration, and DNA spillage, resulting in curbing/reining the growth of cells [80].

- **Zero-valent metal nanoparticles**: Silver and iron nanoparticles are two types of zero-valent metal nanoparticles. Silver nanoparticles are used in water disinfection due to their good antibacterial and antimicrobial properties [80]. Silver nanoparticles are highly poisonous to microorganisms (fungi, viruses, bacteria). The interaction of silver nanoparticles with bacteria generates free radicals, which are used to deteriorate cell membranes, resulting in death of the cells. Also, the breakdown of silver nanoparticles releases Ag^+ ions, which react with enzyme thiol groups, deactivate them, and inhibit the normal functioning of the cells [81,82].
- Al, Fe, and Zn nanoparticles have also been used in wastewater remediation [80]. Fe is considered a premium adsorbent, is comparatively cheap, precipitates, and oxidizes (with the existence of oxygen) [83]. More studies have been conducted on zero-valent iron (Fe) among zero-valent metal nanoparticles. Nano zero-valent Zn powders are also considered an excellent adsorbent toward soluble organic compounds epitomized by COD. The performance of the adsorbent depends on many factors such as contact time, COD concentration, and pH [84].

4.4.5 ECO-FRIENDLY ADSORPTION TECHNOLOGY

Adsorption is viewed as an excellent and promising technology for the elimination of heavy metal contaminants, owing to its conceptual and applicable clarity, and quickness. The working principle is based on capturing the contaminant, heavy metal, from polluted water by chemical and physical processes. The used adsorbents are natural and are obtained from living and non-living matrixes, and biotic and abiotic organisms. To optimize the performance of adsorbents and to achieve an economic profitability, a chemical and a physical reinforcement should be performed. Cow dung powder is one of the adsorbents used for treating wastewater. A previous study [85] demonstrated that the DCP might be an efficient and green adsorbent for the elimination of heavy metal and carcinogenic contaminants in water. In addition, this study pointed out the cow dung powder does not require any pre- or post-remediation, and it is energy- and time-efficient.

A previous study [86] examined the effect of bentonite, chitosan, and chitin on removing contaminants from wastewater. The results revealed that these are excellent cheap natural adsorbents for removing BOD_5 and COD from wastewater.

The contaminants may be adsorbed onto various adsorbents requiring variable degrees of elimination performance and different equilibrium times. pH affects the adsorption ability of the contaminants by the sorbents studied in this study. Chitosan and chitin are deemed as good coagulants, and are green and affordable for smaller wastewater remediation.

4.4.6 Advanced Oxidation

Advanced oxidation is defined as a chemical oxidation escorted by hydroxyl radicals, which are characterized by high reactivity and fugitive oxidants. Radicals should be generated locally in a reactor to enable them to interact with the organics in polluted water. Advanced oxidation has the ability to reduce global organic content, rises bioavailability of resistant organics, remediates sludge, decreases odor, and destroys specific contaminants. The hydroxyl radical (OH) is an unselective oxidant that has an ability to interact rapidly with the organic compounds. The rate constant for the pollutant with hydroxyl radical determines the proportion of devastation of pollutants. Many technologies could be used to generate hydroxyl radicals such as ozone, UV radiation, Fenton reagent, and titanium dioxide/UV radiation.

- **Ozone filtration/disinfection**: Actually, ozone is designated as one of the most effective, safest, eco-friendly, and quickest microbicides, attributable to its powerful oxidizing features. The decomposition of ozone in water forms hydroxyl (OH) and hydrogen peroxyl, which have a cogent oxidizing ability and a great potential in the disinfection operation [87]. In addition, ozone aids the elimination of phenols, cyanides, and tensides from wastewater. In addition, ozonation can be used to eliminate iron and manganese and to restrict the accumulation of manganese and iron on pipelines, and other components, resulting in minimizing the maintenance cost. Ozonation offers many benefits related to contact time, residuals, and contaminants. Ozone removes COD, BOD, and other pollutants in wastewater and blocks/restraints the re-growth of microorganisms, without leaving hazardous residuals and with a short contact time [87]. Also, ozone does not require pre-aeration and filtration. Moreover, less equipment and area, and storage are requested, resulting in reducing their associated cost. The main drawbacks of this technology are high initial capital cost that can be offset in the long term due to needlessness of residual elimination, and lessening of storage and maintenance cost. In recent years, catalytic ozonation gained more attention, attributable to its role in enhancing the performance of ozonation in the process of wastewater treatment [88]. Catalysts are used to boost the decomposition of O_3, produce active free radicals, and used to reinforce the degeneration and mineralization of organic contaminants [88] (Figure 4.6).
- **Fenton technology**: This technology is based on the production of active oxygen species by reacting hydrogen peroxide and iron ions to oxidize organic and inorganic matter. Fenton reagent is the combination of iron ions and hydrogen peroxide. This reagent has the ability to remedy a large quantity

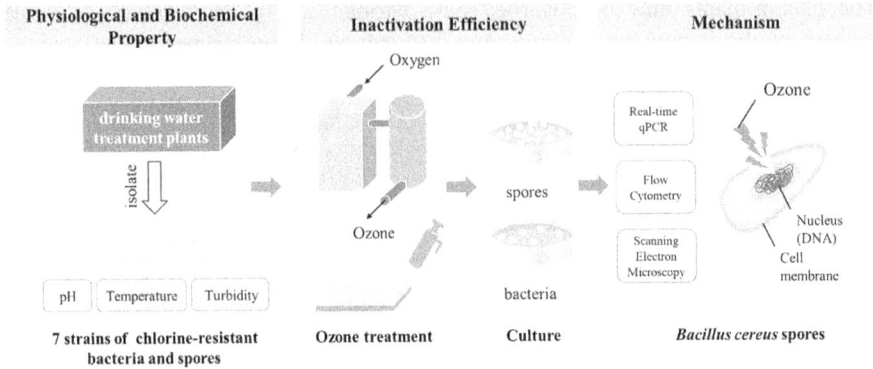

| Physiological and Biochemical Property | Inactivation Efficiency | Mechanism |

FIGURE 4.6 Working principle of ozone disinfection [89].

of polluted water, involving formaldehyde, pesticides, phenols, BTEX, and dyes [90]. This method offers many economic and technic benefits [90]. It can be performed under ambient pressure and temperature, with less energy to achieve fast deterioration and full mineralization of organic compounds. Also, the organic matter is reduced, resulting in ease permitting for more traditional biological remediation. Compared to other advanced oxidation techniques, the Fenton technology is cheaper since hydrogen peroxide and iron are inexpensive and scot-free. On the contrary, the Fenton process has three disadvantages related to pH value, formulation of iron sludge, and remediation and storage of reactants [90]. pH has a significant effect on the treatment performance, where the ideal pH is in the range between 2 and 4. A large amount of chemical is requested to preserve the pH of polluted water since polluted water generated from various industries have different pH values. The reactants hydrogen peroxide and ferrous-ion chemical may explode when stored due to their unstable and reactive properties. In addition, H_2O_2 is deleterious to humans since it is explosive and poisonous [90]. A decrease in catalytic activity and a wastage of iron species can occur due to recurrent reaction of Fe^{2+} with Fe^{3+} (reaction 1) and Fe^{3+} with Fe^{2+} (reaction 2). A formation of iron sludge may also occur, attributable to the accumulation of Fe^{3+} in the aqueous solution, which sediments by means of oxyhydroxide, known as iron sludge [90].

– **UV radiation**: UV is used to disinfect remediated wastewater. It is a physical treatment based on passing wastewater near a UV source. The chemical and physical finesses properties of the wastewater before the disinfection process determine the performance of UV disinfection [91]. The efficiency of UV disinfection increases with the increase in the wastewater quality. This technology is rapid, without generating byproducts and poisoning the water. A combination of H_2O_2 and UV radiation in a synergistic effect to destroy harmful microorganisms and organic chemicals in the wastewater medium [91]. The use of this technology is the production of hydroxyl radicals, which is the main factor that determines the success of this technology. The main disadvantage of this method is the re-growth of microbes during the piping over a long distance and storing for a long time [92]. Also, the performance of UV radiation for virus degradation is less than that of chlorine, and there is no residue for allocation [92].

- **Photocatalysis**: It is an advanced oxidation process, based on the generation of electron–hole pair after shedding light on semiconductor materials to induce chemical reactions such as photocatalytic reactions [93]. The energy of a photon should be equal or largen than the semiconductor band gap [94]. Photocatalysis may be categorized as homogeneous and heterogeneous, as determined by the number of phase and the use of catalysts, such as carbon materials, and semiconductors [93]. Homogeneous photocatalysis is identified by chemical conversions based on the interaction between the chemical reagent and target composite [95] There are two kinds of photocatalytic utilization in the remediation of water: solar photocatalysis and photocatalytic system combined with a source of artificial UV light. These two systems can be used under ambient temperature to destroy many kinds on microbiological and chemical contaminants such as microbes, pesticides, crude oil, inorganic compounds, and metals present in wastewater. The solar photocatalysis process is considered cheap, eco-friendly, and globally actionable due to the use of sunlight. The most widespread photocatalyst is titanium dioxide (TiO_2), attributable to their high reactivity, low cost since it is abundant, and high capability to destroy contaminants by breaking them. The main drawback of this process is indoor applications since the activation of nano-TiO_2 requires UV light, leading to the need of a catalyst to absorb visible light [96]. So, to guarantee a high performance and durability of the catalyst, determination of the most efficient substratum component and set-up system for various utilization is needed. Also, TiO_2 might breakdown any organic matrix where the nanoparticles are enclosed [96]. Therefore, nano-TiO_2 may be used in an inorganic ambiance.

4.5   CONCLUSIONS

Water shortage is considered among the major features of the climate catastrophe. The perpetually increasing population and its rising needs for water resources have led to a requirement of recovering and reusing water to respond demands. Many advantages can be obtained from reusing wastewater such as reduction of waste volume, generation of fertilizer, preventing diseases, ensuring a balance between water supply and demands by reusing the water, and a source of energy. Sustainable liquid waste treatment is defined as a green technology used to treat liquid waste without harming the environment and humans. The present chapter sheds light on the different sustainable technologies used for wastewater treatment. These sustainable technologies use plants, microalgae, natural sorbents, nanoparticles, oxidants, and membrane distillation.

REFERENCES

1. Demirbas, A., Concept of energy conversion in engineering education. *Energy Education Science and Technology Part B-Social and Educational Studies*, 2009. **1**(3–4): p. 183–197.
2. Ward, N.D., T.S. Bianchi, P.M. Medeiros, M. Seidel, J.E. Richey, R.G. Keil, & H.O. Sawakuchi, Where carbon goes when water flows: Carbon cycling across the aquatic continuum. *Frontiers in Marine Science*, 2017. **4**: p. 7.
3. Yoon, T.K., H. Jin, M.S. Begum, N. Kang, & J.-H. Park, CO_2 outgassing from an urbanized river system fueled by wastewater treatment plant effluents. *Environmental Science & Technology*, 2017. **51**(18): p. 10459–10467.
4. Sickman, J., M. Zanoli, & H. Mann, Effects of urbanization on organic carbon loads in the Sacramento River, California. *Water Resources Research*, 2007. **43**: p. 11.
5. Begum, M.S., I. Jang, J.-M. Lee, H.B. Oh, H. Jin, & J.-H. Park, Synergistic effects of urban tributary mixing on dissolved organic matter biodegradation in an impounded river system. *Science of the Total Environment*, 2019. **676**: p. 105–119.
6. Linke, S., E. Turak, & J. Nel, Freshwater conservation planning: The case for systematic approaches. *Freshwater Biology*, 2011. **56**(1): p. 6–20.
7. Fagan, L.L., N. Cakausese, & E. Anderson, *Water Quality in the Ba River and Estuary: Environmental Effects of Multiple Resource Use*. 1995: Institute of Applied Sciences, University of the South Pacific, Suva, Fiji.
8. Cripps, K., Survey of the point sources of industrial pollution entering the port waters of Suva. *For Ports Authority of Fiji*, 1992, p. 74.
9. Syed, S., Solid and liquid waste management. *Emirates Journal for Engineering Research*, 2006. **11**(2): p. 19–36.
10. Morel, A., *Greywater Management in Low and Middle-Income Countries*. 2006: Dubenforf, CH: Swiss Federal Institute of Aquatic Science and Technology.
11. Brandes, M., Characteristics of effluents from gray and black water septic tanks. *Journal (Water Pollution Control Federation)*, 1978. **50**: p. 2547–2559.
12. Wilson, H.M. & H.T. Calvert, *A Text-Book on Trade Waste Waters: Their Nature and Disposal*. 1913: Griffin, Michigan, USA.
13. Tittlebaum, M.E., R.K. Seals, F.K. Cartledge, S. Engels, & H.R. Fahren, State of the art on stabilization of hazardous organic liquid wastes and sludges. *Critical Reviews in Environmental Science and Technology*, 1985. **15**(2): p. 179–211.
14. Koul, B. &P. Taak, *Biotechnological Strategies for Effective Remediation of Polluted Soils*. 2018: Springer, Singapore.
15. Liu, C., S. Subashchandrabose, H. Ming, B. Xiao, R. Naidu, & M. Megharaj, Phycoremediation of dairy and winery wastewater using Diplosphaera sp. MM1. *Journal of Applied Phycology*, 2016. **28**(6): p. 3331–3341.
16. Nagase, H., K.-I. Yoshihara, K. Eguchi, Y. Okamoto, S. Murasaki, R. Yamashita, K. Hirata, & K. Miyamoto, Uptake pathway and continuous removal of nitric oxide from flue gas using microalgae. *Biochemical Engineering Journal*, 2001. **7**(3): p. 241–246.
17. Muñoz, R., C. Rolvering, B. Guieysse, & B. Mattiasson, Photosynthetically oxygenated acetonitrile biodegradation by an algal-bacterial microcosm: A pilot-scale study. *Water Science and Technology*, 2005. **51**(12): p. 261–265.
18. El-Sheekh, M.M., A.A. Farghl, H.R. Galal, & H.S. Bayoumi, Bioremediation of different types of polluted water using microalgae. *Rendiconti Lincei*, 2016. **27**(2): p. 401–410.
19. Brix, H., The use of aquatic macrophytes in water pollution control. *Ambio*, 1989. **18**: p. 100.
20. Vymazal, J. & L. Kröpfelová, *Wastewater Treatment in Constructed Wetlands with Horizontal Sub-Surface Flow*. Vol. 14. 2008: Springer Science & Business Media, Dordrecht.

21. Vymazal, J., Constructed wetlands for wastewater treatment. *Water*, 2010. **2**(3): p. 530–549.
22. Kadlec, R.H., Overview: Surface flow constructed wetlands. *Water Science and Technology*, 1995. **32**(3): p. 1–12.
23. Kadlec, R., Comparison of free water and horizontal subsurface treatment wetlands. *Ecological Engineering*, 2009. **35**(2): p. 159–174.
24. Vohla, C., E. Poldvere, A. Noorvee, V. Kuusemets, & Ü. Mander, Alternative filter media for phosphorous removal in a horizontal subsurface flow constructed wetland. *Journal of Environmental Science and Health*, 2005. **40**(6–7): p. 1251–1264.
25. Jenssen, P. & T. Krogstad, Design of constructed wetlands using phosphorus sorbing lightweight aggregate (LWA). *Advances in Ecological Sciences*, 2003. **11**: p. 259–272.
26. Brix, H. & H.-H. Schierup, Sewage treatment in constructed reed beds—Danish experiences, In: Lijklema, L., Imhoff, K. R., Ives, K. J., Jenkins, D., Ludwig, R. G., Suzuki, M., Toerien, D. F., Wheatland, A. B., Milburn, A. & Izod, E. J. (Eds) *Water Pollution Research and Control Brighton*. 1988, Elsevier, Brighton, p. 1665–1668.
27. Vymazal, J., Removal of nutrients in various types of constructed wetlands. *Science of the Total Environment*, 2007. **380**(1–3): p. 48–65.
28. Mara, D. & H. Pearson, *Design Manual for Waste Stabilization Ponds in Mediterranean Countries*. 1998: Lagoon Technology International Leeds, Leeds.
29. Hager, L.S., *Membrane Systems for Wastewater Treatment*. 2006: WEF Press, USA.
30. Mara, D., *Design Manual for WSP in UK*, Leeds University, UK. Accessible in: http://www.leeds.ac.uk/civil/ceri/water/ukponds/publicat/pdmuk/pdmuk.html, 2003.
31. Crites, R.W., E.J. Middlebrooks, & S.C. Reed, *Natural Wastewater Treatment Systems*. 2010: CRC press, Boca Raton.
32. Mara, D., H. Pearson, J. Oragui, H. Arridge, & S. Silva, Development of a new approach to waste stabilization pond design. *Research Monograph*, 2001. **5**: p. 2–9.
33. Ho, L.T., W. Van Echelpoel, & P.L. Goethals, Design of waste stabilization pond systems: A review. *Water Research*, 2017. **123**: p. 236–248.
34. Muthusaravanan, S., N. Sivarajasekar, J. Vivek, T. Paramasivan, M. Naushad, J. Prakashmaran, V. Gayathri, & O.K. Al-Duaij, Phytoremediation of heavy metals: Mechanisms, methods and enhancements. *Environmental Chemistry Letters*, 2018. **16**(4): p. 1339–1359.
35. Lombi, E., F. Zhao, S. Dunham, & S. McGrath, Phytoremediation of heavy metal–contaminated soils: Natural hyperaccumulation versus chemically enhanced phytoextraction. *Journal of Environmental Quality*, 2001. **30**(6): p. 1919–1926.
36. Assunção, A.G., H. Schat, & M.G. Aarts, Thlaspi caerulescens, an attractive model species to study heavy metal hyperaccumulation in plants. *New Phytologist*, 2003. **159**(2): p. 351–360.
37. Hem, J.D., *Study and Interpretation of the Chemical Characteristics of Natural Water*. 1970: US Government Printing Office, Reston, VA.
38. Stout, L. & K. Nüsslein, Biotechnological potential of aquatic plant–microbe interactions. *Current Opinion in Biotechnology*, 2010. **21**(3): p. 339–345.
39. Zhang, B., J. Zheng, & R. Sharp, Phytoremediation in engineered wetlands: Mechanisms and applications. *Procedia Environmental Sciences*, 2010. **2**: p. 1315–1325.
40. Wang, J., Z. Zhang, Y. Su, W. He, F. He, & H. Song, Phytoremediation of petroleum polluted soil. *Petroleum Science*, 2008. **5**(2): p. 167–171.
41. Khellaf, N. & M. Zerdaoui, Phytoaccumulation of zinc by the aquatic plant, Lemna gibba L. *Bioresource Technology*, 2009. **100**(23): p. 6137–6140.
42. Monferrán, M.V., M.L. Pignata, & D.A. Wunderlin, Enhanced phytoextraction of chromium by the aquatic macrophyte Potamogeton pusillus in presence of copper. *Environmental Pollution*, 2012. **161**: p. 15–22.

43. El-Khatib, A., A. Hegazy, & A.M. Abo-El-Kassem, Bioaccumulation potential and physiological responses of aquatic macrophytes to Pb pollution. *International Journal of Phytoremediation*, 2014. **16**(1): p. 29–45.
44. Agunbiade, F.O., B.I. Olu-Owolabi, & K.O. Adebowale, Phytoremediation potential of Eichornia crassipes in metal-contaminated coastal water. *Bioresource Technology*, 2009. **100**(19): p. 4521–4526.
45. Hadiyanto, M.C., D. Soetrisnanto, & M. Christwardhana, Phytoremediations of palm oil mill effluent (POME) by using aquatic plants and microalgae for biomass production. *Journal of Environmental Science and Technology*, 2013. **6**(2): p. 79–90.
46. Roongtanakiat, N., Vetiver phytoremediation for heavy metal decontamination. *PRVN Technical Bulletin*, 2009. **1**: p. 1–20.
47. Limmer, M. & J. Burken, Phytovolatilization of organic contaminants. *Environmental Science & Technology*, 2016. **50**(13): p. 6632–6643.
48. Gordon, M., N. Choe, J. Duffy, G. Ekuan, P. Heilman, I. Muiznieks, L. Newman, M. Ruszaj, B.B. Shurtleff, & S. Strand, *Phytoremediation of Trichloroethylene with Hybrid Poplars*. 1997, ACS Publications, Washington, DC.
49. Materac, M., A. Wyrwicka, & E. Sobiecka, Phytoremediation techniques of wastewater treatment. *Environmental Biotechnology*, 2015. **11**(1): p. 10–13.
50. Garrison, A.W., V.A. Nzengung, J.K. Avants, J.J. Ellington, W.J. Jones, D. Rennels, & N.L. Wolfe, Phytodegradation of p, p '-DDT and the enantiomers of o, p '-DDT. *Environmental Science & Technology*, 2000. **34**(9): p. 1663–1670.
51. Lin, Y.-P., C.-M. Lin, H. Mukhtar, H.-F. Lo, M.-C. Ko, & S.-J. Wang, Temporal variability in the rhizosphere bacterial and fungal community structure in the melon crop grown in a closed hydroponic system. *Agronomy*, 2021. **11**(4): p. 719.
52. Sheridan, C., P. Depuydt, M. De Ro, C. Petit, E. Van Gysegem, P. Delaere, M. Dixon, M. Stasiak, S.B. Aciksöz, & E. Frossard, Microbial community dynamics and response to plant growth-promoting microorganisms in the rhizosphere of four common food crops cultivated in hydroponics. *Microbial Ecology*, 2017. **73**(2): p. 378–393.
53. Segura, A. & J.L. Ramos, Plant–bacteria interactions in the removal of pollutants. *Current Opinion in Biotechnology*, 2013. **24**(3): p. 467–473.
54. Cunningham, S.D. & W.R. Berti, Remediation of contaminated soils with green plants: An overview. *In Vitro Cellular & Developmental Biology-Plant*, 1993. **29**(4): p. 207–212.
55. Mahmood, T., S.A. Malik, & S.T. Hussain, Biosorption and recovery of heavy metals from aqueous solutions by Eichhornia crassipes (water hyacinth) ash. *BioResources*, 2010. **5**(2): p. 1244–1256.
56. Lopez-Chuken, U.J., S.D. Young, & M.N. Sanchez-Gonzalez, The use of chloro-complexation to enhance cadmium uptake by Zea mays and Brassica juncea: Testing a "free ion activity model" and implications for phytoremediation. *International Journal of Phytoremediation*, 2010. **12**(7): p. 680–696.
57. Kumar, P.N., V. Dushenkov, H. Motto, & I. Raskin, Phytoextraction: The use of plants to remove heavy metals from soils. *Environmental Science & Technology*, 1995. **29**(5): p. 1232–1238.
58. Abdel-Shafy, H.I. & M.S. Mansour, Phytoremediation for the elimination of metals, pesticides, PAHs, and other pollutants from wastewater and soil, In: Kumar, V., Kumar, M., & Prasad, R. (Eds) *Phytobiont and Ecosystem Restitution*. 2018, Springer, Singapore, p. 101–136.
59. Gopal, B., Natural and constructed wetlands for wastewater treatment: Potentials and problems. *Water Science and Technology*, 1999. **40**(3): p. 27–35.
60. Kadlec, R.H. & S. Wallace, *Treatment Wetlands*. 2008: CRC press, Boca Raton.
61. Shahid, M.J., M. Arslan, S. Ali, M. Siddique, & M. Afzal, Floating wetlands: A sustainable tool for wastewater treatment. *Clean–Soil, Air, Water*, 2018. **46**(10): p. 1800120.

62. Cuellar-Bermudez, S.P., G.S. Aleman-Nava, R. Chandra, J.S. Garcia-Perez, J.R. Contreras-Angulo, G. Markou, K. Muylaert, B.E. Rittmann, & R. Parra-Saldivar, Nutrients utilization and contaminants removal. A review of two approaches of algae and cyanobacteria in wastewater. *Algal Research*, 2017. **24**: p. 438–449.

63. Xian, Q., L. Hu, H. Chen, Z. Chang, & H. Zou, Removal of nutrients and veterinary antibiotics from swine wastewater by a constructed macrophyte floating bed system. *Journal of Environmental Management*, 2010. **91**(12): p. 2657–2661.

64. Lawson, K.W. & D.R. Lloyd, Membrane distillation. *Journal of Membrane Science*, 1997. **124**(1): p. 1–25.

65. LaPara, T.M. & J.E. Alleman, Thermophilic aerobic biological wastewater treatment. *Water Research*, 1999. **33**(4): p. 895–908.

66. Cote, P., Z. Alam, & J. Penny, Hollow fiber membrane life in membrane bioreactors (MBR). *Desalination*, 2012. **288**: p. 145–151.

67. Judd, S.J., The status of industrial and municipal effluent treatment with membrane bioreactor technology. *Chemical Engineering Journal*, 2016. **305**: p. 37–45.

68. Komatsu, K., T. Onodera, A. Kohzu, K. Syutsubo, & A. Imai, Characterization of dissolved organic matter in wastewater during aerobic, anaerobic, and anoxic treatment processes by molecular size and fluorescence analyses. *Water Research*, 2020. **171**: p. 115459.

69. Cheng, X., J. Wang, B. Chen, Y. Wang, J. Liu, & L. Liu, Effectiveness of phosphate removal during anaerobic digestion of waste activated sludge by dosing iron (III). *Journal of Environmental Management*, 2017. **193**: p. 32–39.

70. Guesh, K., Á. Mayoral, C. Marquez-Alvarez, Y. Chebude, & I. Diaz, Enhanced photocatalytic activity of TiO_2 supported on zeolites tested in real wastewaters from the textile industry of Ethiopia. *Microporous and Mesoporous Materials*, 2016. **225**: p. 88–97.

71. Yamakata, A. & J.J.M. Vequizo, Curious behaviors of photogenerated electrons and holes at the defects on anatase, rutile, and brookite TiO_2 powders: A review. *Journal of Photochemistry and Photobiology C: Photochemistry Reviews*, 2019. **40**: p. 234–243.

72. Esakkimuthu, T., D. Sivakumar, & S. Akila, Application of nanoparticles in wastewater treatment. *Pollution Research Journal*, 2014. **33**(03): p. 567–571.

73. Lee, K.M., C.W. Lai, K.S. Ngai, & J.C. Juan, Recent developments of zinc oxide based photocatalyst in water treatment technology: A review. *Water Research*, 2016. **88**: p. 428–448.

74. Liu, I.-T., M.-H. Hon, & L.G. Teoh, The preparation, characterization and photocatalytic activity of radical-shaped CeO_2/ZnO microstructures. *Ceramics International*, 2014. **40**(3): p. 4019–4024.

75. Zhou, X., T. Shi, & H. Zhou, Hydrothermal preparation of ZnO-reduced graphene oxide hybrid with high performance in photocatalytic degradation. *Applied Surface Science*, 2012. **258**(17): p. 6204–6211.

76. Dai, K., L. Lu, C. Liang, J. Dai, G. Zhu, Z. Liu, Q. Liu, & Y. Zhang, Graphene oxide modified ZnO nanorods hybrid with high reusable photocatalytic activity under UV-LED irradiation. *Materials Chemistry and Physics*, 2014. **143**(3): p. 1410–1416.

77. Daneshvar, N., D. Salari, & A. Khataee, Photocatalytic degradation of azo dye acid red 14 in water on ZnO as an alternative catalyst to TiO_2. *Journal of Photochemistry and Photobiology A: Chemistry*, 2004. **162**(2–3): p. 317–322.

78. Chatterjee, A. & B. Deopura, Carbon nanotubes and nanofibre: An overview. *Fibers and Polymers*, 2002. **3**(4): p. 134–139.

79. Khin, M.M., A.S. Nair, V.J. Babu, R. Murugan, & S. Ramakrishna, A review on nanomaterials for environmental remediation. *Energy & Environmental Science*, 2012. **5**(8): p. 8075–8109.

80. Singh, J., K. Vishwakarma, N. Ramawat, P. Rai, V.K. Singh, R.K. Mishra, V. Kumar, D.K. Tripathi, & S. Sharma, Nanomaterials and microbes' interactions: A contemporary overview. *3 Biotech*, 2019. **9**(3): p. 1–14.

81. Kalhapure, R.S., S.J. Sonawane, D.R. Sikwal, M. Jadhav, S. Rambharose, C. Mocktar, & T. Govender, Solid lipid nanoparticles of clotrimazole silver complex: An efficient nano antibacterial against Staphylococcus aureus and MRSA. *Colloids and Surfaces B: Biointerfaces*, 2015. **136**: p. 651–658.
82. Borrego, B., G. Lorenzo, J.D. Mota-Morales, H. Almanza-Reyes, F. Mateos, E. López-Gil, N. de la Losa, V.A. Burmistrov, A.N. Pestryakov, & A. Brun, Potential application of silver nanoparticles to control the infectivity of Rift Valley fever virus in vitro and in vivo. Nanomedicine: Nanotechnology, *Biology and Medicine*, 2016. **12**(5): p. 1185–1192.
83. Rivero-Huguet, M. & W.D. Marshall, Reduction of hexavalent chromium mediated by micron-and nano-scale zero-valent metallic particles. *Journal of Environmental Monitoring*, 2009. **11**(5): p. 1072–1079.
84. Mahmoud, A.S., R.S. Farag, M.M. Elshfai, L.A. Mohamed, & S.M. Ragheb, Nano zero-valent aluminum (nZVAl) preparation, characterization, and application for the removal of soluble organic matter with artificial intelligence, isotherm study, and kinetic analysis. *Air, Soil and Water Research*, 2019. **12**: p. 1178622119878707.
85. Barot, N.S. & H.K. Bagla, Eco-friendly waste water treatment by cow dung powder (Adsorption studies of Cr (III), Cr (VI) and Cd (II) using tracer technique). *Desalination and Water Treatment*, 2012. **38**(1–3): p. 104–113.
86. Gaouar-Yadi, M., K. Tizaoui, N. Gaouar-Benyelles, & B. Benguella, *Efficient and Eco-Friendly Adsorption Using Low-Cost Natural Sorbents in Waste Water Treatment.* 2016, NIScPR Online Periodicals Repository, India.
87. Lazarova, V., P.-A. Liechti, P. Savoye, & R. Hausler, Ozone disinfection: Main parameters for process design in wastewater treatment and reuse. *Journal of Water Reuse and Desalination*, 2013. **3**(4): p. 337–345.
88. Wang, J. & H. Chen, Catalytic ozonation for water and wastewater treatment: Recent advances and perspective. *Science of the Total Environment*, 2020. **704**: p. 135249.
89. Ding, W., W. Jin, S. Cao, X. Zhou, C. Wang, Q. Jiang, H. Huang, R. Tu, S.-F. Han, & Q. Wang, Ozone disinfection of chlorine-resistant bacteria in drinking water. *Water Research*, 2019. **160**: p. 339–349.
90. Zhang, M.-H., H. Dong, L. Zhao, D.-X. Wang, & D. Meng, A review on Fenton process for organic wastewater treatment based on optimization perspective. *Science of the Total Environment*, 2019. **670**: p. 110–121.
91. Whitby, G. &G. Palmateer, The effect of UV transmission, suspended solids and photoreactivation on microorganisms in wastewater treated with UV light. *Water Science and Technology*, 1993. **27**(3–4): p. 379–386.
92. Masschelein, W.J. & R.G. Rice, *Ultraviolet Light in Water and Wastewater Sanitation.* 2016: CRC press, Boca Raton.
93. Mansur, A.A., H.S. Mansur, F.P. Ramanery, L.C. Oliveira, & P.P. Souza, "Green" colloidal ZnS quantum dots/chitosan nano-photocatalysts for advanced oxidation processes: Study of the photodegradation of organic dye pollutants. *Applied Catalysis B: Environmental*, 2014. **158**: p. 269–279.
94. Mills, A., C. Hill, & P.K. Robertson, Overview of the current ISO tests for photocatalytic materials. *Journal of Photochemistry and Photobiology A: Chemistry*, 2012. **237**: p. 7–23.
95. Moreira, N.F., C.A. Orge, A.R. Ribeiro, J.L. Faria, O.C. Nunes, M.F.R. Pereira, & A.M. Silva, Fast mineralization and detoxification of amoxicillin and diclofenac by photocatalytic ozonation and application to an urban wastewater. *Water Research*, 2015. **87**: p. 87–96.
96. Al-Rasheed, R.A. Water treatment by heterogeneous photocatalysis an overview, in *Fourth SWCC Acquired Experience Symposium*, Saudi Arabia, Jeddah, 2005.

5 A Review
Medical Wastes Management and Disposals: Cases of MENA Countries

Ghita Amine Benabdallah, Najoua Labjar,
Mohamed Dalimi, and Souad El hajaji

CONTENTS

5.1 Introduction 70
5.2 Categories and Types of Medical Wastes 70
    5.2.1 Sources 70
    5.2.2 Composition 71
5.3 Classification 71
    5.3.1 Categories 71
    5.3.2 Types of Medical Waste 71
5.4 Medical Wastes Management 75
    5.4.1 Stages of Process Management 75
        5.4.1.1 First Step: Waste Sorting and Conditioning 75
        5.4.1.2 Second Step: The Collection 76
        5.4.1.3 Third Step: Storage 77
        5.4.1.4 Fourth Step: Transportation 77
        5.4.1.5 Fifth Step: Treatment 77
        5.4.1.6 Sixth Step: Landfill 77
    5.4.2 Medical Waste Management in the Context of Pandemic COVID-19 77
        5.4.2.1 Example of Waste Generation in Some Countries 78
        5.4.2.2 Management and Disposal of Medical Waste 79
5.5 Treatment Process of Medical Wastes: Cases of MENA Countries 82
    5.5.1 Incineration 83
    5.5.2 Steam/Autoclave Disinfection (STD) 83
    5.5.3 Microwave Disinfection 83
    5.5.4 Reverse Polymerization (RP) 83
    5.5.5 Chemical Disinfection (CHD) 83
5.6 Conclusion 86
References 87

DOI: 10.1201/9781003359784-5

5.1   INTRODUCTION

In recent years, the medical sector has undergone considerable scientific and technological development, enabling it to provide appropriate services for health care (Su et al., 2021). Nevertheless, despite this progress, it is considered one of the sectors that generates a high amount of medical waste. This number has increased sharply due to the pandemic COVID-19 since the declaration of the state of health emergency by the WHO in March 2020 (Fadare and Okoffo, 2020; Yang et al., 2021; You and coll., 2020). Several measures have been put in place to limit and control the spread of the coronavirus such as containment, physical and social distancing, use of personal equipment, rapid test kits, PCR, vaccine. The adoption of these measures has created many management challenges due to the diversity of sources that are no longer limited to healthcare facilities and the potential infectious nature of medical waste. Inappropriate management can lead to risks, direct or indirect impacts on health and the environment. Another challenge is the choice of treatment.

Currently, there are few studies on the management and treatment of solid medical waste in the MENA region, especially after the COVID-19 pandemic. The main objective of this chapter is to present the different types and categories of wastes, the current status of solid medical waste generation during the COVID-19 pandemic, the analysis of the management modalities adopted, and the choice of treatment techniques, especially in some MENA countries.

5.2   CATEGORIES AND TYPES OF MEDICAL WASTES

"Hospital Waste" or "Medical Waste" or "Health Care Waste" or "Health Care Facility Waste" or "Clinical Waste" are some of the terms used to describe special waste generated by healthcare facilities (WHO, 1999a & b).

Waste, solid or liquid, resulting from the activities of all actors involved in the medical sector, such as hospital management staff, doctors, nurses, hospital hygiene officers, environmental protection officers, laboratory technicians, maintenance technicians, cleaners, waste transporters, etc, is considered as medical waste. This waste is mainly the result of diagnostic, monitoring, and treatment activities carried out within healthcare institutions. This heterogeneous waste is very varied in terms of production source and physical and chemical characteristics. However, two sources are considered to be the main sources (Purnomo et al., 2021; Windfeld and Brooks, 2015; Erdogan et al., 2021).

5.2.1   SOURCES

- **Primary sources:** Public hospitals, clinics, scientific research institutions, testing laboratories, and similar institutions: health centres, haemodialysis centres, blood transfusion centres, mortuaries and autopsy centres.
- **Secondary sources:** Private practices in general (dental, veterinary, medical), health training institutions, beauty care centres [7].

COVID-19 pandemic: In the last two years since the outbreak of the COVID-19 pandemic and according to the guidelines of the authorities in the different countries, the sources of healthcare waste are no longer limited to hospitals and health centres, and a wide range of other sources have emerged to cope with the increased number of positive cases affected by the coronavirus, symptomatic or asymptomatic such as newly built temporary hospitals, isolation centres, homes, quarantine units, screening centres, vaccination centres, organization, etc (Yang et al., 2021; Chen and Guo, 2020, Bhakta et al., 2020).

5.2.2 Composition

The composition of harmful hospital waste is characterized by a variety of waste categories: infectious, pathological, pharmaceutical, toxic chemicals and heavy metals, sharps, and may contain genotoxic or radioactive substances (Mbongwe et al., 2008; Cebe et al., 2013; WHO, 2014).

This composition endangers the safety of all health personnel and direct or indirect users of this waste although only 15% of this waste is considered hazardous (10% is infectious and 5% is radioactive or chemical) with risks of infectious, toxic, and radioactive nature. The remaining 85% is classified as household waste and non-hazardous waste with no potential risk (WHO, 2018).

On the other hand, the diversified physicochemical composition of medical waste gives it a high calorific character estimated at 8,820 kcal/Kg. The waste is made up of 54% paper, 20% textile, 26% organic material, 50% plastic as combustible waste, and 10% waste containing metal and 15% glass as incombustible waste. The elemental analysis of this waste is characterized by 35% carbon (C), 15% hydrogen (H), 16% nitrogen (N), 26% oxygen (O), 1% sulphur (S), and 3% chlorine (Cl) (Erdogan et al., 2021). This composition can vary from one hospital to another or even from one department to another depending on their specificities.

5.3 CLASSIFICATION

5.3.1 Categories

According to the World Health Organization and other legislation in different countries, there are different classifications of biomedical waste. Table 5.1 illustrates, as an example, the categories adopted by WHO, France, Spain, Portugal, Morocco, Iran, Turkey, and Lebanon.

5.3.2 Types of Medical Waste

Not all medical and pharmaceutical waste has the same origin and does not cause the same risk and hazard. There are two types of biomedical waste, namely solid waste and liquid waste (WHO, 2011).

  I. **Liquid waste:** This consists of blood residues, liquid chemicals, medical fluids such as gastric lavage fluids, pleural and cardiac puncture fluids, as well as rinsing water from X-ray films, but also household wastewater from kitchens, toilets, and laundry.

TABLE 5.1

Categories Adopted by some Countries

Country/ Organization	Category	References
OMS	* Medical care waste requiring special attention * Non-hazardous healthcare waste * Infectious and highly infectious waste * Other hazardous waste * Radioactive waste	WHO (2014)
Brazil	Group A: residues presenting a risk to public health and the environment due to the presence of biological agents (infectious-biological). Group B: residues presenting a risk to public health and the environment due to physical, chemical, and physicochemical characteristics (hazardous waste). Group C: radioactive waste. Group D: all other residues that are not included in the groups described above.	(8)
France	* DADM: Waste that can be assimilated to household waste * DASRI: Waste from care activities with infectious risks *Toxic and chemical risk waste	
Iran	4 categories of waste are provided for: Non-hazardous waste Drugs/chemicals Infectious waste Sharp waste	Zand and Heir (2020).
Lebanon	Category 1: Non-hazardous waste Category 2: Hazardous waste Category 3: Special waste Category 4: Radioactive waste	Maamari et al. (2015)
Morocco	Category 1: Biohazardous waste Category 2: Chemical hazard waste Category 3: Identifiable human organs and tissues Category 4: Waste similar to household waste	Moroccan Ministry of Health
Spain or Portugal	* Group I: Urban waste *Group II: Non-hazardous sanitary waste *Group III: Hazardous sanitary waste with a biological risk *Group IV: Hazardous sanitary wastes with a chemical hazard	
Turkey	Category 1: Municipal waste Category 2: Medical waste Category 3: Hazardous waste Category 4: Radioactive waste	Birpınar et al. (2009)
United Kingdom	Domestic/municipal waste Offensive waste Anatomical waste Cytotoxic/cytostatic waste Medical waste Infectious waste	Korkut et al.
United States	1 Non-infectious but hazardous waste 2 General waste 3 Radioactive waste 4 Infectious waste	

II. **Solid waste:** This can also be divided into two categories:
   I. Waste similar to household waste, non-hazardous waste produced by health personnel or by those accompanying patients (leftover meals, non-infected paper and packaging, etc.)
   II. Waste from care activities produced in the care services, and hazardous waste. It is made up of:

- **Anatomical waste:** Human organs and tissues, blood bags, placentas.
- **Chemical waste:** There is a wide variety of chemical waste used in health-care facilities: hospital effluents, solvents, expired disinfectants, waste containing heavy metals such as mercury or lead (thermometers, blood pressure gauges), laboratory reagents (Chartier et al., 2014; Ridha et al., 2014) out of these wastes only 20% waste resulting from healthcare activities and remaining are chemical waste (Ilyas et al., 2020).
- **Toxic waste:** Potentially life-threatening substances. Carcinogens, mutagens, toxins, and hazardous chemicals are found in healthcare facilities.
- **Radioactive waste:** This is the action of exposing an organism to the action of radiation. This radiation disturbs the functioning of living cells (breaking of chemical bonds and structural modifications...). Waste is contaminated by radionuclides, produced by care units (radiotherapy, nuclear medicine, etc). The most common examples are syringes, needles, gloves, liquid effluents, and stool from patients who have undergone nuclear medicine tests (Chartier et al., 2014; Ministry of Health, 2004).
- **Sharps waste:** This is another category of solid healthcare waste. Types of needles, broken glasses, ampoules, syringes, blades knives, and pipettes are some examples of sharps that need to be handled and managed with great care and precaution (Askarian et al., 2010; Kalogiannidou et al., 2018; Mato and Kassenga, 1997; Chartier et al., 2014). During the COVID-19 pandemic, these tools may be a source of contamination for waste collectors, knowing that the life span of the virus is a few days on the media.
- **Infectious waste:** Material or preparation containing microorganisms or their toxins that can cause disease in humans such as dressings and swabs, syringes without needles, waste contaminated with blood and physiological fluids, cultures, and stores of infectious agents (Askarian et al., 2010; Chartier et al., 2014). Waste from hospitalized patients placed in isolation, contaminated disposable and medical equipment and devices (such as PPE/masks gloves, goggles, suits and gowns, etc.) used for those patients are infectious waste. This waste has increased during the COVID-19 pandemic (WHO, 2020b). Managing this type of waste during the pandemic is a huge challenge (Rowan and Laffey, 2021).
- **Pathological waste:** This type of waste comes mainly from microbiological or surgical samples (organ, tissue,...), removed from human or animal bodies (Chartier et al., 2014) to perform diagnostics. Pathological waste potentially carrying pathogens, comparable to infectious waste, must be handled with great care to mitigate the risk of transmission of infection (WHO, 2020b).

- **Pharmaceutical waste:** Expired, unused, and contaminated pharmaceuticals; drugs; vaccines; and sera are used for therapy arising from various health facilities, pharmacies, hospitals, and patient isolation sites (Chartier et al., 2014; Malsparo, 2020). This type of waste has seen an exceptional increase as a result of the immense demand for the various protocols by patients and symptomatic and asymptomatic individuals.
- **Non-hazardous health care waste:** This is non-hazardous, non-infected healthcare waste that can be processed by municipal cleaning services in a manner similar to normal household waste processing. These wastes, such as office supplies (paper, newspapers, etc.), packaging and food waste, and plastic water bottles, account for the majority of the total amount of waste generated by healthcare facilities (Askarian et al., 2010; Kalogiannidou et al., 2018) (Algeria thesis).

  Nevertheless, this waste can be at risk of contamination, particularly infectious and bacteriological, when mixed with hazardous waste (Rau et al., 2000, Rushbrook and Zghondi, 2005). During the COVID-19 pandemic, contaminated non-hazardous waste produced by people affected by SARS-CoV-2 is a significant source of contamination.

  All items discarded by a symptomatic or asymptomatic COVID-19-infected person, or by a person in contact with another coronavirus-infected person, generated within a healthcare facility or at home, are considered infectious healthcare waste (Ed and Ym, 2020). This waste must be treated according to COVID-19 waste treatment and disposal rules.
- **Other waste:** The various rapid tests and PCR kits used for the detection of persons affected by the SARS-CoV-2 virus count as additional biomedical waste. This type of waste may present a new source of infectious waste and may be responsible for the transmission of coronavirus. In addition, the measures imposed by the leaders of the various countries (social distancing, conditions of inter-country mobility) have contributed to the increase in the quantity of this waste produced.

The management of hazardous waste must be controlled. It is the responsibility of healthcare institutions to ensure that there are no adverse health and environmental impacts from their waste collection, handling, treatment, and disposal activities (Gayathri et al., 2005). Worldwide, 5.2 million people lose their lives annually due to serious diseases generated from medical waste (Zhao et al., 2022).

However, a large number of diseases can be transmitted through contact with healthcare waste. This is the case for hepatitis B, typhoid, hepatitis C or HIV infection, and COVID-19 (2) (Cebe et al., 2013).

In general, the diversity of sources (hospitals, clinics, laboratories...), the varied nature (glass, plastics, cardboard, metals, etc.), and categories (hazardous or non-hazardous) of biomedical waste require a proper strategy throughout the entire process of its management, from production to the final phase of disposal (production, sorting, collection, packaging, transport, storage, treatment, and disposal). The waste must be collected properly, stored temporarily, and disposed of permanently (Sharma et al., 2020; Cebe et al., 2013).

5.4   MEDICAL WASTES MANAGEMENT

The process of hospital solid waste disposal follows the steps that deserve to be made explicit. In practice, in order to reduce the risk of spreading infections, the modalities and conditions of solid waste management from healthcare facilities must be well controlled and properly followed throughout the process from production to the final disposal phase. Good management requires the collective involvement of all parties concerned (decision makers, doctors, nurses, housekeepers, patients, employees, citizens, …).

However, waste generated in health facilities must always follow an appropriate and well-identified route, from its point of generation to its final disposal (WHO, SCB & UNEP, 2005a & 2005b).

This mainly involves six stages: sorting and packaging, collection, storage, transport, treatment, and disposal.

5.4.1   Stages of Process Management

5.4.1.1   First Step: Waste Sorting and Conditioning

These two activities are complementary to ensure a good management of bio-medical waste. Sorting consists of separating waste according to their categories, properties, specificities, and treatment and elimination methods. Each category is classified according to colour codes and placed in containers or labelled rigid bags, depending on the packaging stage. Sorting is carried out at the production site. It must be carried out with vigilance to avoid mixing the two types of waste, hazardous and non-hazardous waste, whose properties are similar to municipal waste. In addition to the protection of the safety of the personnel, the sorting and the conditioning make it possible to reduce the economic incidence through the reduction of the quantity of waste at risk to treat (Moroccan Ministry of Health, Tunisian Ministry of Health).

Packaging: It is the packaging of the waste followed by the labelling (physical barrier against pathogenic microorganisms) by respecting the some conditions: respect of the limit of filling to the ¾ (verification of the system of closing of the bags, …). It allows to identify the source of the waste (entity, service, department, laboratory of production of the waste), the nature of the waste through the labelling (dangerous, infectious, pathological, …), and facilitates the choice of the appropriate container. There are two types of packaging: (Moroccan Ministry of Health, Tunisian Ministry of Health).

- **Primary packaging:** Consumable packaging such as bags, cartons, etc., used by waste-producing personnel in the departments of healthcare institutions
- **Secondary packaging:** The support containers where the primary packaging is placed

The method of packaging may differ from one country to another depending on the regulations and standards in force in accordance with WHO guidelines and on the type and category of healthcare waste. Table 5.2 shows some examples of colour coding adopted by some countries: Inde, Iran, Morocco, OMS, South Africa, and Turkey.

5.4.1.2 Second Step: The Collection

This is the journey from the site of production of waste previously sorted and packaged, to the central storage area according to conditions of safety and hygiene

TABLE 5.2
Distribution of Colour Codes by Category and by Country

Country/ Organization	Waste Category	Colour Code	References
OMS	Infectious, pathological and sharps waste	Yellow	WHO (2014)
INDE	Chemical and pharmaceutical waste	Brown	BWMHR
	General non-hazardous waste	Black	(2016), Iyer
	Infectious, pathological, chemical, radioactive, and pharmaceutical	Yellow	et al. (2021), MOEFCC
	Contaminated recyclable waste	Red	(2016)
	Sharps waste	White	
	Glassware waste and metal implants	Blue cardboard boxes	
	General biodegradable waste	Green	
Morocco	Solid non-hazardous medical and pharmaceutical waste	Black	Moroccan Ministry of
	Sharps waste	Red or yellow	Health
	Non-hazardous infectious waste	Red	
South Africa Human anatomical waste	Human anatomical waste	Red	SANS (2008).
	Infectious animal waste	Orange	Hangulu and
	Sharps waste	Yellow	Akintola
	Chemical and pharmaceutical waste	Green	(2017)
Iran	Non-hazardous waste	Black	Zand and Heir
	Pharmaceutical/chemical waste	White or brown	(2020)
	Infectious waste	Yellow	
	Sharp waste	Safes	
Turkey	Municipal solid waste	Black	Birpınar et al.
	Recyclable materials	Bleu	(2009)
	Pathological, non-pathological, and infectious medical waste	Red	
	Sharp objects	Plastic containers Yellow	
	Heavy metal waste	Same rule for hazardous waste management	
	Radioactive waste	According to the law of the Turkish Atomic Energy Council (Turkish Ministry of Environment and Forestry, 2005)	

standards required to avoid any accumulation of waste in the care units (Ranjbari et al., 2021).

5.4.1.3 Third Step: Storage

There are two types of storage: intermediate storage (local near the care units) and central storage in situ of the health establishments, with well-closed, ventilated, protected premises inaccessible to animals and all third parties. The objective of this step is to ensure safe and temporary storage of waste pending final disposal (Ranjbari et al., 2021).

5.4.1.4 Fourth Step: Transportation

Two modes of transport of biomedical waste are possible: the internal transport of waste from the intermediate storage place to the central storage through adapted tools (washed and disinfected cart) and the transport outside the hospital by vehicles reserved for this purpose.

5.4.1.5 Fifth Step: Treatment

There is a diversity of techniques for the treatment of biomedical waste. The main ones are incineration, autoclaving, microwave treatment, and reverse polymerization. The main objectives are to reduce the volume, to limit the harmful impact on the environment and on health, and to eliminate the dangerous character of pathogenic germs in the waste (Liang et al., 2021). These techniques for treating biomedical waste will be developed in detail in Chapter 3.

5.4.1.6 Sixth Step: Landfill

This is the installation of the waste by deposition or burial in artificial or natural soil cavities without the intention of subsequent recovery.

The management of medical waste is of great importance because of its potential impact on the environment and, consequently, on human health. In recent years, many efforts have been made by environmental regulators and waste generators to better manage waste from healthcare facilities. In the past, medical waste was often mixed with municipal solid waste and disposed of in residential waste landfills or inappropriate treatment facilities.

5.4.2 MEDICAL WASTE MANAGEMENT IN THE CONTEXT OF PANDEMIC COVID-19

Since the declaration of a health emergency by WHO (Fadare and Okoffo, 2020; Yang et al., 2021; You et al., 2020), the number of positive coronavirus cases has been steadily increasing worldwide. Faced with this unprecedented situation, some measures have been imposed at the level of different countries to reduce the rapid spread of this infectious and contagious disease transmissible through respiratory droplets in the air, direct contact, etc (Yang et al., 2021). Containment, physical distancing, mobility restrictions, travel bans, quarantine isolation, frequent hand washing, etc. are some examples of these initiatives (Klemeš et al., 2020). In addition to these measures, the wearing of masks by citizens and PPE by healthcare workers (masks, coveralls, gowns, plastic gloves, boots,...) have become a necessity to limit the transmission of coronavirus (SARS-CoV2). Although these PPEs allow the protection of the life of several people including those exposed in the first place (doctors, nurses, and agents responsible for the collection and transport of waste,...), they generate a massive amount of waste causing various challenges to ensure effective and appropriate management.

As a result of this COVID-19 pandemic, an increase of 18–425% in medical waste generation was recorded worldwide with a daily variation of 200 t/d to 29,000 t/d during the period from February 22 to the end of September 2020 (Liang et al., 2021). Implications of the COVID-19 pandemic on solid waste generation and management strategies should be considered. In. Environ. Sci. Ing. 15, 115 (2021).

5.4.2.1 Example of Waste Generation in Some Countries

- In February 2020, as the first country affected by COVID-19, 116 million disposable masks were produced daily in China, 12 times the amount produced prior to the COVID-19 pandemic (Chen et al., 2021; Ministry of Industry and Information Technology, 2020) with a 670% increase in biomedical waste (Saxina et al., 2021).
- India as the new epicentre after China, according to available statistics, was producing about 139 tons per day of medical waste between April 2021 and May 2021 at the level of health facilities under the jurisdiction of different member states (Saxina et al., 2021). A clear increase in the volume of biomedical waste has been recorded. It represents six times the usual amount of waste produced before the appearance of COVID-19.
- In the United States, the rate of production of COVID-19 waste increased from 5 million tons/year to 2.5 million tons per month (Ilyas et al., 2020).
- For Tehran, Iran, the estimated amount is 80–110 tons per day during the pandemic or an increase from 18 to 62%, with a production of 5.5 million masks/d (Zand and Heir, 2020). In general, Iranian hospitals, both public and private, noted a significant increase in biomedical waste production estimated at 102.2% and 121% of infectious waste compared to the prepandemic situation (Kalantary et al., 2020; Kalantary et al., 2021).
- In Lebanon, 2.45 kg/lit/d and 0.94 kg/lit/d represent, respectively, the share of infectious hospital waste in private hospitals and other types of health facilities, data collected from a 5-year study of 57 hospitals among the 163 existing in the country. (Maamari et al., 2015).
- For Kuwait, waste generation in two major public hospitals was estimated to be 4.89–5.4 kg/patient/day and 3.65–3.97 kg/bed/day.
- **In Jordan:** The Jordanian King Abdullah University Hospital (KAUH) experienced a daily generation of about 650 kg of biomedical waste – a generation ten times higher than the normal proportion generated before the coronavirus period (Abu-Qdais et al., 2020; Silva et al., 2021).

Similarly, the generation of this waste increased from 0.36–0.5 kg/day to 14.16 kg/day at the Jordanian KAUH. 3.95 kg/bed/day is the rate in terms of beds (Abu Qdais et al., 2007, 2020; Al Shraideh and Abu Qdais 2017).

– **Morocco:** The amount of waste generated at Moulay Abdellah Hospital, located within the Rabat-Salé-Kenitra Region, the second most affected region by COVID-19, experienced a very high flow compared to normal. The quantity of production has increased from 8 to 13 tons/month, respectively, an average increase of 266–433 kg/day. This means an increase from 1.6 to 4 kg/bed/day in the waste rate (Khazraji et al., 2021).

The distribution of waste per bed per day produced by some countries, especially some MENA countries, is as follows : Algéria (1,8), Egypt (1,2), Iran (3,5), Kuwait (3,8), Lebanon (2,45), Libya (1,3), Palestine (1,57), Yemen (2,41), Morocco (4), Jordan (3,95) (Kalantary et al., 2021; Maalouf and Maalouf, 2021; Khazraji et al., 2021; Abu-Qdais et al., 2020; ADEPT, 2020).

Although the pandemic has had positive effects on the soil, climate (reduction of CO_2), water, and the reduction of waste in some tourist areas due to the travel ban, in the areas most affected by the virus, there have been problems with the management and disposal of waste generated by COVID-19.

The countries and regions most affected by the virus have been faced with problems of management and disposal of waste generated by COVID-19. The situation is becoming worrisome and alarming. One of the major problems is the mastery of a quality, adequate, and efficient waste management system (Ansari et al., 2019; Tsakona et al., 2007; Windfeld and Brooks, 2015).

5.4.2.2 Management and Disposal of Medical Waste

The management methods adopted prior to COVID-19 are no longer sufficient and reliable given the critical increase in the quantity of waste and the persistence of virus survival on various contaminated surfaces and media (metal, cardboard, paper, plastic) (Klemeš et al., 2020; Sharma et al., 2020). The management system within healthcare facilities is becoming unable to handle the volume of waste being recovered (Klemeš et al., 2020).

Operators in several countries have been faced with two major challenges. The first challenge is the massive increase in the number of positive cases, the transmissible pathogenicity of the virus, and the exceptional increase in the amount of waste produced, while the second challenge concerns the very limited financial, structural, logistical, and human resources (Morocco-Africa).

Good biomedical waste management practices, from identification to disposal, combined with disinfection and personnel protection measures, are urgently needed and recommended by the authorities (UNEP, 2020).

To cope with the huge amount of waste generated during the COVID-19 pandemic (Das et al., 2021; Sarkodie and Owusu, 2021), limit the spread of the virus, and protect the health of citizens and nature, several countries have instituted and established new strategies and guidelines to effectively manage and treat biomedical waste DM (Agamuthu and Barasarathi, 2021; Sarkodie and Owusu, 2021; Windfeld and Brooks, 2015) (Table 5.3).

TABLE 5.3

Case of some MENA Countries: Morocco (Moroccan Ministry of Health)

Operations	New Waste Management Provisions During COVID-19
Separation and packaging	In Morocco, all COVID-19 waste from isolation units is considered infectious risk waste, including household waste (leftover food, PPE, paper, plastic bottles, disposable plastic utensils, etc). – A new colour code has been adopted. With the exception of sharps, which must be disposed of in yellow single-use containers, all other waste must be packed in red bags. – Labelled red/blue containers were used to differentiate them from medical waste containers from other units. The bags and containers once filled to ¾ must be closed and not exceeded and labelled. Other requirements concerning pre-collection, storage, internal transport (conditions, disinfection, cleaning, precautionary rules to be taken into consideration, etc) have been detailed in these documents.
Pre-collection/ Internal transport	The packaging bags and waste containers are: – pre-collected by the agents in charge of this function, well-protected wearing individual protection equipment guaranteeing their safety (disposable suit or blouse, gloves, glasses, boots, mask,...) – disinfected from the outside with approved disinfectants (bleach diluted 12°) – transported to the temporary storage place using specific reserved carts, while waiting for their final evacuation for treatment.
Stockage	– Specific locked premises not accessible to third parties, meeting hygiene and safety standards, have been dedicated to the storage of disinfected containers. – The premises must be cleaned and disinfected after each waste collection operation.
External transport, treatment, and disposal	– Morocco outsources the three operations of external transport, treatment and disposal of medical waste, including COVID-19 waste, to specialized private companies under specific contracts. – The transport of containers, sprayed with disinfectants, is carried out by vehicles belonging to the subcontracting company authorized by the Ministry of Health, taking into account the necessary hygiene conditions. – These companies treat the COVID-19 waste as a priority as soon as it is received. Once processed, the treated waste goes to the public dump.

Like all countries in the world affected by SARS COV-2, in order to reduce the spread of the virus, Morocco has put in place, in addition to existing laws, decrees and guides, guidelines and procedures relating to the new management conditions (sorting, packaging, storage, and internal transport), and treatment of medical and pharmaceutical waste in isolation units.

These procedures are addressed to all actors involved in the process of waste management and treatment, such as healthcare staff, agents in charge of collection, transport, storage, treatment, and elimination of this waste (Moroccan Ministry of Health).

– Due to COVID-19, in Tehran, the initially adopted guidelines specifying the distribution of biomedical waste (four categories: existing) and treatment of infectious waste within hospitals are no longer respected. The new provisions recommend the collection of hospital waste in double or triple bags, and their direct routing to the disposal site at Aradkouh for Buria (Zand and Heir, 2020).

– In India, due to the pandemic, additional specific rules for the collection and management of biomedical hazardous waste have been issued by the Central Pollution Control Board. (2) These guidelines recommend, in addition to the already existing 2016 rules, the adoption of double-layered yellow bags labelled "COVID-19 waste", considered as hazardous waste generated by healthcare facilities during COVID-19.

Nevertheless, despite all the efforts made, data reveal that in some states, the medical waste treatment centres have been saturated due to the enormous flow of waste generated considering the rapid spread of the virus and the increase in the number of affected cases, far exceeding the available capacity of the treatment system.

Although the pandemic generated positive effects, for the environment the soil, climate (decrease of CO_2), water, decrease of waste in some tourist regions because of the ban on travel, the regions most affected by the virus faced problems with the management and disposal of waste generated by COVID-19 (Klemeš et al., 2020). However, many developing countries have struggled with medical waste management long before the COVID-19 pandemic (Rahman et al., 2020).

Several studies have been conducted in different countries such as Egypt (Abd El-Salam, 2010), Jordan (Abdulla et al., 2008), Africa case of Morocco (Belhadi et al., 2020), Turkey (Alagöz and Kocasoy, 2008), and Iran (Farzadkia et al., 2015; Dastpak et al., 2017) to assess the quality of solid medical waste management practices adopted by various healthcare facilities. The results revealed that there is a great similarity in the findings of these surveys (Abd El-Salam, 2010; Abdulla et al., 2008). The main findings and recommendations are summarized below:

– Insufficient financial and human resources to carry out effective management.
– The management rules relating to the sorting and packaging of different types of waste were not carried out in accordance with the standards in force. The rate of infectious waste and the related costs have increased due to the mixing of infectious hazardous waste with other waste considered as household waste.
– Lack of knowledge and application of sorting standards by health actors: lack of disinfection of waste bins after each use, PPE not used by staff during waste collection, unskilled staff.
– Problem of identification of waste categories and source of waste bags due to lack of labelling.
– Lack of adequate infrastructure: for the storage of waste inside healthcare facilities according to normative requirements and hygiene and safety conditions. Problems with collection bags, PPE.

- Insufficient means of transport and manpower to carry out the collection and transport of waste at the appropriate time (Zhao et al., 2022).
- Insufficient capacity of treatment and disposal facilities for waste domiciled in hospitals (Zhao et al., 2022).

5.4.2.2.1   Recommendations

- Implement appropriate short-, medium-, and long-term strategies, taking into account all stages of waste handling, from production, sorting, packaging, storage, transport, treatment, and disposal, in order to create an appropriate and more efficient management system.
- Implement a policy of continuous improvement of the quality and safety of care activities to anticipate risks and take the necessary preventive measures (DDASS, 2007).
- Implement a national plan for the management of medical waste during health disasters coupled with collection and traceability models based on artificial intelligence.
- Involve all concerned operators (physicians, managers, employees, agents) in the implementation, monitoring, and control of the medical waste management system.
- Provide training and awareness programs for all staff involved in waste management in healthcare facilities: doctors, nurses, employees, agents in charge of waste collection and transport, maintenance staff.

5.5   TREATMENT PROCESS OF MEDICAL WASTES: CASES OF MENA COUNTRIES

In addition to the main medical waste management techniques (sorting, segregation, storage, transport), a variety of methods for the safe disposal of hospital waste have been used by most countries (Belhadi et al., 2020).

There are various disposal and treatment methods used in the management of medical and pharmaceutical wastes, especially hospital wastes, regardless of the solid or liquid nature, such as incineration, autoclaves, pyrolysis, landfill, sterilization by dry or wet steam, microwave radiation, or chemical disinfection (Yong et al., 2009; Ansari et al., 2019; Iyer et al., 2021; Subramanian et al., 2021).

However, due to the lack of universal regulations imposing specific treatment techniques, the choice, with the main objective of mitigating negative impacts on health and nature, is dependent on several qualitative and quantitative parameters: quantity and nature of the waste generated, availability or not of a treatment centre near the health care facility, existence of means of transport, load and capacity of infrastructure, social acceptance of the treatment methods, compliance with existing legislation at the level of each country, and availability of human, material, and financial resources.

In general, the most commonly used techniques for the treatment and disposal of healthcare waste are incineration, autoclave, microwave, chemical disinfection, and land disposal (landfill, pits). Currently, other more sophisticated methods are used such as pyrolysis (Guilherme et al., 2021).

5.5.1 INCINERATION

Is a high-temperature (+ 800 °C) combustion technique used to treat hazardous medical waste (Klemeš et al., 2020). This process properly removes microorganisms and converts the waste into gas and non-combustible residue while reducing its volume and weight (Cobo et al., 2018; Hong et al., 2018). However, there are two types of incineration: the first "in situ incineration" implemented within the medical facility involved, and the second "centralized incineration", external localized and managed by an independent entity (Hervier, 1999; Berrahal, 2001; Daoudi, 2008). According to these authors, incineration is the most recommended solution for the treatment of medical waste. In addition, it contributes to air pollution by producing toxic substances such as chlorinated dibenzodioxins and dibenzofurans (Voudrias, 2016). Currently, the most used method is pyrolysis (Undri et al., 2014; Guilherme et al., 2021).

5.5.2 STEAM/AUTOCLAVE DISINFECTION (STD)

MST or autoclaving is a thermal technology used to disinfect infectious healthcare waste using saturated steam at a temperature above 100 °C for a sufficient amount of time (Veronesi et al., 2005). The treated, non-hazardous waste is subsequently disposed of in sanitary landfills. However, this safe and often recommended method has limitations of use in some areas due to the need for electricity (Belhadi et al., 2020; Evangelos et al., 2016).

5.5.3 MICROWAVE DISINFECTION

This is a rapid, steam-based, 100% environmentally friendly disinfection technology with no odour or toxic residue emissions, used to treat all types of medical waste, solid or liquid, including infectious waste. It is based on electromagnetic waves with a wavelength between $(2,450 \pm 50)$ and (915 ± 25) MHz (Veronesi et al., 2005; Evangelos et al., 2016). However, it is recommended to pay special attention when using advanced microwave instruments (Stolze and Kühling, 2009; Belhadi et al., 2020; Evangelos et al., 2016).

5.5.4 REVERSE POLYMERIZATION (RP)

RP is a pyrolysis that uses microwave energy to treat infectious healthcare waste including metals and plastic waste (Undri et al., 2014). Reverse polymerization allows for 80% volume reduction and 6 log10 inactivation of pathogens (Voudrias, 2016). The residues from this technology are ground up and disposed of in sanitary landfills (Undri et al., 2014). This technique is considered the most expensive of the other techniques studied (Evangelos et al., 2016; Belhadi et al., 2020).

5.5.5 CHEMICAL DISINFECTION (CHD)

This is a system for treating CHD using specific chemicals such as sodium hypochlorite, calcium hypochlorite, chlorine dioxide, etc (Wang et al., 2020a, b).

These disinfectants are mixed with the previously shredded waste to kill and inhibit infectious microorganisms. This type of disinfection is mostly recommended for the treatment of infectious liquid waste such as blood, urine, or hospital drains. Nevertheless, chemical disinfection can only be applicable for small volumes of waste and poses a considerable risk to health (people handling these products) and the environment (liquid waste containing NaOCl) (Hong et al., 2018; Evangelos et al., 2016; Belhadi et al., 2020).

Faced with this diversity of existing techniques, the appropriate choice for the treatment and disposal of waste within different countries is dependent on several factors, political, economic, social, and environmental (Valente and Bueno, 2019; Voudrias, 2016; Wang et al., 2019; Ren and Lützen, 2015).

Unfortunately, the majority of developing countries, especially those under the MENA Region, Saudi Arabia, Egypt, Lebanon, Morocco, are facing several constraints to properly handle medical waste according to the guidelines required by the authorities. In addition, for most MENA countries, the treatment and disposal of waste from some hospitals is usually delegated to private companies. Aware of the importance of the treatment and disposal phases of biomedical waste and its adverse effects and with a view to focus and give priority to the health and safety of patients.

In accordance with the literature, previous studies have analysed and prioritized the methods used for some MENA countries: Saudi Arabia, Egypt, Jordan, Lebanon, and Morocco.

– **Saudi Arabia**: The amount of waste from healthcare facilities has increased massively in recent years in line with the increase in Saudi Arabia's population of about 32 million. The rate of waste generation in hospitals in the Eastern Province is 0.51 kg/bed/day, i.e. 1.66 kg/patient/day. The total amount generated is about 5,781.9 tons/day. In addition, the treatment usually used by hospitals, after sorting, separating, and collecting management techniques, is the dumping of waste in a landfill operated by licensed private companies. The study carried out in the hospitals of this Eastern Province recommended the autoclaving technique as an alternative treatment for biomedical waste, especially plastic waste. This waste can be reused as fuel in waste-to-energy facilities or as an alternative fuel in the cement industry (Alagha et al., 2018).
– **Egypt**: The share of waste generated in Egyptian hospitals in the city of Damanhour is estimated at 1,249 tons/day, with 38.9% of waste considered as hazardous. The rest is similar to household waste. Incineration is the most frequently used method for the treatment of biomedical waste despite the associated pollution risks and the refusal of the population (Abdel-Shafy and Mansour, 2018).
– **Jordan:** The amount of medical waste generated in the Jordanian KAUH hospital is estimated at 0.36–0.5 kg/patient/day. The production of the amount of healthcare waste has increased enormously due to the COVID-19 pandemic, which is 14.16 kg/patient/day, equivalent to 3.95 kg/bed/day in terms of beds. The collected medical waste is incinerated daily in the incinerator available at the Jordan University Science and Technology Campus (Abu Qdais et al., 2007; Al Shraideh and Abu Qdais, 2017).

- **Lebanon**: The biomedical waste stream is 1.0–1.5 kg/bed/day, which represents 9.2–13.8 tons of hazardous healthcare waste per day. The treatment of the majority of medical waste is delegated to a non-governmental organization, with 80%–85% of the waste being treated by this organization, in close collaboration with other stakeholders, using the autoclave technique (MoPH, 2018). In Lebanon, incineration of potential infectious medical waste is prohibited according to current legislation; sterilization remains the recommended method. The remaining parts of the waste (15–20%) are either discreetly incinerated without permission within hospitals or illegally dumped in landfills with the other categories of medical waste (MoE/UNDP/GEF, 2016).
- **Morocco**: An exceptional procedure has been adopted for the treatment of medical and pharmaceutical waste generated from isolation units for patients with COVID-19 the case of Moulay Abdellah Hospital in Salé located within the Rabat Salé-Kenitra Region. Managing medical waste is an essential issue in times of pandemic; in health emergency, health initiatives have been taken to protect the environment by ensuring the health of patients. The hospital has used a company, certified by the Moroccan Ministry of Health, specialized in the transport and treatment of medical waste. This company collects and treats the waste in accordance with the sanitary guidelines and instructions required by the Ministry of Health (Khazraji et al., 2021). For Morocco, also for the other countries listed above, outsources the three operations of external transport, treatment and disposal of medical waste, including COVID-19 waste, to specialized private companies under specific contracts (Moroccan Ministry of Health).

In addition, the results of an evaluation and ranking study of different treatment technologies revealed that the combination of the two methods of incineration and chemical disinfection is the most appropriate treatment for infectious waste in the context of the pandemic. This study was carried out based on the combination of the four criteria: technological, environmental and safety, economic, socio-political, using various approaches: LCA coupled with LCC: LCA-LCC, Analytical Hierarchy Process (AHP), and VIKOR method in an interval value fuzzy environment (IVF) (Zanghelini et al., 2018; Li et al., 2019).

Incineration is the most prevalent technology used by the majority of MENA countries for the treatment of infectious medical waste. This result is also confirmed by several studies conducted within many countries such as South Africa, Mexico, Indonesia, and Bangladesh (Islam et al., 2021a & 2021b; Marome and Shaw, 2021; Singh et al., 2022; UNEP, 2020).

Currently, other more sophisticated techniques are also used for the treatment of medical waste such as disinfection by radiation, disinfection by reverse polymerization, pyrolysis disinfection by plasma, and disinfection by thermal gasification. Unfortunately, these techniques are expensive and require huge investments (Wang et al., 2019; Guangcan et al., 2021).

In addition, the choice of delegating biomedical waste treatment and disposal operations to specific companies is limited only to urban areas and especially to large cities. The issue of treatment poses enormous problems for small towns and rural areas. The massive and rapid increase in infectious medical waste has greatly influenced the treatment and disposal capacity in some countries. Medical waste is dumped or burned in the open (Nzediegwu and Chang, 2020).

In order to face this critical situation, guidelines are still to be suggested and measures are needed to improve the treatment and disposal processes of medical waste by drawing on the experience and good practices of other countries.

China in Wuhan, for example, implemented emergency measures during the pandemic period to increase and strengthen medical waste disposal capacity: the establishment of mobile facilities, the temporary use of municipal waste incinerators, and off-site disposal. In addition, immediately following the pandemic, changes in medical waste management practices took place through increased disposal capacity, implementation of smart management, and development of regulatory policy (Zhao et al., 2022; Yu et al., 2020).

During the pandemic, the diversity of waste sources, including healthcare facilities, hospitals, homes, isolation centres, and interim treatment centres, amplified the difficulty of waste treatment and disposal processes. People in quarantine at home throw their contaminated waste with the household waste. No prior sorting is done. This waste constitutes a risk to the safety and health of waste collection agents and to the environment. Special attention and preventive measures for the treatment of contaminated household waste are necessary to limit the risk of transmission and the direct and indirect impacts of the virus on the environment (Diaz et al., 2005).

The analysis carried out opens a wide field of study and evaluation of the methods used within the MENA region for the treatment and disposal of medical waste. It constitutes a starting point for a common reflection that will allow the elaboration of a management strategy for medical waste, including potentially infectious household waste, in case of a health crisis based on artificial intelligence. Moreover, the diversified physicochemical composition of medical waste gives it an important calorific character; these wastes can be used for energy production if they are well managed and properly treated (Erdogan et al., 2021).

5.6 CONCLUSION

The issue of medical waste has been on everyone's mind and has become increasingly critical in the last two years worldwide since the COVID-19 pandemic. Numerous solid medical waste management initiatives have been presented in this chapter from generation, segregation, collection, treatment to disposal, highlighting the various challenges of the different approaches taken. Because of the pandemic, infectious solid medical waste is no longer limited to healthcare facilities, and special attention must be paid to the management of solid waste generated from other sources, including household waste.

The diversity and infectious nature of medical waste pose a health and environmental risk if not managed effectively. Despite the efforts made by many countries, medical waste management remains an onerous and costly task, but it can be controlled through the efforts and involvement of all stakeholders. The analysis carried out opens a wide field of research study concerning the management of medical waste, particularly infectious waste, based on artificial intelligence and constitutes a starting point for reflection on the optimization of collection and treatment protocols, the possibilities of recycling (paper, plastics, metals, glass), reflection on reduction at source, and the involvement and awareness of the people concerned. To ensure continuous improvement of medical waste management practices, the implementation of a clear and preventive strategy in all countries, especially in the MENA region, is becoming an urgent necessity.

REFERENCES

Magda Magdy Abd El-Salam. Hospital waste management in El-Beheira Governorate, Egypt. *Journal of Environmental Management*, 91, 3, 2010, 618–629, ISSN 0301-4797. https://doi.org/10.1016/j.jenvman.2009.08.012.

Hussein I. Abdel-Shafy, Mona S. M. Mansour. Solid waste issue: Sources, composition, disposal, recycling, and valorization. *Egyptian Journal of Petroleum*, 27, 4, 2018, 1275–1290, ISSN 1110-0621. https://doi.org/10.1016/j.ejpe.2018.07.003.

Fayez Abdulla, Hani Abu Qdais, Atallah Rabi. Site investigation on medical waste management practices in northern Jordan. *Waste Management*, 28, 2, 2008, 450–458, ISSN 0956-053X. https://doi.org/10.1016/j.wasman.2007.02.035.

H. A. Abu-Qdais, M. A. Al-Ghazo, E. M. Al-Ghazo. Statistical analysis and characteristics of hospital medical waste under novel coronavirus outbreak. *Global Journal of Environmental Science and Management*, 6, 4, 2020, 1–10. https://doi.org/10.22034/GJESM.2019.06.SI.03.

H. A. Abu Qdais, M.F. Hamoda, J. Newham. Analysis of residential solid waste at generation sites. *Waste Management & Research*, 15, 4, 1997, 395–406, ISSN 0734-242X. https://doi.org/10.1006/wmre.1996.0095.

H. Abu Qdais, A. Rabi, F. Abdulla. Characteristics of the medical waste generated at the Jordanian hospitals. *Clean Technologies and Environmental Policy*, 9, 2007, 147–152. https://doi.org/10.1007/s10098-006-0077-0.

ADEPT. COVID 19 – Waste Survey Results, Week Commencing 18th May, 2020. Available online at: www.adeptnet.org.uk/ documents/covid-19-waste-survey-results-wc-18th-may (Accessed on 27 September 2020).

P. Agamuthu, J. Barasarathi. Clinical waste management under COVID-19 scenario in Malaysia. *Waste Management & Research*, 39, 1_suppl, 2021, 18–26. https://doi.org/10.1177/0734242X20959701.

P. Aghapour, R. Nabizadeh, J. Nouri, M. Monavari, K. Yaghmaeian. Analysis of hospital waste using a healthcare waste management index. *Toxicological & Environmental Chemistry*, 95, 4, 2013, 579–589.

A.S. Alagha, H. Faris, B.H. Hammo, A.M. Al-Zoubi. Identifying β-thalassemia carriers using a data mining approach: The case of the Gaza Strip, Palestine. *Artificial Intelligence in Medicine*, 88, 2018, 70–83, ISSN 0933-3657. https://doi.org/10.1016/j.artmed.2018.04.009.

Khadija Al-Omran, Ezzat Khan, Nisar Ali, Muhammad Bilal. Estimation of COVID-19 generated medical waste in the Kingdom of Bahrain. *Science of the Total Environment*, 801, 2021, 149642, ISSN 0048-9697. https://doi.org/10.1016/j.scitotenv.2021.149642.

Aylin Zeren Alagöz, Günay Kocasoy. Determination of the best appropriate management methods for the health-care wastes in İstanbul. *Waste Management*, 28, 7, 2008, 1227–1235, ISSN 0956-053X. https://doi.org/10.1016/j.wasman.2007.05.018.

H. Alshraideh, H. Abu Qdais. Stochastic modeling and optimization of medical waste collection in Northern Jordan. *Journal of Material Cycles and Waste Management*, 19, 2017, 743–753. https://doi.org/10.1007/s10163-016-0474-3.

M.M. Alzaabi, R. Hamdy, N.S. Ashmawy et al. Flavonoids are promising safe therapy against COVID-19. *Phytochemistry Reviews* 21, 2022, 291–312. https://doi.org/10.1007/s11101-021-09759-z.

M. Ansari, M.H. Ehrampoush, M. Farzadkia, E. Ahmadi. Dynamic assessment of economic and environmental performance index and generation, composition, environmental and human health risks of hospital solid waste in developing countries; a state of the art of review. *Environment International*, 132, 2019, p.105073.

Mehrdad Askarian, Peigham Heidarpoor, Ojan Assadian. A total quality management approach to healthcare waste management in Namazi Hospital, Iran. *Waste Management*, 30, 11, 2010, 2321–2326, ISSN 0956-53X. https://doi.org/10.1016/j.wasman.2010.06.020.

A. Belhadi, S.S. Kamble, S.A.R. Khan et al. Infectious waste management strategy during COVID-19 pandemic in Africa: an integrated decision-making framework for selecting sustainable technologies. *Environmental Management* 66, 2020, 1085–1104. https://doi.org/10.1007/s00267-020-01375-5.

Z. Bendjoudi, F. Taleb, F. Abdelmalek, A. Addou. Healthcare waste management in Algeria and Mostaganem department. Waste Management, 29, 4, 2009, 1383–1387, ISSN 0956-053X. https://doi.org/10.1016/j.wasman.2008.10.008.

Mehmet Emin Birpınar, Mehmet Sinan Bilgili, Tuğba Erdoğan. Medical waste management in Turkey: a case study of Istanbul. *Waste Management*, 29, 1, 2009, 445–448, ISSN 0956-053X, https://doi.org/10.1016/j.wasman.2008.03.015.

A. Cebe, S. Dursun, H. Mankolli. Hospital solid wastes and its effect on environment. *Journal of International Environmental Application & Science*, 8, 5, 2013, 733–737.

R. Chandra, A. Jain, D. Singh Chauhan. Deep learning via LSTM models for COVID-19 infection forecasting in India. *PLoS ONE*, 17, 1, 2022, e0262708.

Y. Chartier (Ed.). *Safe Management of Wastes from Health-Care Activities*. World Health Organization, 2014.

Y. Chen, C. Guo. Handbook of Emergency Disposal and Management of Medical Waste in China, 2020. http://bcrc.tsinghua.edu.cn/col/1256347643949/2020/06/22/1592834394326.html (Accessed on 21 August 2021).

C. Chen, J. Chen, R. Fang, F. Ye, Z. Yang, Z. Wang, F. Shi, W. Tan. What medical waste management system may cope With COVID-19 pandemic: Lessons from Wuhan. *Resources, Conservation and Recycling*, 170, 2021, 105600, ISSN 0921-3449. https://doi.org/10.1016/j.resconrec.2021.105600.

Derek K Chu, Elie A. Akl, Stephanie Duda, Karla Solo, Sally Yaacoub, Holger J Schünemann. Physical distancing, face masks, and eye protection to prevent person-to-person transmission of SARS-CoV-2 and COVID-19: a systematic review and meta-analysis. *The Lancet*, 395, 10242, 2020, 1973–1987, ISSN 0140-6736. https://doi.org/10.1016/S0140-6736(20)31142-9.

Selene Cobo, Antonio Dominguez-Ramos, Angel Irabien. From linear to circular integrated waste management systems: A review of methodological approaches. *Resources, Conservation and Recycling*, 135, 2018, 279–295, ISSN 0921-3449. https://doi.org/10.1016/j.resconrec.2017.08.003.

Atanu Kumar Das, Md. Nazrul Islam, Md. Morsaline Billah, Asim Sarker. COVID-19 pandemic and healthcare solid waste management strategy – A mini-review. *Science of the Total Environment*, 778, 2021, 146220, ISSN 0048-9697. https://doi.org/10.1016/j.scitotenv.2021.146220.

H. Dastpak, S. Golbaz, M., Farzadkia. Hospital waste minimisation, separation, treatment and disposal in Iran: a mini review study. *Proceedings of Institution of Civil Engineers: Waste and Resource Management*, 170, 2017, 107–118. https://doi.org/10.1680/jwarm.17.00016.

M.A. Daoudi. Évaluation de la gestion des déchets solides médicaux et pharmaceutiques à l'hôpital Hassan-II d'Agadir. Institut national d'administration sanitaire. Neuvième cours de maîtrise en administration sanitaire et santé publique. Institut national d'administration sanitaire, centre collaborateur de l'OMS, Royaume du Maroc, 84, 2008.

L.F. Diaz, G.M. Savage, L.L. Eggerth. Alternatives for the treatment and disposal of healthcare wastes in developing countries. *Waste Management*, 25, 6, 2005, 626–637, ISSN 0956-053X. https://doi.org/10.1016/j.wasman.2005.01.005.

Mehtap Dursun, E. Ertugrul Karsak, Melis Almula Karadayi. Assessment of health-care waste treatment alternatives using fuzzy multi-criteria decision making approaches. *Resources, Conservation and Recycling*, 57, 2011, 98–107, ISSN 0921-3449. https://doi.org/10.1016/j.resconrec.2011.09.012.

Julien S. Frédéric Dutheil. Baker, valentin navel, COVID-19 as a factor influencing air pollution? *Environmental Pollution*, 263, Part A, 2020, 114466, ISSN 0269-7491, https://doi.org/10.1016/j.envpol.2020.114466.

H.H. Eker, M.S. Bilgili. Statistical analysis of waste generation in healthcare services: a case study. *Waste Management & Research*, 29, 8, 2011, 791–796. https://doi.org/10.1177/0734242X10396755.

Samar Elkhalifa, Tareq Al-Ansari, Hamish R. Mackey, Gordon McKay. Food waste to biochars through pyrolysis: a review. *Resources, Conservation and Recycling*, 144, 2019, 310–320, ISSN 0921-3449. https://doi.org/10.1016/j.resconrec.2019.01.024.

Altug Alp Erdogan, Mustafa Zeki Yilmazoglu. Plasma gasification of the medical waste. *International Journal of Hydrogen Energy*, 46, 57, 2021, 29108–29125, ISSN 0360-3199. https://doi.org/10.1016/j.ijhydene.2020.12.069.

Oluniyi O. Fadare, Elvis D. Okoffo. Covid-19 face masks: a potential source of microplastic fibers in the environment. *Science of the Total Environment*, 737, 2020, 140279, ISSN 0048-9697. https://doi.org/10.1016/j.scitotenv.2020.140279.

M. Farzadkia, M. Emamjomeh, S. Golbaz, H. Sajadi. An investigation on hospital solid waste management in Iran. *Global NEST Journal*, 17, 4, 2015, 771–783.

G. Giakoumakis, D. Politi, D. Sidiras. Technologies de traitement des déchets médicaux pour la production d'énergie, de combustibles et de matériaux: une revue. *Energies*, 14, 2021, 8065. https://doi.org/10.3390/en14238065.

L. Hangulu, O. Akintola. Health care waste management in community-based care: experiences of community health workers in low resource communities in South Africa. *BMC Public Health*, 17, 2017. https://doi.org/10.1186/s12889-017-4378-5.

Jingmin Hong, Song Zhan, Zhaohe Yu, Jinglan Hong, Congcong Qi. Life-cycle environmental and economic assessment of medical waste treatment. *Journal of Cleaner Production*, 174, 2018, 65–73, ISSN 0959-6526. https://doi.org/10.1016/j.jclepro.2017.10.206.

N. Huang, P. Pérez, T. Kato et al. SARS-CoV-2 infection of the oral cavity and saliva. *Nature Medicine*, 27, 2021, 892–903. https://doi.org/10.1038/s41591-021-01296-8.

Sadia Ilyas, Rajiv Ranjan Srivastava, Hyunjung Kim. Disinfection technology and strategies for COVID-19 hospital and bio-medical waste management. *Science of the Total Environment*, 749, 2020, 141652, ISSN 0048-9697. https://doi.org/10.1016/j.scitotenv.2020.141652.

S.M.D.U. Islam, P.K. Mondal, N. Ojong, et al. Water, sanitation, hygiene and waste disposal practices as COVID-19 response strategy: insights from Bangladesh. *Environment, Development and Sustainability*, 23, 2021a, 11953–11974. https://doi.org/10.1007/s10668-020-01151-9.

A. Islam, M.A. Kalam, M.A. Sayeed, S., Shano, M.K., Rahman, S. Islam, J. Ferdous, S.D. Choudhury, M.M. Hassan. Escalating SARS-CoV-2 circulation in environment and tracking waste management in South Asia. *Environmental Science and Pollution Research*, 28, 2021b, 61951–61968. https://doi.org/10.1007/S11356-021-16396-8/TABLES/7.

M. Iyer, S. Tiwari, K. Renu, M.Y. Pasha, S. Pandit, B. Singh, N. Raj, S. Krothapalli, H.J. Kwak, V. Balasubramanian, S.B. Jang, G. Dileep Kumar, A. Uttpal, A. Narayanasamy, M. Kinoshita, M.D. Subramaniam, S.K. Nachimuthu, A. Roy, A. Valsala Gopalakrishnan, P. Ramakrishnan, S.G. Cho, B. Vellingiri. Environmental survival of SARS-CoV-2 – a solid waste perspective. *Environmental Research*, 197, 2021, 111015. https://doi.org/10.1016/J.ENVRES.2021.111015.

Òscar Jordà, Sanjay R. Singh, Alan M. Taylor. Longer-run economic consequences of pandemics. *The Review of Economics and Statistics*, 104, 1, 2022, 166–175. https://doi.org/10.1162/rest_a_01042.

R.R. Kalantary, A. Jamshidi, M.M.G. Mofrad et al. Effect of COVID-19 pandemic on medical waste management: a case study. *Journal of Environmental Health Science and Engineering*, 19, 2021, 831–836. https://doi.org/10.1007/s40201-021-00650-9.

Katerina Kalogiannidou, Eftychia Nikolakopoulou, Dimitrios Komilis. Generation and composition of waste from medical histopathology laboratories. *Waste Management*, 79, 2018, 435–442, ISSN 0956-053X. https://doi.org/10.1016/j.wasman.2018.08.012.

G. Kampf, D. Todt, S. Pfaender, E. Steinmann. Persistence of coronaviruses on inanimate surfaces and their inactivation with biocidal agents. *Journal of Hospital Infection*, 104, 3, 2020, 246–251, ISSN 0195-6701. https://doi.org/10.1016/j.jhin.2020.01.022.

B.D. Kevadiya, J. Machhi, J. Herskovitz et al. Diagnostics for SARS-CoV-2 infections. *Nature Materials*, 20, 2021, 593–605. https://doi.org/10.1038/s41563-020-00906-z.

M. Khazraji, L. Mouhir, M. Fekhaoui, L. Saafadi, I. Nassri. Assessment of medical and pharmaceutical waste flows during the coronavirus pandemic in the Rabat-Sale-Kenitra region, Morocco. *Waste Management & Research*. 2021. https://doi.org/10.1177/0734242X211046853.

Jiří Jaromír Klemeš, Yee Van Fan, Peng Jiang. The energy and environmental footprints of COVID-19 fighting measures – PPE, disinfection, supply chains. *Energy*, 211, 2020, 118701, ISSN 0360-5442. https://doi.org/10.1016/j.energy.2020.118701.

Fang Li, Yuan-Yuan Li, Ming-Jin Liu, Li-Qun Fang, Natalie E. Dean, Gary W.K. Wong, Xiao-Bing Yang, Ira Longini, M. Elizabeth Halloran, Huai-Ji Wang, Pu-Lin Liu, Yan-Hui Pang, Ya-Qiong Yan, Su Liu, Wei Xia, Xiao-Xia Lu, Qi Liu, Yang Yang, Shun-Qing Xu. Household transmission of SARS-CoV-2 and risk factors for susceptibility and infectivity in Wuhan: a retrospective observational study. *The Lancet Infectious Diseases*, 21, 5, 2021, 617–628, ISSN 1473–3099. https://doi.org/10.1016/S1473-3099(20)30981-6.

Jiafu Li, Zhiwei Lv, Lei Du, Xiaonan Li, Xuepeng Hu, Chong Wang, Zhiguang Niu, Ying Zhang. Emission characteristic of polychlorinated dibenzo-p-dioxins and polychlorinated dibenzofurans (PCDD/Fs) from medical waste incinerators (MWIs) in China in 2016: A comparison between higher emission levels of MWIs and lower emission levels of MWIs. *Environmental Pollution*, 221, 2017, 437–444, ISSN 0269-7491. https://doi.org/10.1016/j.envpol.2016.12.009.

Fanghua Li, Srikanth Chakravartula Srivatsa, Sankar Bhattacharya. A review on catalytic pyrolysis of microalgae to high-quality bio-oil with low oxygeneous and nitrogenous compounds, *Renewable and Sustainable Energy Reviews*, 108, 2019, 481–497, ISSN 1364-0321. https://doi.org/10.1016/j.rser.2019.03.026.

Y. Liang, Q. Song, N. Wu et al. Implications of the COVID-19 pandemic on solid waste generation and management strategies. *Frontiers of Environmental Science & Engineering*, 15, 2021, 115. https://doi.org/10.1007/s11783-021-1407-5.

Hu-Chen Liu, Jian-Xin You, Chao Lu, Yi-Zeng Chen. Evaluating health-care waste treatment technologies using a hybrid multi-criteria decision making model. *Renewable and Sustainable Energy Reviews*, 41, 2015, 932–942, ISSN 1364-0321, https://doi.org/10.1016/j.rser.2014.08.061.

A. Maalouf, H. Maalouf. Impact de la pandémie de COVID-19 sur la gestion des déchets médicaux au Liban. *Gestion des déchets et recherché*, 2021, 39, 1_suppl, 45–55. https://doi.org/10.1177/0734242X211003970.

Olivia Maamari, Cedric Brandam, Roger Lteif, Dominique Salameh. Health Care Waste generation rates and patterns: the case of Lebanon. *Waste Management*, 43, 2015, 550–554, ISSN 0956-053X. https://doi.org/10.1016/j.wasman.2015.05.005.

Malsparo. Pharmaceutical Waste Management, Pharmaceutical Waste at Healthcare Facilities and the Home. https://www.malsparo.com/pharm.htm, 2020.

W. Marome, R. Shaw. COVID-19 Response in Thailand and its implications on future preparedness. *International Journal of Environmental Research and Public Health*, 18, 3, 2021, 1089. https://doi.org/10.3390/ijerph18031089.

R. Mato, G.R. Kassenga. A study on problems of management of medical solid wastes in Dar es Salaam and their remedial measures. *Resources, Conservation & Recycling*, 21, 1997, 1–16.

Bontle Mbongwe, Baagi T. Mmereki, Andrew Magashula. Healthcare waste management: current practices in selected healthcare facilities, Botswana. *Waste Management*, 28, 1, 2008, 226–233, ISSN 0956-053X, https://doi.org/10.1016/j.wasman.2006.12.019.

Jiahe Miao, Jining Li, Fenghe Wang, Xinyi Xia, Shaopo Deng, Shengtian Zhang. Characterization and evaluation of the leachability of bottom ash from a mobile emergency incinerator of COVID-19 medical waste: a case study in Huoshenshan Hospital, Wuhan, China. *Journal of Environmental Management*, 303, 2022, 114161, ISSN 0301-4797. https://doi.org/10.1016/j.jenvman.2021.114161.

Ministry of Industry and Information Technology. COVID-19 Pandemic Prevention and Control Report (In Chinese), 2020. Available at: https://www.miit.gov.cn/ztzl/rdzt/xxgzbdgrdfyyqfkgz/ gzdt/art/2020/art_cea12b5000b8474da3bd8e85918723f1.html (Accessed on 21 August 2021).

MOEFCC. Biomedical Waste Management and Handling Rules, 2016. https://moef.gov.in/en/division/environment-divisions/hazardous-substances-management-hsm/introduction/

M. Mofijur, I.M. Rizwanul Fattah, Md Asraful Alam, A.B.M. Saiful Islam, Hwai Chyuan Ong, S.M. Ashrafur Rahman, G. Najafi, S.F. Ahmed, Md. Alhaz Uddin, T.M.I. Mahlia, Impact of COVID-19 on the social, economic, environmental and energy domains: Lessons learnt from a global pandemic. *Sustainable Production and Consumption*, 26, 2021, 343–359, ISSN 2352-5509. https://doi.org/10.1016/j.spc.2020.10.016.

C. Nzediegwu, S.X. Chang. Improper solid waste management increases potential for COVID-19 spread in developing countries. *Resources Conservation and Recycling*, 161, 2020, 104947. https://doi.org/10.1016/j.resconrec.2020.104947.

A. Pacheco, S. Vassal, M. Gabarda, M. Jung, M. Hervier, A. Durand, ... M. Vergnenegre. Dossier: rejets et dechets hospitaliers. Partie 2. *Techniques Hospitalières*, 54, 1999, 633.

Vishal Kumar Parida, Divyanshu Sikarwar, Abhradeep Majumder, Ashok Kumar Gupta. An assessment of hospital wastewater and biomedical waste generation, existing legislations, risk assessment, treatment processes, and scenario during COVID-19. *Journal of Environmental Management*, 308, 2022, 114609, ISSN 0301–4797. https://doi.org/10.1016/j.jenvman.2022.114609.

Gayathri V. Patil, Kamala Pokhrel. Biomedical solid waste management in an Indian hospital: a case study. *Waste Management*, 25, 6, 2005, 592–599, ISSN 0956-053X, https://doi.org/10.1016/j.wasman.2004.07.011.

Md Arafatur Rahman, Nafees Zaman, A. Taufiq Asyhari, Fadi Al-Turjman, Md. Zakirul Alam Bhuiyan, M.F. Zolkipli. Data-driven dynamic clustering framework for mitigating the adverse economic impact of Covid-19 lockdown practices. *Sustainable Cities and Society*, 62, 2020, 102372, ISSN 2210-6707. https://doi.org/10.1016/j.scs.2020.102372.

Meisam Ranjbari, Michael Saidani, Zahra Shams Esfandabadi, Wanxi Peng, Su Shiung Lam, Mortaza Aghbashlo, Francesco Quatraro, Meisam Tabatabaei, Two decades of research on waste management in the circular economy: Insights from bibliometric, text mining, and content analyses. *Journal of Cleaner Production*, 314, 2021, 128009, ISSN 0959-6526. https://doi.org/10.1016/j.jclepro.2021.128009.

K. Rebbas, R. Bounar, N. Merniz, M. D. Miara. Management of healthcare waste in the M'Sila region (Algeria). *Journal of EcoAgriTourism*, 14, 1, 2018, 96–104.

Jingzheng Ren, Marie Lützen. Fuzzy multi-criteria decision-making method for technology selection for emissions reduction from shipping under uncertainties. *Transportation Research Part D: Transport and Environment*, 40, 2015, 43–60, ISSN 1361-9209. https://doi.org/10.1016/j.trd.2015.07.012.

Neil J. Rowan, John G. Laffey. Unlocking the surge in demand for personal and protective equipment (PPE) and improvised face coverings arising from coronavirus disease (COVID-19) pandemic – Implications for efficacy, re-use and sustainable waste management. *Science of the Total Environment*, 752, 2021, 142259, ISSN 0048-9697. https://doi.org/10.1016/j.scitotenv.2020.142259.

Ryan M. Samuel, Homa Majd, Mikayla N. Richter et al. Androgen Signaling Regulates SARS-CoV-2 Receptor Levels and Is Associated with Severe COVID-19 Symptoms in Men. *Cell Stem Cell*, 27, 6, 2020, 876–889.e12, ISSN 1934-5909. https://doi.org/10.1016/j.stem.2020.11.009.

S.A. Sarkodie, P.A. Owusu. Global assessment of environment, health and economic impact of the novel coronavirus (COVID-19). *Environ Dev Sustain* 23, 2021, 5005–5015. https://doi.org/10.1007/s10668-020-00801-2.

Parul Saxena, Indira P. Pradhan, Deepak Kumar. Redefining bio medical waste management during COVID-19 in India: a way forward. *Materials Today: Proceedings*, 2021, ISSN 2214–7853. https://doi.org/10.1016/j.matpr.2021.09.507.

Ain Umaira Md Shah, Syafiqah Nur Azrie Safri, Rathedevi Thevadas, Nor Kamariah Noordin, Azmawani Abd Rahman, Zamberi Sekawi, Aini Ideris, Mohamed Thariq Hameed Sultan. COVID-19 outbreak in Malaysia: actions taken by the Malaysian government. *International Journal of Infectious Diseases*, 97, 2020, 108–116, ISSN 1201-9712. https://doi.org/10.1016/j.ijid.2020.05.093.

Hari Bhakta Sharma, Kumar Raja Vanapalli, VR Shankar Cheela, Ved Prakash Ranjan, Amit Kumar Jaglan, Brajesh Dubey, Sudha Goel, Jayanta Bhattacharya. Challenges, opportunities, and innovations for effective solid waste management during and post COVID-19 pandemic. *Resources, Conservation and Recycling*, 162, 2020, 105052, ISSN 0921-3449, https://doi.org/10.1016/j.resconrec.2020.105052.

R. Stolze, J-G. Kühling. Treatment of infectious waste: development and testing of an add-on set for used gravity displacement autoclaves. *Waste Management & Research*, 27, 4, 2009, 343–353. https://doi.org/10.1177/0734242X09335695.

Guangcan Su, Hwai Chyuan Ong, Shaliza Ibrahim, I. M. Rizwanul Fattah, M. Mofijur, Cheng Tung Chong. Valorisation of medical waste through pyrolysis for a cleaner environment: progress and challenges. *Environmental Pollution*, 279, 2021, 116934, ISSN 0269-7491, https://doi.org/10.1016/j.envpol.2021.116934.

A.K. Subramanian, D. Thayalan, A.I. Edwards, A. Almalki, A. Venugopal. Biomedical waste management in dental practice and its significant environmental impact: a perspective. *Environmental Technology and Innovation*, 24, 2021, 101807. https://doi.org/ 10.1016/J.ETI.2021.101807.

Margaret M. Sugg, Trent J. Spaulding, Sandi J. Lane, Jennifer D. Runkle, Stella R. Harden, Adam Hege, Lakshmi S. Iyer. Mapping community-level determinants of COVID-19 transmission in nursing homes: a multi-scale approach. *Science of the Total Environment*, 752, 2021, 141946, ISSN 0048-9697. https://doi.org/10.1016/j.scitotenv.2020.141946.

A.L.P. Silva, J.C. Prata, T.R. Walker, A.C. Duarte, W. Ouyang, D. Barcelo, T. RochaSantos. Increased plastic pollution due to COVID-19 pandemic: challenges and recommendations. *Chemical Engineering Journal*, 405, 2021, 126683. https://doi.org/10.1016/j.cej.2020.126683.

Parteek Singh Thind, Arjun Sareen, Dapinder Deep Singh, Sandeep Singh, Siby John. Compromising situation of India's bio-medical waste incineration units during pandemic outbreak of COVID-19: associated environmental-health impacts and mitigation measures. *Environmental Pollution*, 276, 2021, 116621, ISSN 0269-7491. https://doi.org/10.1016/j.envpol.2021.116621.

E. Singh, A. Kumar, R. Mishra, S. Kumar. Solid waste management during COVID-19 pandemic: recovery techniques and responses. *Chemosphere*, 288, 2022, 132451. https://doi.org/10.1016/J.CHEMOSPHERE.2021.132451.

M. Tsakona, E. Anagnostopoulou, E. Gidarakos. Hospital waste management and toxicity evaluation: A case study. *Waste Management*, 27, 7, 2007, 912–920, ISSN 0956-053X. https://doi.org/10.1016/j.wasman.2006.04.019.

Shu-Feng Tsao, Helen Chen, Therese Tisseverasinghe, Yang Yang, Lianghua Li, Zahid A Butt. What social media told us in the time of COVID-19: a scoping review. *The Lancet Digital Health*, 3, 3, 2021, e175–e194, ISSN 2589-7500. https://doi.org/10.1016/S2589-7500(20)30315-0.

UNEP. Solid Waste Management (Volume II: Regional Overviews and Information Sources), CalRecovery, Inc. and UNEP International Environmental Technology Centre (IETC), 2005a. https://wedocs.unep.org/handle/20.500.11822/30734

UNEP. Integrated Waste Management Scoreboard: A Tool to Measure Performance in Municipal Solid Waste Management, 2005b. https://wedocs.unep.org/handle/20.500.11822/8409

UNEP. Fiche d'information sur la gestion des déchets liés à la COVID-19 | UNEP - UN Environment Programme, 2020. https://www.unep.org › resources › feuillet-dinformation (Accessed on 19 July. 2020).

Andrea Undri, Luca Rosi, Marco Frediani, Piero Frediani. Efficient disposal of waste polyolefins through microwave assisted pyrolysis. *Fuel*, 116, 2014, 662–671, ISSN 0016-2361. https://doi.org/10.1016/j.fuel.2013.08.037.

Andrea Undri, Luca Rosi, Marco Frediani, Piero Frediani, Upgraded fuel from microwave assisted pyrolysis of waste tire. *Fuel*, 115, 2014, 600–608, ISSN 0016-2361. https://doi.org/10.1016/j.fuel.2013.07.058.

Andrea Undri, Marco Frediani, Luca Rosi, Piero Frediani. Reverse polymerization of waste polystyrene through microwave assisted pyrolysis. *Journal of Analytical and Applied Pyrolysis*, 105, 2014, 35–42, ISSN 0165-2370. https://doi.org/10.1016/j.jaap.2013.10.001.

Marica Valente, Matheus Bueno. The effects of pricing waste generation: a synthetic control approach. *Journal of Environmental Economics and Management*, 96, 2019, 274–285, ISSN 0095-0696. https://doi.org/10.1016/j.jeem.2019.06.004.

Balachandar Vellingiri, Kaavya Jayaramayya, Mahalaxmi Iyer, Arul Narayanasamy, Vivekanandhan Govindasamy, Bupesh Giridharan, Singaravelu Ganesan, Anila Venugopal, Dhivya Venkatesan, Harsha Ganesan, Kamarajan Rajagopalan, Pattanathu K.S.M. Rahman, Ssang-Goo Cho, Nachimuthu Senthil Kumar, Mohana Devi Subramaniam, COVID-19: a promising cure for the global panic. *Science of the Total Environment*, 725, 2020, 138277, ISSN 0048-9697, https://doi.org/10.1016/j.scitotenv.2020.138277.

Paolo Veronesi, Cristina Leonelli, Umberto Moscato, Angelo Cappi, Ornella Figurelli. Non-incineration microwave assisted sterilization of medical waste, *Journal of Microwave Power and Electromagnetic Energy*, 40, 4, 2005, 211–218. https://doi.org/10.1080/08327823.2005.11688546.

G. Volpicelli, A. Lamorte, T. Villén. What's new in lung ultrasound during the COVID-19 pandemic. *Intensive Care Med* 46, 2020, 1445–1448. https://doi.org/10.1007/s00134-020-06048-9.

Evangelos A. Voudrias. Technology selection for infectious medical waste treatment using the analytic hierarchy process. *Journal of the Air & Waste Management Association*, 66, 7, 2016, 663–672, https://doi.org/10.1080/10962247.2016.1162226.

Chandra Wahyu Purnomo, Winarto Kurniawan, Muhammad Aziz. Technological review on thermochemical conversion of COVID-19-related medical wastes. *Resources, Conservation and Recycling*, 167, 2021, 105429, ISSN 0921-3449. https://doi.org/10.1016/j.resconrec.2021.105429.

C. Wang, P.W. Horby, F.G. Hayden, G.F. Gao. A novel coronavirus outbreak of global health concern. *The Lancet*, 395, 10223, 2020a, 470–473.

Jiao Wang, Jin Shen, Dan Ye, Xu Yan, Yujing Zhang, Wenjing Yang, Xinwu Li, Junqi Wang, Liubo Zhang, Lijun Pan. Disinfection technology of hospital wastes and wastewater: suggestions for disinfection strategy during coronavirus Disease 2019 (COVID-19) pandemic in China. *Environmental Pollution*, 262, 2020b, 114665, ISSN 0269-7491. https://doi.org/10.1016/j.envpol.2020.114665.

Qiang Wang, Min Su. A preliminary assessment of the impact of COVID-19 on environment – A case study of China. *Science of the Total Environment*, 728, 2020, 138915, ISSN 0048-9697. https://doi.org/10.1016/j.scitotenv.2020.138915.

WHO, 2014. Safe management of wastes from health-care activities, in: Yves Chartier, Jorge Emmanuel, Ute Pieper, Annette Prüss, Philip Rushbrook, Ruth Stringer, William Townend, Wilburn, S., Zghondi, R. (Eds.). Geneva, Switzerland.

Elliott Steen Windfeld, Marianne Su-Ling Brooks. Medical waste management – A review. *Journal of Environmental Management*, 163, 2015, 98–108, ISSN 0301-4797, https://doi.org/10.1016/j.jenvman.2015.08.013.

World Health Organization (WHO). Guidelines for Safe Disposal of Unwanted Pharmaceuticals in and after Emergencies. World Health Organization, 1999a. https://apps.who.int/iris/handle/10665/42238. Available at: (Accessed on 10 January 2011).

World Health Organization (WHO). Safe Management of Wastes from Healthcare Activities. World Health Organization, 1999b. https://apps.who.int/iris/bitstream/handle/10665/42175/9241545259.pdf Available at: (Accessed on 10 January 2011).

World Health Organization (WHO). Wastes from Health-Care Activities. Factsheet No. 253; November, 2011. Available at: http://www.who. int/mediacentre/factsheets/fs253/en (Accessed on 3 May 2017).

World Health Organization. Health-care Waste [Internet]. World Health Organization, 2018 [cited 2020 Apr 15]. Available at: https://www.who.int/news-room/ fact-sheets/detail/health-care-waste (Accessed on 2 August 2020).

Linping Xu, Yan Kong, Mingxue Wei, Yichuan Wang, Minhao Zhang, Benny Tjahjono. Combatting medical plastic waste through visual elicitation: insights from healthcare professionals. *Journal of Cleaner Production*, 329, 2021, 129650, ISSN 0959-6526. https://doi.org/10.1016/j.jclepro.2021.129650.

Lie Yang, Xiao Yu, Xiaolong Wu, Jia Wang, Xiaoke Yan, Shen Jiang, Zhuqi Chen. Emergency response to the explosive growth of health care wastes during COVID-19 pandemic in uhan, China. *Resources, Conservation and Recycling*, 164, 2021, 105074, ISSN 0921-3449. https://doi.org/10.1016/j.resconrec.2020.105074.

Siming You, Christian Sonne, Yong Sik Ok. COVID-19's unsustainable waste management. *Science*, 368, 6498, 26 Jun 2020, 1438. https://doi.org/10.1126/science. abc7778open_in_new.

H. Yu, X. Sun, W.D. Solvang, X. Zhao. Reverse Logistics Network design for effective management of medical waste in epidemic outbreaks: insights from the coronavirus disease 2019 (COVID-19) outbreak in Wuhan (China). *International Journal of Environmental Research and Public Health*, 17, 5, 2020, 1770. https://doi.org/10.3390/ijerph1705177.

M. Yunesian, F. Malekahmadi. *Analysis of the Healthcare Waste Management Status in Tehran Hospitals*. 2014.

Ali Daryabeigi Zand, Azar Vaezi Heir. Emerging challenges in urban waste management in Tehran, Iran during the COVID-19 pandemic. *Resources, Conservation, and Recycling*, 162, 2020, 105051. https://doi.org/10.1016/j.resconrec.2020.105051.

Guilherme Marcelo Zanghelini, Edivan Cherubini, Sebastião Roberto Soares. How multi-criteria decision analysis (MCDA) is aiding life cycle assessment (LCA) in results interpretation. *Journal of Cleaner Production*, 172, 2018, 609–622, ISSN 0959-6526, https://doi.org/10.1016/j.jclepro.2017.10.230.

Hailong Zhao, Hanqiao Liu, Guoxia Wei, Ning Zhang, Haoyu Qiao, Yongyue Gong, Xiangnan Yu, Jianhua Zhou, Yuhang Wu. A review on emergency disposal and management of medical waste during the COVID-19 pandemic in China. *Science of the Total Environment*, 810, 2022, 152302, ISSN 0048-9697, https://doi.org/10.1016/j.scitotenv.2021.152302.

B. Zheng, D. Tong, M. Li, F. Liu, C. Hong, G. Geng, H. Li, X. Li, L. Peng, J. Qi, L. Yan, Y. Zhang, H., Y. Zheng, K. He, Q. Zhang. Trends in China's anthropogenic emissions since 2010 as the consequence of clean air actions. *Atmospheric Chemistry and Physics*, 18, 2018, 14095–14111. https://doi.org/10.5194/acp-18-14095-2018.

6 A Comprehensive Review on Performance Evaluation Methods of Biomass Cook Stoves

Chandrika Samal, Ramesh Chandra Nayak, Manmatha Kumar Roul, Diptikanta Das, and Ramanuj Kumar

CONTENTS

6.1 Introduction 95
6.2 History of Testing Methods 96
    6.2.1 Laboratory-Based Assessment Method: WBT 97
    6.2.2 Test Phases and Performance Parameters of WBT 97
    6.2.3 Emission Testing and Emission Parameters 99
    6.2.4 Limitations of WBT... 101
    6.2.5 Field-Based Assessment Methods 101
6.3 Conclusions.. 101
References... 102

6.1 INTRODUCTION

Cooking is the most vital domestic activity of mankind. Invention of fire conceptualized the idea of cooking, and thus cookstoves. The history of cookstoves is as old as human civilization. Next to open fire, the three-stone fire was the first ever cookstove design developed by mankind with concept of shielding the fire [1]. Gradually, the three-stone fire is replaced by mud stoves, brick stoves, metallic stoves, and technically developed improved cookstoves. As cooking is very much attached to local culture, tradition, and food habits, various cooking cycles along with different cookstove designs are seen across the globe [2]. The majority of cookstove users prefers biomass as fuel. Almost 3 billion people worldwide are using biomasses, such as fuel wood, agricultural residues, and animal wastes as fuels in the traditional biomass cookstoves [3–6]. The major drawbacks associated with traditional biomass cookstoves are low efficiency, high pollutant emission, high fuel consumption, and long activity duration [7–9]. Due to incomplete combustion of solid biomasses, undesirable gaseous products (like CO, NO_x, and SO_2) and solid particles (such as particulate

DOI: 10.1201/9781003359784-6

matters and polycyclic aromatic hydrocarbons) are emitted from traditional stoves [10–12]. These harmful emissions from various sources contribute to serious health hazards, significant climate change, and global warming [13,14]. The primary intention behind designing advanced cookstoves is to produce efficient, clean, and affordable cookstoves to the users [15]. It is essential for researchers to assess the efficiency and emission correctly so that proper modifications can be introduced to achieve a better output. Correct evaluation of cookstove performance during cooking in a kitchen is a critical and complex phenomenon due to involvement of various constraints associated in conducting the tests [16]. Cookstove performance can be varied drastically in field than that in laboratory as a specific cooking procedure is not maintained during cooking. Rigorous testing of biomass cookstoves in different conditions can be considered one of the important aspects to generate accurate and reliable performance results. It was realized by researchers that for assessing the realistic results and for comparing the performance results of different cookstoves used in different regions of world, cookstoves are to be tested using some standardized testing protocols. A standardized cookstove testing method furnishes a better chance of comparing different types of cookstove used in different regions of the globe with decisive results [17]. So, various research community have designed and developed many testing methods and checked their reliability and utility before finalizing any international standard testing protocol for biomass cookstoves. However, all these protocols have been scrutinized from time to time to minimize the uncertainty present in the adopted methodologies. Several recommendations and suggestions have been included in testing protocols to make more reliable and closer-to-actual cooking practices. Many laboratory- and field-based international testing standards and protocols have been designed, evolved, and executed for the testing of cookstove performance.

The present review article provides a detailed study on numerous standardized international cookstove testing protocols along with the detailed evaluation methodology. A modest approach is made to cover various performance parameters associated with the testing methods for proper evaluation and quantification of efficiency and emission of biomass cookstoves.

6.2   HISTORY OF TESTING METHODS

The first initiative on testing of biomass cookstoves (mud stoves) was carried out by Theodorovic of Egypt in 1954, and the next attempt was made in 1961 by Singer of Indonesia [18]. The performance evaluation testing methods were standardized by United States Agency for International Development in the year 1982 by arranging a handful of workshops [19]. In 1982, Volunteer in Technical Assistance (VITA) published the first version of one laboratory-based cookstove testing protocol, named water boiling test (WBT), and later, it was upgraded to many other versions [16]. In 1985, the second version of the WBT was published along with two new field-based testing protocols named controlled cooking test (CCT) and kitchen performance test (KPT) [20]. The WBT is a completely laboratory-based controlled performance testing method, whereas the KPT is exclusively followed for field testing. Although the CCT is conducted in a controlled environment, during testing, real meals are to be cooked in actual households. So, it can be considered a bridge between laboratory- and field-based tests. VITA is a popular publication for cookstove testing protocols, and the WBT is the most preferred testing protocol by researchers [21], accounting for around 73% of the tests conducted [19]; however, the CCT accounts for around 12%, and the KPT is followed by lesser researchers for cookstove testing [19].

6.2.1 LABORATORY-BASED ASSESSMENT METHOD: WBT

The WBT is a simple, widely used, standardized, reproducible, and controlled labo- ratory-based testing protocol used for cookstove performance evaluation [22,23]. It is a systematized and transposable testing standard utilized for comparison of perfor- mance of cookstoves made in different regions for various cooking applications [24]. It emphasizes on how effectively and efficiently a certain quantity of water can boil in a cooking pot by using available fuel energy [25]. The WBT is the most popular test- ing protocol and is followed by many researchers [26–30]. The WBT has been modi- fied continuously into many upgraded versions as per the requirement and demand of the cooking sector. Different versions of the WBT with publishing organizations are presented in Table 6.1. Emphasizing on indigenous culture and regional food practice, some native standards were also developed along with international standards [21]. All the versions of the WBT can be classified into two different categories: two-phase WBT and three-phase WBT. The two-phase WBT has one high-power phase (cold start) and one-low power phase (simmering) [31]. The cold-start phase (stove in the ambient condition) resembles a cooking task that needs a high-power input to achieve a quick increase in temperature; however, the low-power phase is simmering-based. In the three-phase WBT, another phase named a high-start hot start is organized in between the high-power hot start and low-start simmering. Recent versions (4.2.2 and 4.2.3) of the WBT are three-phase-based testing protocols [17].

6.2.2 TEST PHASES AND PERFORMANCE PARAMETERS OF WBT

WBT version 4.2.3 being its latest version, different test phases and performance parameters associated are discussed here in a comprehensive manner. WBT 4.2.3 is proposed for three different phases, immediately followed one after the other: "high- power cold start," "high-power hot start," and "low-power simmering." The high- power cold start is a phase where the fuel combustion starts from a room temperature condition, and the standard quantity of water is allowed to heat up to a local boiling point [26]. Experiments for the high-power hot start are initiated just after the cold start, i.e., when the stove is in a hot condition. This phase is carried out to identify the difference in performance of the stove when it is at room temperature and when it is in hot condition before starting of combustion [31,43]. The pot is replaced by another pot of the same size and same material with the same amount of water at atmospheric temperature. This phase also continues till water reaches the local boiling point, and it provides a comparison in cookstove performance in the room temperature condi- tion and hot condition [44]. The low-power simmering phase represents minimum fuel consumption for simmering of water. The test starts immediately after the hot start. The significance of this phase is to determine the amount of fuel required to

TABLE 6.1

Evolution of WBT into Different Versions by Different Publishing Organizations

Different versions of WBT	Year	Organization	Reference
VITA (version1)	1982	VITA	[16]
VITA (version 2)	1985	VITA	[20]
Solid Biomass Chulha (cookstoves) specifications	1991	Indian Standard Institute, Bureau of Indian Standards	[32]
US761	2007	Uganda National Bureau Standards	[33]
WBT version 3	2007	Aprovecho Research Centre along with University of California at Berkeley	[34]
DB11/T540	2008	Quality and Technical Supervision Bureau of Beijing Municipality, China	[35]
Emissions and performance test protocol	2009	Stove manufacturers (Envirofit International and Philips)	[36]
WBT 4.1.2	2009	Aprovecho Research Centre	[37]
Adapted WBT version 2.0)	2010	Research and Development Unit of Group of Renewable Energy Environment and Solidarity, Cambodia	[38]
Heterogeneous testing protocol	2012	Sustainable Energy Technology Testing and Research Centre, University of Johannesburg	[39]
WBT version 4.2.2	2013	Aprovecho Research Center	[40]
Portable solid bio-mass cookstove (chulha)	2013	BIS: Bureau of Indian Standard	[41]
WBT version 4.2.3	2014	Engineers in Technical and Humanitarian Opportunities of Service, Clean Indoor Air (PCIA) and Global Alliance for Clean Cookstoves.	[42]

WBT, water boiling test; VITA, Volunteer in Technical Assistance.

simmer a certain amount of water just below boiling point (3°C) for 45 minutes with a minimum evaporation of water.

Generally, thermal efficiency is considered a major performance parameter in the WBT. It is the measure of fractional fuel energy utilized for boiling and evaporating of a certain amount of water in each phase. It is calculated as the ratio of the amount of energy utilized for boiling and vaporizing of a certain quantity of water to the amount of heat supplied for complete combustion of a known quantity of feed stock. The heat supplied is calculated theoretically on the basis of the amount of dry fuel used and its calorific value. Another important parameter utilized to show the performance of cookstoves is specific fuel consumption (SFC). SFC is calculated as a ratio of the amount of dry fuel consumed to kilograms of water left at the end of the test. Along with efficiency and SFC, some additional performance parameters are also

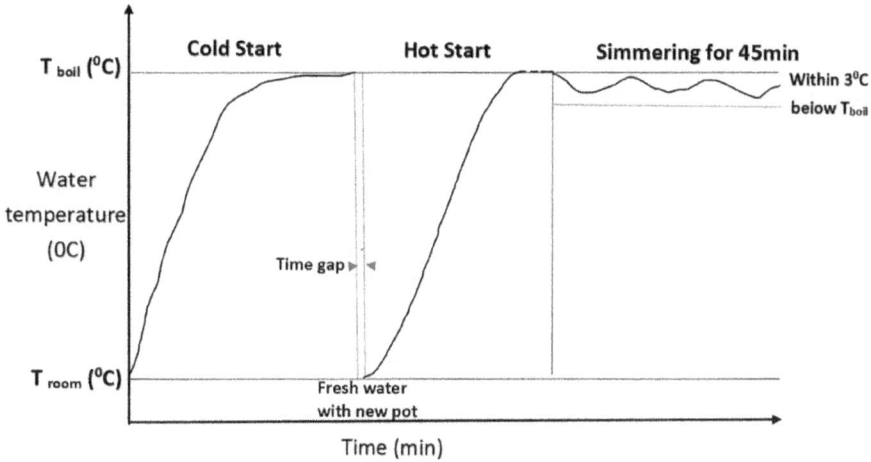

FIGURE 6.1 Three phases of WBT with respect to water temperature and test duration. WBT, water boiling test.

considered for performance evaluation of cookstoves, such as burning rate (dry fuel consumed per unit time to reach the boiling point), fire power (fuel energy consumed per unit time to reach the boiling point), turn down ratio (ratio of high-power hot start to low-power simmering), and time for boiling. Figure 6.1 depicts the three phases of the WBT with respect to water temperature and duration of test.

SFC is the amount of fuel required for 1 kg of water to boil and simmer. Mathematically, it is the ratio of equivalent dry fuel consumed to the amount of water left in the pot at the end of each phase. The burning rate is the ratio of equivalent dry fuel consumed to the time required for boiling. Fire power is the ratio of fuel energy consumed for boiling to the time required for boiling. The turn down ratio indicates the amount of heat adjustment between high- and low-power phases, and mathematically, it is the ratio of fire power of the hot start (high power) to fire power of simmering (low power).

6.2.3 Emission Testing and Emission Parameters

Emission testing has always been one of the integral and important aspects of cookstove research, although initially experimentation on emission was not conducted in a rigorous manner. However, researchers had emphasized on emission testing in a comprehensive manner after publication of WBT 4.2.2. Generally, studies of emission are carried out in a controlled environment. A number of protocols and testing methods were prescribed in the literature to characterize the emission performance of biomass cookstoves. The "hood method" and "chamber method" are the two most preferred emission testing methods by many researchers. According to Arora and Jain [21], the hood method has been used effectively in 70% of studies, whereas 20% of pollutant-monitoring studies followed the chamber method, and the rest 10% of emission studies were carried out by using other field-based plume sampling

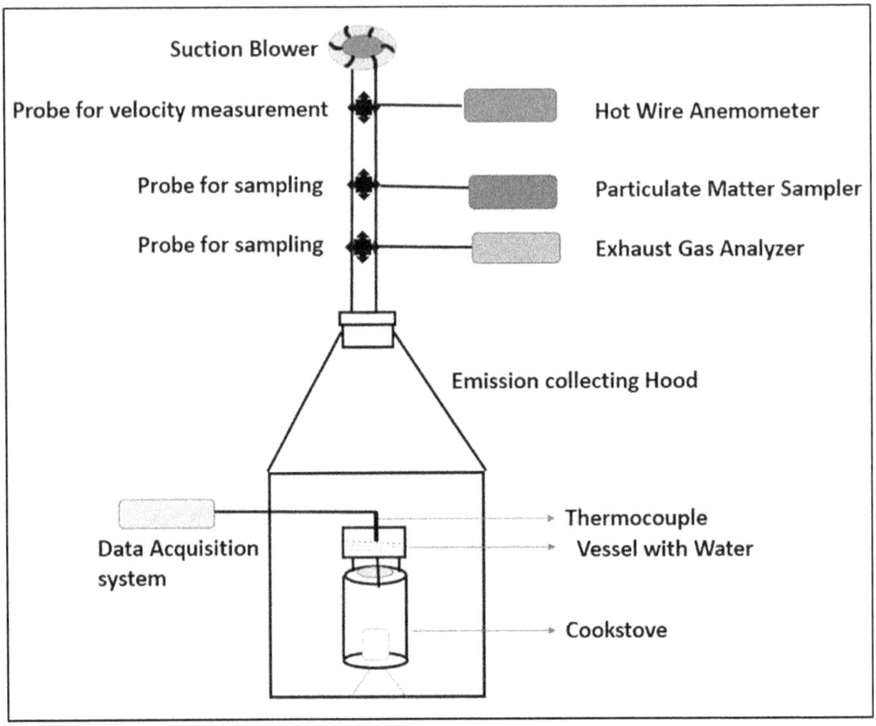

FIGURE 6.2 Experimental setup for a typical laboratory-based hood method for emission evaluation.

methods. Generally, pollutants like carbon monoxide (CO), particulate matter (PM), carbon dioxide (CO_2), methane (CH_4), non-methane hydrocarbons, tetrahydrocannabinol, polycyclic aromatic hydrocarbons, nitrous oxides (NO_X), and sulfur dioxide (SO_2) are monitored using the hood method. During testing, the stove is placed under (1 m below) a conical fume hood (1 m diameter), into which all flue gases are drawn by using a blower without hampering the combustion characteristics [31]. Generally, the exhaust discharge rate is maintained at 0.1 m³/s. Probes are provided in the hood and duct lines to quantify the pollutant emissions by means of different exhaust analyzers. A schematic of a typical experimental setup of the hood method is presented in Figure 6.2. Although this method possesses several limitations because of its economic viability, safe, and easy handling procedure, it is well accepted by the researcher community for laboratory-based testing.

Ahuja et al. [45] proposed the "chamber method" for the assessment of smokiness of unvented biomass cookstoves in the laboratory as well as in field. It is also known as the "simulated kitchen method." In this method, the stove undergoes a cooking cycle in a chamber, and the emission is monitored within that chamber. Duct work and air flow calibrations are not required for this method. A major drawback of this method is the requirement of a huge space for testing, which may not be a suitable practical approach in laboratories. Direct exposure of the tester to exhaust fumes and pollutants during monitoring is another drawback of this method.

6.2.4   LIMITATIONS OF WBT

Although the WBT provides systematic testing procedures and guidelines for controlled laboratory-based tests, the result may vary from user to user, even while performing the experiments with a same cookstove, which necessitates the requirement of trained experts to reduce human error. The testing procedure may vary with local cooking habits, and it may provide contrast results during cooking of normal foods with the locally available fuels. It is also difficult to demonstrate the household SFC for different food items with respect to variations in the number of people to whom food items are to be served.

6.2.5   FIELD-BASED ASSESSMENT METHODS

Open literature reveals that researchers have contributed a lot to verify the findings of field-based evaluation methods of cookstove performance. Although two field-based testing methods, namely, KPT and CCT, were published along with the WBT by VITA in the same year, modified versions of the KPT and CCT are limited compared with the WBT. Moreover, the experiments performed using the CCT and KPT are quite few in numbers compared with experiments conducted using the WBT protocol. In the CCT, the performance of cookstoves is evaluated by measuring the quantity of fuel used for preparing a simple meal using locally available fuel. Cooking time and SFC are two major performance indicators considered in the CCT [46,47]. The KPT is an atypical field-based test, which provides actual indications about stove performance in the real cooking scenario. Generally, the KPT is conducted by stove researchers during stove dissemination programs with actual users or end users. For that reason, the KPT has been used by limited researchers [48]. Along with SFC and cooking time, users' feedback is considered a major performance indicator in the KPT [49]. CO and PM are considered two major emission parameters in field testing [47,50], although works available in open literature on emission tests are not quite large. Field-based tests emphasize on the actual cooking environment, ensuring the accuracy of test results, but these are not economical and are time-consuming [51]. Although some field-based assessment methods like stove use monitors and burning cycle test have been developed in the recent past, these are not emphasized much in open literature, and therefore, these may be included as a future scope of research in this area.

6.3   CONCLUSIONS

A comprehensive review is carried out on globally adopted performance assessment testing procedures, which includes both laboratory- and field-based testing methods. As per the literature, the WBT is a widely used testing method for the performance evaluation of biomass cookstoves. The CCT and KPT are two field-based testing protocols used in the real cooking environment. For proper assessment of the efficiency and emission of biomass cookstoves, and for enhancement of efficiency with clean cooking, there is a need to adopt proper testing methods. The testing methods should be easily assessable and should correlate with real cooking practice, allowing minimal instrumental and human errors. Although handful of testing methods are available in the literature, several important performance parameters for a detailed evaluation of cookstoves during real cooking are yet to be adopted. Laboratory-based tests are useful for enhancement of a stove design, whereas field-based tastings are required for authenticated real kitchen performances. Although the WBT is a widely used popular laboratory-based test, test results may vary during real household cooking. So, a global standardized testing method is yet to be established that will correlate laboratory- and field-based testing methods. Moreover, various social, cultural, and developmental factors across topographical locations, along with affordability and easy accessibility of the method, need to be considered to assess the appropriateness of upgraded technologies in biomass cookstove research.

REFERENCES

1. Samal, C., Mishra, P. C., Mukherjee, S., and Das, D. 2019. Evolution of high performance and low emission biomass cookstoves-An overview. *AIP Conference Proceedings* 2200(1):020021.
2. Samal, C., Mishra, P. C., and Das, D. 2020. Design modifications and performance of biomass cookstoves-A review. *AIP Conference Proceedings* 2273(1):020002.
3. Gupta, A., Mulukutla, A. N., Gautam, S., TaneKhan, W., Waghmare, S. S., and Labhasetwar, N. K. 2020. Development of a practical evaluation approach of a typical biomass cookstove. *Environmental Technology & Innovation* 17:100613.
4. Urmee, T., and Gyamfi, S. 2014. A review of improved cookstove technologies and programs. *Renewable and Sustainable Energy Reviews* 33:625–635.
5. Grabow, K., Still, D., and Bentson, S. 2013. Test kitchen studies of indoor air pollution from biomass cookstoves. *Energy for Sustainable Development* 17(5):458–462.
6. Bonjour, S., Adair-Rohani, H., Wolf, J., Bruce, N.G., Mehta, S., and Prüss-Ustün, A. 1980. Solid fuel use for household cooking: Country and regional estimates for 1980–2010. *Environmental Health Perspectives* 121 (7):784–790.
7. Bhatta, S., Pratap, D., Gakkhar, N., and Rajput, J. P. S. 2021. A comparative experimental investigation of improved biomass cookstoves for higher efficiency with lower emissions. *Proceedings of the 7th International Conference on Advances in Energy Research,* Springer, Singapore, 961–971.
8. Vaccari, M., Vitali, F., and Tudor, T. 2017. Multi-criteria assessment of the appropriateness of a cooking technology: A case study of the Logone Valley. *Energy Policy* 109: 66–75.
9. Harshika, K., Avinash, C., and Kaushik, S. C. 2014. Comparative study on emissions from traditional and improved biomass cookstoves used in India. *International Journal for Research in Applied Science and Engineering Technology* 2(8):249–257.
10. Kim, K. H., Jahan, S. A., and Kabir, E. 2011. A review of diseases associated with household air pollution due to the use of biomass fuels. *Journal of Hazardous Materials* 192(2): 425–431.

11. Benka-Coker, M. L., Peel, J. L., Volckens, J., Good, N., Bilsback, K. R., L'Orange, C., ... and Clark, M. L. 2020. Kitchen concentrations of fine particulate matter and particle number concentration in households using biomass cookstoves in rural Honduras. *Environmental Pollution* 258:113697.

12. Memon, S. A., Jaiswal, M. S., Jain, Y., Acharya, V., & Upadhyay, D. S. 2020. A comprehensive review and a systematic approach to enhance the performance of improved cookstove (ICS). *Journal of Thermal Analysis and Calorimetry* 141:2253–2263.

13. Torres-Rojas, D., Deng, L., Shannon, L., Fisher, E. M., Joseph, S., and Lehmann, J. 2019. Carbon and nitrogen emissions rates and heat transfer of an indirect pyrolysis biomass cookstove. *Biomass and Bioenergy* 127:105279.

14. Carrión, D., Kaali, S., Kinney, P. L., Owusu-Agyei, S., Chillrud, S., Yawson, A. K., ... and Asante, K. P. 2019. Examining the relationship between household air pollution and infant microbial nasal carriage in a Ghanaian cohort. *Environment International* 133:105150.

15. Jetter, J., Zhao, Y., Smith, K. R., Khan, B., Yelverton, T., DeCarlo, P., and Hays, M. D. 2012. Pollutant emissions and energy efficiency under controlled conditions for household biomass cookstoves and implications for metrics useful in setting international test standards. *Environmental Science & Technology* 46(19):10827–10834.

16. Kumar, M., Kumar, S., and Tyagi, S. K. 2013. Design, development and technological advancement in the biomass cookstoves: A review. *Renewable and Sustainable Energy Reviews* 26:265–285.

17. Raman, P., Ram, N. K., and Murali, J. 2014. Improved test method for evaluation of bio-mass cook-stoves. *Energy* 71:479–495.

18. Sharma, S. K. 1993. *Improved Solid Biomass Burning Cookstoves: A Development Manual*. Bangkok, Thailand: Food and Agriculture Organization of the United Nations; Field document No. 44.

19. Kshirsagar, M. P., and Kalamkar, V. R. 2014. A comprehensive review on biomass cookstoves and a systematic approach for modern cookstove design. *Renewable and Sustainable Energy Reviews* 30:580–603.

20. VITA, Volunteers in Technical Assistance. 1985. *Testing the Efficiency of Wood Burning Cookstoves*. Tech. Rep. International Standards. Arlington, Virginia. https://cupdf.com/document/testing-the-efficiency-of-wood-burning-cook-stoves.html.

21. Arora, P., and Jain, S. 2016. A review of chronological development in cookstove assessment methods: Challenges and way forward. *Renewable and Sustainable Energy Reviews* 55:203–220.

22. Mehetre, S. A., Panwar, N. L., Sharma, D., and Kumar, H. 2017. Improved biomass cookstoves for sustainable development: A review. *Renewable and Sustainable Energy Reviews* 73:672–687.

23. Tryner, J., Willson, B. D., and Marchese, A. J. 2014. The effects of fuel type and stove design on emissions and efficiency of natural-draft semi-gasifier biomass cookstoves. *Energy for Sustainable Development* 23:99–109.

24. Yuntenwi, E. A., MacCarty, N., Still, D., and Ertel, J. 2008. Laboratory study of the effects of moisture content on heat transfer and combustion efficiency of three biomass cook stoves. *Energy for Sustainable Development* 12(2):66–77.

25. L'orange, C., DeFoort, M., and Willson, B. 2012. Influence of testing parameters on biomass stove performance and development of an improved testing protocol. *Energy for Sustainable Development* 16(1):3–12.

26. Chica, E., and Pérez, J. F. 2019. Development and performance evaluation of an improved biomass cookstove for isolated communities from developing countries. *Case Studies in Thermal Engineering* 14: 100435.

27. Obi, O. F., Okechukwu, M. E., and Okongwu, K. C. 2019. Energy and exergy efficiencies of four biomass cookstoves using wood chips. *Biofuels* 12:869–878.

28. Saiyyadjilani, S. S., Tewari, P. G., Tapaskar, R., Madival, A. P., Gorawar, M., and Revankar, P. P. 2016. Design of improved biomass cook stove for domestic utility. In: Pawar, P., Ronge, B., Balasubramaniam, R., Seshabhattar, S. (Eds.) *Techno-Societal International Conference on Advanced Technologies for Societal Applications.* Cham: Springer, 53–63.
29. Jetter, J. J., and Kariher, P. 2009. Solid-fuel household cook stoves: Characterization of performance and emissions. *Biomass and Bioenergy* 33(2):294–305.
30. Berrueta, V. M., Edwards, R. D., and Masera, O. R. 2008. Energy performance of wood-burning cookstoves in Michoacan, Mexico. *Renewable Energy* 33(5):859–870.
31. Bailis, R., Berrueta, V., Chengappa, C., Dutta, K., Edwards, R., Masera, O.,... and Smith, K. R. 2007a. Performance testing for monitoring improved biomass stove interventions: Experiences of the household energy and health project. *Energy for Sustainable Development* 11(2):57–70.
32. Smith, K. R., Shuhua, G., Kun, H., and Daxiong, Q. 1993. One hundred million improved cookstoves in China: How was it done. *World Development* 21:941–961.
33. Energy Efficiency Stoves. 2007. Household biomass stoves - Performance requirements and test methods: Uganda standard (First edition). US 761. https://ia801601.us.archive.org/32/items/us.761.2007/us.761.2007.html.
34. Bailis, R., Ogle D., Mac Carty N., Still D., Edwards R., and Smith K. R. 2007b. The water boiling test version 3.0: Cook-stove emissions and efficiency in a controlled laboratory. Technical Report. Berkeley: University of California. https://energypedia.info/images/3/38/Wbt_version_3.0_jan2007.pdf.
35. General Specifications for Biomass Household Stoves. 2008. Quality and technical supervision bureau of Beijing Municipality. DB 11/T 540–2008. https://cleancooking.org/binary-data/DOCUMENT/file/000/000/77-1.pdf.
36. DeFoort, M., L'Orange, C., Kreutzer, C., Lorenz, N., Kamping, W., and Alders, J. 2009. Stove manufacturers emissions & performance test protocol (EPTP). https://cleancooking.org/binary-data/DOCUMENT/file/000/000/73-1.pdf.
37. The Water Boiling Test, Version 4.1.2. 2009. Cookstove emissions and efficiency in a controlled laboratory setting. https://cleancooking.org/binary-data/DOCUMENT/file/000/000/399-1.pdf.
38. Adapted Water Boiling Test (AWBT) V. 2.0. 2010. GERES research & development unit, Cambodia. https://cleancooking.org/binary-data/DOCUMENT/file/000/000/75-1.pdf.
39. The Heterogeneous Testing Procedure for Thermal Performance and Trace Gas Emissions, UJ SeTARcentre Standard Operating Procedure, SOP#1.05. 2012. South Africa: University of Johannesburg. https://cleancooking.org/binary-data/DOCUMENT/file/000/000/74-1.pdf.
40. Water Boiling Test, Version 4.2.2. 2013. Available from: The Water Boiling Taste Aprovecho Research Center. https://cleancooking.org/binary-data/DOCUMENT/file/000/000/399-1.pdf
41. BIS. Indian Standard: Portable Biomass Cookstove (Chulha)-Specification. 2013. First revision. IS 13152 (Part 1). New Delhi, India: Bureau of Indian Standard, 22. https://bis.gov.in/wp-content/uploads/ 2020/05/PM-IS-13152-Part-1-May-2020.pdf.
42. Water Boiling Test V. 4.2.3. 2014. Cookstove emissions and efficiency in controlled laboratory setting. Washington, D.C., USA: Global Alliance for Clean Cookstoves. http://cleancookstoves.org/binary-data/DOCUMENT/file/000/000/399-1. pdf.
43. Obi, O. F., Ezeoha, S. L., and Okorie, I. C. 2016. Energetic performance of a top-lit updraft (TLUD) cookstove. *Renewable Energy* 99:730–737.
44. Sonarkar, P. R., and Chaurasia, A. S. (2019). Thermal performance of three improved biomass-fired cookstoves using fuel wood, wood pellets and coconut shell. *Environment, Development and Sustainability* 21(3):1429–1449.

45. Ahuja, D. R., Joshi, V., Smith, K. R., and Venkataraman, C. 1987. Thermal performance and emission characteristics of unvented biomass-burning cookstoves: A proposed standard method for evaluation. *Biomass* 12(4):247–270.
46. Adkins, E., Tyler, E., Wang J., Siriri D., and Modi, V.2010. Field testing and survey evaluation of house hold biomass cookstoves in rural sub-Saharan Africa. *Energy for Sustainable Development* 14:172–85.
47. MacCarty, N., Still, D., and Ogle, D. 2010. Fuel use and emissions performance of fifty cooking stoves in the laboratory and related benchmarks of performance. *Energy for Sustainable Development* 14(3):161–171.
48. Berkeley Air Monitoring Group. 2012. Prepared for global alliance for clean cookstoves. Stove performance inventory report. Berkeley, CA: United Nations Foundation. http://berkeleyair.com/wp-content/publications/SPT_Inventory_Report_v3_0.pdf.
49. Granderson, J., Sandhu, J. S., Vasquez, D., Ramirez, E., and Smith, K. R.2009. Fuel use and design analysis of improved woodburning cookstoves in the Guatemalan Highlands. *Biomass and Bioenergy* 33(2):306–315.
50. Arora, P., Das, P., Jain, S., and Kishore, V. V. N. 2014. A laboratory based comparative study of Indian biomass cookstove testing protocol and water boiling test. *Energy for Sustainable Development*, 21:81–88.
51. Lee, C.M., and Chandler, C. 2013. Assessing the climate impacts of cookstove projects: issues in emissions accounting. *Chall Sustain* 1(2):5371.

7 Performance Evaluation of an Improved Biomass-Fired Cook Stove
A Comparative Analysis

Chandrika Samal, Ramesh Chandra Nayak, Manmatha Kumar Roul, Diptikanta Das, and Ramanuj Kumar

CONTENTS

7.1 Introduction 107
7.2 Materials and Methods 109
    7.2.1 Cookstoves 109
    7.2.2 Biowastes 110
    7.2.3 Thermo-Physical Characteristics of Biowastes 110
    7.2.4 Testing Protocol 112
    7.2.5 Sampling Setup 112
    7.2.6 Evaluation Parameters 112
7.3 Results and Discussion 114
    7.3.1 Thermal Efficiency 114
    7.3.2 Specific Fuel Consumption 116
    7.3.3 CO Emission 117
7.4 Conclusions 118
References 118

7.1 INTRODUCTION

Domestic cooking practice is one of the vital daily household activities that consume a notable amount of energy, human effort, and time. In most of the developing countries, especially in rural areas, people are very much acquainted with traditional ways of cooking and use traditional biomass cookstoves to meet their daily cooking activities [1,2]. About 3 billion people across the globe use traditional biomass-related sources (fire wood, animal dung, crop residues, etc.) for cooking [3,4]. The major topics of concern for traditional biomass cookstoves are low efficiency and high emission [5,6]. Intending to overthrow these two major disadvantages, many traditional cookstoves are modified by changing the stove configuration, operation,

DOI: 10.1201/9781003359784-7

and design, and a large number of new technically improved stoves are also developed worldwide. Many experiments were also carried out by introducing a variety of new biomass feedstock and feed mechanism methods. Another major challenge the researchers deal with is the reduction of the use of firewood in daily domestic cooking to avoid deforestation. So, it is highly preferable to increase the usage of locally accessible biomass resources, especially plant-based wastes (non-woody) in the development of improved stoves to make users acquainted with biomass fuels other than wood. Due to low cost, abundant availability, and widespread allocation, renewable plant-based biowastes can be considered effective alternatives to wood. Again, depletion of fossil fuels and increased cost of LPG pressurize the modern world to find out some alternative energy sources for daily domestic purposes [5]. It is therefore challenging for researchers to replace non-renewable energy sources by biomass (renewable energy source) in an efficient and clean manner. Most of the researchers have modified existing traditional cookstoves into improved cookstoves, or developed new improved cookstoves based on the availability of the biomass waste in the local region [7]. Joshi and Srivastava [8] designed and fabricated an improved three-pot cookstove, tested its thermal performance using the WBT protocol, compared it with a traditional biomass mud stove, and reported a better performance for the improved one. Parmigiani et al. [9] tested a natural draft cookstove (made of crude earth and mud brick) as per the WBT protocol using agricultural waste (rice husk) as fuel. Although the efficiency was reported to be less than that of many improved cookstoves, this work promoted the efficient use of biowastes with an overall low cost and enlightened a pathway for preventing deforestation caused by household cooking. By conducting a survey on cookstoves, Pande et al. [10] listed some constructional, functional, and environmental flaws in traditional cookstoves. One popular traditional stove was modified by emphasizing on outcomes of the survey report, and the performance parameters were compared with those of previous one. The purpose of this work was to provide high-efficiency and low-emission cooking compared with the traditional cookstove, and the goal was achieved to a considerable extent. Rasoulkhani et al. [11] evaluated the performance of a top-lead-up-draft using apple pruning waste as feedstock, and the result was compared with that of the traditional stove as per WBT 4.2.3. The thermal efficiency (TE) of the top-lead-up-draft was reported to be 35%, whereas that for the traditional stove was 12.6%. Kaur-Sidhu et al. [12] reported the emission factors of five mostly used fuel–stove combinations for cooking in Punjab (India), and pointed out prominence of maximizing energy supply and minimizing emission by exploring new renewable energy strategies and adopting clean biomass fuels.

In this work, the thermal and emission performances of an improved cookstove were evaluated, and the results were compared with those of a traditional cookstove. All the experiments were conducted following the WBT (version 4.2.3) protocols. Three abundantly available plant-based biowastes (palm shells, peanut shells, and acacia pods) in coastal Odisha were selected as fuels. Intention behind the use of plant-based biowastes was to restrict abundant use of firewood in cooking practice, and thus to impede deforestation. The objective of this study is to promote efficient utilization of biowastes to meet daily domestic energy need by implementing improved cookstoves. It is also intended to compare the efficiencies of traditional and improved biomass cookstoves with regionally available plant-based biowastes to promote the concept of "zero waste."

7.2 MATERIALS AND METHODS

7.2.1 COOKSTOVES

Two different types of locally available stoves were selected for evaluation and comparison of efficiency and emission using the WBT. One of the two is a TMS, which was chosen because of its low cost, ease in manufacturing, and use. The TMS was made with a thick layer of mud and was sun-dried for 4–5 days. It was a U-shaped semi-enclosed "Chulla" (locally named) with a circular opening at the front for fuel feeding. Three sharp edges were provided at the top of the rim to hold the pot. These edges also create free space for supplying secondary air and exit of residue gases. Figure 7.1a shows the photograph of TMS used in the study. Another cookstove selected for testing was an IMS as available in the local market. The IMS was designed to fulfill the daily domestic energy demand of a family of 6–7 persons. It was made of a mild steel plate of a thickness of 3 mm cylindrical in shape with a circular fire opening hole. The height of the stove was 270 mm, and the diameter of the combustion chamber was 240 mm. A circular fire hole opening of a diameter of 190 mm was provided at the top of the stove. The stove had a rectangular opening (120×90 mm) at one side for fuel supply. Grates were provided at the bottom of the stove to supply primary air, and small circular holes were incorporated circumferentially throughout the neck to supply secondary air. Two metallic handles were incorporated at both sides of the stove. Figure 7.1b shows a schematic of the IMS used in this experimental study.

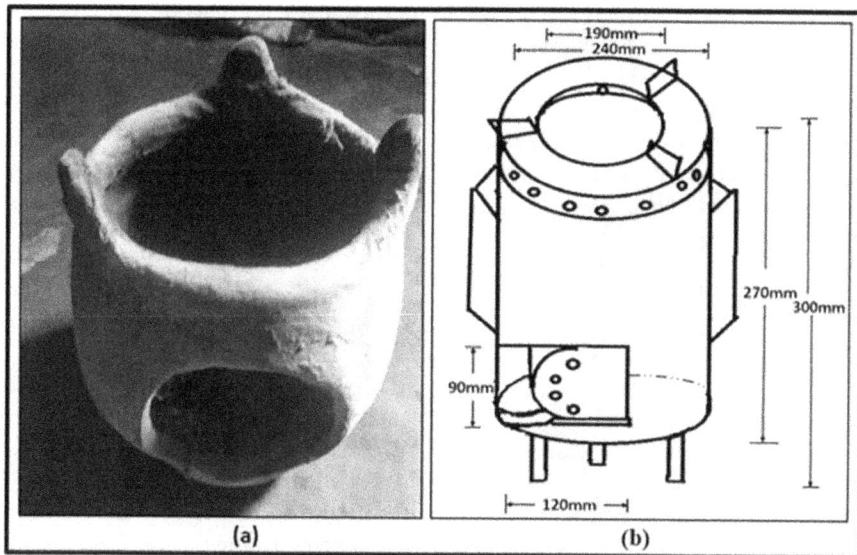

FIGURE 7.1 (a) TMS. (b) IMS. IMS, improved metallic stove; TMS, traditional mud stove.

FIGURE 7.2 Images of plant-based biowaste samples. (a) Palm shells. (b) Peanut shells. (c) Acacia pods.

7.2.2 BIOWASTES

Three different types of plant-based biowastes, i.e., palm shells, peanut shells, and acacia pods, were used as fuels due to their wide availability at the test location. Huge numbers of palm trees are cultivated in coastal Odisha to protect the region from storms, cyclones, and lightning. Usually, palm is used to prepare jaggery and some traditional foods. After extracting juice, the palm shells are thrown as waste. For this experimental investigation, palm shells were collected from a villager who uses palms for making of some traditional foods. Peanut is one of the major oil seed crops cultivated in coastal Odisha. Peanut shells were collected from a local farmer. Acacia pods were collected from a nearby farm where the acacia trees are planted to support forestation. Acacia pods are initially grown straight but become twisted spirals with seeds inside in the fully grown state. Before using acacia pods as fuel, the seeds were taken out from the pods. The motive behind using of all these three plant-based biowastes was to restrict the use of firewood in cooking. After collection of all the biomasses, those are sun-dried for 15 days to reduce excess moisture, cut into required shape and size (if necessary), and stored in air-tight containers. A sample of three plant-based biowastes are depicted in Figure 7.2.

7.2.3 THERMO-PHYSICAL CHARACTERISTICS OF BIOWASTES

All the biowaste fuels used in this study were collected from same source to maintain uniformity in shape and size. The bulk density of the fuel samples was measured as per the ASTM E-873-82 standard using the Archimedes Principle Kit (make: Mettler Toledo). Because of very lightweight and a leafy structure, the acacia pod test sample floated on the water surface, and therefore, its density was unable to be determined. A proximate analysis is employed for determining the amount of moisture content (MC), volatile matter content (VMC), ash content (AC), and fixed carbon content present in biomasses as per the American Society of Testing Materials

(ASTM) [13–15]. A Labotech-BDI72 muffle furnace and a Labotech-BDI50 oven were used for conducting proximate analysis. MC was determined by using the oven dry method, following the ASTM D3173 standard [14,16]. For determination of MC, 1 g of the finely powdered fuel sample was heated in an oven at a constant temperature of 105°C±5°C for 6 hours. The VMC was determined following the ASTM D3175-73 standard [14,16]. One gram of the finely powdered fuel sample was heated in the muffle furnace at 950°±5°C for 7 minutes. The AC of the fuel was measured following the ASTM D3174-73 standard [16]. One gram of the fuel sample was placed in the muffle furnace for heating at 700°C±5°C for 30 minutes. The value of the fixed carbon content was determined using the material balance, i.e., by deducting the sum of percentage of MC, VMC, and AC from total of 100% composition [14,16]. Each experiment was carried out three times, and the mean of the individual result was recorded for further calculations. Using the ultimate analysis, the percentages of carbon (C), hydrogen (H), nitrogen (N), and sulfur (S) were determined following the ASTM D3176 [15,17] by using a CHNS/O elemental analyzer (Model: PerkinElmer-2400). The average results of three runs were recorded for further calculations. Higher heating values of specimens were determined using a digital bomb calorimeter of Parr-6200, following the ASTM D 5468-02 standard [14,15]. Lower heating values of all specimens are measured using Dulong's formula [18] (Table 7.1).

TABLE 7.1

Average results of proximate analysis, ultimate analysis, HHV, and LHV for three plant-based biowastes

Thermo-physical property	Fuel types		
	Palm shell	Peanut shell	Acacia pod
Proximate analysis			
MC (%)	8.376	6.37	10.446
VMC (%)	73.31	67.94	80.643
AC (%)	0.91	5.8	1.49
FCC (%)	17.404	19.88	7.421
Ultimate analysis			
Carbon (%)	47.442	45.96	43.419
Hydrogen (%)	5.450	5.24	5.385
Nitrogen (%)	1.28	1.59	2.25
Sulfur (%)	0.75	0.84	1.44
Oxygen	40.186	40.5	40.642
Heating values			
HHV (MJ/kg)	18.834	17.15	16.593
LHV (MJ/kg)	15.954	14.23	13.747
Bulk density (kg/m³)	0.984	0.887	—

7.2.4 Testing Protocol

The performance of the cookstoves was investigated following protocols of WBT version 4.2.3 [19], using all the bio-fuels under consideration. Before conducting the WBT, the local boiling point of water (boiling point of water in the laboratory, where experiments were carried out) was determined. Once the local boiling point was known, three phases of the WBT (high-power hot start, high-power cold start, and low-power simmering) were conducted one after the other. In high-power cold start phase, a known quantity of water (2.5 L) at ambient temperature was brought to the local boiling point (98°C) using pre-weighed bunch of biomasses. In the next phase, i.e., in the high-power hot start phase, the experiment was started with the hot stove following the same procedure as that of cold start phase. The hot start phase was carried out to characterize the behaviors of cookstoves in a hot environment. In case of the low-power simmering phase, the boil water of the hot start phase was simmered for 45 minutes maintaining 3°C below of the local boiling temperature. The amount of wood burnt, amount of ash produced, amount of water left in the pot, and time taken for boiling (for cold and hot start) were noted for performance evaluation. Three thermocouples were used to measure water temperature, flame temperature, and stove outer body temperature. The temperatures were continuously monitored and recorded by using a CHINO-KR2000 data acquisition system.

7.2.5 Sampling Setup

Emission sampling was conducted using the "hood method" of WBT [20]. A conical hood was fabricated to collect the emissions from the cookstove. The flue duct was connected at the end of emission hood to vent the exhaust out from the laboratory. A blower was installed at the end of the duct to create a proper suction through it. Care was taken to maintain a constant blower speed of $0.1 \, m^3/s$ [21], so the normal combustion of cookstove was not hampered. The sampling probes were provided inside the flue duct for emission sampling and velocity measurement. During the test, CO emission was analyzed by using an automotive emission analyzer (Model: MEXA-584L, Make: Horiba) as CO is one of the most dangerous gases that adversely affect human health. A laboratory-based experimental setup was developed (presented in Figure 7.3) as per the WBT protocol prescribed by Global alliance for clean cookstoves [19]. The setup included a CHINO-KR2000 data acquisition system, an automotive emission analyzer (Model: MEXA-584L, Make: Horiba), a suction blower, a hot wire anemometer, and three K-type thermocouples.

7.2.6 Evaluation Parameters

To study thermal behaviors of the cookstoves with respect to three biowastes such as fuels, TE and SFC were evaluated for each phase of the WBT using experimental data. TE is a major performance parameter, which influences the usability of a stove [22], and is the ratio between the total amount of energy utilized for boiling and vaporizing of water and the amount of energy provided to the cookstove by burning of fuel. The mathematical expression for TE is presented in equation (7.1)

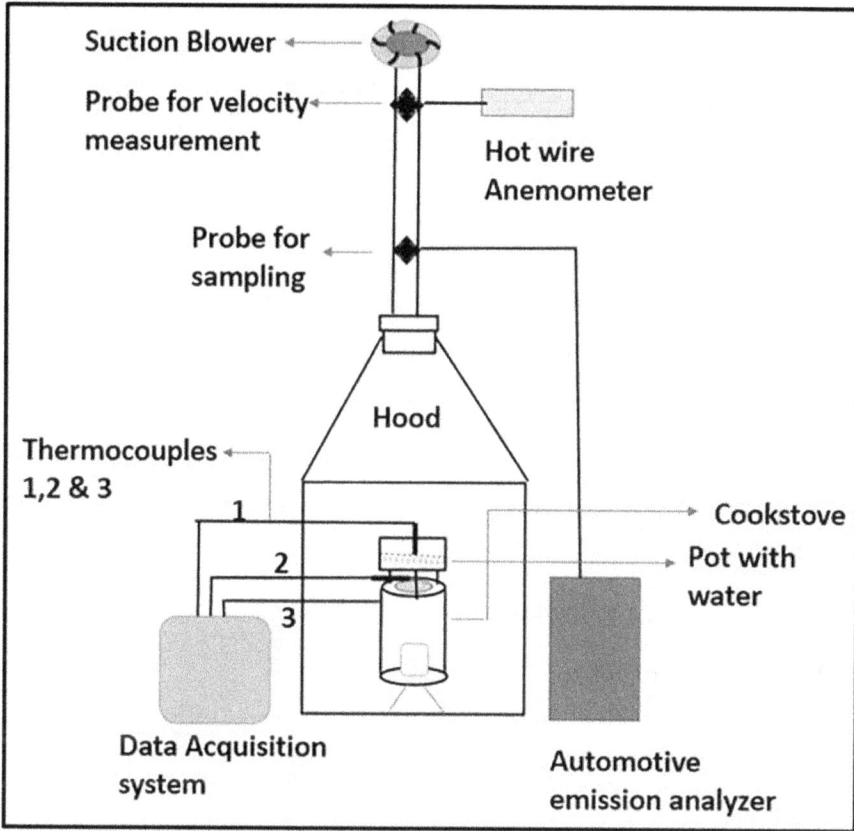

FIGURE 7.3   Schematic of laboratory-based experimental setup for WBT. WBT, water boiling test.

[17,23]. During the calculation of TE, the equivalent dry fuel consumed (DFC) was taken into consideration, rather than the total amount of fuel consumed. DFC represents the sum of the total amount of energy utilized for extracting moisture from fuel and the amount of energy left with char after burning. The mathematical formula used for the calculation of DFC consumed is presented in equation (7.2) [24]. SFC is the amount of fuel consumed by unit kilogram of water to boil and simmer. Mathematically, SFC is the ratio of equivalent DFC to the amount water left in the pot at the end of each phase, as presented in equation (7.3) [17,24].

$$TE = \frac{m_w C_w (T_b - T_w) + (m_{wv} * LH_{fg})}{DFC * LHV} \tag{7.1}$$

$$DFC = m_{fb}[1 - (1.12 * MC)] - 1.5 m_C \tag{7.2}$$

$$SFC = \frac{DFC}{m_{wl}} \tag{7.3}$$

where m_w is the mass of water taken for test (kg), C_w is the specific heat of water at constant pressure (4.186 kJ/kg°C), T_b is the local boiling point of water (°C), T_w is the initial temperature of water in the pot (°C), m_{wv} is the mass of water vaporized after each phase of WBT (kg), LH_{fg} is the latent heat of vaporization, DFC is the dry fuel consumed (kg), LHV is the lower hating value (kJ/kg), m_{fb} is the mass of fuel burned (kg), MC is the moisture content (%), m_c is the mass of net char (kg), and m_{wl} is the Mass of water left in the pot after each phase of WBT (kg).

For emission performance evaluation, CO concentration was continuously monitored throughout the WBT by using a data acquisition system. For comparison of the results, time averaged CO concentration (CO_{tavg}) in ppm was calculated by using equation (7.4) [11,25].

$$CO_{tavg} = \frac{\sum_{i=1}^{i=t} CO_i}{t} \tag{7.4}$$

All the experiments were conducted three times, and arithmetic means ± standard deviations were considered for result analysis.

7.3 RESULTS AND DISCUSSION

7.3.1 THERMAL EFFICIENCY

The TE of TMS and IMS for all the three phases, i.e., cold start (CS), hot start (HS), and simmer (SIM), using palm shells, peanut shells, and acacia pods as fuels are depicted in Figures 7.4 and 7.5, respectively. For TMS, the TE varied from 9.76% ± 2.24% to

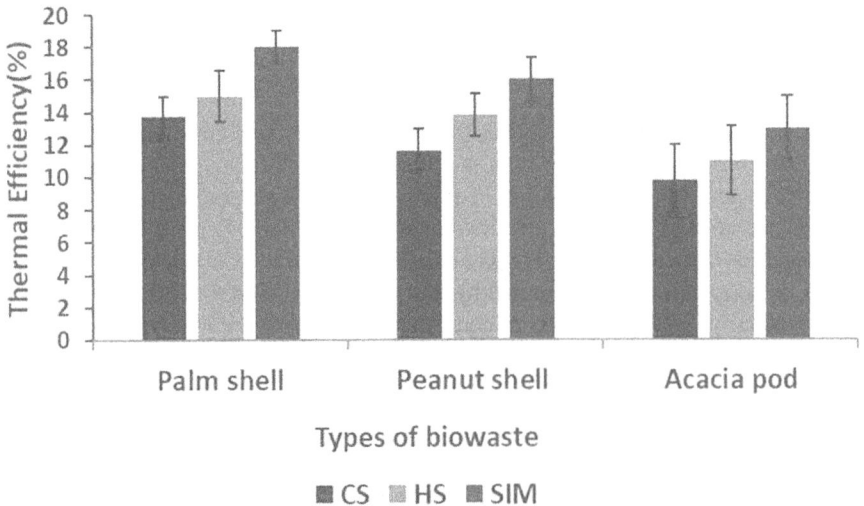

FIGURE 7.4 Mean TE with ± standard deviation for TMS in three phases of WBT. WBT, water boiling test; TMS, traditional mud stove; TE, thermal efficiency; TE, thermal efficiency.

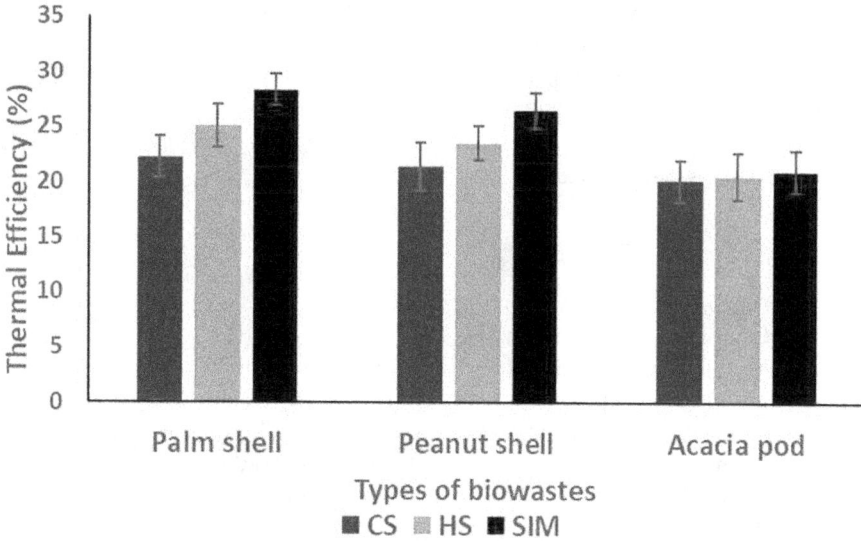

FIGURE 7.5 Mean TE with ± standard deviation for IMS in three phases of WBT. TE, thermal efficiency; WBT, water boiling test; IMS, improved metallic cookstove.

13.74% ± 1.24% in the cold start phase, from 11% ± 2.16% to 15% ± 1.54% in the hot start phase, and from 13% ± 1.94% to 18% ± 1% in the simmering phase. The TE_{avg} of TMS varied from 11.25% ± 2.11% to 15.58% ± 1.26% for all types of biowaste fuels used in the present study, which can be validated with the reports of Berrueta et al. [26]. Further, TE of IMS varied from 20.14% ± 1.87% to 22.28% ± 1.89% in the cold start phase, 20.57% ± 2.12% to 25.13% ± 1.93% in the hot start phase, and 21% ± 1.89% to 28.32% ± 1.42% in the simmering phase. The TE_{avg} of IMS varied from 20.57% ± 1.962% to 25.24% ± 1.74% using all the three biowastes under consideration. These results can be validated with the results provided by Suresh et al. [27] during the evaluation of the TE of one traditional stove (14%–19%) and improved natural draft biomass cookstoves (21%–28%) with different solid biomass fuel types. The average efficiency of IMS was recorded almost 10%–12% greater than the average efficiency of TMS. One of the major factors responsible for the increase in efficiency in IMS as compared to TMS was the use of grates, which were incorporated at the bottom of IMS. The presence of grates ensured the supply of required primary air for mixing with volatiles during the pyrolysis phase. Another important factor was the presence of small circular holes circumferentially at the neck of the stove. Secondary air was provided for combustion in a controlled manner through those holes, which helped complete combustion. TMS required more attention to operate as there was no control on flow of air. For both the stoves, the lowest efficiency was observed in the cold start phase as certain quantity of heat was utilized for warming up the stove, which was not required for the hot start phase. The highest efficiency was observed in the simmering phase due to less heat lost for simmering of water at temperature 3°C lower than the local boiling point. The same trend of variation in TE in three phases of the WBT (highest efficiency in the simmering phase succeeded by the hot start and cold start phases) was observed by Aprovecho Research Center for many biomass cookstoves [28]. The analysis of experimental results on the basis of fuel used revealed that both the cookstoves performed better during burning of palm shells than during burning of peanut shells and acacia pods. This may be attributed to the high energy of palm shells.

7.3.2 SPECIFIC FUEL CONSUMPTION

The mean SFC ± standard deviation for both the stoves with all the three biowastes is depicted in Figure 7.6. The mean SFC for the TMS was $1,154 \pm 178$ g/kg of water, $1,220 \pm 158$ g/kg of water, and $1,446 \pm 160$ g/kg of water while burning palm shells, peanut shells, and acacia pods, respectively, whereas that for the IMS was 543 ± 132 g/kg of water, 619 ± 143 g/kg of water, and 840 ± 162 g/kg of water for the same series of biowaste fuels, respectively. The SFC was observed significantly lower for the IMS than for the TMS, for all the three biowastes. In case of TMS, the biowastes burned under insufficient air, and so less heat was generated. At the same time, blowing of air was required during experimentation with the TMS, which created more availability of oxygen for fuels in the core of the combustion chamber. This led to a higher burning rate, more SFC, and less TE. Again, heat loss from the body of the TMS was observed to be more than that for the IMS. So, more fuel was consumed for boiling and simmering of a certain quantity of water. On the contrary, with the IMS, complete combustion was achieved as biowaste fuels were burnt under sufficient oxygen by supplying primary and secondary air at a controlled rate. Moreover,

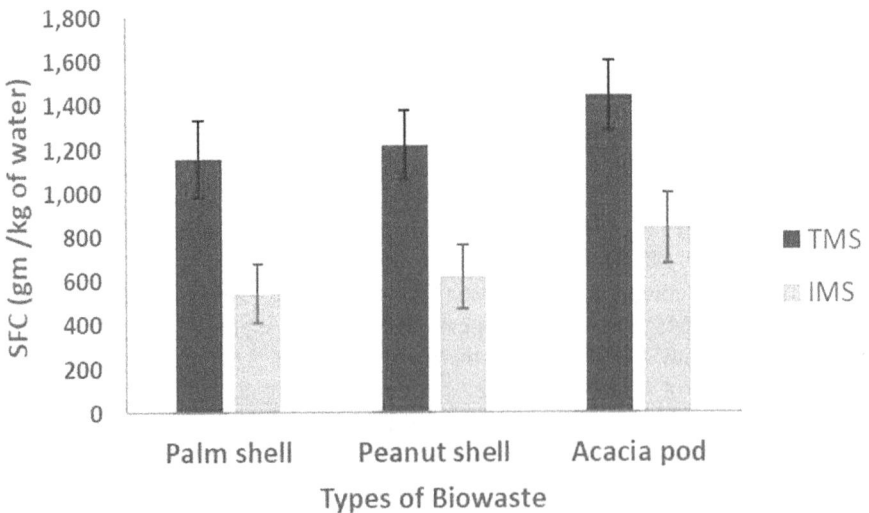

FIGURE 7.6 Comparison of mean SFC with ± standard deviation between TMS and IMS. TMS, traditional mud stove; SFC, specific fuel consumption; IMS, improved metallic cookstove.

as the IMS has low thermal mass, a less amount of heat was lost to the surrounding than that with the TMS. So, a less amount of biowastes was required for boiling and simmering of a unit quantity of water, which reduced the SFC of the IMS.

7.3.3 CO EMISSION

The CO emission was monitored continuously for both the cookstoves with three bio-wastes for all the phases of WBT. However, for easy comparison, time averaged values of CO_{tavg} for each test were calculated for both the stoves with each biowaste, as presented in Figure 7.7. CO_{tavg} emission of the TMS was 64 ± 5.5 ppm, 73 ± 6.21 ppm, and 92 ± 10.28 ppm, whereas for the IMS, it was 14 ± 3.18 ppm, 17 ± 4.89 ppm, and 25 ± 7.32 ppm, respectively, during combustion of palm shells, peanut shells, and acacia pods, respectively. As per indoor air quality guidelines issued by the WHO, the limiting value of CO emission should 100 and 30 ppm for an average of 15 and 60 minutes, respectively [29–31]. The analysis of the above result reveal that the emission from the TMS was far above, and emission from the IMS was within the prescribed limit of the WHO guidelines. Approximately 60% – 70% of reduction in CO emission was noticed in the IMS as compared to the TMS. Higher CO emission in the TMS was mainly due to uncontrolled combustion. While in case of the IMS, control supply of primary and secondary air to the combustion chamber produced high combustion efficiency, and so lower CO emission was observed [11]. Although emission was greatly affected by the type of stoves used, it was also significantly affected by the type of biowastes (fuels) used. The results can be validated with those reported by Tryner et al. [32]. Basically, the variation in thermodynamically fuel parameters like carbon content, oxygen content, and heating values was responsible

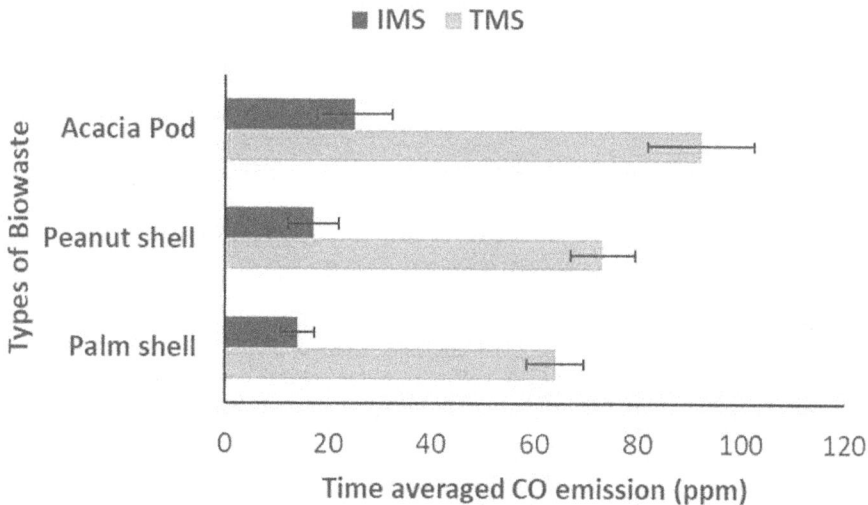

FIGURE 7.7 Comparison of time averaged CO emission between TMS and IMS. TMS, traditional mud stove; IMS, improved metallic cookstove.

for the variation in CO emission for three biowastes [12]. Acacia pods have the lowest carbon content, which obstructed their ability to sustain the flaming phase [33], and so the highest emission was recorded as compared to palm shells and peanut shells.

7.4 CONCLUSIONS

- This comparative study investigated the thermal and emission perfor-mances of a TMS and an improved biomass cookstove using three plant-based biowastes, i.e., palm shells, peanut shells, and acacia pods as fuel. Performances of cookstoves were evaluated following procedures of WBT protocols, version 4.2.3.
- The average TE of TMS was within the range $11.25\% \pm 2.11\%$–$15.58\% \pm 1.26\%$, whereas that for the IMS was within $20.57\% \pm 1.962\%$–$25.24\% \pm 1.74\%$. The TE of the IMS was around 10%–12%, which is more efficient than the traditional cookstove.
- The average SFC of the TMS varied from $1,154 \pm 178$ g/kg of water to $1,446 \pm 160$ g/kg of water, whereas that of the IMS varied from 543 ± 132 g/kg of water to 840 ± 162 g/kg of water. The IMS consumed 50%–60% less fuel than the traditional stove to complete the WBT. The time consumed by the IMS for boiling in the cold start and hot start phases was substantially less (approximate half) than the time consumed by the TMS.
- Around 60%–70% of reduction in CO emission was noticed for the IMS compared to the TMS.
- Palm shells offered a better overall performance in both the cookstoves, followed by peanut shells and acacia pods. However, implementation of the IMS in domestic cooking with these three biowastes could substantially reduce the use of firewood for cooking and could be an initiative to mini-mize deforestation.

MC, moisture content, VMC, volatile matter content, AC, ash content; FCC, fixed carbon content; HHV, higher heating value; LHV, lower heating value.

REFERENCES

1. Samal, C., Mishra, P. C., Mukherjee, S., and Das, D. 2019. Evolution of high perfor-mance and low emission biomass cookstoves-An overview. *AIP Conference Proceedings* 2200(1):020021.
2. Bansal, M., Saini, R. P., and Khatod, D. K. 2013. Development of cooking sector in rural areas in India-A review. *Renewable and Sustainable Energy Reviews* 17:44–53.
3. Arora, P., and Jain, S. 2016. A review of chronological development in cookstove assessment methods: Challenges and way forward. *Renewable and Sustainable Energy Reviews* 55:203–220.
4. Raman, P., Ram, N. K., and Murali, J. 2014. Improved test method for evaluation of bio-mass cook-stoves. *Energy* 71:479–495
5. Samal, C., Mishra, P. C., and Das, D. 2020. Design modifications and performance of biomass cookstoves-A review. *AIP Conference Proceedings* 2273(1):020002.

6. Bhatta, S., Pratap, D., Gakkhar, N., and Rajput, J. P. S. 2021. A comparative experimental investigation of improved biomass cookstoves for higher efficiency with lower emissions. In *Proceedings of the 7th International Conference on Advances in Energy Research*. Springer, Singapore, 961–971.
7. Singh, S. 2014. Comparative study of indoor air pollution using traditional and improved cooking stoves in rural households of Northern India. *Energy for Sustainable Development* 19:1–6.
8. Joshi, M., and Srivastava, R. K. 2013. Development and performance evaluation of an improved three pot cook stove for cooking in rural Uttarakhand, India. *International Journal of Advanced Research* 1(5):596–602.
9. Parmigiani, S. P., Vitali, F., Lezzi, A. M., and Vaccari, M. 2014. Design and performance assessment of a rice husk fueled stove for household cooking in a typical sub-Saharan setting. *Energy for Sustainable Development* 23:15–24.
10. Pande, R. R., Kalamkar, V. R., and Kshirsagar, M. 2019. Making the popular clean: Improving the traditional multi pot biomass cookstove in Maharashtra, India. *Environment, Development and Sustainability* 21(3):1391–1410.
11. Rasoulkhani, M., Ebrahimi-Nik, M., Abbaspour-Fard, M. H., and Rohani, A. 2018. Comparative evaluation of the performance of an improved biomass cookstove and the traditional stoves of Iran. *Sustainable Environment Research* 28(6):438–443.
12. Kaur-Sidhu, M., Ravindra, K., Mor, S., and John, S. 2020. Emission factors and global warming potential of various solid biomass fuel-cook stove combinations. *Atmospheric Pollution Research* 11(2):252–260.
13. Panwar, N. L. 2009. Design and performance evaluation of energy efficient biomass gasifier based cookstove on multi fuels. *Mitigation and Adaptation Strategies for Global Change* 14(7):627–633.
14. Mehetre, S. A., Sengar, S. H., Panwar, N. L., and Ghatge, J. S. 2016. Performance evaluation of improved carbonized cashew nut shell based cookstove. *Waste and Biomass Valorization* 7(5):1221–1225.
15. Ahmed, A., Hidayat, S., Abu Bakar, M. S., Azad, A. K., Sukri, R. S., and Phusunti, N. 2021. Thermochemical characterisation of Acacia auriculiformis tree parts via proximate, ultimate, TGA, DTG, calorific value and FTIR spectroscopy analyses to evaluate their potential as a biofuel resource. *Biofuels* 12(1):9–20.
16. Panwar, N. L. 2010. Performance evaluation of developed domestic cook stove with Jatropha shell. *Waste and Biomass Valorization* 1(3):309–314.
17. Obi, O. F., Ezema, J. C., and Okonkwo, W. I. 2017. Energy performance of biomass cookstoves using fuel briquettes. *Biofuels* 11:1–12.
18. Nzihou, J. F., Hamidou, S., Bouda, M., Koulidiati, J., and Segda, B. G. 2014. Using Dulong and Vandralek formulas to estimate the calorific heating value of a household waste model. *International Journal of Scientific & Engineering Research* 5(1):2229–5518.
19. Global Alliance for Clean Cookstoves, The Water Boiling Test Version 4.2.3. 2014. http://cleancookstoves.org/binary-data/DOCUMENT/file/000/000/399-1.pdf.
20. Arora, P., Das, P., Jain, S., and Kishore, V. V. N. 2014. A laboratory based comparative study of Indian biomass cookstove testing protocol and water boiling test. *Energy for Sustainable Development* 21:81–88.
21. Venkataraman, C., and Rao, G. U. M. 2001. Emission factors of carbon monoxide and size-resolved aerosols from biofuel combustion. *Environmental Science & Technology* 35(10):2100–2107.
22. Shankar, A., Johnson, M., Kay, E., Pannu, R., Beltramo, T., Derby, E., ... and Petach, H. 2014. Maximizing the benefits of improved cookstoves: Moving from acquisition to correct and consistent use. *Global Health: Science and Practice* 2(3):268–274.

23. Sonarkar, P. R., and Chaurasia, A. S. 2019. Thermal performance of three improved biomass-fired cookstoves using fuel wood, wood pellets and coconut shell. *Environment, Development and Sustainability* 21(3):1429–1449.

24. Kshirsagar, M. P., and Kalamkar, V. R. 2014. A comprehensive review on biomass cookstoves and a systematic approach for modern cookstove design. *Renewable and Sustainable Energy Reviews* 30:580–603.

25. Bhanap, I. J., and Deshmukh, R. D. 2012. Development of an improved sawdust gasifier stove for industrial applications. *Journal of Renewable and Sustainable Energy* 4(6):063113.

26. Berrueta, V. M., Edwards, R. D., and Masera, O. R. 2008. Energy performance of wood-burning cookstoves in Michoacan, Mexico. *Renewable Energy* 33(5):859–870.

27. Suresh, R., Singh, V. K., Malik, J. K., Datta, A., and Pal, R. C. 2016. Evaluation of the performance of improved biomass cooking stoves with different solid biomass fuel types. *Biomass and Bioenergy* 95:27–34.

28. Still D, MacCarty N, Ogle D., Bond, and T., Bryden M. 2011. Test results of cookstove performance. Cottage Grove, OR: Aprovecho Research Center; London: Shell Foundation; Washington DC: U.S. Environmental Protection Agency. Available from: http://www.pciaonline.org/resources/test-results-cook-stove-performance.

29. World Health Organization. 2010. *WHO Guidelines for Indoor Air Quality: Selected Pollutants.* Regional Office for Europe: The WHO European center for environmental and Health.

30. Patel, S., Leavey, A., He, S., Fang, J., O'Malley, K., and Biswas, P. 2016. Characterization of gaseous and particulate pollutants from gasification-based improved cookstoves. *Energy for Sustainable Development* 32:130–139.

31. Bartington, S. E., Bakolis, I., Devakumar, D., Kurmi, O. P., Gulliver, J., Chaube, G., ... and Ayres, J. G. 2017. Patterns of domestic exposure to carbon monoxide and particulate matter in households using biomass fuel in Janakpur, Nepal. *Environmental Pollution* 220:38–45.

32. Tryner, J., Willson, B. D., and Marchese, A. J. 2014. The effects of fuel type and stove design on emissions and efficiency of natural-draft semi-gasifier biomass cookstoves. *Energy for Sustainable Development* 23:99–109.

33. Pandey, A., Patel, S., Pervez, S., Tiwari, S., Yadama, G., Chow, J. C., ... and Chakrabarty, R. K. 2017. Aerosol emissions factors from traditional biomass cookstoves in India: Insights from field measurements. *Atmospheric Chemistry and Physics* 17(22):13721–13729.

8 The Glimmers of Hope

Transforming COVID-19 Medical Wastes into Value-Added Products as an Immediate Step to Encounter Medical Waste

R. Sharmila, K. Akila, S. Danushri, and M. Zion Mercy

CONTENTS

8.1 Introduction 122
    8.1.1 Systematic Collection of MW.. 123
    8.1.2 MW Sorting.. 124
    8.1.3 MW Handling... 124
    8.1.4 Transportation to MWT Locations... 125
8.2 Treatment of MWs... 125
    8.2.1 Sanitary Landfill Technology 127
    8.2.2 High-Temperature Incineration Technology.................................. 128
    8.2.3 High-Temperature Pyrolysis Technology 129
    8.2.4 Medium-Temperature Microwave Technology.............................. 130
    8.2.5 Pressure Steam Sterilization Technology 130
    8.2.6 Chemical Disinfection Technology 130
    8.2.7 Plasma Technology 131
    8.2.8 Torrefaction Technology 131
    8.2.9 Acid and Enzymatic Hydrolysis Technology................................. 132
8.3 Energy, Fuels, and Materials as Value-Added Products
Generated by MWT.. 133
    8.3.1 Generation of Value-Added Aromatics from Wasted
COVID-19 Mask through Catalytic Pyrolysis................................ 135
    8.3.2 Production of Butane from the Monthly Production of
COVID-19 Masks 136
    8.3.3 Conversion of COVID-19-Infected Masks to Energy
through Thermochemical Pathway along with Food Waste.............. 137
8.4 Critical Discussion... 137
8.5 Conclusion 139
References... 140

DOI: 10.1201/9781003359784-8

8.1 INTRODUCTION

Biomedical wastes (BMWs) are the potentially contagious waste generated from various hospitals, diagnostic centers, health camps, and treatment establishments involving animals and human research projects. The basic aspect on which the disposal of BMW relies is the classification, evaluation, separation, deposition, transition, treatment, and disposal. The righteous principle on which BMW is managed is based on 3Rs, which refers to reducing, reusing, and recycling. The most effective practice is to avoid the generation of BMW. The disposal of BMW is an extremely laborious and big-budget task.

Coronavirus disease 2019 (COVID-19) caused by severe acute respiratory syndrome coronavirus 2 (SARS-CoV-2) has led to the growing demand for ultra-modern treatment, which, in its, turn has elevated the need for an improvised medical waste (MW) disposal and management strategy. The inability to process such a huge quantity of biohazardous waste has led to contamination of landfills and treating facilities. The high-risk MW generated is harboring a massive negative impact on the environment and human health. The sustainable recovery of MW can tackle the global challenge of MW danger with utmost ease. The COVID-19 BMWs involve PPE gowns, coveralls, surgical masks, head covers, shoe covers, testing kits, cotton, and vials. The unsafe MW generated from different areas seriously contributes to the spread of the disease at a fast rate because of improper collection, segregation, and management processes (Das et al., 2021). The COVID-19 pandemic has acutely surged medical and municipal waste by 350%–500% around the world. India dumps a humongous measure of 90% of hazardous clinical waste in landfills without fitting sanitization treatment (Fadaei, 2021).

According to the World Health Organization (WHO), MW can be classified into seven major types: (i) infectious MW involving bodily fluids, swabs, bandages, culture, autopsies, medical devices, infected animals, blood, and infectious agents; (ii) pathological MW comprising body parts, body fluids, human tissues, animal organs, and carcasses; (iii) sharp MW including blades, disposable scalpels, needles, and syringes; (iv) pharmaceutical MW consisting of vaccines and drugs; (v) cytotoxic MW including genotoxic and highly hazardous MW, mutagenic, carcinogenic, teratogenic, and cancer therapy cytotoxic therapeutics metabolites; (vi) radioactive MW comprising radionuclides, radioactive diagnosis, and radiotherapy materials; and (vii) non-hazardous MW or general medical with no history of contamination with chemical, biological, and radioactive risks.

The Environment Protection Agency in the United States follows a different classification of MW: hazardous MW, general MW, infectious MW, and radioactive MW. Hazardous MW or hazardous MW comprises highly threatening yet non-contagious waste. The general MW consists of a voluminous proportion of household and office MWs. The infectious MW comprises MW generated from radioactive therapy and clinical elements from cancer therapeutics. Finally, the infectious waste contains the contaminated fluids of the body. The research on the sustainability of MW management (MWM) in Africa focused on adaptable approaches to safeguarding public health and the environment (Chisholm et al., 2021). At Lebanon during the COVID-19 epidemic, Maalouf and Maalouf (2021) examined the contagious MW generation rates (MWGRs) and MWM practices. The healthcare workers' knowledge of and use of MWM reduce the incidence of improper administration of national guidelines (Letho et al., 2021). During the COVID-19 pandemic, Mekonnen et al. (2021), investigated the MW in Ethiopia and calculated an average daily MWGR of 493 kg across all medical service facilities.

The classification of MW varies from one region to another. The categorization of MW depends on the models proposed by various agencies. The recent research and literature have categorized MWs depending on the waste discarding operations and the products used by various healthcare and research organizations. The hazardous MW has a high possibility of transmission of infection as the number of foreign pathogens tends to be sufficiently high. The MW generated from hemodialysis, dialysis unit, first aid, autopsies, surgeries, and laboratory functions can cause the transfer of hazardous infections from infected patients to healthy individuals. The country's rules and regulations also play an indispensable role in molding the MW disposal strategy. The by-products from medical laboratories, hospitals, healthcare centers, COVID-19 home isolates, dentist's places, and rehabilitation centers and operation activities contribute a humongous volume of MW to the environment. The number of MWs generated also immensely depends on the economic stability and level of activity of the country. The rich or highly developed nation contributes an enormous quantity of MW as compared to underdeveloped or developing nations. The inappropriate management of MW involves potential infections to patients, MW handlers, healthcare personnel, healthcare assistants, MW handlers, patients, and the public.

Comprehensive evaluations of the technologies used in the management of medical plastic waste treatment pathways were analyzed. This review article also gives a quick overview of the technologies that are available for turning MW into goods with additional value. Researchers should use these waste resources to create value-added products using the currently available technologies as the amount of plastic waste in the world increases, rather than allowing it to pollute the environment. This would turn the current crisis into an opportunity for future industries.

8.1.1 SYSTEMATIC COLLECTION OF MW

The production of functional products begins with the systematic collection of MW. The recent pieces of literature showcase the classification of healthcare waste based on the waste removal rehearses and the materials disposed of as MW (Minoglou et al., 2017). The utilization of reasonable, practical, and harmless collection techniques helps in the management of MW in the ecosystem. Accomplishing higher energy recuperation is fundamental for the innocuous removal of MW. MW primarily comprises radioactive, toxic, and harmful materials, which are related to natural contamination and health threat if researchers are inappropriately treated or discarded. The relationship between various types of MW influences the cycles for the assortment, stockpiling, and obliteration of hazardous MW. The legitimate assortment system for the collection of MW involves prepared specialists, explicit compartments, and in situ pre-treatment (Giakoumakis et al., 2021).

8.1.2 MW SORTING

Healthcare waste is most of the time categorized by the WHO into (i) infectious MW, (ii) pathological MW and sharp MW, (iii) chemical MW, (iv) pharmaceutical MW, (v) cytotoxic MW, (vi) radioactive MW, and (vii) general MW (https://www.who.int/news-room/fact-sheets/detail/health-care-waste). The classification of MWs is lifted by medical practitioners who engage in handling infectious wastes generated from hospitals, COVID care centers, home isolations, vaccination centers, biomedical laboratories, and research facilities. Until now, different classifications of MW are being considered and followed for effective and efficient management of hazardous MW. Primarily, the MWs are categorized as hazardous and non-hazardous portions (Giakoumakis et al., 2021). Various tones (typically yellow, red, or green) signify different kinds of waste, which help in the collection of different categories of MW. Tragically, no worldwide or even local coloring framework has been developed by global or public health specialists to help sort the hazardous MW (Mühlich, 2003).

Moreover, the functional progression of MW is to improve the robotized MW arranging framework issue, utilizing a blended number programming model for the MW task, pre-sorting stations, and robotized directed vehicles optimization. To achieve the objective of sorting, the most ideal option for risky MW is (reducing, arranging, sorting, moving, treatment, until burial) by directing pair correlations with Expert Choice programming for the information handling (Giakoumakis et al., 2021).

8.1.3 MW HANDLING

The classification of MW is followed by MW handling procedures. Various methodologies like incineration, microwave, reverse polymerization, disinfection, and autoclaving are taken into consideration after analyzing various factors like wellbeing risks, social acknowledgment, natural effect, practical expenses, benefits from shelling and reusing, cost of transportation, stockpiling, and the course to the store, enlisted organizations that dispose of MW, date of transport, nature of waste, and weight of MW. The handling of MW involves an insignificant or unplanned working environment to laborers liable for taking care of such irresistible things. Medical care offices are lawfully unreliable for guaranteeing that their faculties do not contact infection from MW (Liu et al., 2021). COVID-19 poses a concern to human health and generates a significant quantity of MW that is hazardous to the environment (Song et al., 2021).

The WHO has stressed the meaning of acting with tact in emergency clinical waste handling. The disposal of MW should likewise be finished in a manner that guarantees insignificant or preferably no unplanned openness of laborers liable for taking care of such irresistible things. Medical care offices are legitimately answerable for guaranteeing that their workforce does not contact irresistible MW put in the fitting waste canister. Nevertheless, even in nations with severe lawful systems, for example, the United Kingdom, reports are proposed that do not furnish adequate safeguards to stay away from contact with hazardous MW. The COVID-19 pandemic raised MW generation (MWG) in health centers in Iran by 102% (Kalantary et al., 2021). However, various protected practices are often overlooked by the concerned authorities. The authoritative ineptitude concerning faulty MWM actions might bring about contamination and ill-ness of either suffering patients or laborers and eventually increase the potential risk for the medical clinic organization. Thus, MW treatment (MWT) offices are becoming overburdened, requiring the utilization of options for taking care of and unloading, for example, co-removal of MW in the general strong waste incinerator, modern broilers, concrete ovens, and profound landfill, to build dealing with ability. The functions of MWT facilities must be updated regularly while handling MW concerning COVID-19 (Giakoumakis et al., 2021).

8.1.4   TRANSPORTATION TO MWT LOCATIONS

The MW piled up is transported to treatment spots either situated within medical facilities or central offsite establishments. The most often utilized techniques for treatment are autoclaving, microwaving, and incineration, which bring about lin-gering ash. The MW is never permitted to be reused or recycled into functional products before appropriate sterilization protocols. The COVID-19 pandemic has overburdened the landfills with soaring medical and municipal wastes, and this has increased the requirement for the incinerators, cement stoves, industrial modern ovens, and profound landfills, to deal with capacity (Hantoko et al., 2021). MWM models used for the assortment and transportation take place by utilizing heuristic and meta-heuristic calculations, taking into consideration "arranging cycles," put-ting in vulnerability and gauging protocols like well-built advancement, and inte-grating arising advances. On-location sanitization forestalls the dangers of tedious COVID-19 MW transportation. The microwave innovation is further accompanied by autoclaving, with sanitization steam at 93°F–350.60°F (Giakoumakis et al., 2021). This third party gathers the loss from the main issues and transports it securely to the last removal office. Sadly, there are a few downsides to this technique since there are lawful holes concerning the obligations of the workers for hire, who can procure a ton. An unbending clinical waste global positioning framework is essential to keep away from or possibly limit illicit unloading, which can, in any case, become ongoing and bring about expanded danger for general wellbeing, which causes more dam-age to the climate in light of microbe discharge. Frail regulation can likewise give third-party individuals one more chance of stashing cash. Researchers can exchange things that ought to be discarded, e.g., sharps on the underground market for subse-quent reuse. Recuperated and non-sterile sharps address a huge danger of patients' contamination through the spread of blood-borne microorganisms. Reusing of pos-sibly irresistible MWs is not permitted, irrespective of the utilization and sanitization process (Giakoumakis et al., 2021) (Figure 8.1).

8.2   TREATMENT OF MWs

The COVID-19 pandemic increased the consumption of personal protective equip-ment (e.g., face masks and gloves), as well as the transmission of infectious MW from clinics, treatment centers, and quarantine households. As a result, MWT installations became inundated, forcing the deployment of alternative management and discarding techniques to boost handling capacities such as co-disposal of MW in general solid

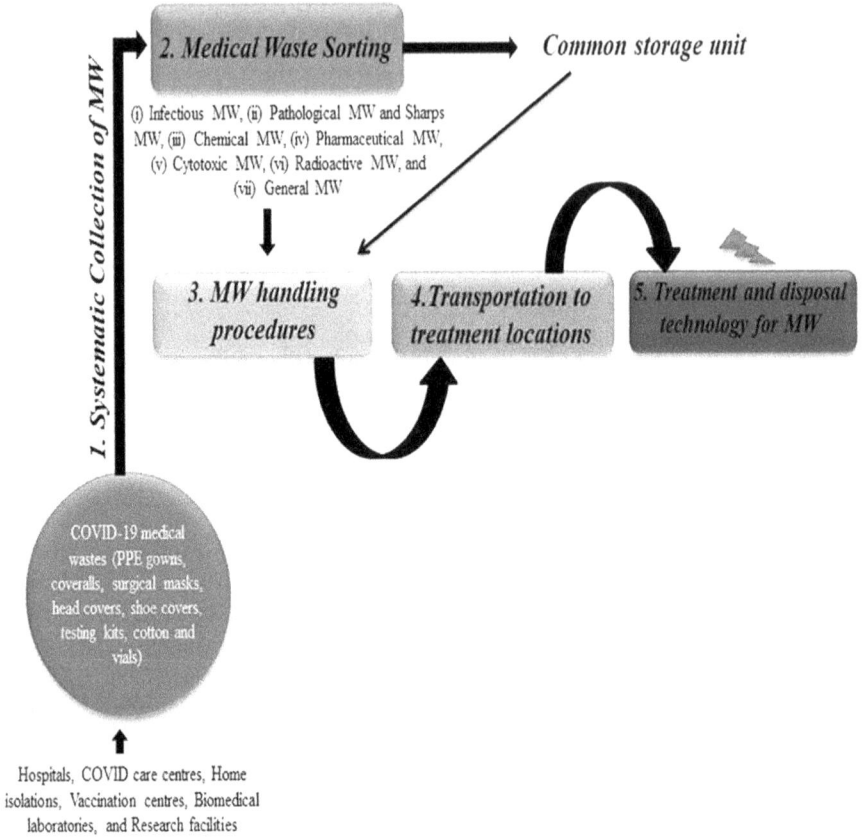

1. Systematic Collection of MW

2. Medical Waste Sorting → **Common storage unit**

(i) Infectious MW, (ii) Pathological MW and Sharps MW, (iii) Chemical MW, (iv) Pharmaceutical MW, (v) Cytotoxic MW, (vi) Radioactive MW, and (vii) General MW

3. MW handling procedures

4. Transportation to treatment locations

5. Treatment and disposal technology for MW

COVID-19 medical wastes (PPE gowns, coveralls, surgical masks, head covers, shoe covers, testing kits, cotton and vials)

Hospitals, COVID care centres, Home isolations, Vaccination centres, Biomedical laboratories, and Research facilities

FIGURE 8.1 A comprehensive evaluation of steps involved in the conversion of MWs generated during the COVID-19 pandemic into functional products by various treatment technologies. MW, medical waste; COVID-19, coronavirus disease 2019.

incineration plants, commercial ovens, and coal ash, and deep dumpsters (Hantoko et al., 2021). Effective MWM formal concept landfill management prevents COVID-19 spread, while on-site remediation and transient storage help mitigate the MWM issue (Das et al., 2021). The WHO's official recognizes that there are almost no ecological, good deals for secure disposal of communicable pollutants and the global society's major concern are over disposal safety at a fair cost and with low environmental impact. It is vital to use trained staff and appropriate containers to collect infectious MW, as well as in situ decontamination. In civilized countries, 50% of MW is burned, 30% is autoclaved, and the remaining has been processed in other forms. Following burning, harmful polychlorinated dibenzo-p dioxins (dioxins) and polychlorinated dibenzofurans (furans) are formed, raising many concerns about air pollution and the generation of poisonous toxins. The necessity for alternate treatment modalities that can safely eradicate any microbe is indisputable, which has led to the usage of autoclaving and microwaving, among other techniques (Figure 8.2).

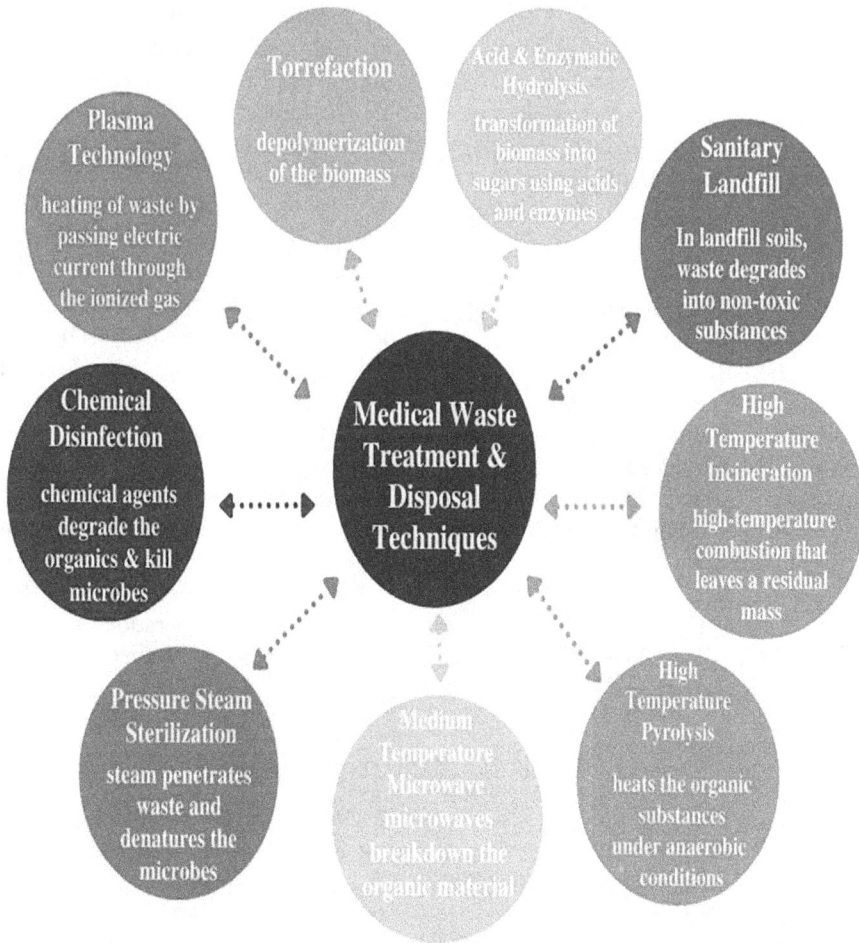

FIGURE 8.2 Insightful analysis of types of MWT and disposal techniques. MWT, medical waste treatment.

8.2.1 SANITARY LANDFILL TECHNOLOGY

The detrimental consequences of landfills on the health and the environment, including the need for appropriate MWM activities to eliminate the effects, were deeply investigated by a wide range of researchers (Ozbay et al., 2021). Inappropriate landfill management causes issues with trash collection and landfill gas generation, resulting in contaminated groundwater and air. Due to the inherent limits of using landfills as a principal disposal option, poor landfill management has a significant negative impact on the environment and people's life. In developing countries, landfilling is the most appropriate method of solid MW disposal (Hereher et al., 2020). One of the most widely used functions of geographical information systems is to locate potential waste sites. Researchers used this technology to locate potential dumpsites in

Oman's Muscat governorate. Furthermore, it is observed that in developing countries, significant non-engineered landfilling techniques have aggravated environmental concerns, while implementing a sanitary landfill appears impracticable due to economic impotence (Nik Ab Rahim et al., 2021). Researchers studied the long-term feasibility of a sanitary landfill strategy in Peninsular Malaysia by incorporating sustainable factors into plan evaluation and merging three policy-related techniques. It was discovered that in the case of An-Najaf city, the optimal sanitary landfill site was determined using a geographical information system and eight acceptable criteria, including the urban area, roads, and other infrastructure, including the urban area, roads, soil conditions, streams, altitude, weather, gradient, and religious/archaeological/historical sites. The earliest approach to MW disposal (MWD) is the sanitary landfill procedure, which is already practiced in certain low-income nations. It emphasizes the garbage's expected degradation into non-toxic substances in landfill soils. As a result, a variety of harmful toxins, such as pathogens and radioactive isotopes, are introduced before the same type of disposal, which will have a substantial impact on the environment and mankind. An effective technique for addressing this problem is to build treatment and output pipelines. The sanitary landfill method is simpler and precise to set up, but it has several shortcomings, including the demand for sterilization and trash reduction before landfill, as well as the necessity for a large number of bits, the need to live in a vast terrestrial zone, and the generation of numerous dangerous gasses in addition to O_2 and H_2. Moreover, soil and groundwater must be consistently monitored (Kareem et al., 2021).

8.2.2 HIGH-TEMPERATURE INCINERATION TECHNOLOGY

After thermochemical processes, researchers examined several disinfection techniques for COVID-19 MW manipulation, sorting, and collecting (Ilyas et al., 2020). As a result, appropriate MWM for COVID-19 limits was investigated utilizing a variety of sanitizing tactics, and even specific concepts were reported to significantly reduce the potential environmental and health threats. Burning medical garbage is a widely approved waste disposal method for all sorts of healthcare waste. However, severe oxidation of the waste under a high-temperature flame causes the compounds to dry and incinerate, resulting in a residual mass that can be treated as an innocuous material and gas because the waste primarily consists of hydrocarbons with a high calorific value, which can still be quickly degraded during cremation, and this method tends to be very efficient.

The use of this approach is limited by the need for a constant high temperature in the furnace, mixing of a high amount of high oxygen, relevant turbulence and integrating degree of the equipment, continuous water content maintenance, ample supply of gas residence time, periodic maintenance of machinery, and finally, sufficient gas contact time, and regulation of the ultimate flue gas. The final product volumes and weights are substantially lowered; the waste is destroyed; the method is stable, formalized, and requires no special knowledge and experience; the produced thermal energy is easy to recycle; the MW is well disinfected/sterilized; and pollutants are eliminated. This technique can be widely employed to all categories of waste irrespective of the quantity of waste.

The process of incineration and the removal of hazardous MW is suitable, according to Greek norms, and significantly reduces their mass. In comparison to alternative treatment technologies, a cremation facility has a high level of automation and versatility, but community support is minimal. Incineration, which involves burning at a very elevated temperature, ensures sterility and creates a small amount of residual ash, which is buried in a landfill. Dioxins, furans, and mercury are the primary poisons generated during incineration. Dioxins are organic molecules made up of two benzene rings linked by two oxygen atoms and containing 4–8 chlorine atoms in place of hydrogen atoms. Their half-lives range from 7 to 11 years, making them extremely persistent and amassing in the atmosphere. Furthermore, researchers have recognized carcinogens and have been linked to reproductive impairment in humans. Furans have a composition similar to dioxins, except perhaps one oxygen atom in between two aromatic rings.

Mercury emissions from the burning of MW account for about 3%–9% of the total overall Hg pollutants. As Hg can concentrate in fatty tissue when breathed, the impact of atmospheric Hg emissions on the environment and public health is profound. Researchers also have identified its adverse effects on the neurological, sexual, and excretory systems. The most effective strategies for decreasing dioxin emissions are textile screens and thorough incineration at temperatures exceeding 800°C (Voudrias et al., 2014).

8.2.3 HIGH-TEMPERATURE PYROLYSIS TECHNOLOGY

Pyrolysis, which runs at 540°C–830°C and involves oxidation/plasma/laser-based and ignition pyrolysis, seems to be a more technically sophisticated solution for MWT. The elevated pyrolysis method burns the organic substances of MW in oxygen-free or oxygen-depleted conditions, disrupting their molecular bonds and converting high-molecular weight organic compounds into flammable gasses and liquids. CO_2, CO, H_2, CH_4, and other compounds and volatile organic compounds are observed in the pyrolysis gas. Cracked gas and coke are burned by the MW high-temperature pyrolysis method. Flammable gasses are employed as pyrolysis fuel, which reduces the cost of maintenance as compared to traditional combustion (Ilyas et al., 2020).

Pyrolysis with an ambient air parameter minimizes the level of flue gas produced while also decreasing the cost of the flue gas filtering device, resulting in a cheaper overall cost than older incineration, and because of the high oxygen combustion during cremation, dioxins are easily formed. In comparison to traditional incineration, pyrolysis can produce minimal dioxins, given the absence of oxygen and the elimination of acid gas. Pyrolysis at high temperatures can directly supply furnaces (Xu et al., 2020). The pyrolysis treatment method for MW offers an alternative strategy for MWT hedging when combined with innocuous solid wastes. Pyrolysis technology has a high rate of energy recuperation, negligible particulate emissions, and adequate finances. Almost fully automated equipment set for MW pyrolysis with synchronized gas retrieval offers a lower workspace, traditional machinery control, industrial operation, few changes in the industry, and effective commercialization. As aforementioned, the pyrolysis technique may be used to circulate energy, reduce energy consumption, low computational costs, and achieve economic feasibility by using the gas formed by the MWT (Ilyas et al., 2020).

8.2.4   MEDIUM-TEMPERATURE MICROWAVE TECHNOLOGY

Due to excessive microwaves for organic material breakdown, the microwave method uses a temperature range of 177°C–540°C for retrograde polymerization. By flapping the bonds of molecules, electromagnetic waves (frequency 300–3,000 MHz, wavelength 1mm–1 m) boost the internal energy. In response to elevated disinfection, a N_2 atmosphere precludes oxygen burning. The reduced energy and temperature used help to cut costs, and because of the harmless residue left over after the disinfection treatment, there are no heat losses and no pollutants. SARS-CoV-2 can be immobilized by a microwave device with a particular design and is suited for decontamination of COVID-19 MW (Wang et al., 2020). On-site decontamination eliminates the hazards of COVID-19 MW shipment, which is time-intensive. Microwave technologies can be used in tandem with autoclaving and sterilization steam at temperatures ranging from 93°C to 177°C (Ilyas et al., 2020). Microwaves kill the vast majority of germs (12.24 cm, 2,450 MHz). Due to the obvious high enough temperature, the infectious particles in the MW water are eliminated. A shredder shreds and breaks down the MW into small pieces. Then, it is saturated, brought to an ultraviolet chamber with microwave emitters, and irradiated for 20 minutes. Eventually, the processed MW is pulverized and blended with municipal solid garbage in containers. It is unsuited for usage in developing nations because of its expensive costs, as well as the high operating and maintenance costs. Comparable approaches are currently being developed.

8.2.5   PRESSURE STEAM STERILIZATION TECHNOLOGY

This technique depends on treating crushed MW for 20 minutes at 121°C below 100 kPa. This method can be used to clean contaminated vestments, needles, and microbiology culturing pieces of equipment, and not tissue or remains. The quantity, temperature, and time of the MW define the productivity of this low-cost/low-operating device. The technique's main flaw is that the finished product has the same volume as the original debris. There are dangerous discharges, and certain toxicants can indeed be handled such as phenol, mercury, and formaldehyde. The use of a large, specialized autoclave for high-pressure steam sterilization is recommended, and the process produces unstable tainted chemicals.

Infectious liquid MW is routinely sterilized with chemical disinfectants. Even yet, large amounts of garbage are challenging to handle. Many innovative technologies could be utilized to replace incineration and eliminate the emission levels it releases. However, these techniques are not frequently used in China since they are thought to be immature and difficult to implement. High-temperature pyrolysis, from the other approach, which may be used multiple times, delivers a higher recycling rate with minimal contaminants and substantial price benefits (Mantzaras and Voudrias, 2017).

8.2.6   CHEMICAL DISINFECTION TECHNOLOGY

Chemical decontamination solutions are feasible for COVID-19 MW after being mechanically shredded (Ilyas et al., 2020). To absorb shredding aerosol, the air is routed through a high-efficiency filter. Chemical disinfectants are added to the crushed MW, which is then placed under a pressure gradient. The infectious micro-organisms are inactivated or terminated, while the organics are digested. Chemical disinfectants leave no lingering danger because researchers eliminate both microbes and microbial spores (Wang et al., 2020). COVID-19 MW chemical disinfection can be split into two categories: chlorine-based and non-chlorine-based technologies (Duarte and Santana, 2020). The chemical disinfection technique utilizes chemi-cal agents to disinfect the MW, such as sodium hypochlorite, ozone, peracetic acid, glutaraldehyde, and others. Not just for liquid waste but also for disposal of solid waste treatment, this strategy is advantageous. The procedure's efficiency is mostly determined by the microorganism type and biological properties, exposure time, pH, chemical composition, and quantity. The chemical modification is a simple, cost-effective, and easily accessible way for achieving speedy disinfection and efficient degumming of the final product, as well as reducing material waste and no waste liquid or waste gas. The method's main drawbacks are the disinfectants' toxicity for people and the method's stringent temperature and pH supervision requirements. The method is not advised for radioactive MW following treatment and volatiles (Ilyas et al., 2020).

8.2.7 PLASMA TECHNOLOGY

Plasma technology is a novel process that relies on the generation of a gas cloud from the ionization of inert gas. When an electrical current passes through this system, the gas is oxidized, resulting in an instantaneous dielectric medium with a very high temperature (up to 3,000°C) and quick-drying and boiling of the waste. A mixture of flammable gasses, such as alkanes, H_2, and CO, is produced as a result of this operation. All harmful bacteria in the garbage are killed after a sec-ond burning. The finished product can be disposed of securely in a landfill. The efficiency of this costly procedure is determined by the instrumentation's power as larger energy output makes temperature conversion easier because no hazard-ous compounds are discharged, the end volumes are greatly reduced, and the heat energy produced can be recovered; this approach is suitable for all types of MW (Aboughaly et al., 2020).

8.2.8 TORREFACTION TECHNOLOGY

Torrefaction is the technique of de-polymerizing biomass. To accomplish the req-uisite degree of de-polymerization of the biomass, a significant length of time is required. The degree of torrefaction is determined by two factors: time and tempera-ture. In the literature, the imposed torrefaction time is sometimes referred to as emit-ter residence time (Cahyanti et al., 2020). The residence time in the reactor begins when the biomass temperature hits 200°C. Formerly, there has been no breakdown of biomass. Torrefaction can be used on medical cotton waste (MCW), a cellulose-based substance with a cellulose content of roughly 95%. To valorize such resources, a variety of treatment procedures are applied. MCW is often contaminated and dangerous. It must be sterilized to exclude any potentially hazardous pathogenic or infectious substances. To achieve this, it is sterilized in an autoclave for 15 minutes at 121°C, 15 bar, to kill any infectious disease-causing organisms. For the sanitization of MCW, this technique is effective (Giakoumakis et al., 2021).

Hanoglu et al. (2019) studied the torrefaction of acrylic and PE textile fibers at temperatures between 300°C and 400°C. They discovered that a key factor in determining the solid product is temperature. Additionally, the torrefaction process could result in a solid product (char) with a high energy density (18–25 MJ/kg) but a low sulfur and ash content (less than 10 wt%) (Mohd Faizal et al., 2018). Additionally, the torrefaction of mixed cotton and PEs may produce high-quality solid fuel that can replace coal. Researchers Matsuzawa et al. (2004) examined the torrefaction of cellulose and polyethylene (PE), polypropylene (PP), polystyrene (PS), and polyvinyl chloride (PVC) mixed materials. Their research indicates that the breakdown patterns of PE, PP, and PS that are mixed with cellulose do not differ noticeably from those that are not. Therefore, it may be said that neither an interaction nor a reply can affect how quickly they degrade. However, when cellulose is combined with polyethylene terephthalate (PET), the rates at which both materials degrade alter, producing more residual char.

8.2.9 Acid and Enzymatic Hydrolysis Technology

Acid hydrolysis is a frequently used method for converting biomass into monosaccharides. H_2SO_4, HNO_3, HCl, and other acids are commonly employed in hydrolysis. When compared to alternative approaches for the same purpose, this technology yields a higher sugar extraction. It is also very reliable. Although acid hydrolysis produces a large amount of sugar, it also generates a large number of breakdown products such as levulinic acid, furfural, formic acid, and hydroxymethylfurfural. To reduce the number of breakdown products, criteria such as temperature, time, type of acid, and pH must be deliberately chosen. For the production of fermentable bio-ethanol sugars, MCW can use acid hydrolysis technology. During this technique, enzymatic hydrolysis might be used as a second approach (Chen et al., 2020).

Enzymatic hydrolysis of cellulose by cellulase converts cellulose to fermentable reducing sugars, whereby yeasts or bacteria convert to ethanol. In a heterogeneous system, this process is a multistep reaction. Endoglucanase and exoglucanase/cell hydrolases separate insoluble cellulose into a solid–liquid phase. In the liquid phase, glucose is generated from glucosidase through intermediate product hydrolysis. Enzymolysis is a less expensive procedure than acid or alkaline hydrolysis. Bacteria and fungi can create hydrolyzed cellulases. Cellulosic enzymatic hydrolysis is a three-step process that includes (i) cellulase adsorption on the cellulose surface, (ii) cellulose hydrolysis to glucose, and (iii) cellulase desorption. Both the amount of yield and the initial rate of enzymatic conversion of cellulose are affected by the substrate concentration. When substrate levels are low, it is necessary to increase the concentration of the substrate to increase the yield and reaction rate of the degradation (Giakoumakis et al., 2021). Additionally Vignesh and Chandraraj (2021) used surfactant as an adjuvant to produce optimal high-solid input enzymatic hydrolysis/fermentation of cotton dust.

8.3 ENERGY, FUELS, AND MATERIALS AS VALUE-ADDED PRODUCTS GENERATED BY MWT

The energy recovery efficiencies (EREs) and environmental effect of the MWT technologies (MWTTs) can be calculated. Energy recovery analysis, life cycle analysis, and life cycle costing methodologies were used to investigate MWTTs such as incineration (rotary kiln or pyrolysis), plasma melting, and sterilizing (steam or microwave). Furthermore, when incineration and sterilization MWTTs were integrated with co-incineration technologies, energy conversion capability and behavioral intention improved. Steam and microwave sterilization + incineration had a high ERE of 83.4%, while plasma melting had a low ERE of 19.2%, according to the energy recovery analysis. The cost of co-incinerating sterilized MW and municipal solid trash was found to be cheap.

Due to a lack of proper heat energy recovery systems, waste heat is not used efficiently enough. MWTTs generate energy and provide heat to North China (Zhao et al., 2021). Although there are limits owing to low power generation efficiency in South China, it is possible to employ electrical power generation to utilize MWTTs that produce heat. Pyrolysis was shown to be a very successful method for the deterioration of COVID-19 MW (Dharmaraj et al., 2021). The majority of COVID-19 MW's plastic component is made up of PET, PE, PP, and PVC, and PS. H_2 and syngas are produced from MW using the plasma gasification technique. The operating settings for a 10-kW microwave air plasma generator were established for the greatest H_2 generation (Erdogan et al., 2021). Despite plasma gasification appearing to be a promising low-cost and long-term solution for MWTTs, more research is required due to a lack of knowledge regarding the toxic compounds produced during the process. The levels of dangerous compounds produced must be below the government's guidelines. Thermal cracking was carried out at 500°C for 40 minutes to create liquefied fuel oil from the polymeric component of MW (Rasul et al., 2021). The production of liquid fuel was 52%, and the higher heating value was 41.32 MJ/kg, which is compatible with a wide range of diesel.

The pyrolysis process MWTT generates environmentally friendly technologies with a high power density and, like traditional pyrolysis, might be used to replace incineration. Researchers used the co-hydrothermal carbonization process to create solid fuel on laboratory- and pilot-scale from PVC containing MW combined with lignocellulosic biomass (woodchips). The addition of woodchips boosted the dechlorinating efficiency of MW in a pilot-scale implementation of the hydrothermal carbonization process. The low-chlorine hydro char product has a higher heating value (24.2 MJ/kg) and might be utilized as a clean coal equivalent (alternative fuel). With the laboratory-scale autoclave reactor, lignocellulosic biomass has a broad array of applications for fuel and energy generation. Combining MWM with the evaluation and production of lignocellulosic biomass is an amazing approach that combines the MWM with the evaluation and production of lignocellulosic biomass. The findings corroborate prior findings, indicating that pyrolysis has the potential to become the most widely used MWTTs shortly (Fang et al., 2020). For MWT and generating power from MW, Manegdeg et al. (2020) used a pyrolyzer–Rankine cycle.

The torrefaction technology on herbal medicine pollutants, with the basic qualification, that torrefied herbal medicine wastes exhibits excellent combustion qualities and is suitable for use as solid fuels, such as co-combustion or pellet manufacture. Torrefaction is a rare MWTT that is utilized in laboratories to produce valuable chemicals with limited MW categories. However, it is a technology that improves the higher heating value feedstock and has a broad range of applications in the management of lignocellulosic biomass. Torrefaction, in combination with lignocellulosic biomass treatment, may become a viable MWTT if MW separation were more organized. Combining the acid pre-treatment with enzymatic saccharification of MCW enhanced the generation of fermentable sugar (Giakoumakis et al., 2021). These sugars were useful in the synthesis of bio-ethanol. The progressive treatment of lignocellulosic biomass is a widespread practice. The originality of that approach was to use this process on MCW and see if it works because MW separation procedures must be redesigned to separate MCW, and this technology has significant limits. The greatest cellulose-to-glucose transformation was found to be 95.6%. Likewise, MCW might be separated from muddled MW and pooled with the rest of the cellulosic biomass MW (papers, textiles, etc.) to be properly handled in future, reinforcing the link between MCW and lignocellulosic MW and the attempt to enhance recycled renewable bio-fuels (Erdogan et al., 2021).

Torrefaction, like acid hydrolysis, is a tried and true biomass treatment process. These technologies have the potential to become part of the MWTTs in future and assist a diverse range of products beyond solid/liquid fuels and adsorbent synthesis. Researchers used thermolysis (pyrolysis at 400°C–550°C) in a stainless steel semi-batch reactor to create liquid fuel from waste disposable syringes. These syringes have PP cores and high-density PE pistons. The pyrolysis oil produced resembled a diesel or petrol mixture in terms of physical qualities. As a result, pyrolysis is a highly effective MWTT for the treatment of MW plastic fractions. This is completely consistent with the literature, which claims that the pyrolyzed MW plastic component yields real worth fuels. MCW was used to make a fibrous cellulose sulfate absorbent (Świechowski et al., 2021). This adsorbent was made by sulfonating MCW with camphor sulfonic acid (CSA) in di-methyl fumarate medium, and it was found to be suitable for the removal of malachite green from aqueous solutions employing batch and column apparatuses. Fast pyrolysis of hazardous MW, produced char and oil, and the creation of polycyclic aromatic hydrocarbons was dependent mostly on control parameters. The MW can be converted to a usable hydrocarbon fuel using fast pyrolysis techniques. Another pyrolytic application is fast pyrolysis, which is used to produce chemicals and fuels from MW mixtures combining paper, textiles, glassware, and plastics, demonstrating the benefits of pyrolysis as a feasible ubiquitous MWTTs.

Researchers used thermophilic bio-digestion conditions to optimize the biogas yield by 92% from industrial recycled MCW. An electrochemical device like a proteolytic fusion reactor could be excellent for transferring chemical energy in the substrate into power via oxidation. The mixed fuel pellets, on the contrary, seemed to have a gross calorific value identical to coal. In reality, when compared to coal, the manufactured fuel pellets had much higher O/C and H/C ratios. Similarly, varied feed fuel pellets improved the quality of the fuel pellets. Finally, low-chlorine pure lignocellulosic materials bio-fuel granules with a higher calorific efficiency could be effectively integrated with hydrothermally processed MW and pyrolytic plastic pollution leftovers. This is an ingenious way of integrating three different types of processed waste, plastic residue, MW, and lignocellulosic biomass, into a single treatment method, pelletizing, for enhanced value in solid fuel generation (Xin et al., 2019).

Researchers discovered that employing a pyrolyzer–Rankine cycling power station to generate energy from MW is economically attractive 400% over just a 5-year cycle. The mission's overall thermodynamic efficacy was 66.8%, CO_2 emissions were considerably decreased, and the net profit margin ratio was 18.6%. This 1-month experiment yielded sufficient data on environmental, economic, and power generation challenges. This has been proven to be an outlay strategy, with consistently low greenhouse emissions and pollutants that are substantially below the national average, while also supplying a lot of energy. During MW incineration, it obtained considerable on-site heat recovery. An incinerator was also utilized for MW, achieving 6.6–8 kW/kg energy, which also approximated 10–12 kg/kg high pressure, with 4.15 kW/kg heat flux as a backup fuel.

The energy efficiency coefficient of the incinerator was 47%–62%. Finally, heat recovery systems evolved by MW thermal treatment. To manufacture carbonized solid fuel from medical peat waste, the torrefaction process was adopted (Swiechowski et al., 2021). Torrefaction at 200°C–550°C increased the higher heating value of peat waste to 21.3 MJ/kg, up from 19.0 MJ/kg in the unmodified mixture. This would be the first attempt to employ torrefied medical peat, a novel technique for lignocellulosic MW usage on a laboratory level, demonstrating the necessity for more research and transformation of MW fraction torrefaction as an alternative to MWTTs. For energy production, an organic Rankine cycle in conjunction with infectious MW incinerator, assessed the system using power, entropy generation, socioeconomic, and hazard assessment. Although Rankine cycle in conjunction with infectious MW incinerator is capable of producing 23.65 kW, and its energetic and energetic efficacy is only 0.91% & 0.89%, respectively (Chaiyat, 2021) (Figure 8.3).

8.3.1   Generation of Value-Added Aromatics from Wasted COVID-19 Mask through Catalytic Pyrolysis

The waste masks can be used as a pyrolysis feedstock as COVID-19 has dramatically expanded its production in recent years. The catalytic rapid pyrolysis can be used to manufacture value-added chemicals from the mask in large quantities, such as aromatic compounds. In order to achieve this, it was determined how zeolite catalyst characteristics affected the effectiveness of the pyrolytic products created from the pyrolysis of used masks. The types of zeolites (HBeta, HY, and HZSM-5), including the mesoporous catalyst Al-MCM-41, and pyrolysis temperature, were used to characterize the compositions and yields of pyrolytic gasses and oils. The mask can be pyrolyzed in a fixed bed reactor, and pyrolysis gasses that were produced can be sent to an additional reactor that include a load of zeolite catalyst because at this temperature, maximum oil yield (80.7 wt%) during the non-catalytic pyrolysis process can be recorded. It was selected as the CFP temperature to compare the catalyst performance for the synthesis of benzene, toluene, ethylbenzene, and xylene

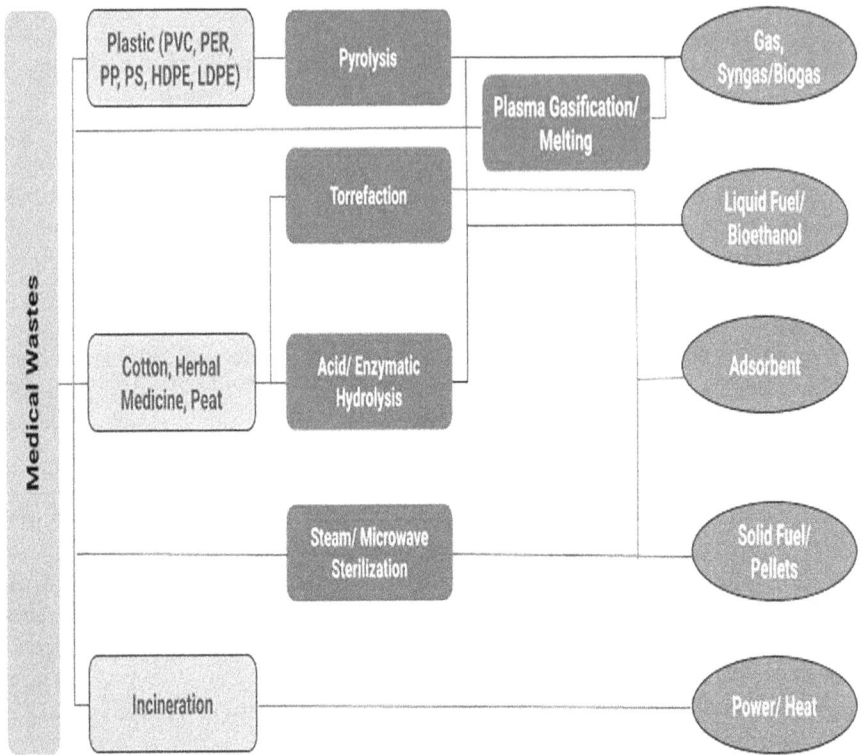

FIGURE 8.3 A structural outline on the value added products processed from various MWTTs. MWTT, medical waste treatment technologies.

(BTEX) because it has larger pores, greater surface areas, and more acid sites than the HZSM-5, the large-pore zeolite groups of HBeta and HY produced BTEX concentrations that were, respectively, 134% and 67% higher than those of the HZSM-5. The impact of zeolite properties on the generation of BTEX via CFP has not been documented before in production of value-added products (Lee et al., 2021).

8.3.2 PRODUCTION OF BUTANE FROM THE MONTHLY PRODUCTION OF COVID-19 MASKS

A fresh thermochemical method to extract butane from the monthly production of billions of useless COVID-19 masks was formulated by researchers to cut down the soaring amount of BMW with the onset of the COVID-19 pandemic. The studies were carried out using thermogravimetry (TGA) with 3-ply face masks (3PFM) over ZSM-5 zeolite with varying proportions of ZSM-5 to 3PFM (w/w: 6, 12, 25, and 50 wt%) at various heating conditions. Furthermore, TG-FTIR and GC-MS analyses were used to examine the impact of ZSM-5 concentration and heating rates. Additionally, using linear and nonlinear isoconversional modeling approaches, the kinetics behavior of the suggested strategy was modeled, allowing the activation energy (Ea) for each conversion area to be determined. Finally, using the distributed activation energy and independent parallel reactions approaches, respectively, the necessary parameters to fit the experimental curves for TGA and differential scanning calorimetry were determined. The findings demonstrated that both aromatic and aliphatic (-C-H) molecules are extremely abundant in the degraded samples. The average Ea at 25% of ZSM-5 (sample enriched with butanol) was estimated in the ranges of 158–187 kJ mol1 (linear methods with R2 > 0.96) and 167–169 kJ/mol (nonlinear methods with R2 > 0.98) based on GC-MS results. However, butanol was the basic component in the generated compounds, with an abundance of 31% at the lowest heating rate (5°C/min). Finally, with a very minor divergence, distributed activation energy and independent parallel reactions were able to reproduce the TGA and differential scanning calorimetry curves of ZSM-5/3PFM samples. Based on that, it is possible to dispose of COVID-19 masks and transform them into a butanol compound using the catalytic pyrolysis technique over ZSM-5 zeolite (Yousef et al., 2022).

8.3.3   Conversion of COVID-19-Infected Masks to Energy through Thermochemical Pathway along with Food Waste

Park et al. (2021) analyzed the objective of valuing the waste materials for energy and resources, and the co-pyrolysis of food waste and single-use face masks (for COVID-19 protection) was examined. This was accomplished by pyrolyzing a disposable face mask (a type of personal protection equipment) to create fuel-range compounds. The majority of the non-condensable permanent hydrocarbons included in the pyrolytic gas that resulted from the pyrolysis of the single-use face mask were CH_4, C_2H_4, C_2H_6, C_3H_6, and C_3H_8. The yields of non-condensable hydrocarbons were improved by raising the pyrolysis temperature. The hind high vacuum of the pyrolytic gas was over 40 MJ kg^{-1}. In addition, the pyrolysis of the disposable face mask resulted in the production of hydrocarbons with wider carbon number ranges, such as those found in motor oil, gasoline, jet fuel, and diesel. The single-use mask produced the maximum outputs of hydrocarbons in gasoline, jet fuel, and diesel at 973 K. The pyrolysis of the single-use face mask produced 14.7 wt% gasoline, 18.4 wt% jet fuel, 34.1 wt% diesel, and 18.1 wt% hydrocarbons in the motor oil range. Through the pyrolysis of the disposable face mask, no solid char was created. Food waste was added to the pyrolysis feedstock to create char; however, the single-use face mask's presence had no effect on the char's characteristics or energy content. Co-feeding food waste with the disposable face mask during pyrolysis resulted in a greater production of H_2 and a decrease in hydrocarbons. The findings of the study can help with thermochemical waste management and usage.

8.4   CRITICAL DISCUSSION

When it comes to MWTT development, the limits in terms of MWTT performance improvement and commercial viability must be considered. It is recommended to use the latest professional technologies for MWT, stating that incineration appears to be the best option in terms of human health and environmental sustainability, but that incomplete plastics burning pollutes the air with furan, dioxin, and poly-chlorinated biphenyls (Okten et al., 2015). Comparing the financial impact of the autoclaving, converting, and ozonation MWTTs, on the contrary, found that the ozonation MW treatment technology was the most cost-effective. The cost investigation was carried out to predict the most environmentally conscious MWTTs, which included (i) autoclave, (ii) microwave, and (iii) lime disinfection technique, accompanied by landfilling/transportation. To restrict the transmission of COVID-19, researchers constructed a safe and efficient infectious MW reverse network infrastructure (Kargar et al., 2020). Researchers were able to reduce the network's operating expenses and hazards by dealing with a variety of infectious MWG healthcare organizations. Researchers devised a linear programming approach to reduce total expenditure, shipping issues, pathogenic MWT threats, and the largest limit of MW left unclaimed in outpatient clinics. Govindan et al. (2021) created a bi-objective mixed-integer linear programming model for MWM, thus reducing overall costs and population exposure to environmental concerns at the same time. The COVID-19 pandemic's MWM issues as well as national and international agencies' recommendations were investigated (Capoor and Parida, 2021). MW operating flow was also employed to solve the automated MW sorting software glitch, employing a mixed-integer programming interface for MW assignment, pre-sorting terminals, and advanced navigation automobile management (He et al., 2021). Creation of an innovative multi-objective was an optimal control strategy to help MW enterprises make better decisions by taking into account the economic, environmental, and social environmental considerations. Researchers hoped to reduce transportation, treatment, and construction costs, along with MW trucking ecological hazard identification, while also increasing work possibilities. While 15% of MW is radioactive, poisonous, or contagious, healthcare organizations have an expanding MWGR (Borowy, 2020). In terms of a multi-objective optimization model solution, researchers studied the scalability and plausibility of an improved the multi-choice goal algorithm (Torkayesh et al., 2021). For the three bottom optimal control problems of sustainably MW acquisition and transportation, Ghannadpour et al. (2021) proposed a unique evolutionary algorithm.

Researchers concentrated on the logistical components of the MWM system, taking into account present legal limits, organizational variables, and financial considerations (Rolewicz-Kalinska, 2016). The aims and limits for full-scale deployment must be included in the framework of the MWM system. A comparison of the various MWM practices in various nations and how they affect the procedures for collecting, storing, and disposing of hazardous MW was found (Bucataru et al., 2021). For recycling stainless steel MW and reusing outdated surgical instruments, Van Straten et al. (2021) studied the viability of the corporate sustainability idea. The MWM system's long-term viability depends on the effective MW logistics and a depth perception among MWT services. Taiwanese researcher (Tsai, 2021) examined the MW generation and the effect of COVID-19 on MWGR and quantity, which grew by 4.17% from 35,747 tons in 2016 to 40,407 tons in 2019. Arun and Wang showed how industry 4.0 may be used to reduce operational MW by studying the mechanism of hierarchical innovation levels. Researchers discovered that implementing the industry 4.0 idea in the healthcare industry helps improve resource efficiency, but additional research is needed to determine its effect on the development of the treatment regimen of MW (Arun and Wang, 2021). Ranjbari et al. (2022) focused on (i) mapping MW research and innovation, (ii) recognizing research concepts and (iii) establishing a MWM literature review in the perspective of the resource efficiency transitions and environment protection notions. The present MW disposal techniques and their consequences on the public health and the environment were examined (Kenny and Priyadarshini, 2021); the results showed that MWTTs rely heavily on basic and moderate MWTTs and that adoption of "greener" MWTTs is lacking due to cost, accessibility, and feasibility. Finding the best locations for the MW disposal centers, according to Yao et al. (2020), is the answer to the complicated connections between the various stakeholders. Researchers emphasized on (i) MW mitigation, protracted administration, and regulation; (ii) MW incineration and its environmental implications; (iii) hazardous MWMTs; and (iv) MW dealing and occupational health and safety and orientation. Finally, an integrated MW logistics operations bring down the cost, the potential for injury, and the time that it takes to dispose of MW safely and reliably, taking into consideration constraints in current premises' disposal functionality as a direct consequence of the COVID-19 pandemic (Mei et al., 2021).

8.5 CONCLUSION

The pyrolyzed plastic waste can be utilized for the arrangement of 13 sorts of fuel pellets. As an end, low-chlorine clean lignocellulosic biomass fuel pellets of high gross calorific worth can be effectively blended in with hydrothermally treated MW and pyrolytic plastic residue. This seems to be a splendid way to deal with joining three various types of pre-treated squander, i.e., MW, plastic build-up, and lignocellulosic biomass, in a typical management procedure, i.e., pelletizing, for solid fuel production (Alam et al., 2019). Energy (heat and power), fuels (gas, fluid, solid), and materials (adsorbents) can be created from MW, further, MW divisions involving plastic, cotton fiber, blood, slaver which can be utilized by different treatment innovations (Giakoumakis et al., 2021). The various processes ensure an alternative to effectively recover MW for commercial fuel and material production.

More research into the improvement of MW collection and management operations, as well as the allocation/location of MWT station deployment problems, should be undertaken. The classification and segregation of MW may be improved greatly, and the isolation of MW at the point of production (healthcare facilities, hospitals, etc.) can substantially boost the economic practicality of the subsequent stages in the production of energy, materials, and fuels. Innovative recycling and recovery strategies can better utilize the thermal energy generated throughout MWT using incineration and analogous technologies, minimizing the environmental impact of thermal pollution. The assessment of higher heating value values and chemical/physical properties of various MW fractions will aid in the continuous improvement and cost reduction. Substantial research needs to be done on the use of MW plastic or lignocellulosic proportion (cardboard/paper, cotton, textiles) conversion technologies in the generation of fuels and materials.

Future research aims at sustainable processing of MW generated from various allocations with effective management of thermal pollution. Innovative recycling and treatment techniques of MW can optimize the cost of production of value-added products like fuel, energy, and material production. The MWT and utilization as a value-added product seem to be economically feasible and most efficient in the production of optimized materialistic output generating zero waste.

REFERENCES

Aboughaly, M., Gabbar, H. A., Damideh, V., & Hassen, I. (2020). RF-ICP thermal plasma for thermoplastic waste pyrolysis process with high conversion yield and tar elimination. *Processes*, 8(3), 281. https://doi.org/10.3390/pr8030281.

Alam, M. T., Lee, J., Lee, S., Bhatta, D., Yoshikawa, K., & Seo, Y. (2019). Low chlorine fuel pellets production from the mixture of hydrothermally treated hospital solid waste, pyrolytic plastic waste residue and biomass. *Energies*, 12(22), 4390. https://doi.org/10.3390/en12224390.

Arun Kumar, P., & Wang, S. J. (2021). The design intervention opportunities to reduce procedural-caused healthcare waste under the industry 4.0 context - A scoping review. *Lecture Notes of the Institute for Computer Sciences, Social Informatics and Telecommunications Engineering*, 446–460. https://doi.org/10.1007/978-3-030-73426-8_27.

Borowy, I. (2020). Medical waste: The dark side of healthcare. *História, Ciências, Saúde-Manguinhos*, 27(suppl 1), 231–251. https://doi.org/10.1590/s0104-59702020000300012.

Bucătaru, C., Săvescu, D., Repanovici, A., Blaga, L., Coman, E., & Cocuz, M. (2021). The implications and effects of medical waste on development of sustainable society-A brief review of the literature. *Sustainability*, 13(6), 3300. https://doi.org/10.3390/su13063300.

Cahyanti, M. N., Doddapaneni, T. R., & Kikas, T. (2020). Biomass torrefaction: An overview on process parameters, economic and environmental aspects and recent advancements. *Bioresource Technology*, 301, 122737. https://doi.org/10.1016/j.biortech.2020.122737.

Capoor, M. R., & Parida, A. (2021). Biomedical waste and solid waste management in the time of COVID-19: A comprehensive review of the national and international scenario and guidelines. *Journal of Laboratory Physicians*, 13(02), 175–182. https://doi.org/10.1055/s-0041-1729132.

Chaiyat, N. (2021). Energy, exergy, economic, and environmental analysis of an organic Rankine cycle integrating with infectious medical waste incinerator. *Thermal Science and Engineering Progress*, 22, 100810. https://doi.org/10.1016/j.tsep.2020.100810.

Chen, F., Lou, J., Hu, J., Chen, H., Long, R., & Li, W. (2021). Study on the relationship between crisis awareness and medical waste separation behavior shown by residents during the COVID-19 epidemic. *Science of the Total Environment*, 787, 147522. https://doi.org/10.1016/j.scitotenv.2021.147522.

Chisholm, J. M., Zamani, R., Negm, A. M., Said, N., Abdel daiem, M. M., Dibaj, M., & Akrami, M. (2021). Sustainable waste management of medical waste in African developing countries: A narrative review. *Waste Management & Research: The Journal for a Sustainable Circular Economy*, 39(9), 1149–1163. https://doi.org/10.1177/0734242x211029175.

Das, A. K., Islam, M. N., Billah, M. M., & Sarker, A. (2021). COVID-19 pandemic and healthcare solid waste management strategy - A mini-review. *Science of the Total Environment*, 778, 146220. https://doi.org/10.1016/j.scitotenv.2021.146220.

Dharmaraj, S., Ashokkumar, V., Pandiyan, R., Halimatul Munawaroh, H. S., Chew, K. W., Chen, W., & Ngamcharussrivichai, C. (2021). Pyrolysis: An effective technique for degradation of COVID-19 medical wastes. *Chemosphere*, 275, 130092. https://doi.org/10.1016/j.chemosphere.2021.130092.

Duarte, P. M., & Santana, V. T. (2020). Disinfection measures and control of SARS-COV-2 transmission. *Global Biosecurity*, 1(3). https://doi.org/10.31646/gbio.64.

Erdogan, A. A., & Yilmazoglu, M. Z. (2021). Plasma gasification of the medical waste. *International Journal of Hydrogen Energy*, 46(57), 29108–29125. https://doi.org/10.1016/j.ijhydene.2020.12.069.

Fadaei, A. (2021). Study of solid waste (municipal and medical) management during the COVID-19 pandemic: A review study. *Reviews on Environmental Health*, 0(0). https://doi.org/10.1515/reveh-2021-0092.

Fang, S., Jiang, L., Li, P., Bai, J., & Chang, C. (2020). Study on pyrolysis products characteristics of medical waste and fractional condensation of the pyrolysis oil. *Energy*, 195, 116969. https://doi.org/10.1016/j.energy.2020.116969.

Ghannadpour, S. F., Zandieh, F., & Esmaeili, F. (2021). Optimizing triple bottom-line objectives for sustainable health-care waste collection and routing by a self-adaptive evolutionary algorithm: A case study from Tehran province in Iran. *Journal of Cleaner Production*, 287, 125010. https://doi.org/10.1016/j.jclepro.2020.125010.

Giakoumakis, G., Karnaouri, A., Topakas, E., & Sidiras, D. (2021). Simulation and optimization of combined acid pretreatment and enzymatic saccharification of medical cotton waste. *Biomass Conversion and Biorefinery*, 11(2), 515–526. https://doi.org/10.1007/s13399-020-00694-1.

Govindan, K., Nasr, A. K., Mostafazadeh, P., & Mina, H. (2021). Medical waste management during coronavirus disease 2019 (COVID-19) outbreak: A mathematical programming model. *Computers & Industrial Engineering*, 162, 107668. https://doi.org/10.1016/j.cie.2021.107668.

Hanoğlu, A., Çay, A., & Yanık, J. (2019). Production of biochars from textile fibres through torrefaction and their characterisation. *Energy*, 166, 664–673. https://doi.org/10.1016/j.energy.2018.10.123.

Hantoko, D., Li, X., Pariatamby, A., Yoshikawa, K., Horttanainen, M., & Yan, M. (2021). Challenges and practices on waste management and disposal during COVID-19 pandemic. *Journal of Environmental Management*, 286, 112140. https://doi.org/10.1016/j.jenvman.2021.112140.

He, X., Quan, H., Lin, W., Deng, W., & Tan, Z. (2021). AGV scheduling optimization for medical waste sorting system. *Scientific Programming*, 2021, 1–12. https://doi.org/10.1155/2021/4313749.

Health-care waste. Who.int. (2022). Retrieved 3 April 2022, from https://www.who.int/news-room/fact-sheets/detail/health-care-waste.

Hereher, M. E., Al-Awadhi, T., & Mansour, S. A. (2020). Assessment of the optimized sanitary landfill sites in Muscat, Oman. *The Egyptian Journal of Remote Sensing and Space Science*, 23(3), 355–362. https://doi.org/10.1016/j.ejrs.2019.08.001.

Ilyas, S., Srivastava, R. R., & Kim, H. (2020). Disinfection technology and strategies for COVID-19 hospital and bio-medical waste management. *Science of the Total Environment*, 749, 141652. https://doi.org/10.1016/j.scitotenv.2020.141652.

Kalantary, R. R., Jamshidi, A., Mofrad, M. M., Jafari, A. J., Heidari, N., Fallahizadeh, S., Hesami Arani, M., & Torkashvand, J. (2021). Effect of COVID-19 pandemic on medical waste management: A case study. *Journal of Environmental Health Science and Engineering*, 19(1), 831–836. https://doi.org/10.1007/s40201-021-00650-9.

Kareem, L., S., Al-Mamoori, S. K., Al-Maliki, L. A., Al-Dulaimi, M. Q., & Al-Ansari, N. (2021). Optimum location for landfills landfill site selection using GIS technique: Al-naja city as a case study. *Cogent Engineering*, 8(1). https://doi.org/10.1080/23311916.2020.1863171.

Kargar, S., Pourmehdi, M., & Paydar, M. M. (2020). Reverse logistics network design for medical waste management in the epidemic outbreak of the novel coronavirus (COVID-19). *Science of the Total Environment*, 746, 141183. https://doi.org/10.1016/j.scitotenv.2020.141183.

Kenny, C., & Priyadarshini, A. (2021). Review of current healthcare waste management methods and their effect on global health. *Healthcare*, 9(3), 284. https://doi.org/10.3390/healthcare9030284.

Lee, S. B., Lee, J., Tsang, Y. F., Kim, Y., Jae, J., Jung, S., & Park, Y. (2021). Production of value-added aromatics from wasted COVID-19 mask via catalytic pyrolysis. *Environmental Pollution*, 283, 117060. https://doi.org/10.1016/j.envpol.2021.117060.

Letho, Z., Yangdon, T., Lhamo, C., Limbu, C. B., Yoezer, S., Jamtsho, T., Chhetri, P., & Tshering, D. (2021). Awareness and practice of medical waste management among healthcare providers in national referral hospital. *PLoS One*, 16(1), e0243817. https://doi.org/10.1371/journal.pone.0243817.

Liu, Z., Liu, T., Liu, X., Wei, A., Wang, X., Yin, Y., & Li, Y. (2021). Research on optimization of healthcare waste management system based on green governance principle in the COVID-19 pandemic. *International Journal of Environmental Research and Public Health*, 18(10), 5316. https://doi.org/10.3390/ijerph18105316.

Maalouf, A., & Maalouf, H. (2021). Impact of COVID-19 pandemic on medical waste management in Lebanon. *Waste Management & Research: The Journal for a Sustainable Circular Economy*, 39(1_suppl), 45–55. https://doi.org/10.1177/0734242x211003970.

Manegdeg, F., Coronado, L. O., & Paña, R. (2020). Medical waste treatment and electricity generation using pyrolyzer-Rankine cycle for specialty hospitals in Quezon City, Philippines. *IOP Conference Series: Earth and Environmental Science*, 463(1), 012180. https://doi.org/10.1088/1755-1315/463/1/012180.

Mantzaras, G., & Voudrias, E. A. (2017). An optimization model for collection, haul, transfer, treatment and disposal of infectious medical waste: Application to a Greek region. *Waste Management*, 69, 518–534. https://doi.org/10.1016/j.wasman.2017.08.037.

Matsuzawa, Y., Ayabe, M., Nishino, J., Kubota, N., & Motegi, M. (2004). Evaluation of char fuel ratio in municipal pyrolysis waste. *Fuel*, 83(11–12), 1675–1687. https://doi.org/10.1016/j.fuel.2004.02.006.

Mei, X., Hao, H., Sun, Y., Wang, X., & Zhou, Y. (2021). Optimization of medical waste recycling network considering disposal capacity bottlenecks under a novel coronavirus pneumonia outbreak. *Environmental Science and Pollution Research*. https://doi.org/10.1007/s11356-021-16027-2.

Mekonnen, B., Solomon, N., & Wondimu, W. (2021). Healthcare waste status and handling practices during COVID-19 pandemic in Tepi General Hospital, Ethiopia. *Journal of Environmental and Public Health*, 2021, 1–7. https://doi.org/10.1155/2021/6614565.

Minoglou, M., Gerassimidou, S., & Komilis, D. (2017). Healthcare waste generation worldwide and its dependence on socio-economic and environmental factors. *Sustainability*, 9(2), 220. https://doi.org/10.3390/su9020220.

MohdFaizal, H., Shamsuddin, H. S., M. Heiree, M. H., Muhammad Ariff Hanaffi, M. F., Abdul Rahman, M. R., Rahman, M. M., & Latiff, Z. (2018). Torrefaction of densified mesocarp fibre and palm kernel shell. *Renewable Energy*, 122, 419–428. https://doi.org/10.1016/j.renene.2018.01.118.

Mühlich, M. (2003). Comparison of infectious waste management in European hospitals. *Journal of Hospital Infection*, 55(4), 260–268. https://doi.org/10.1016/j.jhin.2003.08.017.

Nik Ab Rahim, N. N., Othman, J., Hanim Mohd Salleh, N., & Chamhuri, N. (2021). A non-market valuation approach to environmental cost-benefit analysis for sanitary landfill project appraisal. *Sustainability*, 13(14), 7718. https://doi.org/10.3390/su13147718.

Okten, H. E., Corum, A., & Demir, H. H. (2015). A comparative economic analysis for medical waste treatment options. *Environment Protection Engineering*, 41(3). https://doi.org/10.37190/epe150310.

Ozbay, G., Jones, M., Gadde, M., Isah, S., & Attarwala, T. (2021). Design and operation of effective landfills with minimal effects on the environment and human health. *Journal of Environmental and Public Health*, 2021, 1–13. https://doi.org/10.1155/2021/6921607.

Park, C., Choi, H., Andrew Lin, K., Kwon, E. E., & Lee, J. (2021). COVID-19 mask waste to energy via thermochemical pathway: Effect of Co-feeding food waste. *Energy*, 230, 120876. https://doi.org/10.1016/j.energy.2021.120876.

Ranjbari, M., Shams Esfandabadi, Z., Shevchenko, T., Chassagnon-Haned, N., Peng, W., Tabatabaei, M., & Aghbashlo, M. (2022). Mapping healthcare waste management research: Past evolution, current challenges, and future perspectives towards a circular economy transition. *Journal of Hazardous Materials*, 422, 126724. https://doi.org/10.1016/j.jhazmat.2021.126724.

Rasul, S. B., Som, U., Hossain, M. S., & Rahman, M. W. (2021). Liquid fuel oil produced from plastic based medical wastes by thermal cracking. *Scientific Reports*, 11(1). https://doi.org/10.1038/s41598-021-96424-2.

Rolewicz -Kalinska, A. (2016). Logistic constraints as a part of a sustainable medical waste management system. *Transportation Research Procedia*, 16, 473–482. https://doi.org/10.1016/j.trpro.2016.11.044.

Song, Y., Ye, J., Liu, Y., & Zhong, Y. (2021). Estimation of solid medical waste production and environmental impact analysis in the context of COVID-19: A case study of Hubei province in China. https://doi.org/10.20944/preprints202104.0327.v1.

Śiechowski, K., Leśniak, M., & Białowiec, A. (2021). Medical peat waste upcycling to carbonized solid fuel in the torrefaction process. *Energies*, 14(19), 6053. https://doi.org/10.3390/en14196053.

Torkayesh, A. E., Vandchali, H. R., & Tirkolaee, E. B. (2021). Multi-objective optimization for healthcare waste management network design with sustainability perspective. *Sustainability*, 13(15), 8279. https://doi.org/10.3390/su13158279.

Tsai, W. (2021). Analysis of medical waste management and impact analysis of COVID-19 on its generation in Taiwan. *Waste Management & Research: The Journal for a Sustainable Circular Economy*, 39(1_suppl), 27–33. https://doi.org/10.1177/0734242x21996803.

Van Straten, B., Dankelman, J., Van der Eijk, A., & Horeman, T. (2021). A circular healthcare economy; A feasibility study to reduce surgical stainless steel waste. *Sustainable Production and Consumption*, 27, 169–175. https://doi.org/10.1016/j.spc.2020.10.030.

Vignesh, N., & Chandraraj, K. (2021). Improved high solids loading enzymatic hydrolysis and fermentation of cotton microdust by surfactant addition and optimization of pretreatment. *Process Biochemistry*, 106, 60–69. https://doi.org/10.1016/j.procbio.2021.04.002.

Voudrias, E., & Graikos, A. (2014). Infectious medical waste management system at the regional level. *Journal of Hazardous, Toxic, and Radioactive Waste*, 18(4), 04014020. https://doi.org/10.1061/(asce)hz.2153-5515.0000225.

Wang, J., Shen, J., Ye, D., Yan, X., Zhang, Y., Yang, W., Li, X., Wang, J., Zhang, L., & Pan, L. (2020). Disinfection technology of hospital wastes and wastewater: Suggestions for disinfection strategy during coronavirus disease 2019 (COVID-19) pandemic in China. *Environmental Pollution*, 262, 114665. https://doi.org/10.1016/j.envpol.2020.114665.

Xin, S., Huang, F., Liu, X., Mi, T., & Xu, Q. (2019). Torrefaction of herbal medicine wastes: Characterization of the physicochemical properties and combustion behaviors. *Bioresource Technology*, 287, 121408. https://doi.org/10.1016/j.biortech.2019.121408.

Xu, L., Dong, K., Zhang, Y., & Li, H. (2020). Comparison and analysis of several medical waste treatment technologies. *IOP Conference Series: Earth and Environmental Science*, 615(1), 012031. https://doi.org/10.1088/1755-1315/615/1/012031.

Yao, L., Xu, Z., & Zeng, Z. (2020). A soft-path solution to risk reduction by modeling medical waste disposal center location-allocation optimization. *Risk Analysis*, 40(9), 1863–1886. https://doi.org/10.1111/risa.13509.

Yousef, S., Eimontas, J., Striūgas, N., & Abdelnaby, M. A. (2022). A new strategy for butanol extraction from COVID-19 mask using catalytic pyrolysis process over ZSM-5 zeolite catalyst and its kinetic behavior. *Thermochimica Acta*, 711, 179198. https://doi.org/10.1016/j.tca.2022.179198.

Zhao, H., Wang, L., Liu, F., Liu, H., Zhang, N., & Zhu, Y. (2021). Energy, environment and economy assessment of medical waste disposal technologies in China. *Science of the Total Environment*, 796, 148964. https://doi.org/10.1016/j.scitotenv.2021.148964.

9 Control of Transfrontier Movement of Hazardous Waste
Africa the Final Destination?

M. B. Gasu, G. N. Gasu, and E. J. Mbeng

CONTENTS

9.1 Introduction 145
9.2 Literature Review 146
    9.2.1 What Is Hazardous Waste? 146
9.3 Methodology 148
9.4 Discussions 148
    9.4.1 The Illegal Trade in Hazardous Waste, Poverty, and
            Human Rights Violations: Africa the Destination? 148
    9.4.2 Global Regimes Combating Transfrontier Shipment of Toxic Wastes 151
            9.4.2.1 Global Legal Regimes 151
            9.4.2.2 The Control of Hazardous Waste in Nigeria 154
9.5 Conclusion 157
9.6 Recommendations 158
References 159

9.1 INTRODUCTION

The biosphere is instrumental for life on earth, which is the source of nutrients to plants, forests, mineral resources, animals, and people. Without it, human survival is ultimately at stake. This constitutes the physical environment in which life thrives. The physical elements of the environment include the soil, water, and air (Thornton and Beckwith, 1997; NESREA, 2007). In fulfillment of God's command "to have dominion over all the creatures and subdue the earth" (The Holy Bible King James Version), human activities are now a source of danger to his very existence in most parts of the world. As a result, there has been a rise in air pollution and environmental contamination from hazardous and other wastes.

The world inhabited by humans has become a place of massive production of goods and services with a high level of technology. Waste generation, a by-product of industrial processing and economic development resulting from the production and consumption of industrial and agricultural products, has had extremely negative

DOI: 10.1201/9781003359784-9

consequences on the human environment. The global generation of hazardous waste is believed to have increased 60fold since World War II, from about 5 million tons in 1945 to 300 million tons in 1988 (Hunter et al., 2002) and according to research by UNEP (United Nations Environment Programme), over 400 million tons each year, equating to around 60 kg per person (UNEP, 2019; The world Counts, 2019). It is on record that all over the world, close to 400 million tons of toxic waste is produced every year, and a greater proportion crosses international boundaries (Gasu, 2008). Tons of hazardous substances such as corrosive acids, organic compounds, hazardous metals, and other wastes pose a serious danger to human health by contaminating groundwater, causing leaching and other forms of pollution. The higher the volume of hazardous waste generated, the higher the likely increased the transshipment of such waste and the difficulties of disposing of it in the best practicable mode.

Since the atomic bombing of Japan during World War II, science has found out that creating those bombs for peaceful uses and in the development of more sophisticated military weapons poses threats to the human and their environment. In these instances, the problem of disposal of waste from such sources has remained with manufacturers for the simple reason that nuclear waste, for instance, is just as dangerous to handle as nuclear weapons. More industrialized countries, such as the United States and some European Nations, are the world's largest generators of toxic waste. Nigeria is said to be Africa's leading producer of hazardous waste, with an annual output of 2,469,000 tons (Akpan and Olukanni, 2020). However, it has been estimated that Africa generates a per capita hazardous waste of about 20.1 kg per person per year (Zangina and Ali, 2021). In ancient times, human beings had no problem disposing of their waste, given that their level of economic development used a natural mechanism for waste disposal, which gave them the freehand to manufacture weapons and dispose of the waste (Uchegbu, 1990). It has been observed that because of the repercussions on the environment, socioeconomics, and public health, hazardous waste management is extremely important. Environmental issues resulting from dealing with toxic materials in underdeveloped countries were not brought to the fore until the late 1980s when many incidences of dumping in Africa were documented (Lipman, 2002; Uwagbale, 2016; Akpan and Olukanni, 2020).

9.2 LITERATURE REVIEW

9.2.1 What Is Hazardous Waste?

Black's Law Dictionary defines hazardous waste as "waste that may cause or significantly contribute to an increase in mortality or injury to human health or the environment due to its quantity, concentration, or physical, chemical, or infectious qualities. Toxic waste is defined as dangerous, poisonous compounds such as DDT" (Garner, 2004). Per the OECD (Organisation for Economic Cooperation and Development, 2022), "transfrontier movement of hazardous wastes refers to the transfer of toxic substances from a nation to another whereby the material is classified as hazardous waste in at least one of the nations involved."

"The Resource Conservation and Recovery Act" of 1976 defines toxic waste in the United States as "any solid waste that, when improperly treated, stored, or disposed of, may cause or contribute significantly to an increase in mortality, contribute to irreversible or incapacitating reversible illness, or pose a significant presence or potential hazard on human health or the environment due to its concentration, quantity, physical, chemical, or infectious character" (RCRA, 1976 S.1004(5)). The Clean Air Act defines hazardous air pollutants as risks for adverse health impacts such as carcinogenic, mutagenic, and neurotoxic compounds, or as those that induce adverse environmental effects through bioaccumulation or deposition (Guruswamy and Hendrick, 2001).

The material will be classified as hazardous under the Basel Convention if regulated domestically (Okorodudu-Fubara, 1998). Similarly, the Basel Convention defines hazardous wastes as anything that is explosive, combustible, oxidizing, poisonous, infectious, corrosive, or toxic or any substances that are explosive, ecotoxic, or capable of creating another material with any of the above qualities after disposal. In addition to contaminants that possess these characteristics, several conventions such as the Basel Convention (Article 1 and Annex I) and the Bamako Convention (Article 2 Annex I) contain a list of wastes that have previously been identified as hazardous, including medical waste, organic, chemical, hydrocarbon, or radioactive waste and material that contain traces of heavy metals.

Schedules 12 and 13 of the "National Environmental Protection Management of Solid and Hazardous Waste Regulation in Nigeria" include the criteria for recognizing regulated hazardous waste. Toxic substances such as waste that falls under or exhibits the preset "listed characteristics described in FAC 000-000-99903 and FAC 000-000-99904" (Amokaye, 2004; Sodipo et al., 2017). The unique features of toxic substances include "explosive, corrosive, oxidizing, highly combustible, irritant, harmful, toxic, carcinogenic, infectious, teratogenic, and mutagenic. It also includes substances that release toxic gases or substances capable of yielding another substance or exotoxin" (Uwagbale, 2016; Zangina and Ali, 2021; Kummer 1999). According to the "Harmful Waste (Special Criminal Provisions, etc.) Act 1988", hazardous wastes are flammable, corrosive, chemically reactive, toxic, radioactive, or other properties are considered hazardous (Suleiman et al., 2020).

Several international accords describe hazardous chemicals, and at least four ways can be discerned. The most prevalent approach classifies hazardous substances and activities based on their inherent properties such as toxicity, flammability, explosiveness, or oxidization (Sands, 2003, FEPA, 1991, Uwagbale, 2016; USEPA, 2005; Grasso et al., 2010). A second technique classifies activities by listing specific items expected to have major environmental impacts on their own (USEPA, 2005; Ibrahim et al., 2018). A third approach uses national legislation to define hazardous compounds (FEPA, 1991; USEPA, 2005). Finally, a fourth strategy (which is becoming more popular) is shown in attempts that do not seek to develop broad categories, but rather control individual compounds ("1985 Vienna Convention and the 1987 Montreal Protocol"). In light of this, the research was conceived with the objective of critically examining the global and national regimes on the control of the transfrontier movement of hazardous waste and its effects on the human environment. It also assesses the impact of the illegal trade on hazardous waste in Africa.

9.3 METHODOLOGY

The study made use of primary and secondary sources of data. The primary data employed legislations or statutes such as NESREA, 2007; Waste and Hazardous Regulations 1991; the 1999 Constitution; Treaties and Conventions such as the Basel Convention, 1989; the Bamako Convention; UNCLOS III, 1982; and the Vienna Convention and Montreal Protocol, 1987. Similarly, secondary data were sourced from books, chapters in books, journal articles, periodicals, newspapers, magazines, Internet sources, conference proceedings, published and unpublished thesis/dissertations, and seminar papers that relate to the subject matter. This study, therefore, employed a synergy of the analytical, descriptive, and prescriptive methods. The analytical method was used to analyze the international and national legal frameworks regulating the transfrontier movements of hazardous waste. The objective of this approach was to ascertain the weaknesses inherent in the international system to monitor the movement of hazardous waste correctly. On the contrary, the descriptive method was employed to expose the corrupt practices of authorities of developing countries that have made Africa to be seen as a dumping site. Finally, the prescriptive method sort of advances best practices in managing the transfrontier movement of hazardous waste, which could create a sustainable human environment.

9.4 DISCUSSIONS

9.4.1 The Illegal Trade in Hazardous Waste, Poverty, and Human Rights Violations: Africa the Destination?

The persistent increase in the trade in hazardous waste, despite the public outcry on its implications on the human environment, is multifold. In the early 1980s, the serious efforts toward ecological governance in industrialized nations have increased. Due to strict regulations on the environment in industrialized nations, Africa has become the destination for such trash disposal (Viljoen, 2007; Uwagbale, 2016; Akpan and Olukanni, 2020). The shipment of thousands of tons of mercury into South Africa from the United Kingdom and the United States resulted in the dumping of 1,000–1,900 times higher than WHO-acceptable limits, which was a serious threat to human health (Frey, 1994).

Second, the rising expense of transportation, treatment, and elimination of toxic substances has prompted waste dealers to send the garbage to underdeveloped countries as a less expensive alternative destination. It has been reported that in developing countries, environmental preservation is frequently sacrificed to attain quick economic gains (Barresi, 2020). The countries of the north have taken advantage of the poverty-stricken nature of some countries, such as China, in the early 1990s, and allowed containers transporting waste products for a cost of USD 50 per ton to be dumped in their territory (Sthiannopkao and Wong, 2013). Similar examples in Africa include an Italian company dumping toxic garbage in the Koko fishing town in Nigeria in 1987 (Sthiannopkao and Wong, 2013; Frey, 1994; Perkins et al., 2014; Anyinam, 1991; Gevao et at., 2010; Ibrahim et al., 2018). In the Koko Port scenario, a landowner agreed with an Italian to dump dangerous radioactive waste at the port in exchange for a pitiable stipend (Lipman, 2002). Therefore, the discrepancy in disposal costs between rich and developing countries, according to Hunter et al. (2002) and Akpan and Olukanni (2020), is a primary reason that has fueled the transboundary shipment of trash. Hazardous waste disposal in industrialized countries can cost up to US$ 2,000 for each tonnage compared to $40 per ton in Africa (Viljoen, 2007). The disposal of toxic substances in the OECD nations cost between $100 and $2,000 per ton in 1988 compared to $2.50 and $50 per ton in poverty-emerging nations (Lipman, 2002; Kummer 1999). According to Du Vivier (1988), trash disposal in impoverished countries could cost between US$3 and US$40 per ton compared to US$75 and US$300 in developed countries. In the United Kingdom, the cost of incineration, according to Lipman (2002), might be as high as US$ 10,000 per ton.

Third is the issue of poverty, which is also associated with Africa (Akpan and Olukanni, 2020). According to Kone (2010), Africa is vulnerable to the waste trade's uneven economics since it is home to most poverty-stricken nations, the majority of which are in dire need of foreign exchange. Historically, Africa has been the supplier of raw materials, from which Western nations extrapolate wealth and resources. Even after these resources have served their function, they face the risk of swallowing rubbish generated by their resources (Hunter et al., 2002). In light of the above, Guinea-Bissau in 1988 was paid $600 million to dispose of 15 million tons of toxic garbage for 5 years (Zada, 2019). The contract sum was estimated to be four times higher than that country's GDP (Zada, 2019); although it did not go through due to public concerns, this shows the country is poor and vulnerable.

Fourth, most of the waste is being exported under the disguise of foreign aid, popularly known as cooperation and development aid or, at times, marked as fertilizer. Sometimes, it is arranged contractual terms between two governments of developed and developing countries' states. An example that is still very fresh in our minds is the case between the Republic of Benin and France, where the government of Benin signed an agreement to dispose of hazardous waste for 30 years in exchange for aid. It also causes concern that when rich countries deem hazardous waste treatment too toxic or unprofitable, they ship it to Africa or Asia under recycling. According to Green Peace, "up to 167 million tons of hazardous waste has found a new home in Africa, with a subsidiary of the French corporation Arcelor Mittal suspected of illegally shipping millions of tons of toxic trash (in the form of fuel tankers) between 1993 and 2004" (Bernstorff and Stairs, 2000). As a result, it is becoming commonplace to see ships from the United States and Europe offloading consignments of used electronics spearheaded by corrupt and unscrupulous individuals who accept, despite lack of technical competence and ignore, the environmental and human health consequences of such deals (Cobbling, 1992; Kone, 2010; Akpan and Olukanni, 2020).

The OAU (Organisation of African Unity, now the African Union) overwhelmingly endorsed a resolution on 23 May 1988, proclaiming that dumping radioactive substances on the African continent constitutes an offence against humanity (Kone, 2010). Similarly, in 1968, the United Nations General Assembly recognized the relationship linking fundamental human rights and environmental sanity (UNGA, 1968). The disposal of hazardous substances and other toxic materials over international borders may jeopardize fundamental human rights and a conducive work and habitable environment. The example of Ivory Coast, where toxic waste was dumped in 2006, resulting in adverse effects on the human population, is an example of a grave human rights violation (Kone, 2010).

Over the last two decades, the shipment of hazardous substances into emerging nations has grown to become a profitable venture, where trash dealers and criminal organizations earn up to US$10–12 billion every year. According to Massari and Monzini (2004), a US-based National Intelligence Council estimates the illegal trafficking of hazardous waste to be over 12–15 billion Euros (Akpan and Olukanni, 2020). It has also been noticed that organized criminal gangs have increasingly taken over the booming illegal business of waste peddling to export and import used and worthless materials (SESN, 2009). When the cost of hazardous waste disposal skyrocketed in the 1980s, the nations of the north took advantage of third world nations as cheap dumpsites to dispose of tons of industrial waste (Jennifer, 1994). In the nations of the north, the cost of disposing of a ton of toxic substances is about $2,000, as against $40 a ton on the African continent (Hunter et al., 2002). While developing countries required the revenue generated by taking hazardous trash from the West, they were almost always ill-equipped. They lacked the technical expertise to dispose of such material environmentally responsible. Many developing countries' natural geography and climate rendered hazardous waste more dangerous to handle than industrialized countries; for example, frequent rainfall in sub-Saharan Africa causes landfilled garbage to leach into subterranean water supplies swiftly. LDCs are also more likely than the industrialized world to come into contact with dumpsites. Hundreds of young people scour these dumpsites for goods such as bottles, cans, plastics, metals, and other things to sell to make a living. As a result, landfills are frequently placed near the poorest, who frequent them in quest of these objects to use or sell.

Moreover, the trend for the transborder movement of toxic waste was also a result of a weak legal framework in the LDCs compared with the developed countries. The most notorious are exports to developing countries that are falsely concealed as harmful. Despite their technical and other ineptitude, many underdeveloped nations south of the Sahara have entered the illegal hazardous waste trade where they have imported or received loads of toxic substances at ridiculous prices of $3–40 a ton as against $75–300 per ton in developed nations (Kone, 2010).

Some of the well-documented cases of hazardous waste export to African countries include the Koko waste dump of 1987 in Nigeria by an Italian firm. The French dumped toxic waste in the Republic of Benin in Exchange for 30 years of special financial assistance to the country. The Republic of was paid $1.6 million for receiving hazardous wastes such as radioactive waste and dangerous substances, which were supposed to be buried, burnt, or repurposed (Cobbling, 1992). The Ivory Coast waste dump of 19 August 2006 and the Mombasa Kenya experience of March 2008 are among the well-documented cases. According to BBC News Night's Meirion and Liz (2007), the Ivory Coast incident was the largest hazardous dumping disaster in recent times (Meirion, and Liz, 2007; Comaroff, 2007; CNN.com, 2009; SESN, 2009; Kone, 2010). The tanker ship Probo Koala, hired by Trafigura, was involved in the Abidjan incident, transporting and dumping nearly 500 tons of toxic garbage in the West African city. According to investigations, the hazardous waste, which was initially described as slops but later revealed to be a mixture was supposed to be disposed of in Amsterdam in July 2006 for around $2,50,000. Nonetheless, Trafigura chose the less expensive option and offloaded it in Abidjan. The Abidjan incident resulted in deaths of 20 people, 69 people being hospitalized, over 1,08,000 people seeking medical assistance, and other premature deaths. According to Massari and Monzini in 2004 the unlawful trafficking of hazardous waste generated 15 billion Euros.

9.4.2   GLOBAL REGIMES COMBATING TRANSFRONTIER SHIPMENT OF TOXIC WASTES

Several regimes exist for combating transfrontier shipment of toxic waste at global and national levels.

9.4.2.1   Global Legal Regimes

1. **The Stockholm Conference**
   This Conference, the first coordinated attempt at an international scale to address environmental issues, was held in Stockholm in 1972. The 114 participating states proclaimed that "We have reached a historical turning point in which we must frame our behaviors around the world with greater consideration for their environmental repercussions." The thinking, ideology, and culture of concern for the environment around the world shaped the creation of international environmental law, leading up to the Stockholm Conference (Dokun, 1995; Uwagbale, 2016; Suleiman et al., 2020). The conference deliberated on action plans consisting of 160 recommendations and produced a declaration of 26 principles on the environment. The principles addressed citizens' and government's rights and responsibilities regarding environmental preservation and enhancement. It is worthy of note that virtually all the principles were geared toward protecting the human environment of particular attention to this topic is the control of the transborder shipment and dumping of toxic substances. Sequels with this, principle 6 states as follows:

   > The discharge of toxic substances or other substances and release of heat, in such quantities or concentrations as to exceed the capacity of the environment to render them harmful, must be halted in order to ensure that serious irreversible damage is not inflicted upon ecosystems.

2. **The Rio and Johannesburg Conferences**
   The Rio Conference was held in Rio de Janeiro, Brazil, in 1992 and is widely regarded as the biggest global-level conference ever conducted. It resulted in four important institutional outcomes: the Declaration of Rio de Janeiro on Environment and Development (UN Doc A/Conf. 151/26 1992 154), the ceremonial signing of the Climate Change and Biodiversity Conventions, and Agenda 21, the Soft Authoritative Regulations on Forest Preservation

Chapter 19 on environmentally sound management of toxic chemicals, including prevention of illegal international traffic in toxic and dangerous products, is one of two chapters of Agenda 21 dealing with hazardous substances ecological best practices discussed in Chapter 16 of Agenda 21. By the year 2000, all countries must create hazardous waste treatment and disposal requirements and acquire the ability to monitor the environmental impact of hazardous waste, according to Agenda 21 Chapter 21. The goal for developing countries was to meet the 2005 deadline. By 2025, all countries would dispose of all waste according to international quality guidelines as stipulated by Agenda 21. The 800-page ambitious sustainable development program emphasizes environmental best practices in waste management.

The Plan of Implementation of the WSSD renews these Agenda 21 commitments to use and manufacture chemicals in ways that have the fewest substantial negative impact on human wellbeing and his immediate surroundings by 2020. The general goal of Principle 14 of the Rio Declaration has been expounded on in both Agenda 21 and practicalizing the WSSD line of action. It thus states as follows:

> States should effectively cooperate to discharge or prevent the relocation and the transfer to other states of any activities or substances that cause severe environmental degradation or are found to be harmful to human health.

3. **UN Convention on the Law of the Sea 1982**
   The 1982 UN Convention on the Law of the Sea (UNCLOS III), which became operational on 16 November 1994, is the most powerful and all-encompassing environmental convention currently in force with a foreseeable ambitious plan. It is the oceans' constitution and the characterisitic of each of its 59 environmental protection and conservation provisions. UNCLOS III combines a broad codification of international law and other substantive law rules in a single document, instead of separating issues out by way of additional protocols. UNCLOS III is an umbrella treaty that encompasses a variety of international norms, regulations, and implementing bodies, including the 1972 London Convention, the 1958 UNCOS I, and the MARPOL 73/78 (Birnie and Boyle, 1992).

4. **Basel Convention 1989**
   UNEP began to combat the transboundary movement in the early 1980s, particularly the illegal international trafficking of hazardous waste. After 2 years of talks involving 116 nations, the "Basel Convention on the Control of the Transboundary Movement of Hazardous Waste and their Disposal was adopted in 1989" at a summit called under the auspices of UNEP. On 5 May 1992, the convention went into effect (Suleiman et al., 2020). Like the Cairo Guidelines, the Basel Convention originally adopted a broad management trade approach on hazardous waste, allowing all transboundary movements based on prior informed consent. Opponents of the Basel Convention did little to restrict trade and instead functioned more as a tracking system for continued transfer to developing countries (Hughes, 1996).

According to the convention, toxic substances must be disposed of in an internationally approved sound manner (Uwagbale, 2016), which includes the promotion of the use of clean and innovative technologies (Suleiman et al., 2020). Best international practices in disposing of toxic waste in a sound manner require that human wellbeing and the environment are protected from the harmful effects that such wastes may pose (Article 4(2)). All parties should minimize waste at the point of generation to prevent pollution and its consequences (Article 4(2) (a)).

The transshipment of toxic waste and other waste as defined in Article 9 without notification, consent, deception, or fraud, resulting in the dumping of hazardous waste in violation of the convention, is known as illegal waste trafficking. The statute also states that in the event of an illegal waste deposit, the importing country must take disposition to eliminate such waste through the best practicable means in 30 days of the act of importation of such wastes brought to the importer's attention and that the parties can cooperate as needed. In fact, this Article 9(3) was really a trap for the innocent and vulnerable African countries that do not really possess the technology to deal with such waste. For this reason, Ghana expressed a reservation on Article 9 (3) (Okorodudu-Fubara, 1998).

5. **Bamako Convention**

The Bamako Convention appears as a negation of the policy, principles, and objectives behind the resolution reached at the 48th Ordinary Session meeting of the Council of Ministers of the former OAU (now African Union) in Addis Ababa, Ethiopia, in 1988. The Convention, though born out of disapproval of the Basel Convention managed trade approach, the Bamako Treaty, in fact, closely followed its predecessor in most respects. The Bamako Convention differs from the Basel Convention in that it prevents non-signatories from transporting hazardous waste into Africa. The difference was narrowed considerably, given the recent amendment of the Basel Convention involving OECD and EU nations in 1995.

The convention exhibits weakness in drafting and substance. It lacks a well-articulated focus apart from the hotchpotch expression to foster an intercontinental movement of hazardous waste. It is incomprehensive in several aspects: non-amendable to the rules of legal interpretation and contains several loopholes and inconsistencies. In fact, the Bamako Convention leaves a lacuna, which continues to foster the illegal trade in hazardous waste. The convention has engaged continuous traffic of waste into Africa. The Bamako Convention goes a long way to supplement, rather than restore, the status quo (Okorodudu-Fubara, 1999). The Basel Convention expressly permits a party to impose "additional requirements" to enable it to uphold the ideals of the human environment for healthy living. By contrast, "the Bamako Convention" permits all the stakeholders to be involved in other contractual relationships, provided they are not in conflict with the international acceptable requirements for hazardous waste disposal.

9.4.2.2   The Control of Hazardous Waste in Nigeria

Controlling hazardous waste and substances that pose serious threats to human wellbeing and their immediate milieu is a relatively new phenomenon in Nigeria. Until the public outcry following the dumping of rubbish at the Port of Koko, Edo State, in 1988, hazardous waste management and disposal were not a major public health and environmental concern (Ikhariale, 1989). In Nigeria, the laws, regulations, guidelines, and standards put in place to reduce and/or eliminate hazardous waste generation that marked the beginning of environmental consciousness among others are as follows: "The Harmful Waste Act (Decree 42 of 1988) was preserved as Cap HI LFN 2004; the Environmental Impact Assessment Act was retained as Cap EI2 LFN 2004, and the Harmful Waste Act (Decree 42 of 1988) was retained as Cap HI LFN 2004. Also, the Petroleum Drilling and Production Regulation 1967; National Environmental Protection (Pollution Abatement in Industries and Factories Generating Waste) Regulations S. I. 9 of 1991; Environmental Guidelines and Standards for the Petroleum Industry in Nigeria 2002; National Environmental Protection; Management of Hazardous Wastes S. I. 15; Guidelines on Hazardous Chemicals Management; Guidelines on Pesticides Management; National Implementation Strategy for Chemicals Hazard Communication; National Environmental (Mining and Processing of Coal, Ores, and Industrial Minerals) Regulations, S. I. No. 31 of 2009; The National Oil Spill Detection and Response Agency (NOSDRA) created in 2006; and National Environmental (Electrical/Electronic Sector) Regulations, S. I. No. 23 of 2011" were all enacted to reduce the incidence and/or eliminate hazardous waste (Uwagbale, 2016).

1. **The Special Criminal Provision Act 1988 (Cap H1 LFN 2004)**

   Decree 42 of 1988, which stipulates that the carrying, depositing, transporting, importing, selling, buying, or negotiating in trade of dangerous waste in Nigeria constituted a criminal offence punishable by life imprisonment, was the first step toward the country's effective hazardous waste management (Ikhariale, 1989; Uwagbale, 2016). This act, which is a direct response to the Koko incident, was as time-consuming as before; there was no regulation on waste dumping. The Italian importer Raffaelli was said to have sought through the embassy whether Nigeria had any law barring the dumping of toxic waste from its territory. The Ministry of External Affairs, to which the request was made, never responded to the request presumably because a decree of that sort did not exist at the time.

   The act criminalized the dumping of waste in Nigeria. The act in Section 1 (2) states inter alia that "whosoever: or
   a. Transports or causes to be transported or is in possession to import any harmful waste; or
   b. Imports or causes to be imported any harmful waste on any land, in Nigeria's territorial waters, contiguous zone, or exclusive economic zone, or its internal waters; or
   c. Imports, causes to be imported, or negotiates the importation of any hazardous waste;

d. Anyone who sells, offers for sale, buys, or otherwise deals in hazardous waste commits a crime under the Act" (Suleiman et al., 2020; Zangina and Ali, 2021; Uwagbale, 2016).

The purpose of reproducing Section 1(2) of the act is to show its comprehensiveness. It embodies every commercial activity usually associated with trade in waste.

Section 15 of the decree defines dangerous waste as follows:

> Any injurious, poisonous, toxic or noxious substance and in particular, includes nuclear waste emitting any radioactive substance of the waste in such quantity, whether with another consignment of the same or different substance as to subject a person to the risk of death, fatal injury, or incurable impairment of physical or mental health …

The waste must be not only detrimental but also dangerous, according to the definition. This does not have to cause actual harm, but it must be sufficient to provide a danger of death or injury. A lot of laboratory effort would be required to prove that a certain waste is hazardous and thus illegal to handle. According to Nwufu (2010), Section 6 of the act provides for a life term of prison sentence as well as the seizure of any vessel, car, or property linked to or implicated or connected to the crime of Infringement.

2. **The 1988 (FEPA) Federal Environmental Protection Agency Decree**
   According to Section 20(5), the act states inter alia that the agency has the prerogative to analyze all substances and radioactive chemicals that have the potential to affect public health and welfare, taking into account criteria such as location, quantity, and climatic conditions pertaining to their release. This is an important function entrusted to the agency and poised to have far-reaching effects. The agency's statutory duty is to identify which compounds are hazardous and which hazardous substances, if released, would be injurious to public health or welfare.

   Pursuant to S. 37 of the FEPA 1988, the Minister state as follows "that:
   a. National Environnemental Protection (Effluent Limitation) Regulation, 1991.
   b. Environmental Protection (Pollution Abatement and Facilities Generating Wastes) Regulation, 1991.
   c. Management of Solid and Hazardous Waste Regulations of the National Environmental Protection Agency, 1991. NESREA, in its Section 36, repealed FEPA".

3. **National Environmental Protection Management of Solid and Hazardous Waste S.I. 15**
   There are 13 schedules in addition to the 200-paragraph document. The regulation aims to define the federal minimum criteria for the management of solid and hazardous waste in the country, as well as the design of the country's basic integrated waste management approach. The regulations will

a. declare solid waste to be dangerous or highly hazardous to public health and the environment;
b. establish surveillance and monitoring of dangerous and extremely hazardous waste and substances until they are detoxified or reclaimed;
c. establish a system for manifesting, tracking, reporting, monitoring, record keeping, sampling, and labeling dangerous and extremely hazardous waste by providing the relevant form and guidelines; and
d. identify the target audience.

A lot of fine-tuning still has to be done to this legal instrument, which is supposed to be the springboard of internationally acceptable standards for solid and toxic substances in the country and the minimum standard-bearer for the state environmental protection agency/ministry.

4. **NESREA Act 2007**

Section 2(a) of NESREA Act 2007 is responsible for enforcing legislation, laws, standards, rules, and regulations, and recommendations governing the use of the human environment. This regulation was enacted to ensure that all environmental legislation both at local and global scales signed into law by Nigeria are smoothly implemented. According to Section 1 of the act, "the agency is responsible for environmental protection and development, biodiversity conservation, and the sustainable development of Nigeria's natural resources in general, as well as environmental technology, which includes coordinating and liaising with essential parties both within and outside Nigeria on environmental standards, regulations, rules, laws, policies, and guidelines enforcement". As regards the Control of Hazardous Waste, the Act in Section 27 States inter alia that the following:

1. Except where approved or authorized by Nigerian law, disposal of toxic substances on land, under the waters, air, coastal areas or anywhere in Nigerian territorial area is hereby forbidden.
2. If convicted, a person who violates this law faces a fine of not more than N1,000,000 or a sentence of not more than 5 years in jail.
3. If a corporate entity violates Subsection (1), it will pay a penalty of N1,000,000, plus a daily penalty of N50,000 for the period in which the violation persists.

The act goes on in Subsection (4) to state that where the act of discharge of hazardous wastes was by a company, the highest authority will be held liable to any punishment. However, such a person will not be found guilty if he/she can prove beyond a reasonable doubt that he/she took the necessary precautions to avert such an incidence or that it occurred without his/her consent or his/her pre-knowledge. Finally, Section 27(5) of the act states that "Notwithstanding the provisions of this Section and other sections of the Act, the Harmful Waste (Special Criminal Provisions, etc.) Act shall apply in respect of any hazardous substance constituting harmful waste as defined in Section 37 of the NESREA".

Nigeria has a dismal record of compliance, despite being a signatory to practically all environmental protection treaties necessary for its sustainable development in line with Agenda 21, NEPAD, and NEEDS (Zangina and Ali, 2021). There has been a continuous violation of environmental laws with recurrent national environmental laws and regulations breaches. "The Basel Convention on the Transfrontier Movement of Hazardous Waste and their Disposal, the Montreal Protocol on Substances Depleting the Ozone Layer, signed in 1987, and the Convention on International Trade in Endangered Species, signed in 1973, both deals with large-scale cross-border movement of the substances in question, and enforcement remain one of the most important steps in the implementation of environmental policy".

Some other significant hazardous waste management tools in Nigeria are the economic policy instruments (EPIs) which are contained in the FMIE Policy guide on sound environmental management. Some of the EPIs include "the Polluter Pays Principle, the User Pays Principle, and the Pollution Prevention Pays Principle" (FMEnv., 1999). "The Polluter Pays Principle considers that the polluter should take care of the cost of pollution prevention and remediation, while for the User Pays Principle, the user is expected to cover the price for the use of the resource, which must cover all environmental costs connected with its extraction, transformation, and use, plus the cost of the forgone alternative in the future". Similarly, "the Pollution Prevention Pays Principle, advises the industries to positively invest in pollution prevention, which should include tax holidays and subsidy that promotes pollution abatement; polluted or degraded areas could be restored by utilizing risk management schemes such as the insurance; fines, taxes, penalties, and other charges for defaulters compliance failure to meet up with environmental standards and regulations; and promoting the use of market-based environmental standards and regulations" (FMEnv., 1999).

9.5   CONCLUSION

As can be garnered from this chapter's entire gamut, environmental protection from hazardous wastes is obviously the most pressing task for mankind today. The concept of environment and guaranteeing its stability and balance have been misdirected, a problem which stems from a lack of a proper theoretical framework that necessitates addressing the umbilical tie between human beings and their environment, and vice versa. Contaminants can migrate within and beyond the original medium into which they are introduced since the environment is indivisible, interrelated, and interdependent. When hazardous waste is placed on land after polluting and poisoning the ecosystem, it travels through a variety of environmental pathways, including the atmosphere, rivers, canals, underground watercourses, and outfalls, before eventually reaching the ocean and destroying the marine environment.

As can be discerned from the above discussion on national and international waste control laws, it is visible that their presence provides no cure for the scale of today's environmental pollution from toxic waste. No conventional response is totally adequate; we cannot rely on judicious mixtures of good laws and financial penalties against polluters. We need a positive shift in public perception and behavior and stringent adherence to and implementation of environmental legislation. For smooth implementation of internationally acceptable sound environmental management plans in Africa, technical, particularly in capacity building and technological adaptation, economic, and financial assistance are required. We must restore our environment's dignity, which has been severely harmed by erosion, desertification, and contamination by hazardous and poisonous waste.

9.6   RECOMMENDATIONS

1. Waste should be disposed of in proximity to the place of generation; therefore, in situ or onsite should be encouraged as much as possible (Major and Fitchko, 1992). There should be the minimization/elimination of waste export to regions or even countries beyond the point of generation, sometimes referred to as waste tourism. Policymakers should encourage the location of waste management installations close to or at the points of waste generations in line with the proximity principle. Hazardous waste should be managed close to its point of origin. Thus, inter-state transfer should be outrightly banned as stipulated in the convention on transshipment of toxic substances, Bamako. Therefore, the management and transshipment of toxic substances should make use of the EPIs and best practical environmental option for strategic purposes in waste management only in those countries with technical competence.

2. NESREA should properly organize waste collection and management in the country. It should include the creation of dumpsites for different categories of waste such as biodegradable, non-degradable, and hazardous. This type of development will allow waste that can be recycled to be separated and used for new sets of raw materials to be adequately explored and monitored. Therefore, the hazardous waste dump sites can be fenced to keep out scavengers, protect public health, and ensure public safety.

3. Nigeria needs to be repositioned for environmentally sound waste management practices through appropriate legislations and regulations, realignment with the key premise and waste objective as articulated by Agenda 21. By striving to transform unsustainable patterns of production, consumption, and public information for public engagement, the core cause of the problem must be addressed.

4. Maximizing environmentally sound waste reuse and recycling, and strengthening and increasing national reuse and recycling systems through capacity building are essential. Companies that depend on recycled materials should be encouraged through the development of appropriate infrastructure, and technical and financial assistance to implement environmental plans. A regional team should keep a tight eye on everything, including port facilities.

5. Greenpeace's function as a link between the worldwide economy and the global environment is commendable for our survival on this only life-sustaining planet. Some of these institutions include UNEP, which was at the forefront and instrumental in drafting, facilitating, and negotiating several environmental treaties. In environmental protection, Greenpeace tries to

influence how the sea is used by "bearing witness" and raising attention to what it considers to be environmental violations. Greenpeace developed an effective partnership with developing countries to oppose "free trade" solutions to the global trash trade, and similar organizations should follow suit.

REFERENCES

Akpan, V. E. and Olukanni, D. O. 2020. Hazardous waste management: An african overview. *Recycling*. 5(15): 1–24.

Amokaye, G. O. 2004. *Environmental Law and Practice in Nigeria*, 1st Ed. University of Lagos Press, Lagos, pp. 1, 3, 309, 395, 409.

Anyinam, C. A. 1991. Transboundary movements of hazardous wastes: The case of toxic waste dumping in africa. *International Journal of Social Determinants of Health and Health Services*, 21: 759–777.

Barresi, M. 2020. Position paper dumping of toxic wastes in developing countries. Available online: https://www.academia.edu/14724543/position_paper_dumping_of_toxic_wastes_in_developing_countries (accessed on 15 June 2020).

Basel Convention Country Fact Sheet, Nigeria. 2016. In www.basel.int. Retrieved on 2 September.

Bernstorff, A. and Stairs, K. 2000. Pops in Africa, Hazardous waste trade 1980–2000. Obsolete pesticides stockpiles. Greenpeace inventory. http://www.ban.org/library/afropops.pdf

Birnie, P. W. and Boyle, A. E. 1992. *International Law and the Environment*. Clarendon Press, London, pp. 14, 33, 326.

Cobbling, M. 1992. Europe's toxic colonialism exporting Europe's hazardous waste. *Chemistry and Industry*, 21.

Comaroff, J. 2007. Beyond the politics of bare life: Aids (Bio) Politics, and the Neoliberal order. *Public Culture*, 19(197–219). http://dx.doi.org/10.1215/08992363-2006-030.

Dokun, O. 1995. *Essentials of Environmental Issues: The World and Nigeria in Perspective.* Daily Graphics Publications Ltd., Ibadan, pp. 1–5.

Du Vivier, F. R. 1988. ed. Sang de la Terre. Les vaisseaux du poison- La route des déchets toxiques. Dumping of waste should not be trivialized. Retrieved at http://www.alafrica.com/stories/2008/03/30958.html.

Federal Environmental Protection Agency (FEPA) Act. 1988 No. 58.

Federal Environmental Protection Agency (FEPA) Act. 1992 No. 59.

Federal Environmental Protection Agency (FEPA) Act. 1999 No. 14.

Federal Environmental Protection Agency (FEPA). 1991. The National Guidelines and Standards for Environmental Pollution Control in Nigeria. Federal Environmental Protection Agency.

Federal Ministry of Environment (FMEnv.). 1999. A review of the national policy on the environment. http://environment.gov.ng/index.php/downloads/. Retrieved on September 2nd, 2016.

Frey, R. S. 1994. The international traffic in hazardous wastes. *Journal of Environmental System*, 23, 165–177.

Garner, H. C. 2004. *Black's Law Dictionary*, 8th Ed., West Publishing Co., USA, p. 1621.

Gasu, G. N. 2008. Global and regional regimes on the control of transboundary movement of hazardous waste and their disposal. Being an unpublished Long Essay submitted to the Faculty of Law, Obafemi Awolowo University, Ile-Ife, in partial fulfillment of the requirements for the award of the Masters of Laws (LL.M.) Degree.

Gevao, B., Alegria, H., Jaward, F. M. and Beg, M. U. 2010. *Persistent Organic Pollutants in the Developing World. In Persistent Organic Pollutants*. John Wiley & Sons, Ltd., Chichester, UK, pp. 137–169.

Global Environment Outlook GEO₄. 2007. *Environment for Development, State and Trends of the Environment*, UNEP Publication, Valletta, Malta, pp. 6–9, 38–45.

Grasso, D., Kahn, D., Kaseva, M. E. and Mbuligwe, S. E. 2010. Hazardous waste management. Encyclopedia of life support systems (EOLSS)-UNESCO Sample Chapters.

Guruswamy, L., and Hendrick, B. 2001. *International Law in a Nutshell*. West Group Publishers, pp. 3, 6, 57, 192, 278–279.

Ikhariale, H. 1989. The Koko incident: The environment and the law, In Shyllon, F (ed) The Law and the Environment in Nigeria, Vantage Publishers, Ibadan.

Hughes, N. 1996. *Environmental Law*, 3rd Ed. Butterworths, London, p. 365.

Hunter, D., Salzman, J., Zaelk, D. 2002. *International Environmental Law and Policy*, 2nd Ed. University Casebook Series, New York.

Ibrahim, E., Gushit, J., Salami, S. and Dalen, M. 2018. Accumulation of polychlorinated biphenyls (PCBs) in soil and water from electrical transformers installation sites in selected locations in Jos Metropolis, Plateau State, Nigeria. *Journal of Environmental & Analytical Toxicology*. 8, 1–6.

International Network for Environmental Compliance Enforcement Seaport Security Network SESN. 2009. The international hazardous waste trade through seaports. Working Paper, 24 November 2009, pp. 1–7.

Jennifer, C. 1994. The toxic waste trade with less-industrialised countries: Economic linkages and political alliances. *Third World Quarterly*, 15(3), 505–518. <http//www.jstor.org/journal/taylorfrancis.html. http://dx.doi.org/10.1080/01436599408420393.

Koné, L. 2014. The illicit trade of toxic waste in Africa: The human rights implications of the new toxic colonialism. *SSRN Electronic Journal*. http://dx.doi.org/10.2139/ssrn.2474629.

Kummer, K. 1999. *International Management of Hazardous Wastes: The Basel Convention and Related Legal Rules*. Oxford University Press, Great Britain.

Lipman, Z. 2002. A dirty dilemma: The hazardous waste trade-Proquest. *Harvard International Review*, 23, 67–71.

Massari, M. and Monzini, P. 2004. Dirty businesses in Italy: A case-study of illegal trafficking in hazardous waste. *Global Crime*, 6, 285–304.

Mason, B. (2006) Toxic waste dumping in Ivory Coast. 2006. World Socialist Web Site, wsws. org.

Meirion, J. and Liz, M. 2007. Dirty tricks and toxic waste in Ivory Coast, BBC News night.

National Environmental Protection Management of Solid and Hazardous Waste Regulation. 1991.

National Environmental Standards Regulation Enforcement Agency (Establishment) Act. 2007.

National Environmental Standards Regulatory and Enforcement Agency (NESREA). Corporate Strategic Plan: Building Capacity, Enforcing Compliance. 2009–2012.

National Oil Spill Detection and Response Agency (NOSDRA) (Establishment) Act. 2006.

Nwufu, C. C. 2010. Legal framework for the regulation of waste in Nigeria. *African Research Review: An International Multi-Disciplinary Journal, Ethiopia*, 4(2): 491–501.

OAU Council of Ministers' Resolution on Dumping of Nuclear and Industrial Waste in Africa. 1988. Reproduced in C. Heyns, 2004 Human Rights Law in Africa, p. 342.

Okorodudu Fubara, M. T. 1998. *Law of Environmental Protection: Material and Text*. Caltop Publication (Nigeria) Ltd., Ibadan, pp. 795, 602, 841.

Okorodudu Fubara, M. T. 1999. *Dynamics of a New Environmental Legal Order*. Inaugural Lecture Series 133; Obafemi Awolowo University, p. 32.

Organisation for Economic Cooperation and Development (OECD). 2022. Decision-recommendation of the council on exports of hazardous wastes from the OECD area, OECD/LEGAL/0224. Accessed 3/3/2022.

Perkins, D. N., Drisse, M.-N. B., Nxele, T. and Sly, P. D. 2014. E-waste: A global hazard. *Annals of Global Health*, 80, 286–295.

Report of the International Enquiry Commission on the discharge of toxic waste in the District of Abidjan. 2007.

Sands, P. 2003. *Principles of International Environmental Law*, 2nd Ed. Cambridge University Press, Cambridge, p. 671. http://doi.org/10.1017/CB9780511813511.

Sodipo, E., Omofuma, O. I and Nwachi, V. C. 2017. Environmental law and practice in Nigeria: Overview. Ejide Sodipo & Co, A Q&A Guide to Environment Law Nigeria. www.practicallaw.com/environment-guide. Accessed 28/03/2022.

Sthiannopkao, S. and Wong, M. H. 2013. Handling e-waste in developed and developing countries: Initiatives, practices, and consequences. *Science of The Total Environment*, 463, 1147–1153.

Suleiman, A., Regasa, Y. A. and Kabir, N. 2020. Management of hazardous waste in Nigeria, A proposed strategy for regulators, operators and businesses. *Journal of Scientific and Engineering Research (JSERBR)*, 7(2): 89–96.

The Basel Convention on the Control of Transboundary Movement of Hazardous Waste and their Disposal (Basel Convention). 1989.

The Convention on the Ban of Imports into Africa and Control of Transboundary Movement of Hazardous Waste into Africa (Bamako Convention). 1991.

The Federal Environmental Protection Agency (FEPA). 1988.

The Harmful Waste (Special Criminal Provision etc.) Act. 1988 (Cap H1 LFN 2004)

The Holy Bible King James Version. 15 March, 2019.

The Rio Declaration. 1992.

The Stockholm Declaration. 1972.

The United Nations Convention on the Law of the Sea (UNCLOS III). 1982.

The World Counts Hazardous Waste Statistics. 2019. The world counts. Available online: https://www.theworldcounts.com/counters/waste_pollution_facts/hazardous_waste_statistics (accessed on 22 May 2019).

Thornton, J. and Beckwith, S. 1997. *Environmental Law Sweet & Maxwell*. London, pp. 1, 2, 169.

Uchegbu, J. 1990. The legal regulation of environmental protection and enforcement in Nigeria. *Journal of Private and Property Law*, 8/9: 58.

UN Doc A/Conf. 151/26 1992 154 Agenda 21 Para. 19.1, 19.11 & 21.

UN Doc. Conf/A/48/Pc.

UNGA Res. 2398 (XXII)(1968).

United States Environmental Protection Agency. 2005. Introduction to hazardous waste identification (40 CFR Parts 261). *Solid Waste and Emergency Response*,(5305W) EPA530-K-05-012, 1–26.

Uwagbale, E. D. 2016. Hazardous waste management and challenges in Nigeria. *Public Health International*, 1(1): 1–5.

Viljoen, F. 2007. *International Human Rights Law in Africa*. Oxford University Press, pp. 290–292.

Zada, L. 2019. Trade-in hazardous waste: Environmental justice versus economic growth environmental justice and legal process. Available online: edisciplinas.usp.br/pluginfile.php/4060252/mod_resource/content/1/AULA%208%20-%20DEBATE%20Trade%20in%20Hazardous%20Waste.pdf. (accessed on 26 January 2023).

Zangina, A. S. and Ali, A. F. 2021. An overview of hazardous waste management in Nigeria. *Dutse Journal of Pure and Applied Sciences (DUJOPAS)*, 7(4b): 214–221.

10 E-waste Management Strategies and its Opportunities
Toxic but Beneficial

*Falguni Shinde, Priyanka Vibhandik,
and Khalid Alfatmi*

CONTENTS

10.1 Introduction 164
10.2 Inventory of E-Waste 164
    10.2.1 Types of E-waste: Below are Different Types of E-waste 165
    10.2.2 E-waste Composition 165
10.3 E-waste Generation in India 166
    10.3.1 Stakeholder Involvement in E-waste Generation 167
10.4 Disposal of E-waste 168
    10.4.1 Physical Disposal Processes 168
        10.4.1.1 E-waste Recycling.. 168
        10.4.1.2 Landfill Disposal... 169
    10.4.2 Technologies for Disposal Processes.. 169
        10.4.2.1 Dismantle/Sort/Grind/Dissolve Strategy
                (CEA, JCP Gabriel) DES/ISEC/DMRC Marcoule)........... 169
        10.4.2.2 Microscopic Piping Strategy... 170
        10.4.2.3 Recovering Rare-earth Materials 171
10.5 An Economic Opportunity with Toxicity.. 171
    10.5.1 Major Determinants of the Global Waste Trade Sector 171
    10.5.2 E-waste Economy in Unorganized Sector 172
10.6 India's Rules for E-waste Management and Handling 172
    10.6.1 Applicability 172
    10.6.2 Rules 173
    10.6.3 Responsibilities.. 174
10.7 Technologies Used in E-waste Management 179
    10.7.1 E-waste Management Using Deep Learning Object Classifier 179
        10.7.1.1 E-waste Image Recognition 180

DOI: 10.1201/9781003359784-10

    10.7.2 Solution for Manual E-waste Retrieval and Collection
            Route Planning 181
            10.7.2.1 Concept of GPS and Dijkstra's Algorithm........................ 182
            10.7.2.2 Algorithm for Image Matching.. 183
            10.7.2.3 Image-matching Algorithm... 184
10.8 Conclusion 186
References.. 187

10.1 INTRODUCTION

India is one of the top five nations in the fields of space exploration and scientific research [1], and it is ranked third among the world's most attractive investment destinations for technology transactions. Science and technology are very important in modern India since they are a key component of economic success. The growth of technology has resulted in a huge increase in e-waste generation in India, which is increasing at a rate of 10% per year [1].

After the United States and China, India is the third largest producer of e-waste [2], with about 3.23 million metric tonnes produced each year. Since 2011 [2], India has been the only country in Southeast Asia to have a legal framework for dealing with e-waste [2].

As a result, a large volume of e-waste winds up in landfills or is burned in the open air in India's rural and urban areas, which prefer traditional garbage processing and disposal methods. By leaching harmful chemicals into soil and water, this results in soil and water contamination, posing major health and environmental risks.

The pandemic-induced increase in the use of electronic devices is expected to exacerbate the problem of e-waste generation soon. Until now, metropolitan regions have been the primary generators of e-waste [2]. Because e-waste generation is continually increasing, India would benefit from a multi-pronged strategy for e-waste management.

This study looks at how a "deep learning object classifier can be used to improve e-waste management." This method uses a deep learning CNN and a faster R-CNN to detect residential e-waste pickup [3]. The user uploads a photograph of the equipment to be retrieved, and the collection firm, with the help of CNN and R-CNN, detects the image, and plans a garbage retrieval route, assigning the appropriate truck and personnel. This system has various flaws, such as the fact that some of the trash retrieval requests and collection planning is done manually, which raises the mobile cost, negatively impacting the collection firms' appraisal of this approach. As a result, this study proposes a technical solution to this challenge by combining an AI-based system with GPS to automate waste retrieval and route planning.

10.2 INVENTORY OF E-WASTE

E-waste is created by a variety of factors. E-waste is defined as electrical equipment that has reached the end of its useful life. People typically choose to buy new equipment rather than using old ones, which contributes to an increase in e-waste. Computer accessories (68%) are the most common type of e-waste, followed by telecommunications equipment (12%), electrical apparatus (8%), health gear (7%), and

TABLE 10.1
Examples of E-waste

Home Appliances	Information Communication Devices	Entertainment Devices	Electronic Utilities	Office and Medical Equipment
Microwave	Cell phones	DVDs	Massage chairs	Cords and cables
Fans	Computers	Stereos	Remotes Control	IT servers and racks
Cookers	Laptops	Televisions	Treadmill	Wi-Fi dongles
Heaters	Hard drives	Fax machines	Lamps	Phone and PBX
	Circuit boards	Printers	Smart watches	Systems
		Copiers	Smart lights	Audio and video
		Video games		equipment

home e-scrap (5%) [4]. E-waste contains valuable metals such as gold, silver, platinum, lanthanum, neodymium, palladium, and others that need to be commoditized because "urban mining" of e-waste, including gold and silver, can create $21 billion in revenue [4]. Hazardous compounds in e-waste include hexavalent chromium, mercury, cadmium, and lead, which can harm the human reproductive, renal, and respiratory systems [4].

10.2.1 Types of E-waste: Below are Different Types of E-waste

Electronic and electrical goods are divided into three categories (Table 10.1):

1. **White goods:** Air conditioners, dishwashers, refrigerators, and washing machines.
2. **Brown goods:** TVs, camcorders, and cameras.
3. **Gray goods:** Computers, printers, fax machines, and scanners. Because of their hazardous content, gray items are more difficult to recycle.

The percentage creation of "Waste of Electrical and Electronic Equipment (WEEE)" is shown in the graph (Figure 10.1):

10.2.2 E-waste Composition

Electronic and electrical equipment contain a wide range of chemical components and compounds. Electronic waste is made up of a complex mix of different components, many of which contain toxic substances. Lead, cadmium (Pd), mercury (Hg), polyvinyl chloride, brominated flame retardants, chromium (Cr), beryllium (Be), and other chemicals are found in most electronic devices; additionally, some instruments, such as televisions, video, and computer monitors, use CRTs, which contain significant amounts of lead [5]. Long-term exposure to these chemicals can harm the brain, kidneys, and bones, as well as the reproductive and endocrine systems [5].

Table 10.2 shows toxic metals present in e-waste and their adverse effects on humans.

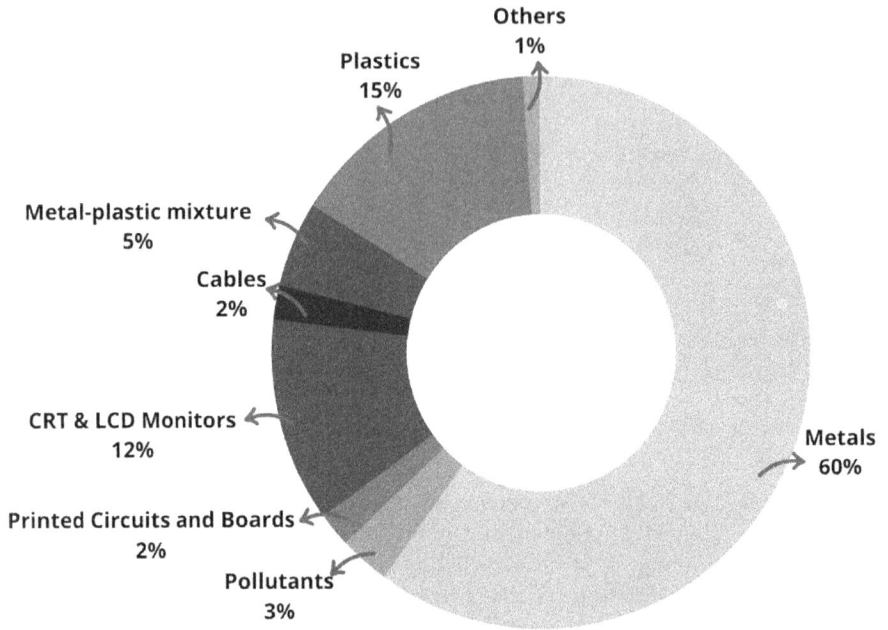

FIGURE 10.1   The distinctive contents of a WEEE.

TABLE 10.2
Toxic Metals in E-waste & their Consequences on Humans

Metals	Present in	Consequences
Lead	Acid battery, CRT	Kidney failure
Cadmium	Battery, CRT, housing	Long-term accumulated toxicity, arthritis
Mercury	Batteries, switches, housing.	Chronic brain damage, liver damage, damage to the central nervous system, and lateral and fetal nervous systems
Chromium VII	Decorative hardener, corrosion protection	DNA damage, lung cancer
Plastic	Computer molding, cabling	Generation of dioxins and furans

10.3 E-WASTE GENERATION IN INDIA

After China and the United States, India has surpassed China as the world's third largest producer of e-waste, with 3 million tonnes produced annually [2]. India is ranked third after China and the United States. These three countries accounted for 38% of the total 53.6 million tonnes of e-waste generated globally in 2019, producing 10,14,961.2 tonnes of rubbish, an increase of 31.6% from the previous year. These rising numbers may result in dangerous changes in the ecosystem [6].

In India, ten states account for 70% of the waste, while 65 cities account for more than 60% of the waste; The states of Maharashtra, Tamil Nadu, Uttar Pradesh,

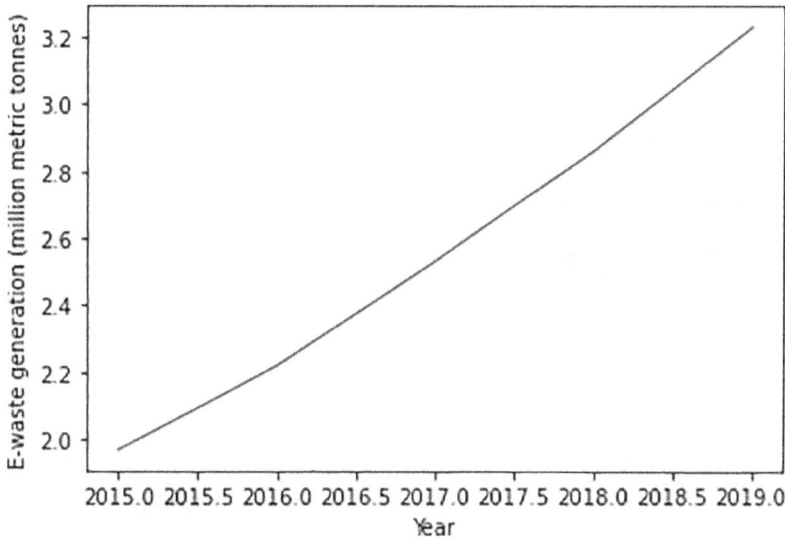

FIGURE 10.2   Yearwise e-waste generation.

West Bengal, Delhi, Karnataka, Gujarat, Madhya Pradesh, and Punjab top the list of e-waste-producing states in India [7]. Although there is no large-scale facility for recycling waste in the country, Mumbai, Delhi, Bengaluru, Chennai, Kolkata, Ahmadabad, Hyderabad, Pune, Surat, and Nagpur all contribute significantly to e-waste in India. Mumbai is the most populous of these cities, followed by others. There are no applicable laws regarding e-waste. As a result, there are no environmental laws or guidelines in India that specifically address the issue of e-waste (Figure 10.2).

10.3.1   STAKEHOLDER INVOLVEMENT IN E-WASTE GENERATION

In India, a variety of parties engage in the management of e-waste. People in India frequently replace older items such as laptops and televisions with newer versions, features, or alternatives to fit their current needs. Because the initial user sells the used equipment to a family member or friend in India, an EEE may have several users. Several significant stakeholders in India's e-waste business are depicted in this diagram.

According to the paper, stakeholders in underdeveloped nations operate on the three levels of the e-waste generation hierarchy as follows:

1. Primary
2. Secondary
3. Tertiary

Formal market importers, manufacturers, and market organizations that receive e-waste directly from domestic consumers, either through exchange programs or

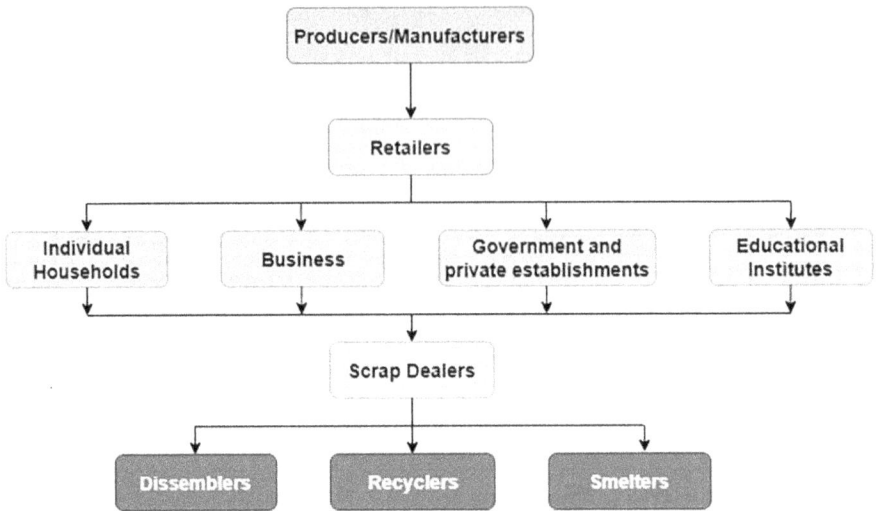

FIGURE 10.3 Different stakeholders involvement in the e-waste flow.

in abandoned form, are referred to as "primary e-waste generators" [8]. The most important stakeholders are scrap dismantlers, who buy e-waste in bulk from the first level. These stakeholders trade e-waste with "secondary e-waste generators" due to their inadequate dismantling capabilities [8]. Between the first- and second-level waste generators, the market is semi-formal (somewhat formal), whereas between the third- and fourth-level trash generators, the market is fully informal. "Tertiary-level stakeholders" include metal, plastic, and electronic item extractors [8]. These are the most important stakeholders at the second and third levels. They employ extraction techniques that are inherently dangerous (Figure 10.3).

10.4 DISPOSAL OF E-WASTE

The production of e-waste is unavoidable. It can be minimized but not eliminated. Recycling and landfill disposal are the most well-known and extensively used practices for disposing of e-waste internationally.

10.4.1 Physical Disposal Processes

10.4.1.1 E-waste Recycling

E-waste is electronic garbage made up of unused electronic devices. Improper e-waste recycling might result in harmful fumes or compounds being released into the environment [9]. Inadequate recycling of this e-waste might result in the discharge of harmful gases or compounds into the environment [9]. The term "e-waste recycling" refers to the process of processing and repurposing electronic waste. Components of e-waste that can be recycled include the following:

TABLE 10.3

Comparison of E-waste Disposal

Developed Countries	Underdeveloped Nations
Incineration with municipal solid waste	Open burning
Landfill disposal	Open dumping

1. **Mercury:** Devices containing mercury are recycled using specialized technology to remove the mercury, resulting in end items such as metric instruments, dental amalgams, and fluorescent lighting.
2. **Circuit boards:** Smelting is used by some specialist companies to recover resources like tin, gold, silver, copper, palladium, and other valuable metals.
3. **Toner and Ink cartridges:** These toner and ink cartridges are recycled and remanufactured by recyclers in various manufacturing businesses. The plastic and metals that are recovered are then used as raw materials for other goods.
4. **Batteries:** Batteries are recycled to recover cadmium, steel, nickel, and cobalt, which can then be reused in new batteries.

10.4.1.2   Landfill Disposal

Landfilling is the most common method of e-waste disposal. E-waste is frequently disposed of in landfills alongside other garbage. It is sometimes burned openly, sending poisonous and carcinogenic chemicals into the atmosphere [9].

The differentiation between developing and industrialized countries for e-waste disposal is displayed in Table 10.3.

10.4.2   Technologies for Disposal Processes

10.4.2.1   Dismantle/Sort/Grind/Dissolve Strategy (CEA, JCP Gabriel) DES/ISEC/DMRC Marcoule)

The sorting process seeks to reduce the amount of variability and chemical complexity in the mixture to be handled. It is carried out at all scales, including device (generation, kind), module (frames, batteries, printed circuits, exterior envelopes, etc.), basic electronic components (resistances, cables, capacities, bare boards, chips, etc.), and even grinding powder [10]. The most effective method is to completely disassemble the devices. This phase is difficult to automate because of the variety and complexity of equipment. Disassembly is still done by hand, which makes sorting down to the elementary component-level expensive [10] (Figure 10.4).

Before any chemical treatment, the most typical process employed by recyclers is device or module-scale grinding, followed by particle separation using physical methods based on magnetic characteristics and density differences. Depending on the purity of the resulting powders, chemical or thermal treatment is performed to refine the final product's composition [10]. The liquid–liquid extraction technique

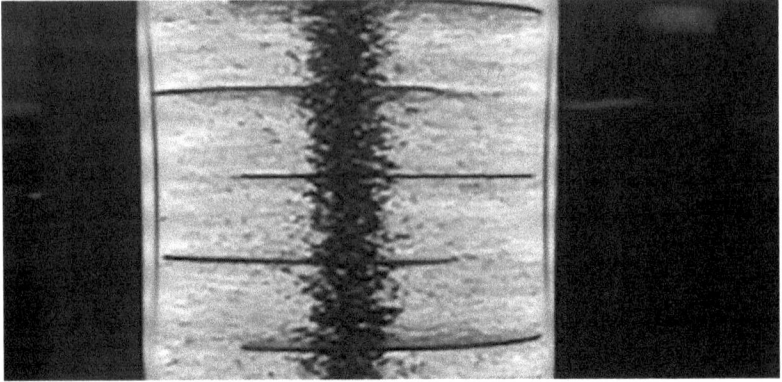

FIGURE 10.4   5 cm-diameter pulsed extraction column.

FIGURE 10.5   Microscopic piping.

is commonly used to separate chemical components from solutions. This method includes dissolving metals or metal oxides in acid (e.g., nitric acid), which produces an emulsion like a French vinaigrette. In the extraction column, the acid solution is aggressively combined with an organic solvent and one or more molecules that promote the transfer of specific metals from acid to solvent. This procedure is repeated until the required purity level is achieved [10].

10.4.2.2 Microscopic Piping Strategy

For the extraction process, an integrated device called microfluidics is being created, which is fully automated and equipped with all the necessary equipment. The piping in this device is modest, 100 m thick in this case, allowing for the utilization of a small amount of material. The fundamental component of the 5-cm side-extraction microfluidic chip, chemical elements flow between two such components through a membrane placed between them, while fluids flow in a zigzag manner through the half-pipe. The addition of pipelines, pumps, and analysis modules, such as infrared, is made [10]. JCP Gabriel, CEA, A. El Mangaar (Figure 10.5).

The advantage of microfluidics over traditional devices is that the phenomena of chemical element transport at the oil–water interface are well understood. A porous membrane regulates the exchange of water and soil surface, as well as the time of contact between the two phases, which are driven into microfluidic channels by computer-controlled syringe pumps. Material flows can then be calculated precisely [10].

10.4.2.3 Recovering Rare-earth Materials

This method permits researchers to investigate the extraction of key metals contained in mobile phones. These metals, which are primarily manufactured in China, are crucial to modern technologies. They are currently recycled at an extremely low rate—less than 5%. This is especially problematic because their production is extremely costly and can result in social and environmental issues [10].

The extraction of rare earths is made possible by combining two specialized extracting molecules, which is 100 times more efficient than extracting the molecules alone [10].

10.5 AN ECONOMIC OPPORTUNITY WITH TOXICITY

E-waste is the world's fastest-growing trash stream, highlighting its twin identities as an environmental hazard and a potential economic resource. Toxic compounds such as lead and mercury are frequently found in e-waste, whereas valuable materials such as gold, silver, and copper can be found in laptops and phones. Only about 20% of the world's e-waste gets collected and transferred to professional recyclers, while the destiny of the rest is unknown [11]. The developing world imports the majority of used electronics to give the devices a second and third life and help close the digital divide. The pretense of "repair and reuse," on the other hand, is commonly used to hide unlawful e-waste exports. Even though this is trash, the rise of hazardous e-waste activities in cities such as Pakistan and Ghana, for example, attests to the wealth hidden in abandoned electronics. The raw materials contained in e-waste were worth $61 billion in 2016, according to researchers at the United Nations University, which was more than the gross domestic product of middle-income countries like Croatia or Costa Rica [11]. Circuit boards contain more valuable metals than even the most productive mines. In 2016, gold retrieved from e-waste accounted for more than a tenth of all gold mined that year. Unfortunately, much of this valuable stuff is either reburied in landfills or openly burned. According to e-waste disposal statistics, Americans alone throw away phones, worth $60 million in gold and silver each year [11].

10.5.1 MAJOR DETERMINANTS OF THE GLOBAL WASTE TRADE SECTOR

Global e-waste creation is expected to increase by 17%, to around 57.4 million tonnes per year by 2021, according to estimates [6].

A fundamental hurdle to estimation is the fact that secondary and waste products are invisible in national statistics on production, sales, and trade in goods. Furthermore, due to various definitions of what constitutes e-waste, there are disparities in the quantification of e-waste generation and the identification of e-waste flows [6]. Despite a lack of statistical data on e-waste and hazardous waste movements in general, it is commonly acknowledged that shipments are changing away from final disposal and toward recovery and recycling operations [6]. Recycling rates increase by 18% on average each year. E-waste commerce has expanded not just between rich and poor countries, but also inside developing countries, indicating continuous expansion and tolerance for cross-border migration. Because most of the trade is regionalized and intra-regional, the flow of e-waste is more than just moving from developed to underdeveloped countries. Furthermore, the transaction is unrestricted and is determined by the value assigned to trash at the time of the exchange. The flow of e-waste is a complex topic; for example, in 2001, Africa shipped most of its e-waste to Korea and Spain [6]. Internal markets and Asia's rise as the principal recipient of global exports have both seen significant growth in global trade since 2006 [6].

10.5.2 E-waste Economy in Unorganized Sector

The unorganized sector, primarily scrap dealers, manages over 95% of the booming economy's e-waste. We could see drastic environmental damage and health hazards in the coming decades if e-waste is not disposed of safely and according to established environmental standards. As a result, the government must regulate e-waste management, recycling, and dismantling facilities to ensure that they use technologically advanced methods. In addition, the personnel in the informal industry should be streamlined.

Fifty million tonnes of e-waste are thrown into landfills yearly, with just 10%–18% of it, making it into the recycling process. India's figures are even more discouraging. The only way-out is to construct a proper recycling chain and establish e-waste processing units.

10.6 INDIA'S RULES FOR E-WASTE MANAGEMENT AND HANDLING

Hazardous compounds used in the creation of electronic products have a maximum limit set by the e-waste management rules. The laws also detail how to acquire approval from the "Pollution Control Board" to handle e-waste.

10.6.1 Applicability [12]

These restrictions apply to all manufacturers, wholesalers, and retailers who deal with the listed items:

1. Mainframes, minicomputers, central processing units (CPUs), input and output devices used with CPUs, laptop, notebook, and notepad computers, printers, printer cartridges, copying equipment.
2. Phones, cordless phones, cell phones, and answering machines.
3. Television sets [LCD and LED technologies].
4. Household appliances include things like air conditioners (without centralized air-conditioning plants), washing machines, and refrigerators.
5. Mercury-based bulbs, and other electrical, fluorescent lights, and electronic consumer items.

These rules also apply to E-retailers and stockists.

These regulations should not be considered in the following circumstances:

1. Micro, Small, and Medium Business (MSMB) assessee.
2. Generation of e-waste containing radioactive material (as defined in Section 2(1)(i) of the Atomic Energy Act of 1962) by the assessee.

10.6.2 RULES [12]

Every company that makes electrical or electronic products should follow the following rules:

The company must make sure that the following chemical concentrations do not exceed the following limits:

1. Up to 0.1% of the total weight of the product can be added of substances such as mercury, lead, polybrominated diphenyl ethers, hexavalent chromium, and polybrominated biphenyls.

Up to 0.01% of the overall weight of the product, cadmium—a chemical, can be used. The company should make plans to gather obsolete electrical and technological equipment. Customers should be contacted, and these items should be collected and taken to a dismantler's or recycler's facility.

The e-waste management rules consist of "The Extended Producer Responsibility Plan." In accordance with the notion, businesses are encouraged to set up a deposit scheme. According to the program, the consumer who purchases the goods should be required to pay a deposit. When the product's usefulness has passed, the customer can return it to the company and obtain a refund of the deposit plus interest. If the company forms a separate department named the e-waste collection, handling, and disposal department, customers should be allowed to use the scheme. The following are some of the responsibilities of this department:

1. Providing a platform for clients who want to return old electrical devices to the company.
2. Collecting and preserving the artifacts in an environmentally friendly manner.
3. Sending the objects to be recycled or dismantled.
4. A toll-free phone number is given to the customers to call for the return of outmoded electronic devices. A specific address and e-mail address are also provided to the customers.
5. Make sure that after the goods have been returned, the deposits made by clients at the time of purchase are refunded.

Make sure that mercury-containing items are appropriately immobilized before transferring them to a dismantler or recycler. Mercury is transformed into a safer form from a hazardous state when compounds containing mercury are immobilized. As a result, there is less chance of environmental pollution.

TABLE 10.4
E-waste Collection Target

Sr. No.	Economical Year	Generation of e-waste that needs to be conveyed to the dismantling or recycling centers (minimum percentage)
1)	2019–20	30
2)	2020–21	40
3)	2021–22	50
4)	2022–23	60
5)	2023 onwards	70

E-waste can only be held for 180 days at the most. The time restriction can be increased by up to 365 days by "The State Pollution Control Board (SPCB)." If the e-waste needs to be handled before it is conveyed to a dismantler or recycler, an extension of time will be made.

The below table lists the e-waste collection targets, as stated in the E-waste (Management) Amendment Rules of 2018 (Table 10.4).

If the company is not able to satisfy the conditions, it should cease manufacturing electrical products, and only after the conditions indicated above have been met, it will be able to resume production.

10.6.3 RESPONSIBILITIES

Manufacturer Responsibilities [13]

1. E-waste produced during the manufacturing of any electrical or electronic appliance must be gathered and transported for recycling or disposal.
2. Application should be given for an authorization in Form 1 (a) to the relevant SPCB, which will provide the authority in Form 1 after following the procedure indicated in sub-rule (2) of rule 13. (bb).
3. Ensure the safety of the environment while e-waste is being stored or transported.
4. Maintain records of Form 2 of e-waste generated, handled, and disposed of. Make these records available to the SPCB for examination.
5. Annually submit Form 3 returns by 30th June following the budget year in question to the SPCB.

Producer's Responsibilities [13]
The following is an important task for the manufacturer of the electrical and electronic appliances listed in schedule:

1. Using the frameworks outlined below, implement Extended Producer Responsibility Authorization:
   a. E-waste generated from old equipment or equipment that is at the end of its life cycle must be collected and channelized with the same electrical and electronic appliance code as old waste available on the date these

rules take effect, in accordance with the targets prescribed in Schedule III, in accordance with Schedule I.

b. The Responsibility of the Extended Producer—Authorization process must be followed for channeling e-waste from "end-of-life" items, including those from their service centers, to an authorized dismantler or recycler. Fluorescent and other mercury-containing bulbs may be routed from a collection center to a treatment, storage, and disposal facility if recyclers are unavailable.

c. A pre-treatment is necessary for disposal in a treatment, storage, and disposal facility to immobilize the mercury and reduce the volume of garbage to be disposed of.

d. Extended Producer Responsibility Authorization should include a general scheme for collecting waste electronic and electric appliances from previously sold electronic and electric appliances, whether directly or through any authorized agency, and channeling the items collected to authorized agencies, such as through dealers, collection centers, producer responsibility organizations, buyback arrangements, exchange schemes, deposit refund systems, and other methods.

e. Providing contact information such as address, e-mail address, toll-free telephone numbers, or helpline numbers to consumers or bulk consumers via their website and product user documentation to assist in the return of end-of-life electrical and electronic equipment.

f. Through advertisements, media, posters, publications, and any other means of communication, public awareness of end-of-life electrical and electronic equipment as well as product user documentation accompanying the equipment should be increased.

  i. Address, e-mail address, toll-free phone numbers or helpline lines, and information about the company's website.

  ii. Information on hazardous ingredients as described in sub-rule 1 of rule 16 in electrical and electronic equipment.

  iii. Information on the risks associated with improper e-waste management, disposal, unintentional breakage, damage, or recycling.

  iv. Do's and Don'ts, as well as instructions on how to manage and dispose of the equipment once it has been used.

  v. Adding the following visible, legible, and indelible emblem to objects or product user manuals to prevent e-waste from being thrown away with other garbage.

  vi. The methods and processes accessible to their clients for returning e-waste for recycling, including, if applicable, the Deposit Refund Scheme's details.

g. On an individual or collective basis, the producer may choose to accept Extended Producer Responsibility. Individual producers can accomplish Extended Producer Responsibility by setting up their own collection center, implementing a take-back system, or doing both. In a communal system, producers can join a producer responsibility organization, an e-waste exchange, or both. In all cases, the individual producer must apply to the Central Pollution Control Board for Extended Producer Responsibility Authorization using Form 1 and the method described in rule 13 sub-rule (1).

2. To provide information on the implementation of a Deposit Refund Scheme, if one is included in the Extended Producer Responsibility Plan, to ensure the collection of end-of-life items and their channelization to authorized dismantlers or recyclers. However, at the time of take-back of the end-of-life goods, the manufacturer shall reimburse the deposit amount obtained from the consumer or bulk consumer at the time of sale, along with interest at the prevailing rate for the period of the deposit.

3. Electric and electronic equipment can only be imported by companies that have been granted Extended Producer Responsibility.

4. Maintain records in the Form 2 of all handled e-waste. Those records should be made available to the Central Pollution Control Board or the appropriate SPCB for examination.

5. Submit the annual returns by June 30th following the financial year in question on Form 3 to the Central Pollution Control Board. If a producer has many offices in a state, he/she must file a single yearly return that includes data from all of them.

6. The producer must apply for authorization in Form 1 to the Central Pollution Control Board, after which, the Extended Producer Responsibility Authorization in Form 1 will be granted (aa).

7. Production without Extended Producer Responsibility Authorization is banned under this regulation.

Collection Centers' Responsibilities [13]

1. Collect e-waste for the producer, dismantler, recycler, or refurbisher, including orphaned items; however, the producer's collection centers can also collect e-waste for the dismantler, refurbisher, and recycler, including orphaned products.

2. Assist the facilities in adhering to the Central Pollution Control Board's rules and guidelines, as they are issued from time to time.

3. Ensure that the e-waste they gather is safely stored until it is sent to an authorized dismantler or recycler.

4. Ensure that no environmental damage happens because of e-waste storage and transportation.

5. Keep Form 2 records of all e-waste handled in compliance with the Central Pollution Control Board's standards and ensure their availability for inspection by the Central Pollution Control Board or the applicable SPCB as needed.

Dealers' Responsibilities [13]

1. If the producer has delegated collection to the dealer, the dealer shall collect the e-waste by providing the consumer with a clearly marked box, bin, or container in which to deposit e-waste; or a take-back mechanism that sends

the e-waste to be recycled as directed by the producer; or collected to a collection center, dismantler, or recycler.
2. If applicable, the dealer, retailer, or e-retailer will reimburse the amount in accordance with the take-back procedure. For e-waste depositors, the producer's deposit refund system.
3. It is the responsibility of each dealer to ensure that the e-waste generated is safely moved. Recyclers or dismantlers have been authorized.
4. Make certain that the environment is not harmed during the storage and transit of e-waste products.

The Refurbisher's Responsibilities [13]

1. During the refurbishment process, the generated e-waste must be collected and sent to an authorized dismantler or recycler via the company's collection center.
2. Apply for a one-time authorization on Form 1(a) to the competent SPCB, following the procedure indicated in sub-rule [14] of rule 13. (a) In accordance with Form 1 (bb), the concerned SPCB shall approve the refurbisher on a one-time basis, and the authorization will be deemed approved if no objections are received within thirty days. (b) The approved refurbisher must report the amount of e-waste generated to the applicable SPCB on a yearly basis.
3. Ensure the safety of the environment during the storage and transportation of e-waste.
4. Verify that the process of refurbishment has no harmful effects on human health or the environment.
5. Ensure safe transportation of e-waste to collection centers, dismantlers, or recyclers that are authorized.
6. Prior to or in anticipation of the 30th June following the budget year to which the return relates, file the yearly returns on Form 3 with the competent SPCB.
7. Records of e-waste handled should be maintained on Form 2 and made available for inspection by the appropriate authority.

Consumer's/Bulk consumer's Responsibilities [13]

1. Consumers and bulk consumers of electrical and electronic equipment listed in Schedule I must ensure that any e-waste they generate is routed through an authorized producer's collection center, dealer, dismantler, or recycler, or the producer's designated take-back service provider to an authorized dismantler or recycler.
2. Large consumers of Schedule I electrical and electronic equipment must preserve Form 2 records of e-waste generated and make them available to the SPCB for review.
3. End-of-life electrical and electronic equipment listed in Schedule I may not be mixed with e-waste containing radioactive material, as defined by the Atomic Energy Act of 1962 (33 of 1962) and its regulations.

4. In Form 3, by the 30th of June following the budget year to which the return applies, the majority of consumers of electrical and electronic appliances listed in Schedule I must file the yearly returns with their appropriate SPCB. If a bulk consumer has multiple offices in a state, the SPCB must receive a single annual return containing information from all the locations on the same day.

Dismantler's Responsibilities [13]

1. Ascertain that the facility and dismantling operations comply with the Central Pollution Control Board's rules and guidelines as provided from time to time.
2. Obtain approval from the appropriate SPCB, as described in rule 13's sub-rule (3).
3. Make certain that no environmental harm happens to the e-waste products while being stored and transported.
4. Make certain that the decommissioning methods have no harmful consequences for human health or the environment.
5. Make sure that deconstructed e-waste is separated and transported to a recycling facility that can recover the materials.
6. Check that non-recyclable or non-recoverable components are properly treated, stored, and disposed of at the appropriate recycling or recovery facility.
7. Maintain a Form 2 record of all e-waste collected, dismantled, and transferred to an authorized recycler, and make that record accessible for examination by the Central Pollution Control Board or SPCB in question.
8. By June 30th following the budget year in question, Form 3 return must be submitted to the appropriate SPCB.
9. Unless approved as a recycler for material refining and recovery by the SPCB in charge, a person cannot process any e-waste for material recovery or refining.
10. Any operation without the permission of a dismantler, as stated in this article, is unlawful.

The Recycler's Responsibilities [13]

1. Shall make sure that the facility and recycling procedures follow the Central Pollution Control Board's rules and guidelines as they are updated.
2. Shall receive permission from the SPCB, using the method outlined in rule 13 sub-section (3).
3. This will ensure that no harm is done to the environment while e-waste is stored and transported.
4. This will ensure that recycling methods have no negative environmental consequences.
5. Provide the Central Pollution Control Board or the appropriate SPCB with all records for examination.

6. Ensure that any fractions or materials that cannot be recycled in its facilities are delivered to authorized recyclers.

7. Ensure that any residue formed during the recycling process is properly disposed of in a licensed treatment storage and disposal facility.

8. Maintain a Form 2 record of any e-waste that is collected, deconstructed, recycled, or delivered to an approved recycler, and make that record available for examination by the concerned Central Pollution Control Board or SPCB.

9. In Form 3 by June 30th following the end of the budget year to which the return pertains, file annual returns with the appropriate SPCB.

10. E-waste that does not contain any radioactive material can be accepted for recycling by non-Schedule I electrical and electronic appliances or components must be declared while seeking a license from the appropriate SPCB.

11. Any action taken without the authorization of a recycler, as specified in this article, is considered environmentally detrimental.

State Government Responsibilities for Environmentally Appropriate E-waste Handling [13]

1. In existing and prospective industrial parks, estates, and industrial clusters, the State Department of Industry, or any other government body authorized by the State Government in this regard, must ensure that industrial space or sheds are set aside or allocated for disassembly and recycling of e-waste.

2. The State Department of Labour, or any other government body, The State Government, has authorized the following:
   a. Ensure that workers involved in dismantling and recycling are recognized and registered.
   b. Assist in the formation of groups of such personnel to improve the efficiency with which dismantling facilities are established.
   c. Organize industrial skill development programs for dismantling and recycling workers.
   d. Undertake annual inspections to safeguard the safety and well-being of deconstruction and recycling personnel.

3. The state government must create an integrated plan for executing these requirements, and a report must be submitted to the Ministry of Environment, Forestry, and Climate Change annually.

10.7 TECHNOLOGIES USED IN E-WASTE MANAGEMENT

10.7.1 E-WASTE MANAGEMENT USING DEEP LEARNING OBJECT CLASSIFIER

In order to properly dispose of e-waste, most garbage collection organizations rely on public engagement. Various garbage collection techniques have been proposed, including collection at supermarkets, electrical and electronic equipment stores, and municipal centers [3]. Mobile collection, such as pavement collection, and on-demand collection, when people request that materials be collected from their homes, are two further approaches [3]. On-demand collection, particularly in city centers, might be a good option for waste collection that is well planned. For efficient garbage collection, the right number of vehicles and optimized routing are required to avoid extra collection costs and a potential loss of profit. Furthermore, in order to increase the quantity of waste collected legally, households and waste collection firms require a flexible and dependable communication system to act as a bridge between them. India and China have previously investigated using IoT to provide rubbish collection services. E-commerce and new technologies are currently popular in the field of e-waste recycling [3]. After doing research in 2019, Zhang et al. discovered that e-commerce was not widely embraced by Chinese citizens [3]. Residents' views, subjective norms, and the acknowledged advantage of e-commerce were all positively related to their intention to use e-commerce for e-waste recycling. We use a myriad of apps daily, and mobile applications and smart phones are ubiquitous. With the right use of these apps, the waste collection process can be made considerably easier [3]. Object detection and identification are used in artificial intelligence systems in robotics, electronics, road safety, autonomous driving, intelligent transportation systems, and content identification [3]. A customized adjustable frame is used to identify the object's location and size after it has been identified. Image processing (often employed in sorting objects with specified shapes and transparencies) and recognition systems are utilized extensively in trash management and waste processing applications [3]. Many solutions based on machine learning algorithms have been developed for the treatment of e-waste. One method uses support vector machine (SVM) classifiers to find garbage containers and check the level of waste in each container. E-waste comes in all shapes and sizes, and vertical garbage bins can be empty, partially filled, or completely full of it [3]. In 2016, Hanna et al. proposed a content-based image retrieval method to examine the likelihood of extracting texture from pictures to recognize the amount of solid trash accumulated in containers [3]. Neural networks with deep learning are used in an e-waste separation approach. R-CNN was used to automate the identification and size of waste equipment detected in photographs, while CNN was utilized to classify the trash equipment [3].

10.7.1.1 E-waste Image Recognition

For the identification of household e-waste collection, deep learning CNN and quicker R-CNN are employed in this method. The collection company and individuals seeking waste removal could employ digital image sharing to bridge the information gap. Depending on the capabilities of the smartphone, this system could possibly work as a mobile application. Individual images of the equipment to be disposed of are taken and sent to a server that uses image recognition software to identify the items. The unique characteristics of these objects are used in the learning process. For the identification procedure, images of various models of televisions or monitors, refrigerators, and washing machines are gathered. This approach has an average accuracy of 90–96%, but it can only identify one object in an image. Automated visual e-waste recognition benefits both garbage collection firms and residents, who can request waste recovery by simply emailing a photo of the waste equipment. This approach works on both a server and a mobile app [3].

$$y = \left[y_{i,j,q} \right] = \left[b_k + \sum_{i=0}^{k_w} \sum_{j=0}^{k_w} \sum_{p=o}^{c} \left(w_{i,j,p,q} \times x_{i+i,j+j,p} \right) \right] \tag{10.1}$$

Convolutional and pooling layers alternate in the feature-learning module. On the input picture x, the convolution layer performs a 2D convolution with m different filters.
   Max pooling,

$$y = \left[y_{i,j,q} \right] = \max_{1 \le i \le w_p, \ 1 \le j \le h_p} \left(x_{i+1,j+1,p} \right) \tag{10.2}$$

This strategy, which takes use of the widespread usage of smart phones with the ability to click and upload photographs, as an essential instrument for collection planning, allows the potential recovery of secondary raw material content. The company must estimate how many items are to be collected through resident communication in order to minimize any possible collection complications and to send the appropriate truck and crew. At this time, requesting e-waste collection and arranging the collection process are not automated. This results in high mobile collection costs, which has a negative impact on collection companies' evaluation of this strategy [3] (Figure 10.6).
   We have suggested an approach that can possibly overcome the drawbacks of the above system. The manual e-waste retrieval is automated by using AI and GPS algorithms, discussed below.

10.7.2 SOLUTION FOR MANUAL E-WASTE RETRIEVAL AND COLLECTION ROUTE PLANNING

As we have shown, the implementation of a deep learning object classifier using image recognition has a problem, in that, the process of waste retrieval request and collection route planning is done manually, which raises mobile costs and negatively impacts the evaluation of this approach by businesses. We'll look at a theoretical way of automating this process here. Initially, the user (customer) uploads an image of the equipment to the company's server, together with the number of equipment (if relevant) and the location. The image is then detected using a machine learning method that employs CNN and R-CNN. Following image recognition, an image-matching algorithm is used to match this image with an image stored in the company's database. When a match is identified, the information about that piece of equipment is taken from the database (model, size, weight, etc.) and then transferred to the system. The system then recommends the sort of vehicle that will be needed to retrieve the equipment, as well as the number of people who will be needed. The system then uses the GPS mechanism, which is the Dijkstra's shortest path method, to calculate the shortest distance between the source and the firm. The management sees all this data. The corporation may simply cut mobile costs and the time necessary for waste collection route design with the use of this information.

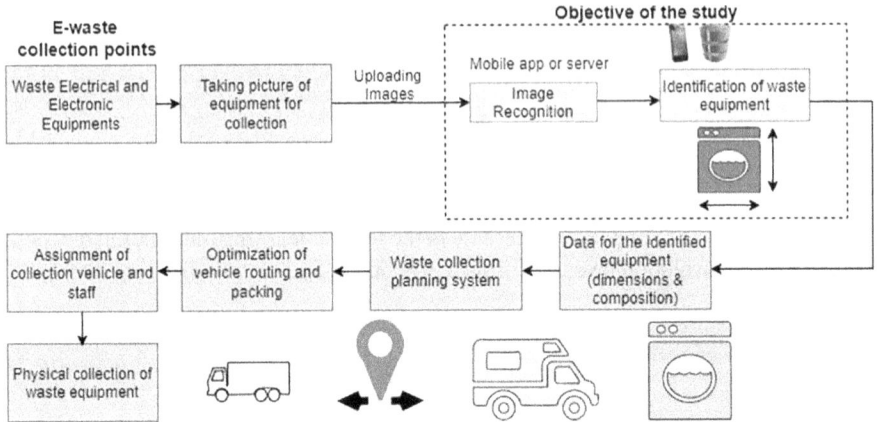

FIGURE 10.6 E-waste collection using image recognition and visual classification of waste equipment.

10.7.2.1   Concept of GPS and Dijkstra's Algorithm

The global positioning system (GPS) is a satellite-based navigation system established and administered by the United States Department of Defence [15]. It is made up of satellites, control and monitoring stations, and GPS receivers that receive data from satellites. To determine a user's precise location, GPS triangulation is used [15].

GPS is used in a variety of ways, like the following [15]:

1. Position and location determination.
2. Navigate from one location to another.
3. Creation of digitized maps.
4. Determination of distance between any two points.

Dijkstra's algorithm (like Prim's method for minimal spanning tree) is a graph-based search algorithm. The single-source shortest path problem is solved by this algorithm. Only positive weighted graphs are used with it. This algorithm is frequently used in routing to discover the shortest path with the least amount of expense.

A. Dijkstra's Shortest Path Algorithm using GPS [15]:
   Step 1: Make all the node distances infinite.
   Step 2: Using GPS, determine the current location of the source node.
   Step 3: Make the set of nodes empty at first. After that, it's a matter of queuing and determining the vertices mid-distance.
   Step 4: Use GPS to determine the location of the visited nodes.
   Step 5: If the shortest path has been discovered, set it to the desired trackback.
   Step 6: After determining the shortest path, use GPS to determine the location of the visited nodes.

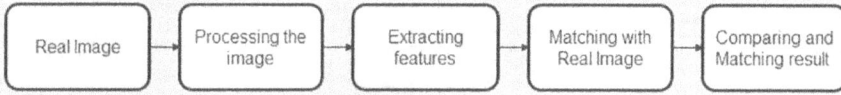

FIGURE 10.7 Image-matching process.

10.7.2.2 Algorithm for Image Matching

Though the principles of different image-matching algorithms are different, their main processes are the same [16] (Figure 10.7).

1. Pre-processing of original image.
2. Extraction of image-matching information from the pre-processed original image.
3. Matching the image.

The image-matching algorithm consists of four elements, namely

1. Similarity measurement [16]
2. Feature space [16]
3. Search space [16]
4. Search strategy [16]
   1. Similarity measurement:

A distance or correlation function is used to calculate the degree of similarity between the original and target images [16]. Absolute difference, square difference, mean absolute difference, and mean square difference are the primary distance functions, whereas the covariance and normalized cross-correlation (NC) functions are the primary correlation functions [16]. For two images A mxn and B mxn, their AD is defined by

$$AD(A,B) = \sum_{i=0}^{m}\sum_{j=1}^{n}\left|A_{i,j} - B_{i,j}\right| \tag{10.3}$$

This SD is given by

$$SD(A,B) = \sum_{i=0}^{m}\sum_{j=1}^{n}\left(A_{i,j} - B_{i,j}\right)^2 \tag{10.4}$$

Their MAD and MSD are defined by $\dfrac{AD(A,B)}{M \times N}$ and $\dfrac{SD(A,B)}{M \times N}$, respectively.

The NC function is defined by

$$NC(A,B) = \frac{\sum_{i=1}^{m}\sum_{j=1}^{n}\left(A_{ij} - B_{i,j}\right)}{\sqrt{\sum_{i=1}^{m}\sum_{j=1}^{n}\left(A_{i,j}\right)^2}\sqrt{\sum_{i=1}^{m}\sum_{j=1}^{n}\left(B_{i,j}\right)^2}} \quad (10.5)$$

1. Feature space: Includes grayscale values, image edges, corners, textures, and shapes [16].
2. Search space: It is a set of image-matching parameters [16].
3. Search strategy: It employs the proper search method to determine the optimal estimation of transformation parameters such as translation and rotation in the search space, resulting in the greatest possible similarity between the transformed images [16].

10.7.2.3 Image-matching Algorithm

Presently, many image-matching algorithms are in existence. Based on these principles, algorithms can be classified as follows:

1. Gray value-based image-matching method [16]
2. A feature-based image-matching algorithm [16]
3. Frequency domain analysis-based image-matching method [16]
4. Neural network-based image-matching method [16]
5. A semantic recognition-based image-matching technique [16]

Among the options listed above, feature and semantic recognition algorithms can be used most effectively, as demonstrated below.

A. **Image-matching algorithm based on features:**
   Principle: To achieve image matching based on the features of the target image, this type of image-matching algorithm searches for the image with the highest similarity of the feature in the search space [16] (Figure 10.8).
   Step 1: From the image, extract image features such as corner points, inflection points, edge points, line features, surface features, and moment features [16].
   Step 2: To describe the image's attributes, a set of parameters were employed [16].
   Step 3: Align the parameters with the image you want to use as a target [16].
B. **Image-matching algorithm based on semantic recognition:**
   Principle: In this type of image-matching algorithm, the semantic information contained in the picture is used to generate image matching [16]. The image's underlying visual features, such as colors, lines, and contours, are extracted first; then, by analysis of the underlying features, high-level semantic features are constructed; and finally, the semantic features are

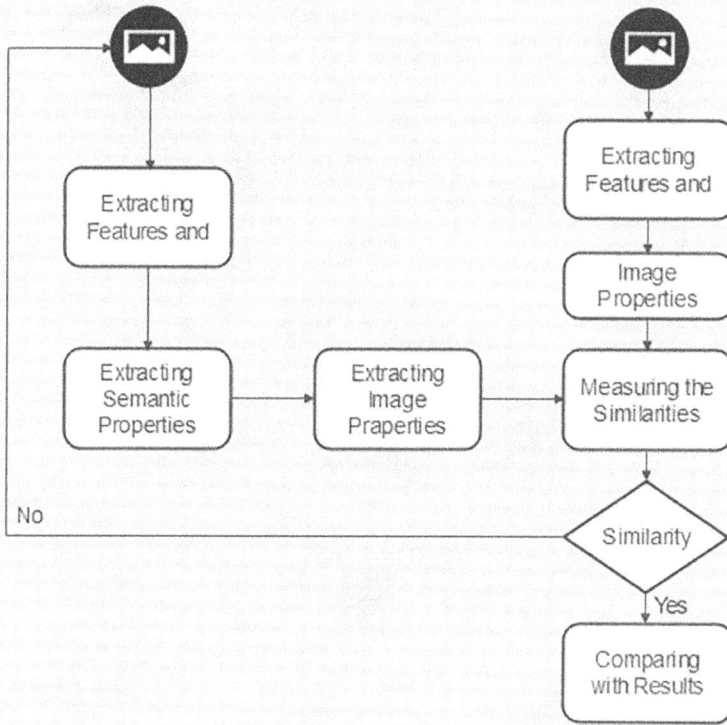

FIGURE 10.8 Flowchart based on features.

used to judge the similarity of the images to locate the target image and complete the image matching [16] (Figure 10.9).

Characteristics: The semantic properties indicate how well the visual content is comprehended. This method of image matching is akin to a human's intuitive comprehension of an image's meaning, which is a crucial step toward intelligent image matching [16]. These methods are used in a variety of fields, including artificial intelligence, computer vision, and machine learning, to correlate image data with its abstract meaning. The goal of this research is to create a high-precision, quick semantic matching method [16]. However, there are a few fundamental issues that must be addressed. First, how to extract high-level semantic features from the underlying visual characteristic processing; second, image data are huge, and semantic features are abundant; third, image meaning is subjective; and finally, how to extract semantic features reliably and make them consistent with image meaning. Alternatively, the hunt for such algorithms is still in its initial stages [16].

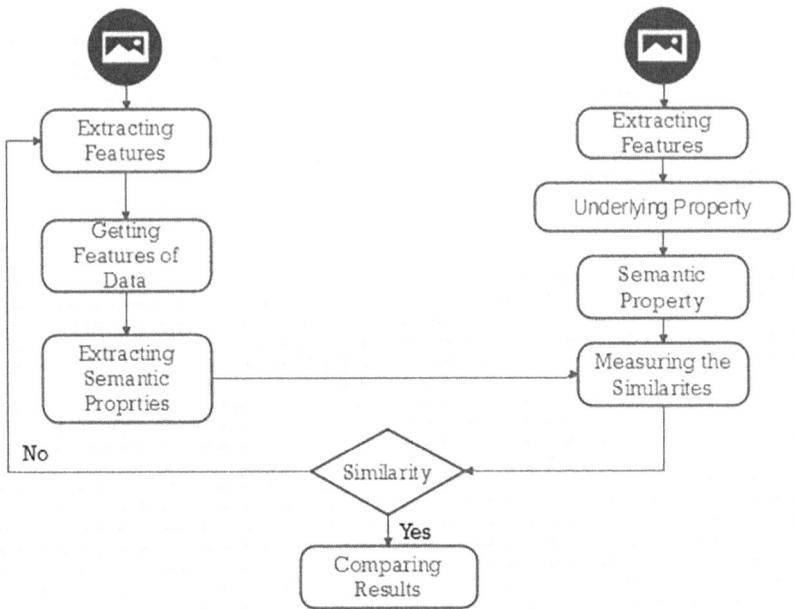

FIGURE 10.9   Flowchart based on semantic recognition.

10.8   CONCLUSION

This chapter examines the causes of e-waste development in India, the threats related to inappropriate processing and ejection of e-waste, and the technology used to handle e-waste. After conducting the research, it was discovered that, despite the existence of legal guidelines for e-waste disposal in India, people tend to ignore them, resulting in increasing mishaps and health risks. Export of e-waste to developing and underprivileged countries is done by several developed countries. It offers advantages like the extraction of several rare metals from e-waste and downsides like the remaining portions of e-waste becoming rubbish and adding to the trash. The most crucial aspect of e-waste management is segregation. This can be done easily by the application of ML (e.g., CNN, RCNN, SVM, etc.) and IOT.

This chapter has reviewed the Indian rules, their applicability, and the responsibilities of various stakeholders for handling and managing e-waste. This chapter also looked at how deep learning object classifiers can help with e-waste management. Following the general overview, an e-waste collection solution based on AI and GPS algorithms is presented. As a result, this lecture offers a technical answer to the problem of laborious and expensive e-waste collection. In AI, the image-matching algorithm is used, and in GPS, the Dijkstra shortest path algorithm is used.

REFERENCES

[1] Roy, E. (2021). 31.6% rise in e-waste generation last year: Ashwini Choubey to Rajya Sabha. The Indian Express.

[2] Rao, R. (2021). E-waste: A growing problem. Business Today.

[3] Piotr Nowakowski, T. P. (2020). Application of deep learning object classifier to improve e-waste collection planning. *Waste Management*, 109, 1–9.

[4] Rahul S Mora, K. S. (2021). E-waste management for environmental sustainability: An exploratory study. *Science Direct*, 98, 193–198.

[5] Daniel Mmereki, B. L. (2016). The generation, composition, collection, treatment and disposal system, and impact of E-waste. IntechOpen Book Series, London, SW7 2QJ, UK.

[6] Lundgren, K. (2012). The global impact of e-waste Addressing the challenge. *International Labour*, 1–72.

[7] Pinto, V.N. (2008). E-waste hazard: The impending challenge. *Indian Journal of Occupational and Environmental Medicine*, 12(2), 65.

[8] Kunal Sinha, A. B. (2013). Electronic waste management in India: A stakeholder's perspective. *Electronic Green Journal*, 36. https://doi.org/10.5070/G313618041

[9] Sivakumar, V. (2016). Global and Indian e-waste management - methods & effects. *13th International Conference on Science, Engineering and Technology*, Vellore Institute of Technology, Vellore: Research Gate.

[10] New Technologies to Recycle Electronic Waste. (2020). The Conversation. Retrieved from https://theconversation.com/amp/new-technologies-to-recycle-electronic-waste-133288.

[11] Larmer, B. (2018). E-waste offers an economic opportunity as well as toxicity. The New York Times Magazine.

[12] E-Waste Management. (n.d.). India Fillings. Retrieved from https://www.indiafilings.com/learn/e-waste-management/.

[13] Government of India Ministry of Environment, F. A. (2016). Retrieved from https://cpcb.nic.in/displaypdf.php?id=RS1XYXN0ZS9FLVdhc3RlTV9SdWxlc18yMDE2LnBkZg==.

[14] D, Z. (2021). What is E-waste? Definition and why It's Important. Great Lakes Electronics Corporation. https://www.ewaste1.com/what-is-e-waste/

[15] Singal, P. and Chhillar, R.S., 2014. Dijkstra shortest path algorithm using global positioning system. *International Journal of Computer Applications*, 101(6), 12–18.

[16] Zhang, X. and Feng, Z., 2018, April. New development of the image matching algorithm. In *Ninth International Conference on Graphic and Image Processing (ICGIP 2017)* (Vol. 10615, pp. 509–516). SPIE.

11 Performance Evaluation of Concrete Incorporating Admixtures and Stone Waste Powder

Kiran Devi, Babita Saini, and Paratibha Aggarwal

CONTENTS

11.1 Introduction 189
11.2 Experimental Detail 190
11.3 Results and Discussion 191
    11.3.1 Compressive Strength 191
    11.3.2 Split Tensile Strength 192
    11.3.3 Electrical Resistivity 192
    11.3.4 Correlation between Compressive Strength and
            Electrical Resistivity 194
    11.3.5 Chemical Attack 194
            11.3.5.1 Change in Strength 194
            11.3.5.2 Electrical Resistivity 196
            11.3.5.3 Microstructural Analysis 197
            11.3.5.4 Prediction Model 199
            11.3.5.5 Environment and Economic Analysis 199
            11.3.5.6 Performance Evaluation 200
11.4 Conclusions 201
References 202

11.1 INTRODUCTION

The expansion in the construction industry has increased cement consumption globally. The cement manufacturing process is considered a high-energy process and contributes to global warming by generating hazardous carbon dioxide (CO_2). The fuel cost to burn the clinker is an additional burden on the cement-making unit. This predicament entails the development of cost-effective and environment-friendly alternatives to cement on a worldwide scale (Singh et al. (eds) 2019; Devi et al. 2018b). On the other hand, the solid waste generated from various industries has become an environmental threat globally. Direct dumping of wastes, for example, dust and slurry, from stone factories pollutes waterways, damages soil fertility owing to blockage by

DOI: 10.1201/9781003359784-11

extremely fine dust particles, impairs the human and animal respiratory systems, and degrades the scenic view of nearby regions. Scientific and technical remedies for direct disposal of wastes are required. Utilization of such dumping wastes as cementitious material substitutes in concrete production could be a good option (Hussain 2015; Ashish 2018; Devi et al. 2018c). Sustainable concreting comprises the reduction of CO_2 emissions through the use of industrial by-products in cement and concrete production resulting in conservation of the natural resources (Bjegovic et al. 2012). The use of 20% granite sludge as cement substitution for cement produced favorable results for mortar and concrete in terms of strength and durability (Mashaly et al. 2018). Marble waste powder and silica fume both increased the mechanical performance of concrete at an early and later age. Electrical resistivity was improved by silica fume, whereas marble waste powder lowered it marginally (Khodabakhshian et al. 2018). Kota stone dust in mortar improved the early-age strength (Meena et al. 2018). The addition of stone powder as a substitution to fine aggregates improved the compressive strength (CS) and elastic modulus of concrete with manufactured sand (Zheng et al. 2020). The replacement of fine aggregates with fine stone powder up to 15% in mortar enhanced the strength and chemical resistance significantly (Rajagopalan and Kang 2021). The compressive and flexural strength of concrete with the inclusion of marble powder and granite powder increased upto a certain extent afterward reduction in strength was observed (Singh and Aggarwal 2022). The incorporation of fine granite waste powder up to 15% improved the performance of geopolymer concrete along with minimizing the embodied energy and embodied carbon dioxide (Saxena et al. 2022). Stone waste up to 10% has reduced water absorption, density, and mechanical strength insignificantly at later age. The resistance of concrete specimens against sulfuric acid attack was significant (Rashwan et al. 2022).

Apart from industrial wastes, chemical admixtures are being used in concrete to improve its certain properties. Among various admixtures, accelerators are employed to speed up the setting and hydration rate, in order to provide early-age strength for ease in removal of exterior formworks, as well as to increase construction pace and save time. Calcium nitrate (CN) and triethanolamine (TEA) have been employed in reinforced concrete structures as accelerating admixtures (Aggoun et al. 2008; Devi et al. 2018a,b). TEA accelerates C_3A hydration while delays the C_3S hydration, and this effect is stronger when TEA content is increased (Han et al. 2015).

The effects of stone waste powder (SWP) and accelerators on strength and durability characteristics of normal strength concrete at 180 and 365 days have been studied. The environmental and economic assessment of different concrete mixtures has been carried out. The purpose of incorporating additives in concrete was to evaluate the long-term performance of concrete specimens under different laboratory or field conditions. The aim is the development of eco-friendly concrete end products by utilizing industrial waste.

11.2   EXPERIMENTAL DETAIL

Locally available, ordinary Portland cement of 43 Grade confirming to IS: 8112-1988 and aggregates (fine and coarse) confirming to IS: 383–2016 were used to prepare the specimens for testing (Devi et al. 2020). Calcium nitrate and triethanolamine as

TABLE 11.1
Detail of Various Mix Proportions of Concrete (Devi et al., 2019)

Mix designation	Cement, kg/m³	Sand, kg/m³	Aggregates, kg/m³	Water, kg/m³	CN, kg/m³	TEA, kg/m³	SWP, kg/m³
N1	390	505	1,090	168	0	0	0
N2	360.75	505	1,090	168	0	0	29.25
N3	390	505	1,090	168	0	0.195	0
N4	390	505	1,090	168	3.9	0	0
N5	390	505	1,090	168	3.9	0.195	0
N6	360.75	505	1,090	168	0	0.195	29.25
N7	360.75	505	1,090	168	3.9	0	29.25
N8	360.75	505	1,090	168	3.9	0.195	29.25

accelerating admixtures along with or without SWP were used in the concrete mixtures (Devi et al. 2020). Eight various concrete mixtures consisting of CN (1%), TEA (0.05%), and SWP (7.5%) individually and in combination were casted (Devi et al. 2019) and have been designated in Table 11.1.

CS of the concrete cubes having each side 150 mm and split tensile strength of cylinder having dimension 150×300 mm was determined at 180 and 365 days as per IS: 516–1959 (Devi et al. 2020). Electrical resistivity (ER) of concrete cubes was determined at 180 and 365 days, using two-point method (Khodabakhshian et al. 2018; Devi et al. 2019). The influence of magnesium sulfate ($MgSO_4$) and sodium chloride (NaCl) solution on CS and ER was investigated experimentally at $(28^* + 152^\#)$ and $(28^* + 337^\#)$ days (* indicate water curing and $^\#$ represents the specimens under chemical solution attack) (ASTM C1012/C1012 M-12, 2012). Scanning electron microscopy (SEM) and energy dispersive X-ray spectroscopy (EDS) techniques were used to examine the microstructure of concrete specimens (Devi et al. 2019, 2020). The chemical-calculations-based criteria proposed by Mohammed et al. (2013) were used to examine the formation of hydration products, that is, CSH, CH, and ettringite phases (Devi et al. 2020). The ecological and economic feasibility of additives in concrete mixtures was also evaluated (Devi et al. 2021).

11.3 RESULTS AND DISCUSSION

11.3.1 Compressive Strength

The consequences of calcium nitrate (CN), triethanolamine (TEA), and SWP on the long-term strength behavior of concrete have been examined, and the results are depicted in Figure 11.1. The addition of CN and SWP alone and in combination improved the CS of concrete in proportion to the control mix, which may be due to better bonding in concrete matrix, higher lime concentration in CN and pore-filling action of fine particles of SWP, respectively (Neville and Brooks 1997; Singh et al. 2017; Kumar et al. 2018). The particles of SWP provide a nucleus for hydration and

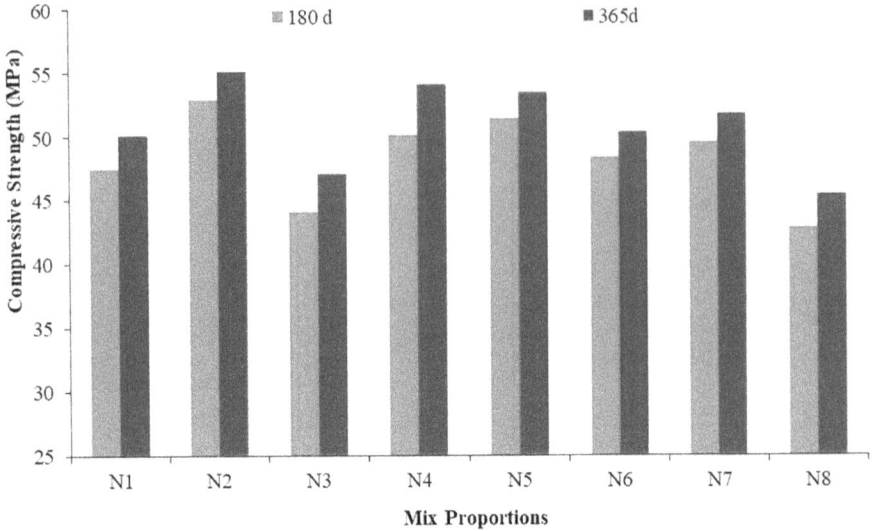

FIGURE 11.1 Variation of CS of different concrete mixtures.

accelerate the hydration, resulting in formation of dense matrix. The previous studies conducted by Al-Akhras et al. (2010), Ergun (2011), and Khodabakhshian et al. (2018) observed comparable results to the present study. The inclusion of TEA reduced the CS of concrete except for mixes N3 and N8, which may be due to its dominating effect. At 180 and 365 days, the CS ranged from 42.79 to 52.89 MPa and 45.47 to 55.15 MPa, respectively, for various mixes. The maximum percentage variation was 11%–10% and 10%–9% after 180 and 365 days of curing for mixes N2 and N8, respectively.

11.3.2 Split Tensile Strength

The tensile strength is taken into account while designing heavy concrete structures such as dams, subjected to earthquakes, pavement slabs, and airfield runways. Thus, the tensile strength of the structure is equally important. The variation in STS of different concrete mixes has been illustrated in Figure 11.2 at 180 and 365 days. Results showed that incorporation of CN and SWP enhanced the STS, whereas TEA declined the strength. The previous studies by Almeida et al. (2007) and Khodabakhshian et al. (2018) reported the same trend for STS. STS varied from 4.52 to 6.5 MPa and 5.16 to 6.91 MPa, respectively, at 180 and 365 days. The maximum increase and reduction in STS were 12%, 10% at 180 days and 22%, 12% at 365 days. The mix N2 had the maximum strength among all mix proportions.

11.3.3 Electrical Resistivity

ER of concrete specimens containing accelerators and SWP has been investigated and is depicted in Figure 11.3. The figure shows that resistivity is increased after mixing CN and SWP in concrete, and, strong bonding due to CN and pore-filling

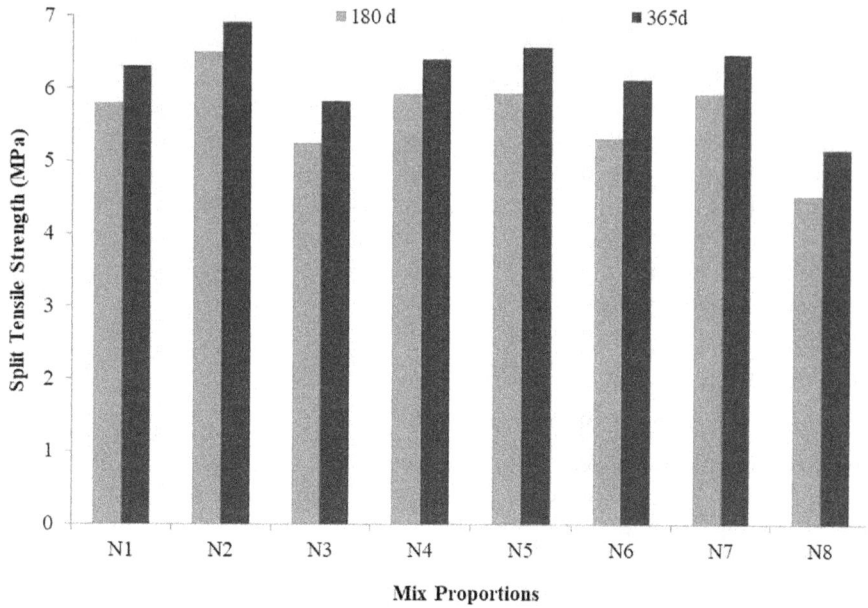

FIGURE 11.2 Variation of split tensile strength of different concrete mixes.

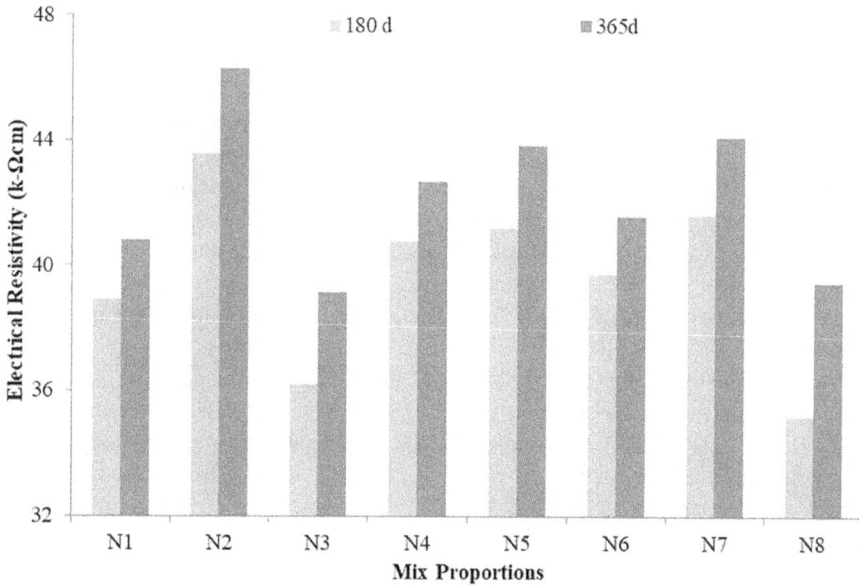

FIGURE 11.3 Variation of electrical resistivity of different concrete mixes.

effect of SWP are the possible reasons. However, TEA decreased ER. ER of different concrete mixtures followed the same trend as CS. The addition of SWP in concrete increased the ER by forming a thick matrix that restricts the ion transport inside the concrete specimens and gel formation increase during the hydration process (Xiao et al. 2007; Silva and Brito 2013).

The electrical resistivity of various concrete mix proportions was found in the range of 35.19–43.57 kΩ-cm at 180 days and 38.47–46.29 kΩ-cm at 365 days, respectively. The maximum increase as 25%, 13%, and decrease in ER was 10%, 4% at 180 and 365 days, respectively. Higher electrical resistivity (>20 kΩ-cm) is an indicator of a lower corrosion rate.

11.3.4 Correlation between Compressive Strength and Electrical Resistivity

The experimental results were used to derive the relationship between CS and ER. An exponential equation was developed for the correlation of CS and ER at a later age, as depicted in Figure 11.4a and 11.4b.

Table 11.2 consists of equations and regression coefficients obtained from the correlations. At 180 and 365 days, good correlation between CS and ER was derived (Khodabakhshian et al. 2018).

11.3.5 Chemical Attack

The changes in strength and electrical properties of various mixes after exposing them to chemicals have been studied at 152 and 337 days, and have been explained as mentioned below.

11.3.5.1 Change in Strength

The changes in the values of ER of various concrete mixes against magnesium sulfate (S) and calcium chloride (C) solution at 180 and 365 days have been illustrated in Figure 11.5. It has been found that ER increased with exposure duration for all mixes due to excellent bonding of CN and SWP's pore-filling effect and pozzolanic reaction except mix N8. The consumption of portlandite and gypsum, as well as the formation of hydrous products during pozzolanic reactivity and the development of compact matrix, may account for the increase in CS during sulfate attack (Cheng et al. 2018; Devi et al. 2020). A continuous hydration process develops the ettringite in confined voids and reduces the continuity of voids/pores which restrains the sulfate ion penetration and consequently improves the strength properties of the concrete product (Singh and Siddique 2014; Singh et al. 2016; Devi et al. 2020). CS under sulfate attack reached up to 25.62–42.31 MPa at 180 days and 21.01–44.79 MPa at 365 days which was 11.5%, 1.4%, and 6.6% higher at 180 days and 14.5%, 2.6%, and 7.6% higher after 365 days for the mix proportions N2, N4, and N7 than control mix. CS of the concrete mixes having no additives and exposed to sulfate attack dropped by 9.9%, 20.4%, 13.5%, and 32.4% at 180 days and by 6%, 16.7%, 11.4%, and 43.7% after 365 days for mix proportions N3, N5, N6, and N8.

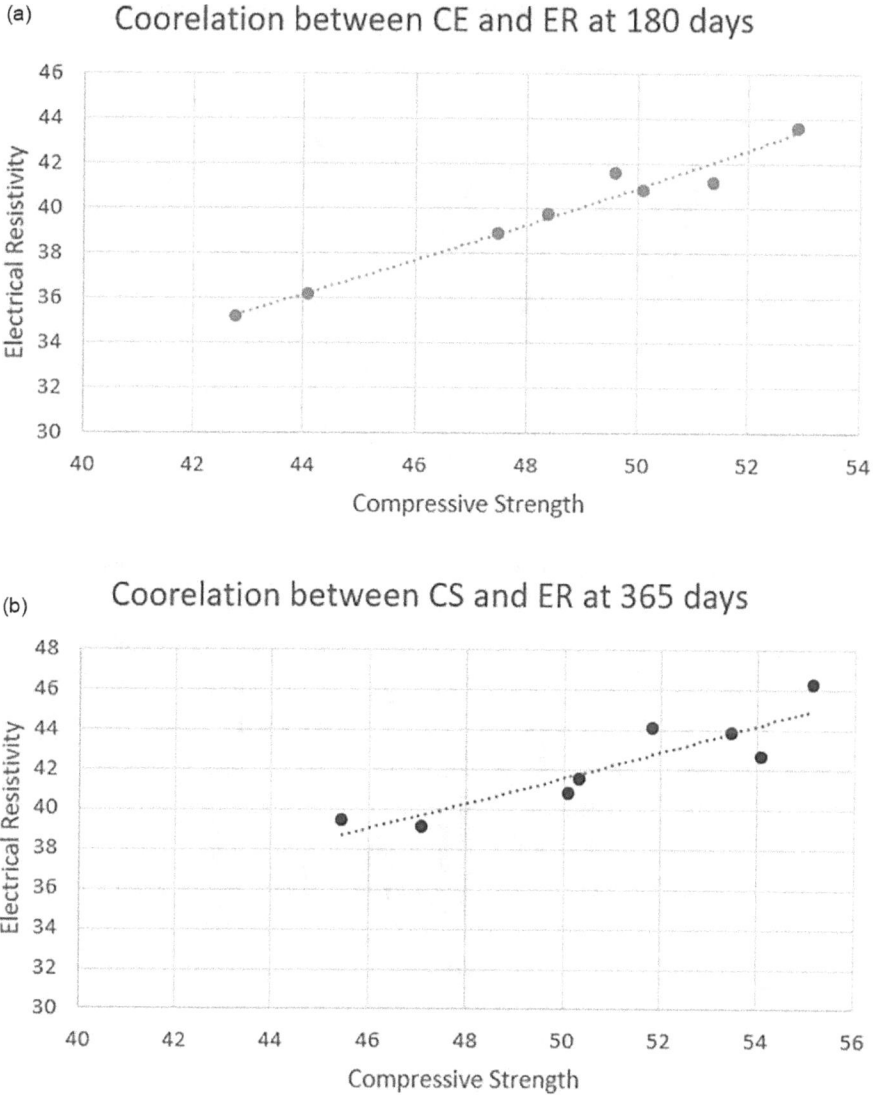

FIGURE 11.4 Relation between CS and ER after (a) 180 and (b) 365 days.

TABLE 11.2
Correlation between CS and ER of Different Concrete Mixes

Property 1	Property 2	Curing period (days)	Equation	Regression coefficient
CS	Electrical resistivity	180	$14.669e^{0.0205}{}_x$	0.96
		365	$19.144e^{0.0155}{}_x$	0.82

■ 180 D-WC   365 D-WC ■ 180 D-SA ■ 365 D-SA   180 D-CA   365 D-CA

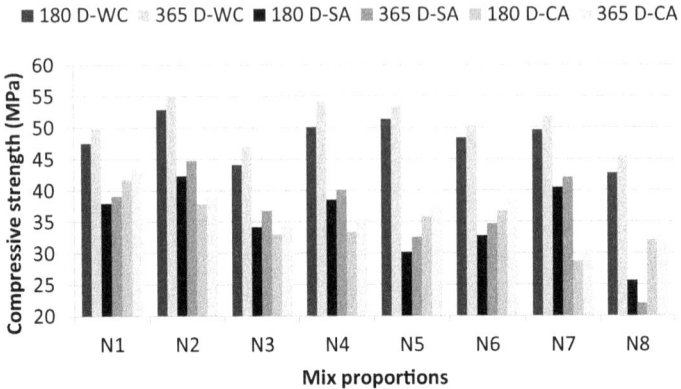

FIGURE 11.5   Variation in CS of different mixes under chemical attack.

■ 180 D-WC   365 D-WC ■ 180 D-SA ■ 365 D-SA   180 D-CA   365 D-CA

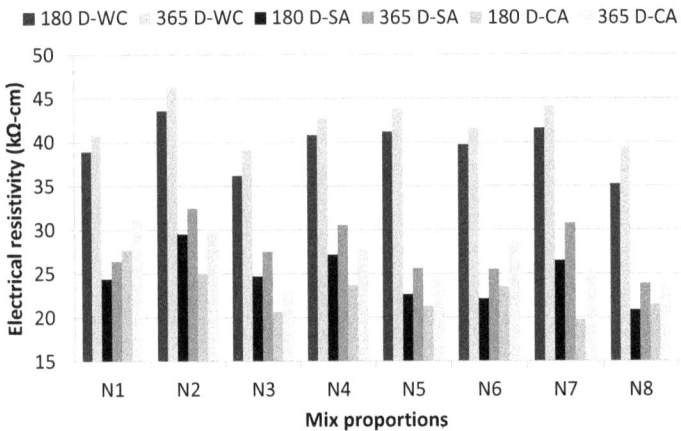

FIGURE 11.6   Electrical resistivity of mix proportions under chemical solution exposure.

Figure 11.5 shows that, when compared to the reference mix, CS decreased with age for all mix proportions under the exposure to chloride attack. Among all additives, SWP mixed specimens showed the highest equivalent strength. Under chloride attack, CS of different mix proportions ranged from 32.14 to 41.82 MPa at 152 days and 30.46 to 43.49 MPa at 335 days against chloride attack.

11.3.5.2 Electrical Resistivity

Figure 11.6 shows the ER of concrete specimens with various mix proportions against sulfate and chloride attack at later age. Resistance against electrical current was found to be increased, and the reasons could be same as discussed in the previous section. The mix proportions attained a range of 20.87–29.47 kΩ-cm at 180 days and 23.88–32.4 kΩ-cm at 365 days, respectively, after sulfate attack. Electrical resistivity of concrete increased by 20.8%, 1.2%, 11.3%, and 8.5% after 180 days and 22.7%, 4.1%, 15%, and 16.5% after 365 days, for mix proportions N2, N3, N4, and N7 after sulfate immersion. In comparison with control mix, electrical resistivity of concrete mixes declined by 7.2%, 9.3%, and 14.5% after 180 days and 3%, 3.4%, and 9.5% after 365 days of sulfate attack.

Under chloride attack, the resistivity of different mix proportions ranged from 18.49 to 27.65 kΩ-cm at 180 days and 22.69 to 31.2 kΩ-cm after 365 days. When comparing plain mix to mix proportions N2, N3, N4, N5, N6, N7, and N8, ER decreased by 9.8%, 25.2%, 14.3%, 23%, 15%, 28.6%, and 22.3% at 180 days and 4.6%, 25.9%, 11.1%, 21%, 7.8%, 18.6%, and 24.1% at 365 days.

The higher values of ER under the exposure of sulfate attack indicate a low corrosion rate, but increase in exposure time accelerated corrosion rate from low to moderate. The ER value of concrete specimens under the chloride solution exposure did not show any significant trend as sulfate attack produced for corrosion rate. However, the increase in exposure time to chloride attack decreased the ER value which showed acceleration in corrosion rate as suggested by Broomfield (2007).

11.3.5.3 Microstructural Analysis

The microstructural analysis of different water-cured mix proportions was carried out after 365 days using SEM and EDS techniques. Microscopic images of relative mix proportions have been shown in Figure 11.7a–h, and values of hydrous components ratios calculated from EDS analysis have been given in Table 11.3. The SEM micrographs of concrete specimens at 365 days of water curing have been shown in Figure 11.7a–h. Figure 11.7c shows the needle-type crystal, ettringite, which is randomly distributed and caused a looser structure with porosity. The strength reduction takes place and can be observed from lab results. Figure 11.7b, d–g showed dense and compacted matrix which decreased the porosity of matrix and enhanced the concrete strength. Similar trend was observed for the CS. Figure 11.7g showed plate-like structure and dense matrix, that is, calcium hydrate (CH). The experimental results of concrete strength properties and ER were confirmed by the SEM micrographs and EDS analysis, and found to be in the same trend.

Table 11.4 shows the values of CS, hydrous ratios, calculated using EDS analysis. It was found that hydration product calcium silicate hydrate (CSH) was formed due to consumption of CH during the pozzolanic reaction of stone powder, formed dense matrix, and aided in strength of concrete. The lesser amount or absence of sulfur (S) also indicates the lesser amount of mono-sulfate and ettringite. Ca/Si ratio was higher in mix proportions N3, N4, N6, and N8 than control mix, while mix N4 and N7 has an absence of sulfur content. A low Ca/Si ratio indicates the improvement in CS by the formation of CSH and consumption of CH during pozzolanic reactions (Cheng et al. 2018). The mix N3 had the highest value of Ca/Si which signifies the lowest strength as confirmed from CS results. Mixes having higher value of Ca/Si but still have improvement in strength (as obtained from macrostudy) were due to either absence or lower value of sulfur or (Al + Fe)/Ca ratio. The values of hydrous ratios were in the specified range which confirmed the formation of hydration products accordingly as reported by Mohammed et al. (2013).

FIGURE 11.7   SEM image at 365 days for mixes (a) N1, (b) N2, (c) N3, (d) N4, (e) N5, (f) N6, (g) N7, and (h) N8.

TABLE 11.3

Formation of Hydrous Products in Different Concrete Mixes at 365 Days

Mix Designation	CS	Hydrous Ratio		
		Ca/Si	(Al+Fe)/Ca	S/Ca
N1	50.12	1.849 (CSH)	0.348 (CSH)	0.004 (---)
N2	55.15	1.758 (CSH)	0.115 (CSH)	0 (CSH)
N3	47.1	6.145 (AFm)	0.083 (---)	0.026 (CH)
N4	54.1	2.077 (CSH)	0.174 (CSH)	0 (CSH)
N5	53.49	0.436 (---)	0.929 (AFm)	0 (CSH)
N6	50.35	0.821 (CSH)	0.422 (AFm)	0 (CSH)
N7	51.83	2.648 (CSH)	0.313 (AFm)	0 (CSH)
N8	45.47	3.583 (AFm)	0.121 (CSH)	0 (CSH)

TABLE 11.4

Statistical Value from Predicted Model

Property	R	R²	MAE	RMSE	Equation
CS	0.9607	0.9228	1.366	1.2102	$y = 0.9609x + 2.2503$
ER	0.8627	0.7496	2.2448	2.2495	$y = 28.682e^{0.0089x}$

11.3.5.4 Prediction Model

Figure 11.8a and b shows the graphical presentation of correlation between predicted and experimental CS and electrical resistivity at 365 days of water curing. Coefficient at 95% confidence level with minimal error value for both has been found satisfactory. Table 11.4 shows the statistical parameters and obtained equations for CS and ER values.

11.3.5.5 Environment and Economic Analysis

The ecological and economic analysis of various concrete mix proportions has been given in Table 11.5. The values of EE, ECO_2, and cost of different concrete mixtures varied from 1,896.2 to 2,037.1 MJ/m³, 343.4 to 372.5 kg CO_2e/m³, and 4,664 to 6,487 INR/m³. The inclusion of SSP reduced the value of environmental parameters with production cost by 7%, 7%, and 4%, whereas CN insignificantly enhanced all slightly. Since the TEA quantity was very low, its effect was not considered.

The EE and ECO_2 per unit strength reduced for mix proportions N2, N4, N5, N6, and N7, and for the rest of mixes, that is, N3 and N8 increased in reference to plain mix as shown in Figure 6.3. The percentage increase in EE and ECO_2 were 0.03% and 0.51%; cost varied from 1.3% to 39.16%; reduction in EE was 6.9%; and ECO_2 and cost varied from 6.83% to 7.34% and 3% for all mix proportions.

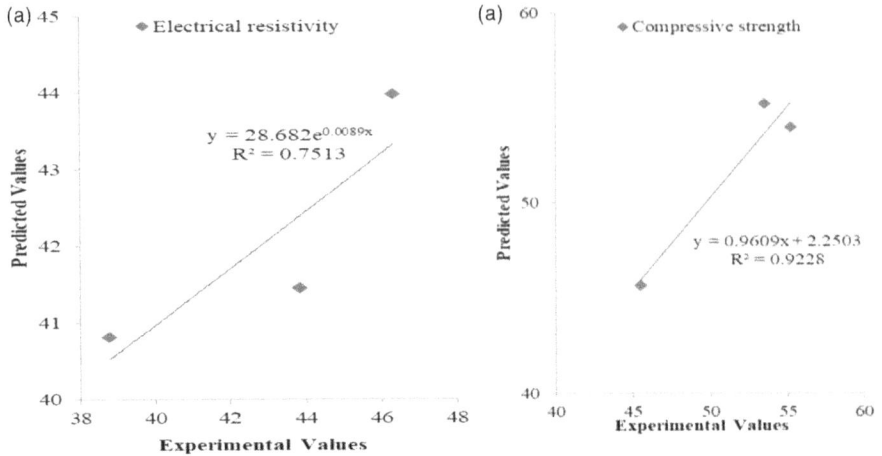

FIGURE 11.8 Correlation between predicted results and experimental results for (a) CS and (b) ER.

TABLE 11.5
Strength, EE, ECO$_2$, and Cost of Concrete Mixtures

Mix designation	EE (MJ/m³)	ECO$_2$ (kgCO$_2$e/m³)	Cost (INR/m³)
N1	2,036.6	370.6	4,810
N2	1,896.2	343.4	4,664
N3	2,036.6	370.6	5,017
N4	2,037.1	372.5	6,487
N5	2,037.1	372.5	6,694
N6	1,896.2	343.42	4,871
N7	1,896.7	345.3	6,341
N8	1,896.7	345.3	6,548

11.3.5.6 Performance Evaluation

The performance index of concrete for different mix proportions according to the requirements was evaluated and has been given in Table 11.6. The highlighted mixes had the optimum performance according to the properties, that is, mechanical strength, ER, resistance against chemical attack at various curing ages for the individual criteria as given in the table, respectively. The mix N2, that is, SSP at 7.5%, is the mix that can be preferred in various requirements such as economic, environmental, and strength.

TABLE 11.6
Individual Performance Indices of Different Concrete Mixes

Mix designation	365D CS	365D STS	365D ER	365D SCS	365D SER	365D CCS	365D CER	EE	ECO$_2$	Cost
N1	4.54	4.57	4.41	4.37	4.07	5.00	5.00	5.00	4.97	3.59
N2	5.00	5.00	5.00	5.00	4.99	4.50	4.77	4.65	4.61	3.48
N3	4.27	4.21	4.23	4.10	4.24	3.97	3.71	5.00	4.97	3.75
N4	4.90	4.64	4.61	4.48	4.70	4.08	4.45	5.00	5.00	4.85
N5	4.85	4.75	4.73	3.63	3.95	4.27	3.95	5.00	5.00	5.00
N6	4.56	4.43	4.49	3.87	3.93	4.39	4.61	4.65	4.61	3.64
N7	4.70	4.69	4.77	4.70	4.74	3.50	4.07	4.66	4.63	4.74
N8	4.12	3.73	4.26	2.46	3.68	3.78	3.80	4.66	4.63	4.89

11.4   CONCLUSIONS

Due to the continuous consumption of natural resources and generation of greenhouse gases during the production of cement, it is necessary to employ innovative and inexpensive materials such as industrial wastes. Accelerators as additives and Kota SWP as cement replacement were used in this work to explore strength, electrical resistivity, chemical resistance, and microstructural qualities at later ages. The following are the primary conclusions of this research:

- The strength and electrical resistivity of concrete improved with the introduction of calcium nitrate and SWP individually and in combination at later stage due to better bonding within the matrix and SWP filling effect forming dense matrix. Moreover, the same also has been confirmed from SEM-EDS analysis.
- Addition of TEA in concrete compromises the concrete properties, and same also has been confirmed by microstructural analysis.
- The chemical resistance of concrete against sulfate solution attack was found improved by using SWP and calcium nitrate whereas TEA reduced it.
- The correlations between CS and electrical resistivity at both 180 and 365 days were found acceptable with high correlation coefficient.
- Microstructural analysis of various samples verified the experimental results. Low value of Ca/Si ratio signifies the enhancement in strength, and presence of sulfur diminishes the strength and durability.
- The mix N2 (SWP), N4 (CN), and N7, that is, (CN+SWP), performed better among all the mix proportions at later age.

According to the findings of the study, SWP and calcium nitrate, both alone and in combination, provide good results in terms of strength and chemical resistance at 180 and 365 days. The use of SWP in cementitious composites helps to solve some of the challenges related to its disposal and storage, as well as produce environment-friendly and cost-effective concrete products. The N2, N4, and N7 concrete mixes may be utilized in field applications under both typical and hard environmental conditions, including sulfate attack.

The highlighted values show the best mix among all mixes (N1, N2…N8) for the respective properties (CS, STS, ER, etc.).

REFERENCES

Aggoun, S., Cheikh-Zouaoui, M., Chikh, N. and Duval, R. 2006. Effect of some admixtures on the setting time and strength evolution of cement paste at early ages. *Construction and Building Materials* 22:106–110.

Al-Akhras, N.M., Ababneh, A. and Alaraji, W.A. 2010. Using burnt stone slurry in mortar mixes. *Construction and Building Materials* 24:2658–2663.

Almeida, N., Branco, F., de Brito, J. and Santos, J.R. 2007. High-performance concrete with recycled stone slurry. *Cement and Concrete Research* 37: 210–220.

Ashish, D.K. 2018. Feasibility of waste marble powder in concrete as partial substitution of cement and sand amalgam for sustainable growth. *Journal of Building Engineering* 15:236–242.

ASTM C1012/C1012M-12. 2012. *Standard Test Method for Length Change of Hydraulic-cement Mortars Exposed to a Sulphate Solution*. ASTM International, West Conshohocken, USA.

Bjegovic, D., Stirmer, N. and Serdar, M. 2012. Durability properties of concrete with blended cements. *Materials and Corrosion* 63:1087–1096.

Broomfield, J. P. 2007. *Corrosion of Steel in Concrete*, 2nd edition, Taylor & Francis, New York.

Cheng, S., Shui, Z., Sun, T., Huang, Y. and Liu, K. 2018. Effects of seawater and supplementary cementitious materials on the durability and microstructure of lightweight aggregate concrete. *Construction and Building Materials* 190:1081–1090.

Devi, K., Acharya, K.G. and Saini, B. 2018c. *Significance of Stone Slurry Powder in Normal and High Strength Concrete*. Springer Nature Switzerland AG 2019, Ludhiana, India.

Devi, K., Saini, B. and Aggarwal, P. 2018a. Effect of accelerators with waste material on the properties of cement paste and mortar. *Computers and Concrete* 22:153–159.

Devi, K., Saini, B. and Aggarwal, P. 2018b. *Combined Use of Accelerators and Stone Slurry Powder in Cement Mortar*. Springer Nature Switzerland AG 2019, Ludhiana, India.

Devi, K., Saini, B. and Aggarwal, P. 2019. Utilization of Kota stone slurry powder and accelerators in concrete. *Computers and Concrete* 23:189–201.

Devi, K., Saini, B. and Aggarwal, P. 2020. Long term performance of concrete using additives. *International Journal of Microstructure and Materials Properties*. https://doi.org/10.1504/IJMMP.2020.104612.

Devi, K., Saini, B. and Aggarwal, P. 2021. Impact of high temperature on mortar mixes containing additives. *Journal of Engineering Research*. https://doi.org/10.36909/jer.10477.

Ergun, A. 2011. Effects of the usage of diatomite and waste marble powder as partial replacement of cement on the mechanical properties of concrete. *Construction and Building Materials* 25:806–812.

Han, J., Wang, K., Shi, J. and Wang, Y. 2015. Mechanism of triethanolamine on Portland cement hydration process and microstructure characteristics. *Construction and Building Materials* 93:457–462.

Hussain, A. 2015. Kota stone slurry problem and possible solutions. https://www.slideshare.net/AkhtarHUssain10/kota-stone-slurry-problem-and-possible-solutions (Accessed on 23th January 2023).

IS 383. 2016. *Indian Standard Specification for Coarse and Fine Aggregates from Natural Sources for Concrete*. Bureau of Indian Standard, New Delhi, India.

IS 8112. 1989. *Specification for 43-Grade Ordinary Portland Cement.* Bureau of Indian Standard, New Delhi, India.

Khodabakhshian, A., de Brito, J., Ghalehnovi, M. and Shamsabadi, E.A. 2018. Mechanical, environmental and economic performance of structural concrete containing silica fume and marble industry waste powder. *Construction and Building Materials* 169:237–251.

Kumar, M.P., Mini, K.M. and Rangarajan, M. 2018. Ultrafine GGBS and calcium nitrate as concrete admixtures for improved mechanical properties and corrosion resistance. *Construction and Building Materials* 182:249–257.

Mashaly, A.O., Shalaby, B.N. and Rashwan, M.A. 2018. Performance of mortar and concrete incorporating granite sludge as cement replacement. *Construction and Building Materials* 169:800–818.

Meena, Y.K., Meena, D., Chaudhary P., and Sharma A. 2018. Partial replacement of fine aggregate with Kota stone dust & fly ash in cement mortar. *2nd International Conference on New Frontiers of Engineering, Management, Social Science and Humanities*, Mahratta Chamber of Commerce, Industries and Agriculture Tilak Road, Pune (India), 96–102.

Mohammed, M.K., Dawson, A.R. and Thom, N. H. 2013. Production, microstructure and hydration of sustainable self-compacting concrete with different types of filler. *Construction and Building Materials* 49:84–92.

Neville, A.M. and Brooks, J.J. 1997. *Concrete Technology.* 2nd Edition, Longman Publishing Limited, England.

Rajagopalan, S. R. and Kang, S.T. 2021. Evaluation of sulfate resistance of cement mortars with the replacement of fine stone powder. *Journal of Material Cycles and Waste Management* 23:1995–2004.

Rashwan, M.A., Basiony, T.M. Al., Mashaly, A.O. and Khalil, M.M. 2021. Self-compacting concrete between workability performance and engineering properties using natural stone wastes. *Construction and Building Materials* 319:126132. https://doi.org/10.1016/j.conbuildmat.2021.126132.

Saxena, R., Gupta, T., Sharma, R.K. and Siddique, S. 2022. Mechanical, durability and microstructural assessment of geopolymer concrete incorporating fine granite waste powder. *Journal of Material Cycles and Waste Management.* https://doi.org/10.1007/s10163-022-01439-0.

Silva, P. and de Brito, J. 2013. Electrical resistivity and capillarity of self-compacting concrete with incorporation of fly ash and limestone filler. *Advances in Concrete Construction* 1:65–84.

Singh, C. and Aggarwal, V. 2022. Experimental investigation of concrete strength properties by partial replacement of cement-sand with marble-granite powder. *Materials Today Proceedings.* https://doi.org/10.1016/j.matpr.2022.04.438.

Singh H., Garg P., Kaur I. (Eds.). 2019. *Proceeding of 1st International Conference on Sustainable Waste Management through Design.* Springer Nature, USA.

Singh, M. and Siddique, R. 2014. Strength properties and micro-structural properties of concrete containing coal bottom ash as partial replacement of fine aggregate. *Construction and Building Materials* 50:246–256.

Singh, M., Srivastava, A. and Bhunia, D. 2017. An investigation on effect of partial replacement of cement by waste marble slurry. *Construction and Building Materials* 134:471–488.

Singh, S., Nagar, R. and Agrawal, V. 2016. Performance of granite cutting waste concrete under adverse exposure conditions. *Journal of Cleaner Production.* https://doi.org/10.1016/j.jclepro.2016.04.034.

Xiao, L., Li, Z., Wei and X. 2007. Selection of superplasticizer in concrete mix design by measuring the early electrical resistivities of pastes. *Cement and Concrete Composites* 29:350–356.

Zheng, S., Liang, J., Hu, Y., Wei, D., Lan, Y., Du, H. and Rong, H. 2021. An experimental study on compressive properties of concrete with manufactured sand using different stone powder content. *Ferroelectrics* 579:189–198.

12 Feasibility Study on Utilization of Waste Materials in Concrete

R. Padmapriya, J. S. Sudarsan, and N. Sunmathi

CONTENTS

12.1 Introduction 205
12.2 Materials 207
    12.2.1 Cement...207
    12.2.2 Fine Aggregate...207
    12.2.3 Coarse Aggregate 207
    12.2.4 Water...208
    12.2.5 Class F Fly Ash..208
    12.2.6 Tannery Waste and Copper Slag..208
    12.2.7 Steel slag and E waste..209
12.3 Methodology and Mix Proportioning...210
    12.3.1 Mix Proportioning 213
12.4 Experimental Investigation..213
    12.4.1 Test on Fresh Concrete 213
    12.4.2 Compressive Strength Test 213
    12.4.3 Split Tensile Test..214
    12.4.4 Flexural Strength Test..214
    12.4.5 Durability of Concrete Test 216
12.5 Result and Discussion..216
    12.5.1 Compressive Strength..216
    12.5.2 Split Tensile Strength...218
    12.5.3 Flexural Strength 220
12.6 Conclusion 220
References..222

12.1 INTRODUCTION

Concrete has the ability to play a major and positive role in the environment as one of the most important building materials for infrastructure and growth. No construction activity is currently possible without the use of concrete. It is the most widely utilized building material on the planet because of its exceptional strength, durability, and workability [1]. The construction of key infrastructure projects in India,

DOI: 10.1201/9781003359784-12

such as highways, airports, nuclear power plants, bridges, and dams, is expanding year after year as a result of privatization and globalization. Such development operations require a significant amount of valuable natural resources. This not only accelerates the loss of natural resources but also raises the expense of building structures. Waste materials can be utilized as cement, fillers, fibers, or aggregate replacement because cement is a hazardous contaminant. Waste materials can be used as a source of energy in the environment. When compared to steel-confined concrete, fiber-reinforced polymer-confined concrete has a significant increase in ductility and strength [2,3]. Despite the fact that concrete is susceptible to the environment, our reliance on it is greater. Fine and coarse aggregates are two types of aggregate that play an important role in concrete. Many industrial processes generate the use of alternative materials in concrete. Different types of waste had been tried to find alternatives to cement, fine aggregates, and coarse aggregates as a part of the concrete to achieve sustainability and 3R principle.

The expansion of fly ash debris to ground waste concrete improves the quality of ground waste concrete-based geopolymer up to 75% fly fiery remains substance. Because of its poor tensile strength, geopolymer concrete is fragile. The addition of fibers transforms the material from brittle to ductile [4]. Using tannery waste as fine aggregate in concrete lowers the cost of the concrete, and because of its low density, it can be utilized in earthquake-prone areas. The bulking percentages for tannery waste and fine aggregate are practically the same [5,6]. The average strength on splitting tensile and flexure increased in 7 and 28 days when fine aggregate was replaced with tannery sludge, while compressive strength decreased when the sludge percentage was increased above 15%. The use of slags in concrete has further environmental and practical benefits for a variety of businesses, particularly in places where a large amount of CS (Compressive Strength) is generated. Concrete mixes contain 25%–100% of copper slag as an FA substitute, and CS can be used as a partial or complete replacement for FA in concrete mixes of grade M20 to M50 without causing significant workability changes [7,8]. Electrical and electronic devices has become an integral part of our daily lives, providing greater convenience, security, and ease of purchase. Electronic equipment has a high rate of obsolescence as a result of technological advancements, resulting in one of the world's fastest-growing waste streams. It has been found that the strength of concrete with E waste plastic as an aggregate is good [9,10].

Class F fly ash is derived from anthracites and bituminous loads, according to ASTM C618 standards. It contains more alumina and silica than class C fly ash and has a higher loss on ignition. Fly ash from class F has a lower calcium content than fly ash from class C. Steel slag is a by-product of smelting ores and used materials, and it is an industrial by-product obtained from the steel making industry. In many areas, using steel slag as an aggregate is considered conventional practice. Where tannery waste is a major cause of pollution, it can be recycled into concrete. Copper slag is a discarded material extracted from copper during heat up process, where impurities turn into slag that floats atop the molten metal. E waste is a word used to describe electronic goods that have outlived their usefulness. Where waste materials were replaced for aggregate, replacement percentages of zero, ten, twenty, and thirty were used. These scraps are ideal for concrete preparation.

12.2 MATERIALS

To substitute cement and aggregates, five alternative components were employed in concrete. Cement is substituted with class F flay ash. Tannery waste and copper slag were used in place of fine aggregates. For coarse aggregates, steel slag and E trash were used instead.

12.2.1 CEMENT

The physical properties of the cement used in this study were evaluated using standard tests such as setting time, standard consistency, compressive strength, and specific gravity, and the results are reported in Table 12.1 [11].

12.2.2 FINE AGGREGATE

Fine aggregates are one ordinary silt particles that have been mined from the ground. The physical properties were studied and are tabulated in Table 12.2.

12.2.3 COARSE AGGREGATE

Coarse aggregates are particles with a diameter of more than 4.75 mm. 9.5–37.5 mm in diameter is the most typical size range. The physical properties were tested and are tabulated in Table 12.3 [12,13].

TABLE 12.1
Physical Properties of Cement

S .No	Description of test	Results	Limit as per IS 269/456
1	Standard consistency (%)	28	-
2	Initial setting time	180	31 (Min)
3	Final setting time	290	600 (Max)
4	Specific gravity	3.12	3.17
5	Compressive strength (N/mm^2)	52.44	42 (Min)
			58 (Max)
6	Fineness of cement I 90mic sieve retained (%)	1.89	10 (Max)

TABLE 12.2
Physical Properties of Fine Aggregate

S .No	Description of test	Results	Limit as per IS 383: 2386
1	Material passing 75 micron by mass (%)	3.85	10 (Max)
2	Specific gravity	2.664	-
3	Water absorption (%)	1.64	-

TABLE 12.3
Physical Properties of Coarse Aggregate

S .No	Description of test	Results		Limit as per IS 383 : 2386
		12.5 mm	20 mm	
1	Specific gravity	2.782	2.752	-
2	Water absorption (%)	0.44	0.64	-
3	Flakiness (%)	3.24	5.9	25 (Max)
4	Elongation (%)	6.09	8.17	25 (Max)
5	Crushing value (%)	14.98	14.98	30 (Max)
6	Impact value (%)	13.46	13.45	30 (Max)
7	Abrasion value (%)	13.4	13.4	30 (Max)

TABLE 12.4
Properties of Water

S. No	Description of test	Results	Limit as per IS 456-2000
1	pH value	8.69	6.0 (Min)
2	Organic solids (mg/L)	45	200 (Max)
3	Inorganic solids (mg/L)	283	3000 (Max)
4	Suspended matter (mg/L)	1.0	2000 (Max)
5	Chloride content (mg/L)	194	2000 (Max for PCC)
			500 (Max for RCC)
6	Sulfate content (mg/L)	1.0	400 (Max)

12.2.4 WATER

Table 12.4 shows the results of water testing according to IS: 3025-1964, which confirms the requirements of IS: 456-2000.

12.2.5 CLASS F FLY ASH

Class C fly ash comes from subbituminous and lignite coals, according to ASTM C 618. It has a smaller loss on ignition than class F fly ash due to its composition, which is primarily calcium, alumina, and silica. The sp. gr of class F fly ash is 2.10 is used for this analysis.

12.2.6 TANNERY WASTE AND COPPER SLAG

Tannery waste was collected in Pallavaram, Chennai, for this study. Fine aggregate is used instead of tannery sludge. The quality of tannery sludge as well as chemical removal or reduction strategies, was investigated. Tables 12.5 and 12.6 show the physical and chemical properties of sludge and test values.

TABLE 12.5
Physical Properties of Tannery Waste

S. No	Test	Result
1	Form	Well powder
2	Color	Gray
3	Odor	Yes
4	Fineness modulus	6.18%
5	Water absorption	1.05%
6	Specific gravity	1.69

TABLE 12.6
Chemical Properties of Tannery Waste

S. No	Test	Result (%)
1	SiO_2	6.3
2	Al_2O_3	62.22
3	Cl-	1.33
4	Cr_2O_3	0.87
5	P_4O_{10}	0.09
6	Na_2O	0.20
7	CaO	24.43
8	K_2O	0.06

TABLE 12.7
Physical Properties of Copper Slag

S. No	Description of test	Results	Limit as per IS 383 : 2016
1	Specific gravity	3.15	2.1 to 3.2
2	Hardness	7 Mohr scale	-
3	Conductivity m S/M	4.8	-
4	Chloride content	<0.002	0.04 (Max)

The Star Grid in Tuticorin has been recognized as a copper slag source, and samples have been analyzed according to technical criteria. Copper slag is a discarded creation that results from copper heat up. The physical and chemical properties of copper slag are tabulated Tables 12.7 and 12.8.

12.2.7 STEEL SLAG AND E WASTE

JSW Steel in Hyderabad was identified as a source of steel slag, and samples were analyzed in accordance with IS 12089: 1987 and technical specifications.

TABLE 12.8
Chemical Properties of Copper Slag

S. No	Description of test	Results	Limit as per IS 383: 2016
1	Aluminum oxide (Al_2O_3) in %	2.51	-
2	Titanium dioxide (TiO_2) in %	0.54	-
3	Iron oxide (Fe_2O_3) in %	55.2	70.00% (Max)
4	Silicon dioxide (SiO_2) in %	35.17	-
5	Calcium oxide (CaO) in %	0.20	12.0% (Max)
6	Magnesium oxide (MgO) in %	0.98	-
7	Potassium oxide (K_2O) in %	1.22	-
8	Sodium oxide (Na_2O) in %	0.85	0.3% (Max)
9	Copper (CU) in %	0.44	-

TABLE 12.9
Physical Properties of Steel Slag

S.No	Description of test	Results	Limit as PER IS 16714 : 2018
1	Specific gravity	2.91	-
2	Fineness (m2/kg)	383	320 (Min)
3	Slag activity index	66.80	60% (Min)
	a. 7 days	76.17	75% (Min
	b. 28 days		

GGBS (Ground granulated blast furnace slag) is a by-product of iron production in blast furnaces. Tables 12.9 and 12.10 show the properties of steel slag.

Crushed wastes were employed in concrete in the size of coarse particles clarified in Figure 12.1. The specific gravity of discarded garbage that has been used is 1.01.

12.3 METHODOLOGY AND MIX PROPORTIONING

The following approach was developed based on the outcomes of the early investigation. A total of 300 specimens were cast for three trials based on ASTM standards. IS 456-2000 was adopted to conduct cube, cylinder, and prism testing. A cube specimen was also produced for durability testing. The process adopted in the entire research study is represented in the following flowchart. Mix proportioning adopted and specimen details for three trials for this study are tabulated in Tables 12.11 and 12.12, respectively (Figure 12.2).

The above process has been adopted in executing the trial study, and the optimization of data was also carried out by using different tools.

TABLE 12.10
Chemical Properties of Steel Slag

S. No	Description of test	Results	Limit as per IS 16714: 2018
1	Manganese oxide (MnO) %	0.12	5.5%(Max)
2	Magnesium oxide (MgO) %	7.57	17%(Max)
3	Sulfide sulfur (S) %	0.68	2.0%(Max)
4	Sulfate (as SO_3)) %	0.24	3.0%(Max)
5	Insoluble residue %	0.26	3.0%(Max)
6	Chloride (Cl)content %	0.003	0.1%(Max)
7	Glass content %	96	85%(Min)
8	Loss on ignition %	0.06	3.0%(Max)
9	Moisture content %	0.03	1.0%(Max)
10	$\frac{CaO+MgO+1/3Al_2O_3}{SiO_2+2/3\,Al_2O_3}$	1.49	1.0%(Min)
11	$\frac{CaO+MgO+Al_2O_3}{SiO_2}$	1.94	1.0%(Min)

FIGURE 12.1 E waste.

TABLE 12.11

Mix Proportion

Material mix	Replacement 1		Replacement 2		Replacement 3		Replacement 4		Replacement 5	
	Cement (%)	Class F fly ash (%)	FA (%)	Tannery waste (%)	FA (%)	Copper slag (%)	CA (%)	Steel slag (%)	CA (%)	E waste (%)
Nominal mix	100	0	100	0	100	0	100	0	100	0
Mix 1	90	10	90	10	90	10	90	10	90	10
Mix 2	80	20	80	20	80	20	80	20	80	20
Mix 3	70	30	70	30	70	30	70	30	70	30

TABLE 12.12

Specimen Data for Three Trails

S. No	Test	Materials/replaced material	0%	10%	20%	30%	Total
1	Compressive strength	Cement/fly ash	9	9	9	9	36
		Fine aggregate/tannery waste	9	9	9	9	36
		Fine aggregate/copper slag	9	9	9	9	36
		Coarse aggregate/steel slag	9	9	9	9	36
		Coarse aggregate/E waste	9	9	9	9	36
2	Split tensile strength	Cement/fly ash	3	3	3	3	12
		Fine aggregate/tannery waste	3	3	3	3	12
		Fine aggregate/copper slag	3	3	3	3	12
		Coarse aggregate/steel slag	3	3	3	3	12
		Coarse aggregate/E waste	3	3	3	3	12
3	Flexure strength	Cement/fly ash	3	3	3	3	12
		Fine aggregate/tannery waste	3	3	3	3	12
		Fine aggregate/copper slag	3	3	3	3	12
		Coarse aggregate/steel slag	3	3	3	3	12
		Coarse aggregate/E waste	3	3	3	3	12

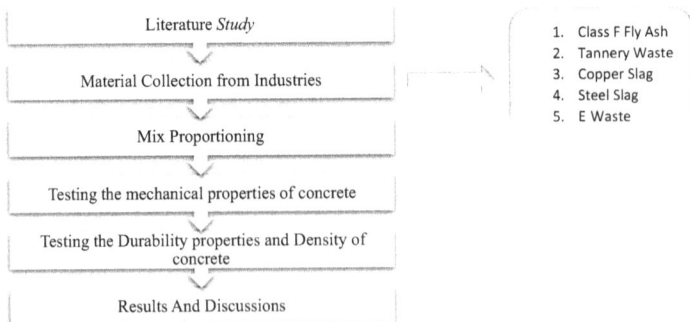

Literature *Study*

↓

Material Collection from Industries

↓

Mix Proportioning

↓

Testing the mechanical properties of concrete

↓

Testing the Durability properties and Density of concrete

↓

Results And Discussions

1. Class F Fly Ash
2. Tannery Waste
3. Copper Slag
4. Steel Slag
5. E Waste

FIGURE 12.2 Methodology.

12.3.1 Mix Proportioning

Grade : M_{20}
Ratio : 1: 1.5:3
W/C Ratio: 0.5

12.4 EXPERIMENTAL INVESTIGATION

12.4.1 Test on Fresh Concrete

Various approaches can be used to determine the workability of fresh concrete. The need for the slump cone test is to decide the flow of concrete. According to the results of the study, both the slump and compaction factor tests show that concrete has a medium workability, as shown in Table 12.13.

12.4.2 Compressive Strength Test

A compressive strength test limits how considerable compressive load a solid can stand before cracking. A gently applied load compresses a cube, prism, or cylinder

TABLE 12.13
Test on Wet Concrete

S. No	Replacement					
	Material	Replaced material	%	Slump cone	Compaction factor	Workableness
1	Cement	Fly ash	0	69	0.86	Medium
			10	62	0.86	Medium
			20	64	0.89	Medium
			30	63	0.91	Medium
2	Fine aggregate	Tannery waste	0	52	0.85	Medium
			10	53	0.87	Medium
			20	57	0.87	Medium
			30	60	0.89	Medium
3	Fine aggregate	Copper slag	0	66	0.87	Medium
			10	69	0.88	Medium
			20	71	0.90	Medium
			30	75	0.92	Medium
4	Coarse aggregate	Steel slag	0	62	0.85	Medium
			10	64	0.87	Medium
			20	69	0.90	Medium
			30	69	0.91	Medium
5	Coarse aggregate	E waste	0	55	0.85	Medium
			10	57	0.86	Medium
			20	59	0.86	Medium
			30	61	0.87	Medium

TABLE 12.14
Compression Strength for Waste Replace Concrete

	Replacement			Compressive strength		
S. No	Material	Replaced material	%	7th day	14th day	28th day
1	Cement	Fly ash	0	15.99	19.23	23.21
			10	18.98	23.38	26.09
			20	16.42	18.20	25.50
			30	15.83	20.50	20.08
2	Fine aggregate	Tannery waste	0	14.79	16.23	22.31
			10	16.82	24.63	25.89
			20	18.42	24.20	27.15
			30	13.67	22.85	25.68
3	Fine aggregate	Copper slag	0	13.73	18.35	21.46
			10	16.68	20.69	25.68
			20	18.60	23.73	28.35
			30	17.89	19.78	23.10
4	Coarse aggregate	Steel slag	0	14.78	16.23	21.21
			10	17.75	20.25	22.24
			20	19.67	23.46	24.79
			30	21.78	24.78	27.36
5	Coarse aggregate	E waste	0	15.78	17.64	20.35
			10	16.57	19.54	22.43
			20	14.28	20.56	21.43
			30	13.08	15.50	20.08

between the platens of a compression-testing equipment. The test values for five various replacements are tabulated in Table 12.14.

12.4.3 SPLIT TENSILE TEST

Split tensile strength is important for concrete, which can be used to indirectly calculate its tensile force. On the 28th day of curing, a specimen cylinder with a diameter of 150 mm and a height of 300 mm was cast. The test values for five different replacements are tabulated in Table 12.15.

12.4.4 FLEXURAL STRENGTH TEST

The flexural strength of concrete was tested to see how well it could withstand breaking forces. The test was carried out on specimen prisms of diameters of 100, 100, and 500 mm, which had been cast following a 28-day cure period. The test values for five different replacements are tabulated in Table 12.16.

TABLE 12.15

Split Tensile Test for Waste Replacement Concrete

	Replacement			Split tensile strength
S. No	Material	Replaced material	%	28th day
1	Cement	Fly ash	0	2.41
			10	2.68
			20	2.91
			30	1.79
2	Fine aggregate	Tannery waste	0	2.75
			10	3.48
			20	3.76
			30	3.80
3	Fine aggregate	Copper slag	0	3.61
			10	3.81
			20	4.12
			30	3.89
4	Coarse aggregate	Steel slag	0	2.62
			10	2.86
			20	3.52
			30	3.15
5	Coarse aggregate	E waste	0	2.46
			10	2.98
			20	3.15
			30	3.56

TABLE 12.16

Flexural Test for Waste Replacement Concrete

	Replacement			Flexural test
S. No	Material	Replaced material	%	28th day
1	Cement	Fly ash	0	3.5
			10	3.93
			20	4.12
			30	4.56
2	Fine aggregate	Tannery waste	0	3.50
			10	3.81
			20	3.90
			30	4.03
3	Fine aggregate	Copper slag	0	3.74
			10	4.10
			20	4.69
			30	4.86
4	Coarse aggregate	Steel slag	0	4.34
			10	4.38
			20	5.04
			30	5.67
5	Coarse aggregate	E waste	0	3.17
			10	3.46
			20	3.98
			30	4.18

TABLE 12.17
Durability of Concrete

S. No	Replacement Material	Replaced material	%	Water absorption (%)	Acid test (%)	Alkaline test (%)
1	Cement	Fly ash	0	1.38	2.56	4.74
			10	1.42	2.51	4.69
			20	1.59	2.48	4.16
			30	1.80	2.42	4.00
2	Fine aggregate	Tannery waste	0	1.30	2.76	4.69
			10	1.34	2.71	4.56
			20	1.39	2.68	4.44
			30	1.40	2.64	4.23
3	Fine aggregate	Copper slag	0	1.42	2.55	4.52
			10	1.54	2.50	4.38
			20	1.57	2.46	4.10
			30	1.77	2.39	3.99
4	Coarse aggregate	Steel slag	0	1.39	2.59	4.78
			10	1.56	2.56	4.47
			20	1.62	2.49	4.28
			30	1.69	2.46	4.08
5	Coarse aggregate	E waste	0	1.33	2.62	4.89
			10	1.38	2.60	4.45
			20	1.45	2.59	4.21
			30	1.58	2.54	3.78

12.4.5 DURABILITY OF CONCRETE TEST

Concrete's durability refers to how long it can survive various conditions over time. Specimens are tested for moisture content, effects of acids, and the assessment effects, as listed in Table 12.17.

12.5 RESULT AND DISCUSSION

12.5.1 COMPRESSIVE STRENGTH

Five different test specimens are tested for compression. Figures 12.3–12.7 depict compression tests on specimens. For replacement materials class F fly ash and E waste, the compression value is increased until 10% replacement and then shows variations or decline in strength. When tannery waste is replaced, strength grows until it reaches 20% replacement, after which it begins to decline. The graph demonstrates that the strength of copper slag grows up to 30% replacement, and the strength of steel slag improves as well.

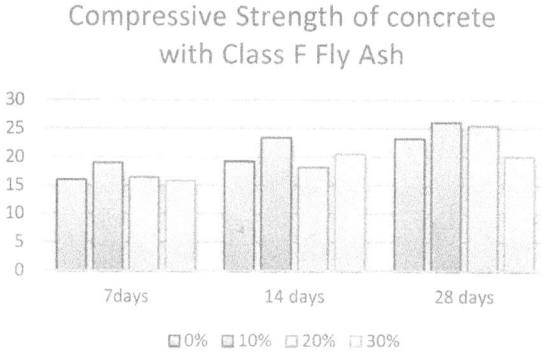

FIGURE 12.3 Compressive strength of concrete with class F fly ash.

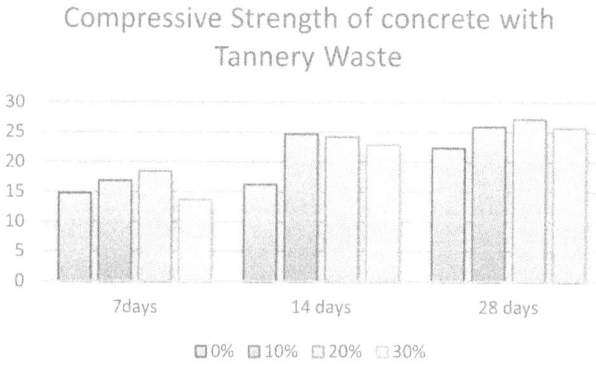

FIGURE 12.4 Compressive strength of concrete with tannery waste.

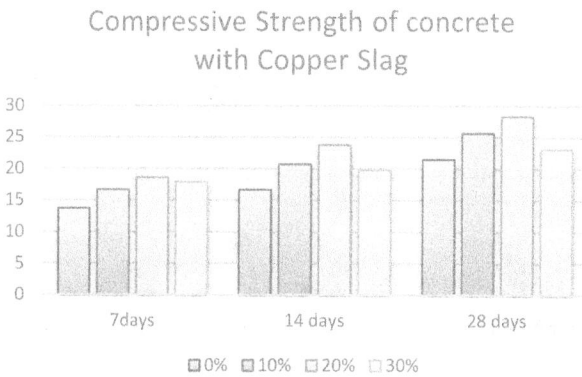

FIGURE 12.5 Compressive strength of concrete with copper slag.

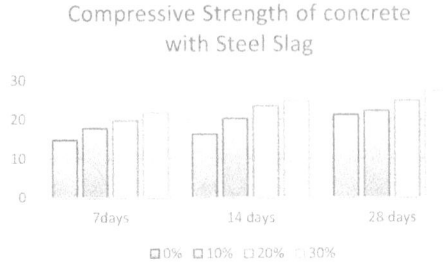

FIGURE 12.6   Compressive strength of concrete with steel slag.

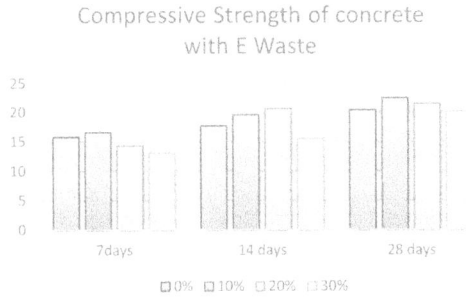

FIGURE 12.7   Compressive strength of concrete with E waste.

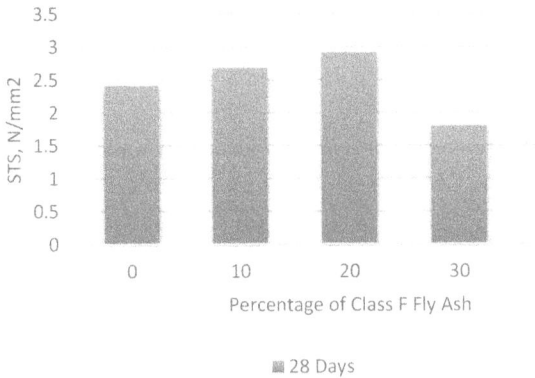

FIGURE 12.8   Split tensile strength of concrete with class F ash.

12.5.2   SPLIT TENSILE STRENGTH

Five different test specimens were tested for tension. Figures 12.8–12.12 depict split tensile tests on specimens. For replacement of class F fly ash, copper slag and steel slag strength increased till 20% replacement, and on further replacement, strength decreases. On replacement of tannery waste and E waste, strength increases gradually.

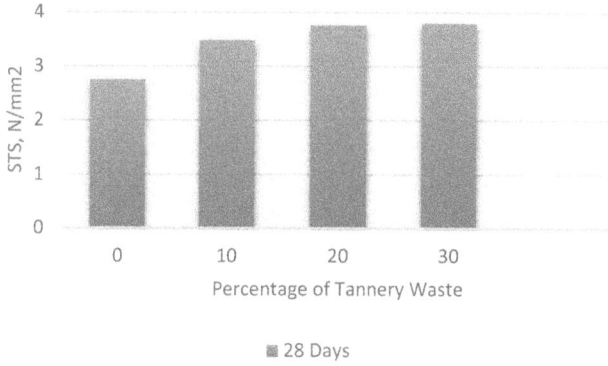

FIGURE 12.9 Split tensile strength of concrete with tannery waste.

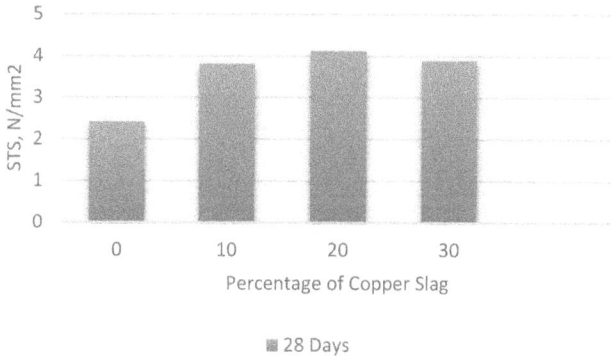

FIGURE 12.10 Split tensile strength of concrete with copper slag.

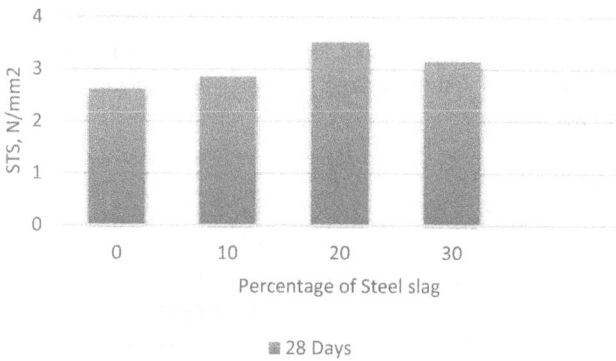

FIGURE 12.11 Split tensile strength of concrete with steel slag.

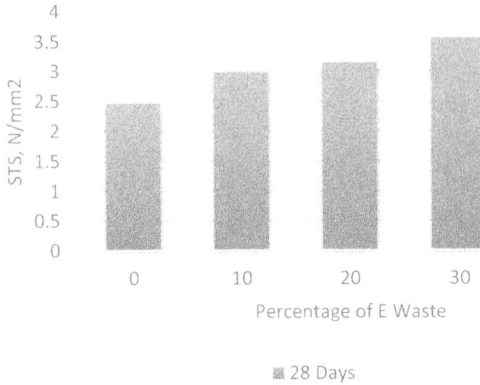

FIGURE 12.12   Split tensile strength of concrete with E waste.

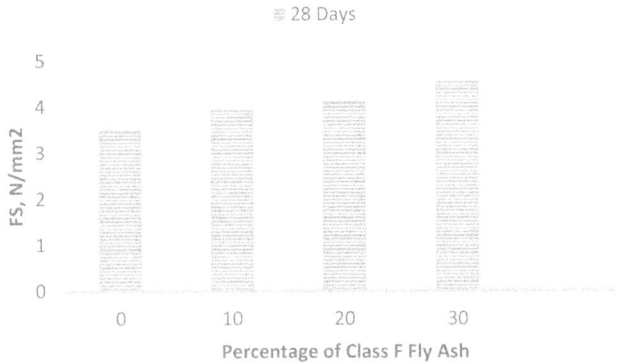

FIGURE 12.13   Flexure strength of concrete with class F ash.

12.5.3 FLEXURAL STRENGTH

Five different test specimens are tested for tension. Figures 12.13–12.17 depict split tensile tests on specimens. On increasing the waste percentage, flexure strength also increases in all percentages.

12.6 CONCLUSION

Experiments were conducted to assess the effect and qualities of concrete when all conventional ingredients were replaced by various waste items. The following are the conclusions drawn from the test results: The all-strength graph displays the ups and downs of varied values for all replacement materials, indicating that adding waste material increases strength when compared to nominal concrete. On all percentages of waste material replacements, the concrete's durability is adequate. The cost of

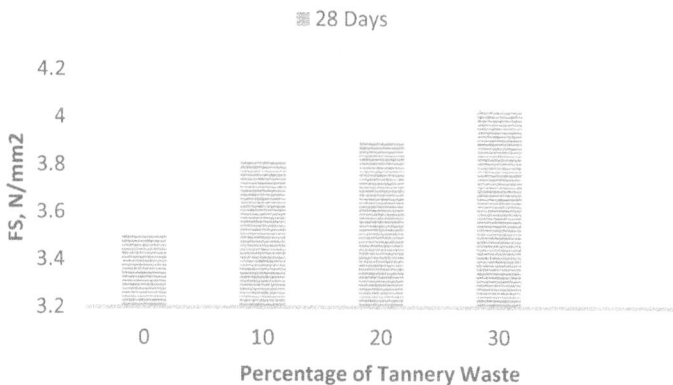

FIGURE 12.14   Flexure strength of concrete with tannery waste.

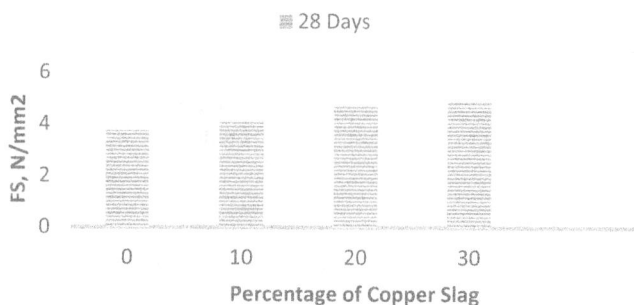

FIGURE 12.15   Flexure strength of concrete with copper slag.

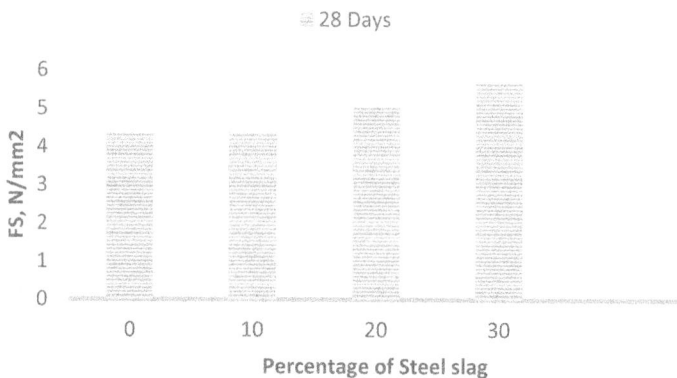

FIGURE 12.16   Flexure strength of concrete with steel slag.

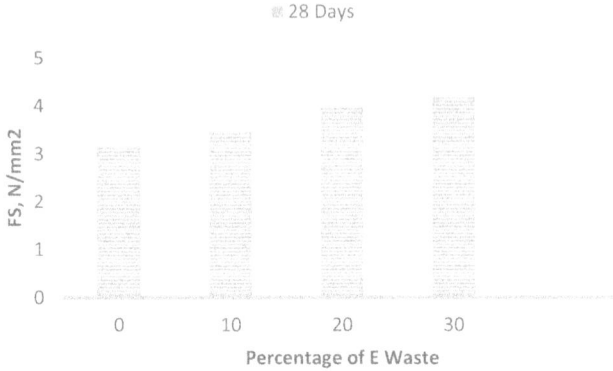

FIGURE 12.17   Flexure strength of concrete with E waste.

producing concrete is reduced by incorporating waste resources. Utilization of the waste material has proven the concept of sustainability. Based on the comparison of the preceding data, a different waste replacement for M20 concrete is suggested. This revision was conducted solely for the purpose of replacing M20 grade concrete; however, it might be expanded to include greater grades of concrete in future.

REFERENCES

[1] P. Guo, W. Meng, H. Nassif, H. Gou, and Y. Bao, "New perspectives on recycling waste glass in manufacturing concrete for sustainable civil infrastructure," *Constr. Build. Mater.*, vol. 257, p. 119579, 2020, doi: 10.1016/j.conbuildmat.2020.119579.

[2] V. K. B. Raja et al., "Geopolymer green technology," *Mater. Today Proc.*, vol. 46, pp. 1003–1007, 2021, doi: 10.1016/j.matpr.2021.01.138.

[3] B.S. Shekhawat, and D.V. Aggarwal, 2014. Utilisation of waste glass powder in concrete–A Literature Review. *International Journal of Innovative Research in Science, Engineering and Technology*, 3(7), pp. 1–5.

[4] W. Lokuge, D. Eberhard, and K. Karunasena, "Geopolymer concrete with FRP confinement," In *6th International Composites Conference (ACUN 6): Composites and Nanocomposites in Civil, Offshore and Mining Infrastructure*, 14–16 November, Melbourne, Australia, 2012.

[5] N. Sunmathi, R. Padmapriya, and J.S. Sudarsan, Feasibility study of tannery waste as an alternative for fine aggregate in concrete. *Sustainable Constr. Mater.- Lect. Notes Civ. Eng.*, vol. 194, pp. 397–405, 2021.

[6] K. Muniraj, B. Asha, and P. Raja Ramachandran, "Application of fine aggregate by replacement of tannery dry sludge in concrete," *Int. J. Appl. Environ. Sci.*, vol. 9, pp. 2805–2815, 2019.

[7] C. K. Madheswaran, P. S. Ambily, J. K. Dattatreya, and N. P. Rajamane, "Studies on use of Copper Slag as Replacement Material for River Sand in Building Constructions," *J. Inst. Eng. Ser. A*, vol. 95, no. 3, pp. 169–177, 2014, doi: 10.1007/s40030-014-0084-9.

[8] Q. Dong, G. Wang, X. Chen, J. Tan, and X. Gu, "Recycling of steel slag aggregate in portland cement concrete: An overview," *J. Clean. Prod.*, vol. 282, p. 124447, 2021, doi: 10.1016/j.jclepro.2020.124447.

[9] B. T. A. Manjunath, "Partial replacement of e-plastic waste as coarse-aggregate in concrete," *Procedia Environ. Sci.*, vol. 35, pp. 731–739, 2016, doi: 10.1016/j.proenv.2016.07.079.

[10] R. Widmer, H. Oswald-Krapf, D. Sinha-Khetriwal, M. Schnellmann, and H. Böni, "Global perspectives on e-waste," *Environ. Impact Assess. Rev.*, vol. 25, no. 5, pp. 436–458, 2005, doi: 10.1016/j.eiar.2005.04.001.

[11] IS 456–2000-2000, "Concrete, plain and reinforced," *Bur. Indian Stand. Delhi*, pp. 1–114, 2000.

[12] IS 2386- Part III, "Method of test for aggregate for concrete. Part III- Specific gravity, density, voids, absorption and bulking," *Bur. Indian Stand. New Delhi*, p. 1963, 2002.

[13] IS2386, "IS : 2386 (Part I)-1963- Indian Method of test for aggregate for concrete. Part I - Particle size and shape," *Indian Stand.*, p. 1963, 2002.

13 Decontamination of Pollutants Present in the Total Environment Using Microorganisms

M. Supreeth, Siddesh V. Siddalingegowda,
H. G. Lingaraju, and Shankramma Kalikeri

CONTENTS

13.1 Insert Graphical Abstract 225
    13.1.1 Pollution in Aquatic Environment..225
    13.1.2 Pollution in Terrestrial Environment...226
    13.1.3 Pollutants in Air...227
13.2 Removal of Pollutants by Microorganisms 228
13.3 Microbial Remediation Mechanism 228
    13.3.1 Decontamination of Heavy Metals...231
    13.3.2 Decontamination of Dyes..232
    13.3.3 Decontamination of Pesticides 232
    13.3.4 Decontamination of Petroleum Hydrocarbons 233
    13.3.5 Decontamination of Air Pollutants..233
13.4 Conclusion and Future Scope 234
References...234

13.1 INSERT GRAPHICAL ABSTRACT

13.1.1 POLLUTION IN AQUATIC ENVIRONMENT

Water (surface and groundwater) gets polluted if invaded by contaminants from point sources (factories, refineries, power plants, mines, wastewater treatment plants, etc.) and non-point sources (watershed, cars, buses, and trains) (Schweitzer and Noblet 2018). Most of surface water pollutants come from construction, municipalities, agriculture, resource extraction production, and industry (Walker et al. 2019). In aquatic lentic ecosystems, about 50%–70% of pollutants entering water bodies comes from domestic sewage (Dash et al. 2020). As a result, the flora and fauna are affected, and humans also encounter numerous problems due to the water systems.

Excess nutrients such as nitrates and phosphates commonly originate from domestic sewage and stimulates the growth of microorganisms mainly algae, which often increases biological oxygen demand and reduces the amount of dissolved oxygen available for aquatic life. In recent years, lakes are becoming the victim of eutrophication; as a result, many important lakes are deteriorating and shrinking. Eutrophication in fresh water has caused several problems all over the world. In about 53% in Europe, 50% in Asia, 48% in North America, 41% in South America, and 28% in Africa, lakes are eutrophicated (Wang et al. 2013).

Toxic sludges used as coloring dyes from industries are being disposed into rivers without prior treatment. These dyes cause cancer and mutation in genetic sequences and suppress enzyme activities. Antimicrobial drugs such as antibiotics, the usage of which has crossed 1,00,000 tons per year, have become ubiquitous in the environment and have become an aquatic pollutant in fresh water, affecting non-target fresh water organisms (Danner et al. 2019). Other pollutants such as pharmaceuticals and personal care products are released enormously into aquatic systems due to their widespread usage in recent times, causing several health problems (Zhang et al. 2014; Xiong et al. 2018).

Another important source of drinking water to a large population is groundwater, which serves as long-term water reserve for irrigation purposes. Globally 2.5 billion people are dependent on groundwater for drinking purposes (Qian et al. 2020). Due to anthropogenic activities, several emerging contaminants like pharmaceuticals and personal care products, perfluorinated compounds, endocrine disruptors, pesticides, heavy metals, and biological agents are being detected in groundwater across the globe (Kurwadkar 2017). The consumption of contaminated groundwater leads to several diseases like dysfunction of the liver, kidney problems, and cancer (Singh et al. 2018).

Marine water, which covers a larger part of the Earth, has been contaminated by activities carried out by humans. Marine water gets contaminated by the release of liquid petroleum hydrocarbons by oil spills due to human activities (Zhang et al. 2018). Oceans are being used as household dustbins to dispose of waste created by human activities. Various scientific studies showed that ocean water contains more than 150 Mt of plastics or more than 5 trillion plastic particles (macro- and micro-), thus affecting the food web (Hahladakis, 2020; Sharma and Chatterjee, 2017).

13.1.2 Pollution in Terrestrial Environment

Most of the organisms including humans on the Earth are dependent on soil for their survival. Hence, it is important to protect soil from contamination (Chae and An 2018). However, due to modern agricultural practices, unscientific mining and over-exploitation of non-renewable resources, the surface soil is contaminated by polychlorinated biphenyls, total petroleum hydrocarbons, and heavy metals.

Aromatic hydrocarbon contamination occurs both naturally and due to anthropogenic activities, and these contaminants are known for toxic, mutagenic, and carcinogenic properties (Phale et al. 2019). To meet the food needs of the ever-expanding global population, agriculture practices are expanding at a rapid rate by converting forests into agricultural lands. This conventional agricultural practice demands excess use of pesticides by farmers. Excessive and persistent usage of pesticides deteriorates the soil quality and environment (Morillo and Villaverde, 2017). Exposure to pesticides results in the development of various diseases in humans, which is documented in a review done by Kim et al. (2017). Another source of soil contamination is through non-ferrous metal smelting (Ettler 2015). Due to industrial operations, heavy metals such as cadmium (Cd-II), chromium (Cr-VI), copper (Cu-II), lead (Pb-II), and zinc (Zn-II) are widely spread in the environment. These heavy metals are not toxic to the environment because of their chelating reactions. However, combination of heavy metals and polycyclic aromatic hydrocarbons has caused increased concerns (Liu et al. 2017, Raj et al. 2018). Higher concentrations of heavy metals in agricultural soils are of global concern because of their association with food production and security (Kowalska et al. 2018). Cd, a primary soil pollutant, is known for its toxic health effects on organisms including humans even on minute concentration. Cd finds its routes into humans via the food chain when vegetables are grown in soil contaminated with Cd. In Japan, people suffered from the disease Itai-Itai when consumed rice grown in Cd-polluted soil (Khan et al. 2017). Contamination of xenobiotic compounds in the terrestrial environment leads to a decrease in soil activity and loss of biodiversity affecting the humification process, resulting in alteration in ecosystem functionality (Ceci et al. 2019). Certain soil contaminants like cadmium, dioxin, lead, mercury, and herbicides derived from trinitrotoluene are of major public health concern directly in relation to soils and human health impacts.

13.1.3 Pollutants in Air

Air pollution is one of the important environmental risk factors to human health. Ambient air pollution is causing 3 million deaths annually all over the globe, where Western Pacific and Southeast Asia are most affected. One in every nine human deaths in 2012 was due to air pollution (Dangi et al. 2019). The atmospheric air is polluted due to rapid economic growth by industrialization which leads to consumption of more energy and results in increased concentration of sulfur oxides (SO_x), nitrogen oxides (NO_x), carbon monoxide (CO), carbon dioxide (CO_2), suspended particulate matter (PM), and ozone (O_3) (Agarwal et al. 2019). The increase in air pollution decreases immunity in humans (Glencross et al. 2020), and even a small increase in $PM_{2.5}$ in the atmosphere will result in increased death rates associated with viral diseases. Studies reported that a 10 μg/m^3 increase in long-term exposure to $PM_{2.5}$ was associated with an 11% increase in cardiovascular mortality (Bourdrel et al. 2017).

Indoor air is polluted due to occupational activities, tobacco smoking, materials, household products, combustion of fossil fuels, pets, underground garages, outdoor air sources, and chemical reactions (Soreanu et al. 2013). Indoor air pollutants include benzene, toluene, ethylbenzene, xylene, polyaromatic hydrocarbons (PAHs), formaldehyde, and volatile organic compounds (VOCs) (Wei et al. 2017, Weyens et al. 2015). On average, a person inhales 6–10 L of air per minute, which amounts to 15,000 L/day (Gawrońska and Bakera, 2015). Indoor air at homes, schools, offices, and other indoor air spaces where more than 85% of time spent is reported to have 5%–10% more pollutants than the outdoor environment, which is a major health, economic, and social concern (Kim et al. 2018). Breathing contaminated air may cause respiratory and cardiovascular diseases, leading to sick building syndrome and

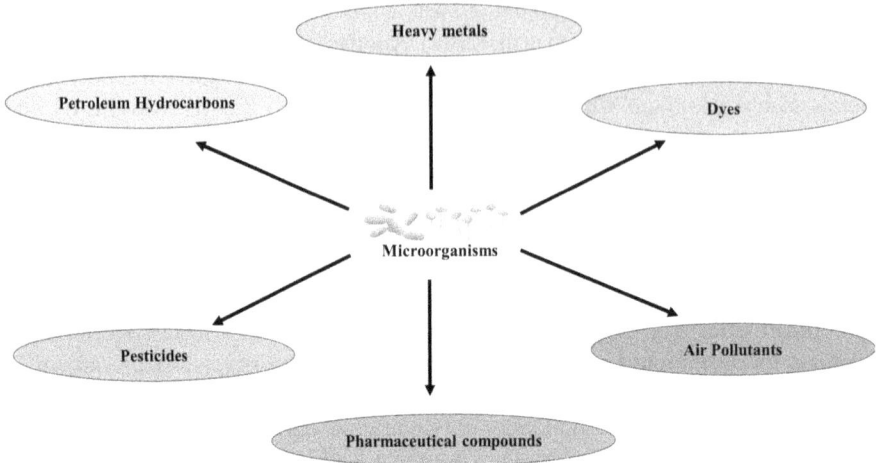

FIGURE 13.1   Microbial applications in removal of different pollutants.

building-related illnesses (Brilli et al. 2018). Along with this, there are other emerging group of pollutants arising in newer forms, posing severe concerns to human health and environment (Agrawal et al. 2020).

13.2   REMOVAL OF POLLUTANTS BY MICROORGANISMS

Research has been concerted on removing pollutants from air, water, and soil using physical, chemical, and biological methods. However, the latter method involving microorganisms is the most accepted method because of its efficiency and eco-friendly and cost-effective nature. Microorganisms (bacteria, fungi, and algae) are involved in removing several recalcitrant pollutants irrespective of biotic and abiotic stresses they come across (Kour et al. 2021). Due their metabolic diversity, microorganisms can be employed to remove any recalcitrant pollutants such as heavy metals (Jacob et al. 2018; Rahman 2020; Zhang et al. 2020; Tang et al. 2021), petroleum hydrocarbons (Li et al. 2020; Ławniczak et al. 2020; Miri et al. 2019; Partovinia and Rasekh 2018), pharmaceutical products (Kumar et al. 2019; Akerman-Sanchez and Rojas-Jimenez 2021; Naghdi et al. 2018; Ethica et al. 2018), pesticides (Bhatt et al. 2021; Bose et al. 2021; Sarker et al. 2021), dyes (Srivastava et al. 2022; Ajaz et al. 2020; Vikrant et al. 2018; Ihsanullah et al. 2020), and other emerging pollutants (Sutherland and Ralph 2019; Roccuzzo et al. 2021; Ratnasari et al. 2021; Figure 13.1). Table 13.1 gives an example of microbial applications in removal of different kinds of pollutants from total environment.

13.3   MICROBIAL REMEDIATION MECHANISM

Microorganisms use pollutants as their carbon source, thereby mitigating pollutant concentrations. The pollutants present in the environment is taken up by indigenous microorganisms, resulting in natural biodegradation. However sometimes, due to

TABLE 13.1

Bioremediation of pollutants by microorganisms

S. No.	Pollutant	Technique	Microorganisms	Efficiency	Time	References
1	Mixed contaminants (poly aromatic hydrocarbons, diesel oil, and heavy metal)	Biodegradation	*Acinetobacter* spp.	60% of 1% w/w of diesel, 150 mg/L of PAHs and 100–40 mg/L of heavy metals	0–7 days	Czarny et al. (2020)
2	Phenanthrene and heavy metal	Biodegradation	*Burkholderia fungorum* FM-2	99.67% of 300 mg/L	0–3 days	Liu et al. (2019)
3	Heavy metal - chromium (VI)	Photo-assisted microbial electrosynthesis	$WO_3/MoO_3/g$-C_3N_4 + *Serratia marcescens*	1.7 time of 60 mg/L	0–14 days	Huang et al. (2021)
4	Co-contaminant (heavy metal and total petroleum hydrocarbons)	Natural attenuation	Mixed cultures	------		Khudur et al. (2018)
5	Fungicide: carbendazim	Biodegradation	*Bacillus subtilis* CB2	98% of 1,000 mg/L	0–7 days	Singh et al. (2019)
6	Drug: paracetamol	Biodegradation	*Pseudomonas moorei* KB4	50 mg/L	24 hours	Żur et al. (2018)
7	Congo red dye	Bioreactor	*Bacillus* sp. MH587030.1	95.7% of 50 mg/L	136 hours	Sonwani et al. (2020)
8	Malachite green dye	Biodegradation	*Lasiodiplodia* sp.	96.9% of 50 mg/L	24 hours	Arunprasath et al. (2019)
9	Methylene blue dye	Biodegradation	*Galactomyces geotrichum* KL20A	76.6% of 50 ppm	48 hours	Contreras et al. (2019)
10	Anthraquinone Dye Green 3	Biodegradation	*Hortaea* sp.	100% of 100 mg/L	72 hours	Al Farraj et al. (2019)
11	Organophosphate pesticides: chlorpyrifos, dichlorvos, dipterex, phoxim, and triazophos	Enzymatic biodegradation	*Bacillus amyloliquefaciens* YP6	80 % of 2.03 mg/mL	8 hours	Meng et al. (2019)
12	Organophosphate pesticides: chlorpyrifos and dichlorvos	In situ bioremediation	*Pseudomonas aeruginosa*, *Taonella mepensis*, and *Methylobacterium zatmanii*	90% of 1,000 mg/L of dichlorvos 100% of 100 mg/L of chlorpyrifos	20 days for dichlorvos 12 days for chlorpyrifos	Gaonkar et al. (2019)

(Continued)

TABLE 13.1 (*Continued*)

Bioremediation of pollutants by microorganisms

S. No.	Pollutant	Technique	Microorganisms	Efficiency	Time	References
13	Organophosphate pesticide: diazinon	Biodegradation	*Bacillus* sp.	3.79%–58.52% of 20 mg/L	14 days	Nasrollahi et al. (2020)
14	Organophosphate pesticides: methyl parathion, paraoxon, chlorpyrifos, and coumaphos	Bioreactor	*Sphingomonas* sp. and *Brevundimonas* sp.	150 mL of 0.15 mM of chlorpyrifos of 1.5 mL/min	17 hours	Santillan et al. (2020)
15	Cypermethrin	Biodegradation and detoxification	*Bacillus subtilis* strain 1D	95% of 450 ppm	15 days	Gangola et al. (2018)
16	Organophosphate pesticides: profenofos and quinalphos	Biodegradation	*Kosakonia oryzae* strain -VITPSCQ3	82% of up to 800 mg/L of Profenofos 92% of up to 800 mg/L of Quinalphos	48 hours	Dash and Osborne (2020)
17	Antibiotic: sulfamethoxazole	Biodegradation	*Acinetobacter* sp.	100% of 5–240 mg/L	7 hours	Wang and Wang (2018)
18	Polyethylene	Biodegradation	*Aspergillus flavus*	----	28 days	Zhang et al. (2020b)
19	Polyethylene terephthalate	Biodegradation	*Bacillus cereus SEHD031MH Agromyces mediolanus PNP3*	17% of 2.63 g/L	17 days	Torena et al. (2021)
20	Pharmaceutical micropollutant: clofibric acid	Biodegradation	*Pseudomonas aeruginosa*	35% of 0.25–2 mg/L	168 hours	Hemidouche et al. 2018)
21	Xylene	Microbe-assisted phytoremediation	*Enterobacter cloacae LSRC11, Staphylococcus* sp. *A1*, and *Pseudomonas aeruginosa*	>53% of 100 ppm	120 hours	Sangthong et al. (2016)
22	Benzene	Microbe assisted phytoremediation	*Enterobacter* sp. *EN2*	5 ppm	67 days	Setsungnern et al. (2018)
23	Formaldehyde	Analog device method	*Methylobacterium* sp. *XJLW*	100 % of 15 g/L	24 hours	Shao et al. (2019)

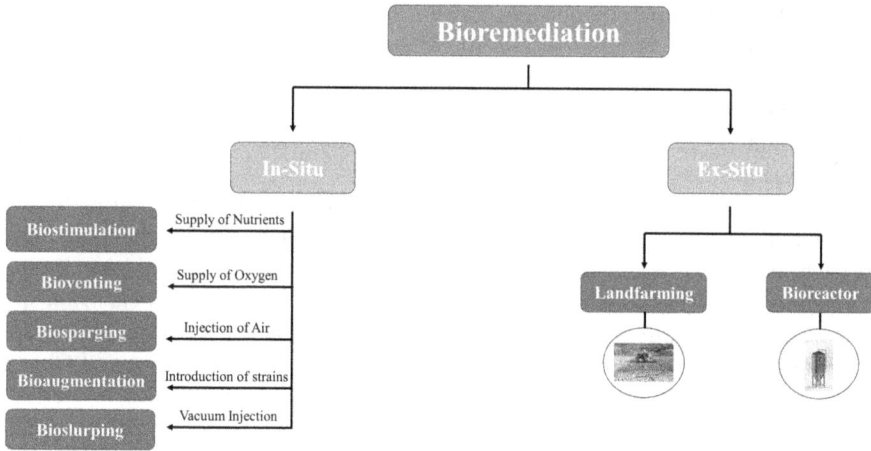

FIGURE 13.2 Different methods of bioremediation.

certain environmental factors, the normal microbial flora present in the contamination site cannot utilize the pollutants as carbon source. In such cases, the bioremediation technique can be applied, where microorganisms will be able to remove contaminants. Bioremediation techniques include in situ and ex situ methods (Figure 13.2). In the in situ method, the contaminants are removed from the site itself by several methods, such as biosparging, bioventing, biostimulation, bioaugmentation, and bioslurping. In the ex situ method, the contaminants are transferred from the contamination site to the treatment plant to remove contaminants. Land farming and bioreactors can be applied to remove the pollutants in the ex situ method (Das and Dash 2014).

13.3.1 Decontamination of Heavy Metals

Elements that have a relatively heavy density (>5 g/cm^3) and are toxic to humans and other organisms even in minute concentrations are considered heavy metals. Examples are arsenic (As), cadmium (Cd), chromium (Cr), copper (Cu), lead (Pb), mercury (Hg), nickel (Ni), platinum (Pt), silver (Ag), tin (Sn), titanium (Ti), vanadium (V), and zinc (Zn). These heavy metals are present in the environment naturally and through anthropogenic activities. Due to rapid urbanization and industrialization, these heavy metals are present beyond the recommended levels in environmental samples. Conventional methods like precipitation, adsorption, reverse osmosis, cementation, ion exchange coagulation/flocculation, and electrochemical removal methods are employed to remove heavy metals. However, these methods result in incomplete removal, high expenses, and secondary pollution (Shrestha et al. 2021). Over a decade, scientists have looked for alternates by using biological methods involving microorganisms to remove heavy metals. Due to their previous exposure to heavy metals, microorganisms have developed resistance and can be employed in remediation. Through the process of oxidation, reduction, methylation, and dealkylation, microorganisms convert metals into metals and metalloids. *Curtobacterium* sp., *Microbacterium* sp., *Kocuria* sp., *Klebsiella* sp., *Enterobacter, Leifsonia* sp., *Klebsiella* sp., *Bacillus* sp., *Rhodococcus* sp., *Flavobacterium* sp., *Chryseobacterium humi, Ralstonia eutropha, Azospirillum* sp., *Mesorhizobium huakuii,* and *Pseudomonas putida* are few bacterial strains having high tolerance toward various heavy metals (Sharma et al. 2021).

Microalgae is another group that shows strong tolerance and fast growth during bioremediation studies of heavy metals. Microalgae remove the pollutants through biosorption and bioaccumulation. *Parachlorella* sp., *Nannochloropsis* sp., *Scenedesmus* sp., *Spirulina* sp., *Chlorella* sp., *Chlamydomonas* sp., *Coelastrum* sp., *Monoraphidium* sp., diatoms, *Pithophora* sp., *Phormidium* sp., and *Oscillatoria* sp. are potential microalgae known to bioremediate heavy metals (Leong and Chang 2020). Mycoremediation of heavy metals from contaminated sites is also very effective. Through extracellular and intracellular sequestration, fungi show resistance toward heavy metals. *Aspergillus* Sp., *Pleurotus* sp., *Rhizopus* sp., *Curvularia* sp., *Acrimonium* sp., and *Pythium* sp. are some of the genera of fungi known for bioremediation of various heavy metals (Arora et al. 2017; Deshmukh et al. 2016).

13.3.2 DECONTAMINATION OF DYES

Effluents of tannery, textile, paper, food, pharmaceutical, and other related industries release huge amounts of untreated synthetic dyes into the surrounding water ecosystem, posing serious threats to humans and the environment. Human health is affected when these dyes come in contact, affecting vital organs like the brain, kidney, heart, and liver along with respiratory, immune, and reproductive systems (Rovira and Domingo 2019). When compared to physical and chemical methods (oxidation, ozonation, adsorption, membrane separation, ion exchange, and coagulation) for removal of dyes, bioremediation with bacteria, algae, and fungi exhibit metallic versatility, which can be employed for effective dye removal, and it is also cost-effective (Kalia and Singh 2020).

Bacterial biodegradation of dyes is more effective and faster in removing dyes than fungal strains. Bacterial genera capable of removing dyes are *Bacillus* sp., *Pseudomonas* sp., *Nocardiosis* sp., *Enterococcus* sp., *Alcaligenes* sp., *Klebsiella* sp., *Vibrio* sp., *Neisseria* sp., *Proteus* sp., and *Staphylococcus* sp. However, fungal species are more efficient and capable of removing a wide variety of synthetic dyes that cannot be degraded by bacteria. *Phanerochaete* sp., *Penicillium* sp., *Myrothecium* sp., *Aspergillus* sp., *Trametes* sp., *Pleurotus ostreatus* sp., and *Pleurotus* sp. Photosynthetic microalgae have also been investigated for dye removal. *Chlorella* sp., *Aphanocapsa* sp., *Spirogyra* sp., *Oscillatoria* sp., and *Scenedesmus* sp. have successfully removed different widely used dyes. Compared to single microbial culture, mixed consortium has shown more efficiency (Srivastava et al. 2022).

13.3.3 DECONTAMINATION OF PESTICIDES

Environmental pollution by pesticides is ubiquitous, and pesticides are the most common contaminants across the globe. Several types of pesticides (insecticides, herbicides, weedicides, fungicides, and more) are applied to agricultural crops to protect them from invading pests. However, the applied pesticides reach the target organism by just 1%, and the remaining 99% come in contact with non-target organisms, causing soil, water, and air pollution. These pesticides undergo natural degradation by indigenous microorganisms and results in the formation of biotransformed products. These products may be non-toxic or even be more toxic than the parent compound (Supreeth and Raju 2017).

Bioremediation of pesticides has turned out to be the gold standard method for decontamination (Jiang and Li 2018). Several strains of bacteria, actinomycetes, fungi, and algae have successfully removed various pesticides from soil and water. Bacterial genera known to degrade pesticides include *Pseudomonas* sp., *Bacillus* sp., *Burkholderia* sp., *Klebsiella* sp., and *Streptomyces* sp. Fungal genera include *Trichoderma* sp., *Aspergillus* sp., *Phanerochaete* sp., and white rot fungi (Sehrawat et al. 2021).

13.3.4 DECONTAMINATION OF PETROLEUM HYDROCARBONS

Pollution caused by crude oil-based hydrocarbons in oceans and sea water is the largest environmental pollution affecting aquatic flora and fauna. Crude oil spill consisting of petroleum hydrocarbons and phenolic substances are the recalcitrant compounds that resist degradation. Physicochemical and biological methods are employed across the globe to remove the pollutants from water. However, biological methods consisting of bacteria, fungi, and algae are cost-effective and eco-friendly for removing petroleum hydrocarbons (Sayed et al. 2021). Compared to conventional techniques such as physical and chemical methods, bioremediation of PAHs using microorganisms by production of biosurfactants is highly successful. Microorganisms (Protobacteria, Firmicutes, Actinobacteria, and fungal strains) possessing alkane monooxygenases and ring-hydroxylating dioxygenases genes can be employed for removing these hydrocarbons from the contaminated water ecosystem (Fuentes et al. 2014)

13.3.5 DECONTAMINATION OF AIR POLLUTANTS

Air pollution is the major and most severe global threat currently faced by mankind. Due to anthropogenic activities, nitrogen dioxide (NO_2), ammonia, and VOC concentration has increased tremendously, resulting in 1,800,000 premature deaths in 2018 (Vohra et al. 2022). Air pollutants such as PM, VOCs, inorganic air pollutants, persistent organic pollutants, heavy metals, and black carbon concentration are beyond the threshold limit, and the concentration of these pollutants must be brought down. Currently, filtration, electrostatic precipitation, membrane separation, ozonation, thermal oxidation, catalytic oxidation, UV photolysis, photocatalytic oxidation, enzymatic oxidation, botanical purification, biofiltration, non-thermal plasma, dry scrubbing, wet scrubbing, non-selective catalytic reduction, and selective catalytic reduction are the methods involved in removing air pollutants. However, due to expensive, high maintenance, and likelihood for pollution remission, these conventional techniques are not recommended. Alternative to these techniques, phytoremediation of pollutants using plants is a cost-effective, energy-saving, and eco-friendly technique in remediating air pollutants (Xin et al. 2020). The plant can be combined with microorganisms (bacteria, fungi, and algae) for enhanced remediation of air pollutants (Supreeth 2021).

13.4   CONCLUSION AND FUTURE SCOPE

Conventional remediation techniques involving physical and chemical methods are not completely achievable. Bioremediation using microorganisms has become an alternative method. It is evident that several bacterial, algal, and fungal cultures can successfully utilize pollutants as their carbon source and become excellent candidature for bioremediation programs. However, studies on degradation of several kinds of pollutants by microorganisms are limited to the laboratory scale. Microorganisms are successful in removing several recalcitrant pollutants completely within a short period of time, thereby proving to become excellent tools for decontamination. However, there are not much data regarding remediation of pollutants by microorganisms at actual contamination sites. Hence, more studies are required at the field level.

Research should be focused on development of microbial consortium to remove pollutants completely in the presence of co-contaminants without formation of any toxic secondary by-products in wastewater treatment plants, textile dye industries, pharmaceutical industries, and many more. These microorganisms can be further improved to remove pollutants by microbial engineering, strain improvement, microbial fuel cells, and other biotechnological approaches. Apart from microorganisms, plants can also be combined for enhanced remediation, especially air pollutants. Except few challenges and drawbacks, decontamination of pollutants using microorganisms is an excellent alternative to remove pollutants from the environmentally polluted world.

REFERENCES

Agarwal P, Sarkar M, Chakraborty B, Banerjee T (2019) *Phytoremediation of Air Pollutants.* Amsterdam: Elsevier Inc.

Agrawal K, Bhatt A, Chaturvedi V, Verma P (2020) *Bioremediation : An Effective Technology toward a Sustainable Environment via the Remediation of Emerging Environmental Pollutants.* Amsterdam: Elsevier.

Ajaz M, Shakeel S, Rehman A (2020) Microbial use for azo dye degradation—a strategy for dye bioremediation. *Int Microbiol* 23:149–159. https://doi.org/10.1007/s10123-019-00103-2.

Akerman-Sanchez G, Rojas-Jimenez K (2021) Fungi for the bioremediation of pharmaceutical-derived pollutants: A bioengineering approach to water treatment. *Environ Adv* 4:100071. https://doi.org/10.1016/j.envadv.2021.100071.

Al Farraj DA, Elshikh MS, Al Khulaifi MM, et al (2019) Biotransformation and detoxification of antraquione dye green 3 using halophilic hortaea sp. *Int Biodeterior Biodegrad* 140:72–77. https://doi.org/10.1016/j. ibiod.2019.03.011.

Arora S, Singh AK, Singh YP (2017) *Bioremediation of Salt Affected Soils: An Indian Perspective.* 1–313. https://doi.org/10.1007/978-3-319-48257-6.

Arunprasath T, Sudalai S, Meenatchi R, et al (2019) Biodegradation of triphenylmethane dye malachite green by a newly isolated fungus strain. *Biocatal Agric Biotechnol* 17:672–679. https://doi.org/10.1016/j. bcab.2019.01.030.

Bhatt P, Bhatt K, Sharma A, et al (2021) Biotechnological basis of microbial consortia for the removal of pesticides from the environment. *Crit Rev Biotechnol* 41:317–338. https://doi.org/10.1080/07388551.2020.1853032.

Bose S, Kumar PS, Vo DVN, et al (2021) Microbial degradation of recalcitrant pesticides: A review. *Environ Chem Lett* 19:3209–3228. https://doi.org/10.1007/s10311-021-01236-5.

Bourdrel T, Bind MA, Béjot Y, et al (2017) Cardiovascular effects of air pollution. *Arch Cardiovasc Dis* 110:634–642. https://doi.org/10.1016/j.acvd.2017.05.003.

Brilli F, Fares S, Ghirardo A, et al (2018) Plants for sustainable improvement of indoor air quality. *Trends Plant Sci* 23:507–512. https://doi.org/10.1016/j.tplants.2018.03.004.

Ceci A, Pinzari F, Russo F, et al (2019) Roles of saprotrophic fungi in biodegradation or transformation of organic and inorganic pollutants in co-contaminated sites. *Appl Microbiol Biotechnol* 103:53–68. https://doi.org/10.1007/s00253-018-9451-1.

Chae Y, An YJ (2018) Current research trends on plastic pollution and ecological impacts on the soil ecosystem: A review. *Environ Pollut* 240:387–395. https://doi.org/10.1016/j. envpol.2018.05.008.

Contreras M, Grande-Tovar CD, Vallejo W, Chaves-López C (2019) Bio-removal of methylene blue from aqueous solution by Galactomyces geotrichum KL20A. *Water (Switzerland)* 11:1–13. https://doi.org/10.3390/w11020282.

Crini G, Lichtfouse E (2018) *Green Adsorbents for Pollutant Removal: Innovative Materials.* Switzerland: Springer Nature, p 131.

Czarny J, Staninska-Pięta J, Piotrowska-Cyplik A, et al (2020) Acinetobacter sp. as the key player in diesel oil degrading community exposed to PAHs and heavy metals. *J Hazard Mater* 383. https://doi.org/10.1016/j.jhazmat.2019.121168.

Dangi AK, Sharma B, Hill RT, Shukla P (2019) Bioremediation through microbes: Systems biology and metabolic engineering approach. *Crit Rev Biotechnol* 39:79–98. https://doi.org/10.1080/07388551.2018.1500997.

Danner MC, Robertson A, Behrends V, Reiss J (2019) Antibiotic pollution in surface fresh waters: Occurrence and effects. *Sci Total Environ* 664:793–804. https://doi.org/10.1016/j. scitotenv.2019.01.406.

Das S, Dash HR (2014) *Microbial Bioremediation: A Potential Tool for Restoration of Contaminated Areas.* Waltham, MA: Elsevier Inc.

Dash DM, Osborne WJ (2020) Rapid biodegradation and biofilm-mediated bioremoval of organophosphorus pesticides using an indigenous Kosakonia oryzae strain -VITPSCQ3 in a vertical-flow packed bed biofilm bioreactor. *Ecotoxicol Environ Saf* 192:110290. https://doi.org/10.1016/j.ecoenv.2020.110290.

Dash S, Borah SS, Kalamdhad AS (2020) Study of the limnology of wetlands through a one-dimensional model for assessing the eutrophication levels induced by various pollution sources. *Ecol Modell* 416:108907. https://doi.org/10.1016/j.ecolmodel. 2019.108907.

Deshmukh R, Khardenavis AA, Purohit HJ (2016) Diverse metabolic capacities of fungi for bioremediation. *Indian J Microbiol* 56:247–264. https://doi.org/10.1007/ s12088-016-0584-6.

Ethica SN, Saptaningtyas R, Muchlissin SI, Sabdono A (2018) The development method of bioremediation of hospital biomedical waste using hydrolytic bacteria. *Health Technol (Berl)* 8:239–254. https://doi.org/10.1007/s12553-018-0232-8.

Ettler V (2015) Soil contamination near non-ferrous metal smelters: A review. *Appl Geochemistry* 64:56–74. https://doi.org/10.1016/j. apgeochem.2015.09.020.

Fuentes S, Méndez V, Aguila P, Seeger M (2014) Bioremediation of petroleum hydrocarbons: Catabolic genes, microbial communities, and applications. *Appl Microbiol Biotechnol* 98:4781–4794. https://doi.org/10.1007/s00253-014-5684-9.

Gangola S, Sharma A, Bhatt P, et al (2018) Presence of esterase and laccase in Bacillus subtilis facilitates biodegradation and detoxification of cypermethrin. *Sci Rep* 8:1–11. https://doi.org/10.1038/s41598-018-31082-5.

Gaonkar O, Nambi IM, Suresh Kumar G (2019) Biodegradation kinetics of dichlorvos and chlorpyrifos by enriched bacterial cultures from an agricultural soil. *Bioremediat J* 23:259–276. https://doi.org/10.1080/10889868.2019.1671791.

Gawrońska H, Bakera B (2015) Phytoremediation of particulate matter from indoor air by Chlorophytum comosum L. plants. *Air Qual Atmos Heal* 8:265–272. https://doi.org/10.1007/s11869-014-0285-4.

Glencross DA, Ho TR, Camiña N, et al (2020) Air pollution and its effects on the immune system. *Free Radic Biol Med.* https://doi.org/10.1016/j.freeradbiomed.2020.01.179.

Hahladakis JN (2020) Delineating and preventing plastic waste leakage in the marine and terrestrial environment. *Environ Sci Pollut Res* 27:12830–12837. https://doi.org/10.1007/s11356-020-08139-y.

Hemidouche S, Favier L, Amrane A, et al (2018) Successful biodegradation of a refractory pharmaceutical compound by an indigenous phenol-tolerant pseudomonas aeruginosa strain. *Water Air Soil Pollut* 229:1–16. https://doi.org/10.1007/s11270-018-3684-6.

Huang L, Song S, Cai Z, et al (2021) Efficient conversion of bicarbonate (HCO_3-) to acetate and simultaneous heavy metal $Cr(VI)$ removal in photo-assisted microbial electrosynthesis systems combining $WO_3/MoO_3/g-C_3N_4$ heterojunctions and Serratia marcescens electrotroph. *Chem Eng J* 406:126786. https://doi.org/10.1016/j.cej.2020.126786.

Ihsanullah I, Jamal A, Ilyas M, et al (2020) Bioremediation of dyes: Current status and prospects. *J Water Process Eng* 38:101680. https://doi.org/10.1016/j.jwpe.2020.101680.

Jacob JM, Karthik C, Saratale RG, et al (2018) Biological approaches to tackle heavy metal pollution: A survey of literature. *J Environ Manage* 217:56–70. https://doi.org/10.1016/j.jenvman.2018.03.077.

Jiang J, Li S (2018) Microbial degradation of chemical pesticides and bioremediation of pesticide-contaminated sites in China. *Twenty Years Res Dev Soil Pollut Remediat China* 655–670. https://doi.org/10.1007/978-981-10-6029-8_40.

Kalia A, Singh S (2020) Myco-decontamination of azo dyes: Nano-augmentation technologies. *3 Biotech* 10. https://doi.org/10.1007/s13205-020-02378-z.

Khan MA, Khan S, Khan A, Alam M (2017) Soil contamination with cadmium, consequences and remediation using organic amendments. *Sci Total Environ* 601–602:1591–1605. https://doi.org/10.1016/j.scitotenv.2017.06.030.

Khudur LS, Gleeson DB, Ryan MH, et al (2018) Implications of co-contamination with aged heavy metals and total petroleum hydrocarbons on natural attenuation and ecotoxicity in Australian soils. *Environ Pollut* 243:94–102. https://doi.org/10.1016/j.envpol.2018.08.040.

Kim KH, Kabir E, Jahan SA (2017) Exposure to pesticides and the associated human health effects. *Sci Total Environ* 575:525–535. https://doi.org/10.1016/j.scitotenv.2016.09.009.

Kim KJ, Khalekuzzaman M, Suh JN, et al (2018) Phytoremediation of volatile organic compounds by indoor plants: A review. *Hortic Environ Biotechnol* 59:143–157. https://doi.org/10.1007/s13580-018-0032-0.

Kour D, Kaur T, Devi R, et al (2021) Beneficial microbiomes for bioremediation of diverse contaminated environments for environmental sustainability: Present status and future challenges. *Environ Sci Pollut Res* 28:24917–24939. https://doi.org/10.1007/s11356-021-13252-7.

Kowalska JB, Mazurek R, Gąsiorek M, Zaleski T (2018) Pollution indices as useful tools for the comprehensive evaluation of the degree of soil contamination–A review. *Environ Geochem Health* 40:2395–2420. https://doi.org/10.1007/s10653-018-0106-z.

Kumar M, Jaiswal S, Sodhi KK, et al (2019) Antibiotics bioremediation: Perspectives on its ecotoxicity and resistance. *Environ Int* 124:448–461. https://doi.org/10.1016/j.envint.2018.12.065.

Kurwadkar S (2017) Groundwater pollution and vulnerability assessment. *Water Environ Res* 89:1561–1577. https://doi.org/10.2175/106143017x15023776270584.

Ławniczak Ł, Woźniak-Karczewska M, Loibner AP, et al (2020) Microbial degradation of hydrocarbons—basic principles for bioremediation: A review. *Molecules* 25:1–19. https://doi.org/10.3390/molecules25040856.

Leong YK, Chang JS (2020) Bioremediation of heavy metals using microalgae: Recent advances and mechanisms. *Bioresour Technol* 122886. https://doi.org/10.1016/j.biortech.2020.122886.

Li Q, Liu J, Gadd GM (2020) Fungal bioremediation of soil co-contaminated with petroleum hydrocarbons and toxic metals. *Appl Microbiol Biotechnol* 104:8999–9008. https://doi.org/10.1007/s00253-020-10854-y.

Liu SH, Zeng GM, Niu QY, et al (2017) Bioremediation mechanisms of combined pollution of PAHs and heavy metals by bacteria and fungi: A mini review. *Bioresour Technol* 224:25–33. https://doi.org/10.1016/j. biortech.2016.11.095.

Liu X xin, Hu X, Cao Y, et al (2019) Biodegradation of phenanthrene and heavy metal removal by acid-tolerant burkholderia fungorum FM-2. *Front Microbiol* 10:1–13. https://doi.org/10.3389/fmicb.2019.00408.

Meng D, Jiang W, Li J, et al (2019) An alkaline phosphatase from Bacillus amyloliquefaciens YP6 of new application in biodegradation of five broad-spectrum organophosphorus pesticides. *J Environ Sci Heal - Part B Pestic Food Contam Agric Wastes* 54:336–343. https://doi.org/10.1080/03601234.2019.1571363.

Miri S, Naghdi M, Rouissi T, et al (2019) Recent biotechnological advances in petroleum hydrocarbons degradation under cold climate conditions: A review. *Crit Rev Environ Sci Technol* 49:553–586. https://doi.org/10.1080/10643389.2018.1552070.

Morillo E, Villaverde J (2017) Advanced technologies for the remediation of pesticide-contaminated soils. *Sci Total Environ* 586:576–597. https://doi.org/10.1016/j.scitotenv.2017.02.020.

Naghdi M, Taheran M, Brar SK, et al (2018) Removal of pharmaceutical compounds in water and wastewater using fungal oxidoreductase enzymes. *Environ Pollut* 234:190–213. https://doi.org/10.1016/j.envpol.2017.11.060.

Nasrollahi M, Pourbabaei AA, Etesami H, Talebi K (2020) Diazinon degradation by bacterial endophytes in rice plant (Oryzia sativa L.): A possible reason for reducing the efficiency of diazinon in the control of the rice stem–borer. *Chemosphere* 246:125759. https://doi.org/10.1016/j.chemosphere.2019.125759.

Partovinia A, Rasekh B (2018) Review of the immobilized microbial cell systems for bioremediation of petroleum hydrocarbons polluted environments. *Crit Rev Environ Sci Technol* 48:1–38. https://doi.org/10.1080/10643389.2018.1439652.

Phale PS, Sharma A, Gautam K (2019) *Microbial Degradation of Xenobiotics Like Aromatic Pollutants from the Terrestrial Environments*. Elsevier Inc.

Qian H, Chen J, Howard KWF (2020) Assessing groundwater pollution and potential remediation processes in a multi-layer aquifer system. *Environ Pollut* 263:114669. https://doi.org/10.1016/j.envpol. 2020.114669.

Rahman Z (2020) An overview on heavy metal resistant microorganisms for simultaneous treatment of multiple chemical pollutants at co-contaminated sites, and their multi-purpose application. *J Hazard Mater* 396:122682. https://doi.org/10.1016/j.jhazmat.2020.122682.

Raj K, Sardar UR, Bhargavi E, et al (2018) Advances in exopolysaccharides based bioremediation of heavy metals in soil and water: A critical review. *Carbohydr Polym* 199:353–364. https://doi.org/10.1016/j. carbpol.2018.07.037.

Ratnasari A, Syafiuddin A, Kueh ABH, et al (2021) Opportunities and challenges for sustainable bioremediation of natural and synthetic estrogens as emerging water contaminants using bacteria, fungi, and algae. *Water Air Soil Pollut* 232:1–23. https://doi.org/10.1007/s11270-021-05183-3.

Roccuzzo S, Beckerman AP, Trögl J (2021) New perspectives on the bioremediation of endocrine disrupting compounds from wastewater using algae-, bacteria- and fungi-based technologies. *Int J Environ Sci Technol* 18:89–106. https://doi.org/10.1007/s13762-020-02691-3.

Rovira J, Domingo JL (2019) Human health risks due to exposure to inorganic and organic chemicals from textiles: A review. *Environ Res* 168:62–69. https://doi.org/10.1016/j.envres.2018.09.027.

Sangthong S, Suksabye P, Thiravetyan P (2016) Air-borne xylene degradation by Bougainvillea buttiana and the role of epiphytic bacteria in the degradation. *Ecotoxicol Environ Saf* 126:273–280. https://doi.org/10.1016/j.ecoenv.2015.12.017.

Santillan JY, Rojas NL, Ghiringhelli PD, et al (2020) Organophosphorus compounds biodegradation by novel bacterial isolates and their potential application in bioremediation of contaminated water. *Bioresour Technol* 317:124003. https://doi.org/10.1016/j.biortech.2020.124003.

Sarker A, Nandi R, Kim JE, Islam T (2021) Remediation of chemical pesticides from contaminated sites through potential microorganisms and their functional enzymes: Prospects and challenges. *Environ Technol Innov* 23:101777. https://doi.org/10.1016/j.eti.2021.101777.

Sayed K, Baloo L, Sharma NK (2021) Bioremediation of total petroleum hydrocarbons (Tph) by bioaugmentation and biostimulation in water with floating oil spill containment booms as bioreactor basin. *Int J Environ Res Public Health* 18:1–27. https://doi.org/10.3390/ijerph18052226.

Schweitzer L, Noblet J (2018) Water Contamination and Pollution. *Green Chem An Incl Approach* 261–290. https://doi.org/10. 1016/B978-0-12-809270-5.00011-X.

Sehrawat A, Phour M, Kumar R, Sindhu SS (2021) Bioremediation of pesticides: An eco-friendly approach for environment sustainability BT. In: Panpatte DG, Jhala YK (eds). *Microbial Rejuvenation of Polluted Environment*, vol. 1. Springer, Singapore, pp 23–84.

Setsungnern A, Treesubsuntorn C, Thiravetyan P (2018) Chlorophytum comosum–bacteria interactions for airborne benzene remediation: Effect of native endophytic Enterobacter sp. EN2 inoculation and blue-red LED light. *Plant Physiol Biochem* 130:181–191. https://doi.org/10.1016/j. plaphy.2018.06.042.

Shao Y, Wang Y, Yi F, et al (2019) Gaseous formaldehyde degrading by methylobacterium sp. XJLW. *Appl Biochem Biotechnol* 189:262–272. https://doi.org/10.1007/s12010-019-03001-5.

Sharma P, Pandey AK, Kim SH, et al (2021) Critical review on microbial community during in-situ bioremediation of heavy metals from industrial wastewater. *Environ Technol Innov* 24:101826. https://doi.org/10.1016/j.eti.2021.101826.

Sharma S, Chatterjee S (2017) Microplastic pollution, a threat to marine ecosystem and human health: A short review. *Environ Sci Pollut Res* 24:21530–21547. https://doi.org/10.1007/s11356-017-9910-8.

Shrestha R, Ban S, Devkota S, et al (2021) Technological trends in heavy metals removal from industrial wastewater: A review. *J Environ Chem Eng* 9:105688. https://doi.org/10.1016/j.jece. 2021.105688.

Singh S, Kumar V, Singh S, Singh J (2019) Influence of humic acid, iron and copper on microbial degradation of fungicide Carbendazim. *Biocatal Agric Biotechnol* 20:101196. https://doi.org/10.1016/j.bcab.2019.101196.

Singh UK, Ramanathan AL, Subramanian V (2018) Groundwater chemistry and human health risk assessment in the mining region of East Singhbhum, Jharkhand, India. *Chemosphere* 204:501–513. https://doi.org/10.1016/j. chemosphere.2018.04.060.

Sonwani RK, Swain G, Giri BS, et al (2020) Biodegradation of Congo red dye in a moving bed biofilm reactor: Performance evaluation and kinetic modeling. *Bioresour Technol* 302:122811. https://doi.org/10.1016/j.biortech.2020.122811.

Soreanu G, Dixon M, Darlington A (2013) Botanical biofiltration of indoor gaseous pollutants - A mini-review. *Chem Eng J* 229:585–594. https://doi.org/10.1016/j. cej.2013.06.074.

Srivastava A, Rani RM, Patle DS, Kumar S (2022) Emerging bioremediation technologies for the treatment of textile wastewater containing synthetic dyes: a comprehensive review. *J Chem Technol Biotechnol*, 97:26-41. https://doi.org/10.1002/jctb.6891.

Supreeth M (2021) Enhanced remediation of pollutants by microorganisms–plant combination. *Int. J. Environ. Sci. Technol.* 19, 4587–4598. https://doi.org/10.1007/s13762-021-03354-7.

Supreeth M, Raju N (2017) Biotransformation of chlorpyrifos and endosulfan by bacteria and fungi. *Appl Microbiol Biotechnol* 101:5961–5971. https://doi.org/10.1007/s00253-017-8401-7.

Sutherland DL, Ralph PJ (2019) Microalgal bioremediation of emerging contaminants - Opportunities and challenges. *Water Res* 164:114921. https://doi.org/10.1016/j. watres.2019.114921.

Tang X, Huang Y, Li Y, et al (2021) Study on detoxification and removal mechanisms of hexavalent chromium by microorganisms. *Ecotoxicol Environ Saf* 208:111699. https://doi.org/10.1016/j.ecoenv. 2020.111699.

Torena P, Alvarez-Cuenca M, Reza M (2021) Biodegradation of polyethylene terephthalate microplastics by bacterial communities from activated sludge. *Can J Chem Eng* 99:S69–S82. https://doi.org/10.1002/cjce. 24015.

Vikrant K, Giri BS, Raza N, et al (2018) Recent advancements in bioremediation of dye: Current status and challenges. *Bioresour Technol* 253:355–367. https://doi.org/10.1016/j. biortech.2018.01.029.

Vohra K, Marais EA, Bloss WJ, et al (2022) Rapid rise in premature mortality due to anthropogenic air pollution in fast-growing tropical cities from 2005 to 2018. *Sci Adv* 8(14):4435.

Walker DB, Baumgartner DJ, Gerba CP, Fitzsimmons K (2019) *Surface Water Pollution*, 3rd edn. Elsevier Inc.

Wang S, Wang J (2018) Biodegradation and metabolic pathway of sulfamethoxazole by a novel strain Acinetobacter sp. *Appl Microbiol Biotechnol* 102:425–432. https://doi. org/10.1007/s00253-017-8562-4.

Wang Z, Zhang Z, Zhang Y, et al (2013) Nitrogen removal from Lake Caohai, a typical ultra-eutrophic lake in China with large scale confined growth of Eichhornia crassipes. *Chemosphere* 92:177–183. https://doi.org/10.1016/j.chemosphere.2013.03.014.

Wei X, Lyu S, Yu Y, et al (2017) Phylloremediation of air pollutants: Exploiting the potential of plant leaves and leaf-associated microbes. *Front Plant Sci* 8:1–23. https://doi. org/10.3389/fpls.2017.01318.

Weyens N, Thijs S, Popek R, et al (2015) The role of plant–microbe interactions and their exploitation for phytoremediation of air pollutants. *Int J Mol Sci* 16:25576–25604. https://doi.org/10.3390/ijms161025576.

Xin B, Lee Y, Hadibarata T (2020) Phytoremediation mechanisms in air pollution control : A review. *Wat Air Soil Poll* 231:437.

Xiong JQ, Kurade MB, Jeon BH (2018) Can microalgae remove pharmaceutical contaminants from water? *Trends Biotechnol* 36:30–44. https://doi.org/10.1016/j.tibtech.2017.09.003.

Zhang B, Matchinski EJ, Chen B, et al (2018) *Marine Oil Spills-Oil Pollution, Sources and Effects*, 2nd Edn. London: Elsevier Ltd.

Zhang D, Gersberg RM, Ng WJ, Tan SK (2014) Removal of pharmaceuticals and personal care products in aquatic plant-based systems: A review. *Environ Pollut* 184:620–639. https://doi.org/10.1016/j.envpol.2013.09.009.

Zhang H, Yuan X, Xiong T, et al (2020a) Bioremediation of co-contaminated soil with heavy metals and pesticides: Influence factors, mechanisms and evaluation methods. *Chem Eng J* 398:125657. https://doi.org/10.1016/j.cej.2020.125657.

Zhang J, Gao D, Li Q, et al (2020b) Biodegradation of polyethylene microplastic particles by the fungus Aspergillus flavus from the guts of wax moth Galleria mellonella. *Sci Total Environ* 704. https://doi.org/10.1016/j.scitotenv. 2019.135931.

Żur J, Wojcieszyńska D, Hupert-Kocurek K, et al (2018) Paracetamol – toxicity and microbial utilization. Pseudomonas moorei KB4 as a case study for exploring degradation pathway. *Chemosphere* 206:192–202. https://doi.org/10.1016/j. chemosphere.2018.04.179.

14 State of Agro-wastes Management in Nigeria
Status, Implications, and Way Forward

Toyese Oyegoke, Ayandunmola Folake Oyegoke, Opeoluwa Olusola Fasanya, and Abdul-Alim Gambo Ibrahim

CONTENTS

14.1 Introduction 242
14.2 Methodology Deployed in the Survey 243
14.3 Agricultural Produces & Their Potential Wastes 244
    14.3.1 The Trend of Agricultural Production and Its Contribution to
        National Economy 244
    14.3.2 Agro-wastes Generation in Nigeria 245
14.4 Current Practice and Its Implications 248
    14.4.1 Open Burning 248
        14.4.1.1 Background Information and Concept of the Practice 248
        14.4.1.2 Health and Environmental Implications of Open
                Burning of the Agro-wastes 248
    14.4.2 Dumping of Agro-wastes on Traditional Landfill (Open Dumpsite).....249
        14.4.2.1 Background Information and Concept of the Practice 249
        14.4.2.2 Health and Environmental Implications of Using
                Traditional (Undesigned) Landfills.................................249
    14.4.3 Use of Agro-wastes for Cooking, Warming, and Lighting...............250
        14.4.3.1 Background Information and Concept of the Practice 250
        14.4.3.2 Health and Environmental Implications of Using it for
                Cooking, Warming, and Lighting.....................................250
    14.4.4 Random Dumping of the Agro-wastes within the Communities.....251
        14.4.4.1 Background Information and Concept of the Practice 251
        14.4.4.2 Health and Environmental Implications of Randomly
                Dumping Wastes in Communities 251
14.5 Ways Forward to Improve the Management Practice................................252
    14.5.1 Conversion of Agro-wastes to Biofuels 252

DOI: 10.1201/9781003359784-14

        14.5.1.1 Concept of Approach 252

        14.5.1.2 Principle of Agro-waste Transformation to be Biofuels 252

        14.5.1.3 A Review of Existing Biofuels Study Reports in
the Literature...253

        14.5.1.4 SWOT Analysis of Agro-waste Transformation to
be Biofuels 255

    14.5.2 Biopower and Bioenergy Generation.................................256

        14.5.2.1 Concept of Approach 256

        14.5.2.2 Principle of Deploying Waste to Bioenergy Approach......256

        14.5.2.3 Report of Bioenergy Studies in the Literature.................257

        14.5.2.4 SWOT Analysis of Deploying Waste to Bioenergy
Approach...257

    14.5.3 Use of Agro-wastes in Adsorbents Productions 259

        14.5.3.1 Concept of Approach 259

        14.5.3.2 Principle & Report of Studies for the Use of
Carbonization...259

        14.5.3.3 Principle and Report of Studies for the Non-char-based
Adsorbents 261

        14.5.3.4 SWOT Analysis of Transforming Agro-wastes
into Adsorbent..262

14.6 Conclusions & Recommendations...263

    14.6.1 Conclusions...263

    14.6.2 Recommendations 263

Acknowledgement 263

References..263

14.1 INTRODUCTION

For many years, agriculture has been the dominant form of occupation across the sub-Sahara African countries from the pre-colonial era to date (Lutz et al., 2019; Robert & Akinlawon, 2022). In countries like Nigeria, Ghana, and many other African countries, agriculture has played a vital role in the sustenance of their people and economies. For some countries, focus is on the cultivation of food crops essentially while some invest more in cash crops (Gareth, 2009; Ireti Olamide, 2015). The literature further indicates that some of these African countries are engaged in exporting crops like palm oil, groundnut, tobacco, cocoa, rubber, rice, maize, and many others (David, 2021; FAO, 1997; Nicolas & Guido, 2011). Historically, agricultural produce claimed a significant portion of the annual government revenue in Nigeria before the discovery of crude oil and other solid minerals (Bjornlund et al., 2020; Ekenta et al., 2017). Even though the attention of the government has primarily shifted to the oil and gas sector of her economy, the investment in the agricultural sector did not stop even though it could be reported to be low when compared to investments in the oil and gas sector of Nigeria economy (Ekenta et al., 2017; Okotie, 2018). This trend is however changing as there are renewed calls, strategies, and actions for an agricultural revolution in the country.

During harvest after the cultivation of agricultural produce across developing countries like Nigeria, wastes are always generated. Some of these wastes include rice husk, groundnut shell, maize cob, maize stalk, tomato wastes, sorghum stalk, bean shells, sugarcane bagasse, and many others (Odimegwu, 2014; Oyegoke et al., 2022). The accumulated waste continuously generated nationwide has been contributing nuisance to our environment resulting in different forms of land, air, and water pollution, which could lead to the habitation of deadly microbes, render our surroundings unclean, and release unfriendly and harmful gases to neighboring residences and a lot more (Alemma-Ozioruva, 2017; Karshima, 2016; Ogundele et al., 2018).

Going by the sustainable development goals of the United Nations (UNDP, 2022), which tend to promote better environmental sustainability and good public health care, it is essential to devise a better approach to managing agro-waste in such a way that the waste can be safely disposed of or transformed into valuable products. The concerned goals include good health and well-being (Goal 3); clean water and sanitation (Goal 6); affordable and clean energy (Goal 7); and sustainable cities and communities (Goal 8) (UNDP, 2022). The current approach to managing agro-wastes has been largely reported to be unsafe and a threat to having good healthy living and an eco-friendly environment both in rural and in urban cities (Eneji et al., 2016). This has been contributing to the difficulty in achieving the set goals of UNDP mentioned earlier.

Moreover, the significance of agriculture to humanity and its sustenance cannot be overemphasized both in developed and in developing nations like Nigeria. The negative implications of how its wastes are currently managed cannot be overlooked for the sake of eco-friendly and good public health. It is therefore vital to correct these current practices and unfold alternative approaches to better manage the waste.

As a step forward to educating the public on a better approach to advancing the management of agro-wastes, this report attempts to present the trend of Nigeria's annual agricultural production, estimate the corresponding volume of wastes generated annually, showcase the current practice deployed in the management of the wastes generated, health safety, and environmental implication of the existing practice deployed in the management of wastes in our communities, and to review alternative measure for the existing ones used as a possible way forward to having better health, safety, and environment across our nation, along with a possible recommendation to guide policy-making and future studies.

14.2 METHODOLOGY DEPLOYED IN THE SURVEY

The study would primarily rely on secondary data, gathered essentially from existing published research articles reported in reputable peer-reviewed journals and some other resources like statistical databases collated by organizations such as the Food and Agriculture Organization of United Nations (FAOSTAT, 2022) and Nigeria Bureau of Statistics (NBS, 2022). Reputable government agencies' online resources and that of some reputable newspapers and magazines were also consulted. Figure 14.1 diagrammatically presents the framework deployed in this report's survey.

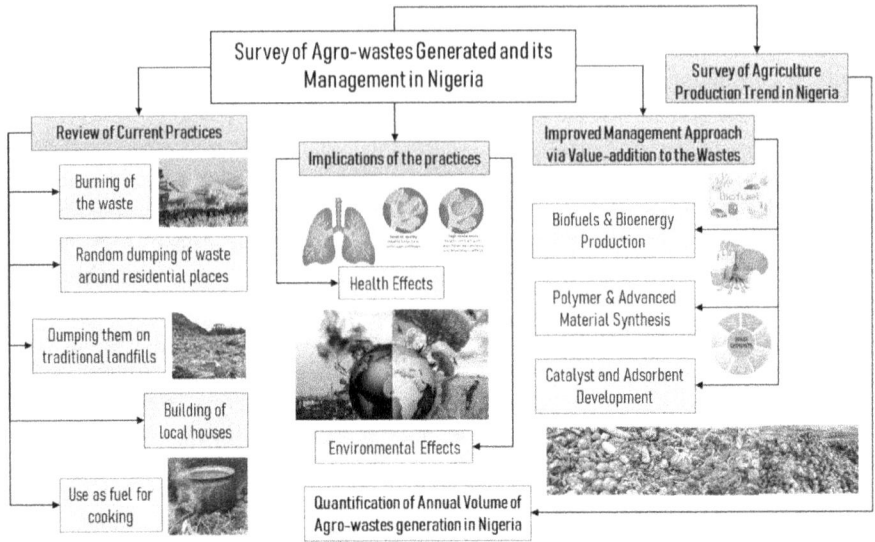

FIGURE 14.1 A schematic diagram showcases the framework deployed in this report's survey.

The report features sections like agricultural produces and its potential wastes with the aid of the existing primary research reports and databases in the literature; current practice and its implications with the aid of the existing primary research reports in the literature; ways forward to improve the management practice with the aid of the existing primary research reports in the literature using strengths, weaknesses, opportunities, and threats (SWOT) analysis; and proposition of relevant recommendations to government, entrepreneurs, scholars, and residents of Nigeria on the subject.

14.3 AGRICULTURAL PRODUCES & THEIR POTENTIAL WASTES

14.3.1 The Trend of Agricultural Production and Its Contribution to National Economy

The practice of cultivating the soil for the sustenance of humans across Africa, especially in Nigeria in the pre-colonial and early post-independence era, has often become the nominal occupation of most residents, especially in the rural and sub-urban regions of the country. According to the report of the FAOSTAT (2022), the national agriculture production has continuously grown which is evident in Figure 14.2. The trend was found to be inconsistently linear with an R-squared value of 0.6833, but rather found to be polynomial as shown in Figure 14.2 with an R-squared value of 0.924.

The treemap plot (Figure 14.3) presents the average fraction of the respective crops produced annually in Nigeria indicating cassava (34%), yams (27%), maize (6%),

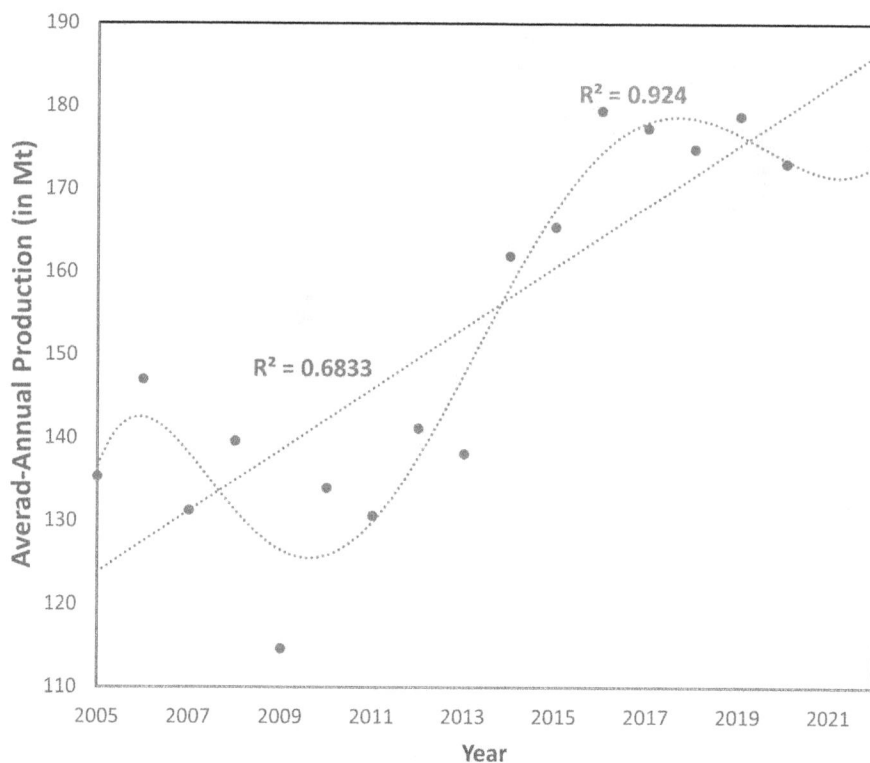

FIGURE 14.2 The trend of agricultural production (Annual-averaged) in Nigeria from the report of FAOSTAT (2005–2020).

sorghum (5%), oil palm fruit (6%), rice (4%), groundnut (2%), sweet potatoes (2%), cowpeas (2%), plantains (2%), tomatoes (2%), millet (2%), and other crops like sugarcane, mango, pineapple, and potatoes. FAOSTAT's (2022) report indicates cassava claimed the largest fraction of the total averaged-annual agriculture production within 2005–2021.

A further survey of the literature (NBS, 2022; Statista, 2022) also indicates that agriculture has continued to contribute significantly 21%–31% to Nigeria's gross domestic product (GDP) which is evident in Figure 14.4, according to the 2019–2021 reports.

14.3.2 AGRO-WASTES GENERATION IN NIGERIA

To further understand the volume of agro-wastes annually generated in Nigeria on an average basis, relevant ratios from existing literature were used in estimating the quantity of waste. In absence of the relevant ratios, a realistic ratio (or factor) was deployed for such a case as "estimated." The report of the analysis carried out is presented in Table 14.1 showcasing the crop, potential class of wastes, ratio deployed (from relevant reference or assumption in absence of reference), an overall fraction

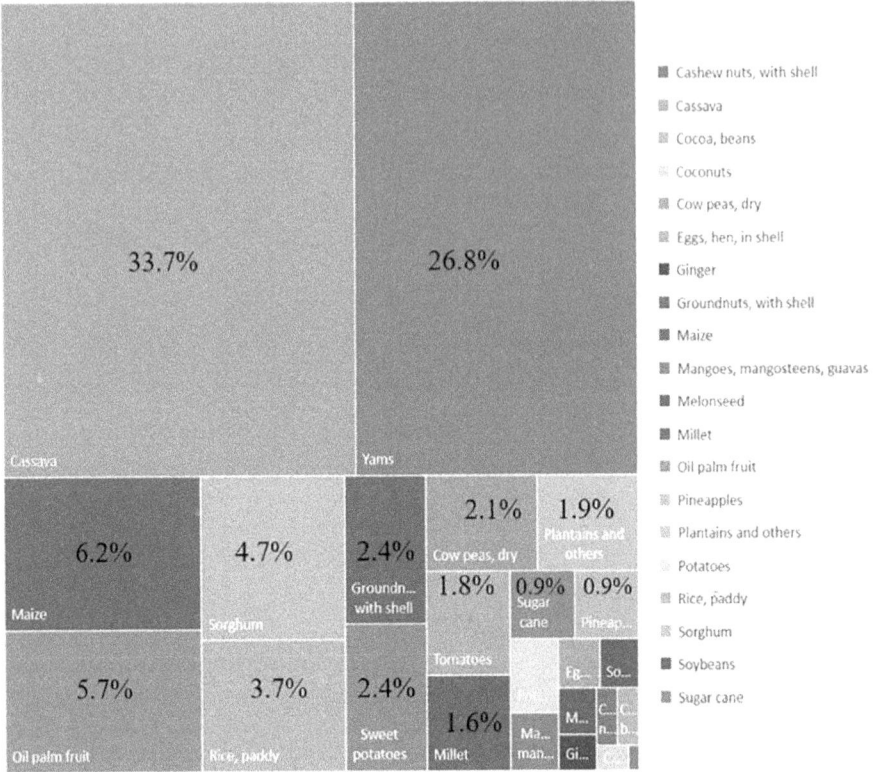

FIGURE 14.3   Treemap plot showing the average fraction of respective annual crop production from the report of FAOSTAT (2005–2020) in Nigeria.

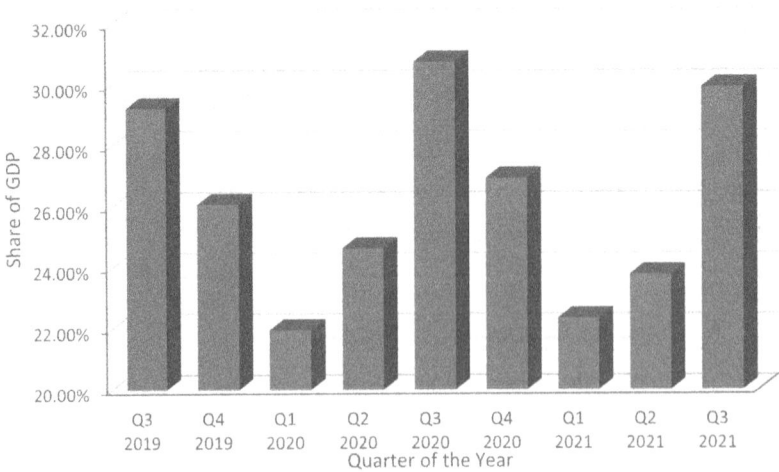

FIGURE 14.4   Contribution of agriculture to GDP in Nigeria from the 3rd quarter of 2019 to the 3rd quarter of 2021 (NBS, 2022; Statista, 2022).

TABLE 14.1

Analysis of Agro-wastes Annually Generated in Nigeria Farms using the produces

Crops	1. Potentials waste	Crop-to-waste mass ratio (source), CW	Overall wastes fraction, $\sum \dfrac{1}{CW}$	Annual averaged crop production[e], (Mt)	Wastes generated (Mt)
Cassava (CS)	Peel, stem	CS:peel = 100/17 [a] CS:stem = 9:1	0.0671	50.80	3.41
Yam	Peel, leaves, stem	Yam:peel = 91/9 Yam:leaves = 95/5 Yam:stem = 95/5	0.0208	40.42	0.84
Maize	Cob, stalk, straw	Maize:straw = 9/1 Maize:cob = 6/5 Maize:stalk = 7/4	0.0837	9.28	0.78
Palm oil fruit (POF)	Kernel shell (KS), leave, empty fruit bunch (EFB)	POF:KS = 4/6 POF:leaves = 9/1 POF:EFB = 4/6	0.0968	8.58	0.83
Sorghum (SG)	Stalk, bagasse, straw	SG:stalk = 7/3 SG:bagasse = 6/4 SG:straw = 9/1	0.0779	7.10	0.55
Rice	Husk, straw	Grain:straw = 3/5 [b] Grain:husk = 9.7/1.9 [c]	0.1753	5.65	0.99
Groundnut (GN)	Shell, leaves, stem	GN:hull = 0.89/0.30 [d] GN:stem = 8/2 [d]	0.1435	3.61	0.52
Sweet potatoes (SP)	Peels, leaves	SP:peel = 8/2 SP:leaves = 9/1	0.0769	3.56	0.27
Millet	Leaves, stalk	ML:leaves = 6/4 ML:stalks = 4/6	0.4615	2.42	1.12
Tomatoes (TM)	Leaves, stem	TM:leave = 6/4 TM:stem = 5/5	0.4000	2.65	1.06
Cowpeas (CW)	Shell, stem, leaves	CW:shell = 6/4 CW:stem = 6/5 CW:leave = 9/1	0.08547	3.23	0.28
Plantain	Peels	Plaintain:peels = 8/3	0.3750	2.91	1.09
Sugarcane (SG)	Bagasse, leaves	SG:bagasse = 9/3 SG:leave = 9/1	0.0833	1.32	0.11
Pineapple (PA)	Peel, stem, leaves	PA:stem = 9/3 PA:peel = 9/1 PA:leaves = 9/1	0.0476	1.31	0.06
Potatoes (PO)	Peel, leaves	PO:peel = 9/3 PO:leave = 9/1	0.0833	1.07	0.09
Mango	Peel, seed	Mango:peel = 9/1 Mango:seed = 9/3	0.0833	0.84	0.07
Total	-	Crop:waste (source)	2.3615	144.75	12.07

(using the summation of the waste fractions) of the wastes, averaged crop production, and wastes generated.

Report from the estimation of the average annual agro-waste generated from the largest cultivated crops in Nigeria (as shown in Table 14.1) via the use of data adapted from FAOSTAT and NBS databases (FAOSTAT, 2022; NBS, 2022) indicated that the total waste fraction to be 2.36, that is, 2.36 wt of wastes would be averagely generated from 1wt crops cultivated. In all, the overall average of the total waste generated was estimated to be 12.06 Mt (i.e., 12.06 mega-tonnes).

Data source: All ratios are estimated ratios sourced from literature (Nguyen et al., 1991; Akcali et al., 2007; Javier et al., 2019; Ebrayi et al., 2007; FAOSTAT, 2022).

This volume of agro-wastes annually generated in our respective communities which are often wrongly managed can be visibly seen in our surroundings road-side, dumpsites, including residential places where they are being used as fuel. Consideration of alternative safer approaches to better managing the wastes could further promote the eco-friendliness of our communities.

14.4 CURRENT PRACTICE AND ITS IMPLICATIONS

In practice, many of the rural, suburban, and a fraction of urban cities still engage in ineffective ways of managing their agro-wastes which tend to pose some level of threat to their public health and environmental sustainability. Some of these practices have been shown to have detrimental health and environmental implications. The current popular practices deployed in Nigeria such as open burning, random disposal around residential places, and a lot more are presented here with the implications of promoting the practices.

14.4.1 OPEN BURNING

14.4.1.1 Background Information and Concept of the Practice

This method is identified commonly as an ancient, and cheap mode of agricultural waste management which is unapproved and sanctioned by the government and considered a disadvantage to the economy. Also, harmful and dangerous chemicals such as polychlorinated dibenzo-p-dioxins and dioxins are known to be emitted during the burning of agricultural waste (Folorunsho & Mamman, 2016). Poor knowledge on how effective conversion of agro-wastes will boost the economy and the negative effect of burning agro-wastes on the environment and health also contribute to the high occurrence of the method, especially among farmers.

14.4.1.2 Health and Environmental Implications of Open Burning of the Agro-wastes

The burning of wastes like agro-wastes produces greenhouse gases (such as methane, carbon dioxide, nitrogen) known to deplete the ozone layer (Elbasiouny et al., 2020). Other reported emissions during the burning of such wastes including stench, soot, noxious gases, and the release of dioxins and furan (Daffi et al., 2020) have been suspected to be involved in some types of cancer, reproductive damage, and birth defect (Rim-Rukeh, 2014). Continuous pollution of air in this manner can cause respiratory symptoms such as chronic bronchitis, allergic rhinitis, acute exacerbation of asthma, scarring of the lungs, and lung cancer in long term (Nkwachukwu et al., 2010; Owoyemi et al., 2016). Many a times, soot and fly ash from the burning process tends to also end up in water bodies leading to water pollution. The practice of agro-wastes burning's impact on the environment and human health is negative, thereby requiring an urgent need for the practice to be abolished and discouraged.

14.4.2 DUMPING OF AGRO-WASTES ON TRADITIONAL LANDFILL (OPEN DUMPSITE)

14.4.2.1 Background Information and Concept of the Practice

Landfills are lands allocated and separated for disposal of refuse to prevent pollution of the environment, and they are divided into sanitary landfills, semi-controlled landfills, and open dumping (Ezechi et al., 2017). Most landfills in Nigerian communities are largely traditional and not engineered, otherwise known as dump sites. The sites are randomly situated around residential places and with some located close to the water bodies without government approvals or regulations. It is a usual practice that when dumpsites get overloaded with wastes, volunteers sometimes organize open burning of the wastes amidst residential places while some transport the decayed component of the dumpsite to their farms to be deployed as manure. The decayed components have high nutrients like potassium and nitrogen that can improve soil yield but they are still yet grossly underutilized (Sabiiti, 2011). The disposal and dumping of agro-waste on open land or sea, and open dumping of refuse have been identified as the prevalent method for waste disposal (Adogu et al., 2015), which are not subjected to any waste treatment measure at all.

14.4.2.2 Health and Environmental Implications of Using Traditional (Undesigned) Landfills

Some of the major environmental impacts include toxicity of the water for those dumpsites located close to water bodies, the habitation of dangerous rodents within residential places, and the generation of landfill gases such as methane, carbon dioxide, organosulfur gases, carbon monoxide, nitrate, and many other unsafe pollutants. These pollutants pose a serious threat to the nature of air in such residential places, and also its accumulation contributes to global warming and the depletion of the ozone layer (Ogunrinola & Adepegba, 2012; Nkwachukwu et al., 2010; Owoyemi et al., 2016). This happens as a result of the continuous process of anaerobic decomposition of the agro-wastes if disposed of at landfills. This is particularly common when it is not an engineered landfill which causes pollution of land, water, and air leading to environmental nuisance and public health issues.

Moreover, the traditional landfills (open dumpsites) which are improperly designed can also lead to underground water pollution. This is due to the slow seepage of water that can contain cyanide toxins which if not channeled or drained appropriately can find their way to groundwater and contaminate drinking water. This occurs if the landfill is located close to a water body causing pollution of underground and surface waters through either chemicals or pathogens (Idowu et al., 2012; Nkwachukwu et al., 2010). An increase in the number of rodents, pests, insects, and birds which are disease vectors (cockroach, rat, mosquito) as well as filthy and messy environments can affect the health of the residents. Residents have a higher probability of falling victim to illnesses like typhoid fever, malaria, river blindness, diarrhea diseases, skin irritation and dermatology diseases, and cholera which can lead to death (Ogunrinola & Adepegba, 2012; Idowu et al., 2012).

An environment free of pollution and indiscriminate dumping of refuse is a step to the attainment of sustainable and robust health (Idowu et al., 2012). The clearance of agro-waste mostly in the developing world is mostly by burning, landfills, and open dumping at unsanctioned places causing a nuisance to the environment. Deterioration of soil, and quality of air and water are the attestation to poorly managed treatment of agro-waste, especially organic waste generated by animals leading to acidic rain and production of ammonia which is an odorous gas (Obi et al., 2016).

14.4.3   Use of Agro-wastes for Cooking, Warming, and Lighting

14.4.3.1   Background Information and Concept of the Practice

The use of agro-wastes as fuel for cooking is an ancient approach commonly deployed in largely the rural part of developing nations like Nigeria. Agro-wastes like stalks, stems, shells, leaves, and straws (Ajibade, 2007) are deployed as fuel for cooking food, warming bodies during winter/cold seasons, and lighting in the night via the burning of the wastes (Folorunsho & Mamman, 2016; Yevich & Logan, 2003). Many of those who practice this approach are widely found in the rural communities (Yusuf et al., 2020) and are largely found to be of poor financial background without the means to purchase alternative energy sources like solar, electricity, gas, or kerosene which are safer to use (Folorunsho & Mamman, 2016; Yevich & Logan, 2003).

14.4.3.2   Health and Environmental Implications of Using it for Cooking, Warming, and Lighting

In using agro-waste as cooking fuel, small particles and other constituents are released into the air due to incomplete combustion causing a detrimental effect on human health due to toxic emissions during its usage (Nwaokocha & Giwa, 2016). Some of these toxic emissions include hydrocarbons, carbon monoxide, and nitrites which are freely released into the environment. Indoor air pollution can lead to a long spectrum of respiratory ailments and disorders ranging from bronchitis, asthma, allergic rhinitis, chronic obstructive pulmonary disease (Gujba et al., 2015), childhood diseases, infant death, infectious disease, cardiovascular diseases (Kersten et al., 1998), which also occurs as a result of these emissions. It was established in a study done among rural women in southern Nigeria who use biomass fuel in cooking that there is a significant relationship between biomass fuel usage and markers of cardiovascular risk. The use of cleaner fueled to a reduction in blood pressure and a 4.5% reduction of being at risk of hypertension (Ofori et al., 2018). Further, looking into the environmental effect indicates that the use of agricultural waste in cooking can also help reduce deforestation and forest degradation (Gujba et al., 2015; Japhet, 2020). It still remains a less effective approach to managing the agro-wastes largely generated, especially during the harvest period.

14.4.4 Random Dumping of the Agro-wastes within the Communities

14.4.4.1 Background Information and Concept of the Practice

This is similar to open dumping mostly seen in communities and areas where humans reside wherein the movement of man from one point to another and activities of man as regard to its environment plays a role in the disposal of agro-waste around residential areas. In developing countries, most of the agro-wastes are dumped openly (Adeyemi & Adeyemo, 2007) usually in fallow lands, uncompleted buildings, market places, roadsides, and drainages. Individuals can dispose of waste anywhere at any time (instinctively waste disposal) mostly at inappropriate places that cause an eyesore and environmental nuisance (Nkwachukwu et al., 2010). The absence of ideal waste storage containers and inefficient and regular waste transportation to sanitary landfills designated by the government could also contribute to the waste littering in residential areas. In some cases, the wastes are dumped illegally at the roadside, at the entrance of the market, on the bank of the river, and inside the drainage and the lagoon (Elbasiouny et al., 2020; Sangodoyin, 1993), which is unsafe approach to disposing of wastes.

14.4.4.2 Health and Environmental Implications of Randomly Dumping Wastes in Communities

Disposal of agro-wastes in residential areas encourages pest breeding ground for mosquitoes, flies which cause intestinal worms (Olukanni & Akinyinka, 2012), hepatitis, cholera, diarrhea airborne disease like measles, tuberculosis, whooping cough asthma, waterborne disease like dysentery, typhoid fever, and cholera. Inset-borne disease like the plague, and Lassa fever (Jatau, 2013; Sridhar & Hammed, 2014). There is a multiplication of microorganisms (fungi, bacteria) as untreated waste dumping into the river causes accumulation of toxic substances and seep into the food chain through fish and aquatic creatures which feed on it (Jatau, 2013; Nagendran, 2011). There is a vicious cycle of human food chain contamination process going on from agro-waste causing an environmental hazard to the ecosystem and depletion in the health of community members (Pandey, 2020).

Furthermore, due to the proximity to a residential area, the environmental pollution caused becomes a fetal problem healthwise. Wind blowing of litter around the community leaving it dirty and causing an eyesore with stent odor occurs as a result of the putrefaction process of manure organic matter from the agro-waste and production of methane, hydrogen sulfide, and nitrogen which cause greenhouse gases and depletion of the ozone layer. Indiscriminate disposal of used pesticide containers leads to leachates; surface water contamination which contains heavy metals, for example, mercury, zinc, lead, and chromium is released into the environment; and the end products and degradation process are not without harmful effects (Obi et al., 2016; Maton et al., 2016; Pandey, 2020; Taiwo, 2011). The ammonia released from agricultural waste affects the atmospheric air and contributes to acidic rain leading to soil sedimentation and soil erosion (Pandey, 2020). The practice of randomly disposing wastes especially those ones dumped along the drainage also contributes to promoting flooding if it eventually leads to blockage of the drain and reduction in water quality with an increase in microbes (Elbasiouny et al., 2020; Sangodoyin, 1993).

The health and well-being of humans have a direct correlation with waste production, segregation, and disposal in relation to the environment. Diseases predominant in a community or acquired from the community by residents of the community have direct (and indirect) linkage with the nature of the environment. The hygienic disposal or conversion of agro-waste will go a long way in correcting negative implications or effects on the environment and health status of residents living in the area. A study done in Calabar, Nigeria showed that there is a significant relationship between waste disposal and environmental health which further buttress the point that for the environment to be healthy and be in good shape, a proper and good strategy for waste management cannot be overlooked (Eneji et al., 2016). The health and environmental effect of indiscriminate waste disposal might not be felt until much later, and these effects are usually not linked or attributed to inefficient agricultural waste management unless meticulous investigations are carried out (Ndidi et al., 2009).

14.5 WAYS FORWARD TO IMPROVE THE MANAGEMENT PRACTICE

With a view of the implications of the current practices deployed in the management of agro-wastes in Nigeria, it has largely been established that most of the current approaches have been posing a threat to public health and environmental sustainability. The need to proffer a way forward to better manage this waste cannot be overemphasized. The promotion of biofuels to complement the current use of fossil fuels, bioenergy generation to reduce the pressure on the energy demand, alongside the development of adsorbents for pollution management and control via the use of agro-wastes are been proposed here.

14.5.1 CONVERSION OF AGRO-WASTES TO BIOFUELS

14.5.1.1 Concept of Approach

Fuels that are generally obtained via a chemical approach from biomass rather than the slower geological approach are referred to as biofuels (Ajadi et al., 2020; Ganesan et al., 2020; Molino et al., 2018; Nigam & Singh, 2011). This type of fuel is mostly in liquid and gaseous forms of fuels and finds wide application in automobile and portable power plants.

> Examples of fuels in this class include bioethanol, bio-oil, biodiesel, biomethane, biokerosene, biohydrogen and among others (Aisien & Aisien, 2020; Akhihiero et al., 2021; Oyegoke et al., 2022; Oyegoke & Ibraheem, 2021; Tongshuwar & Oyegoke, 2021; Xing et al., 2021). They are largely produced from green or bioresources and have been proven to be eco-friendly and sustainable for transforming the generated agro-wastes into valuable biofuels using the different approaches in Figure 14.5, diagrammatically presented.

14.5.1.2 Principle of Agro-waste Transformation to be Biofuels

Biofuels are produced via different approaches which vary from the type of feedstock available for the processing and the desired kind or grade of biofuels (as shown

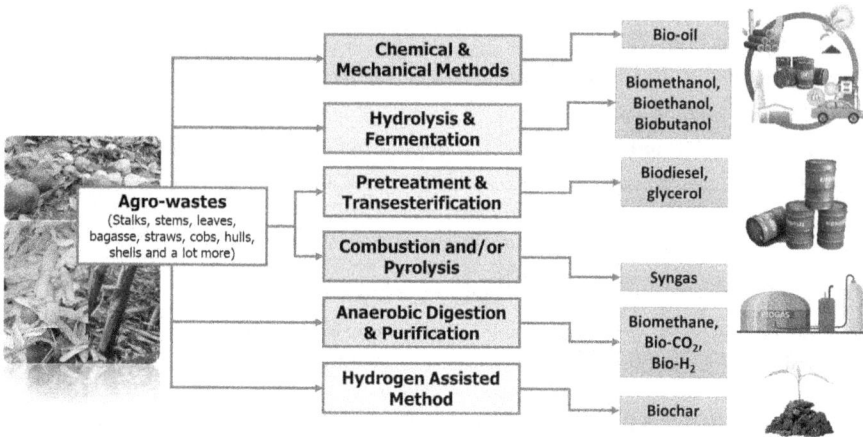

FIGURE 14.5 Different agro-wastes to biofuel conversion approaches.

in Figure 14.5). Bioalcohols like bioethanol, biomethanol, and biobutanol are largely processed via a process known as fermentation which would only take place in presence of simple sugars. In absence of simple sugars, other preliminary processes like pretreatment and hydrolysis would have to be carried out to have the feedstock transformed into simple sugar like glucose (Ebabhi et al., 2019; Isah et al., 2019; Oyegoke et al., 2022).

Another class of biofuels is biodiesel produced via the use of oil-rich agro-wastes in a transesterification process. The esterification and neutralization methods are normally used in the pretreatment processes for the agro-wastes before subjecting the resulting processed wastes to transesterification in presence of either homogenous or heterogeneous catalysts. Such a process helps in reducing the free fatty acid content of the feedstocks (Ajadi et al., 2020; Akhihiero et al., 2021; Funmilayo Aransiola et al., 2012; Nwoko et al., 2019).

Biogas like biomethane is generated via the anaerobic digestion of the feedstock normally in presence of enzymes at high temperatures in presence of moisture. Purification processes are used in separating the carbon dioxide, hydrogen, sulfide, and other materials present in the raw methane gas (Aisien & Aisien, 2020; Chinwendu et al., 2019; Khotmanee & Pinsopon, 2021). Other valuable biofuels like bio-oil and biochar are produced via chemical/mechanical methods and hydrogen-assisted processes, respectively (Ganesan et al., 2020).

14.5.1.3   A Review of Existing Biofuels Study Reports in the Literature

A survey of works reported in the literature indicated that a lot of studies have been carried out on the development of biofuels (biogas, bioethanol, biodiesel, biochar, syngas, and many other ones). A few of the recent reports found the literature on the use of agro-wastes in the production of biofuels are presented in Table 14.2.

Here, diverse reports have established the feasibility of obtaining syngas from rice husk, bioethanol from corncob, cassava peels, sorghum bagasse, groundnut peels and

TABLE 14.2
Some Existing Biofuel Reports are Studied in the Literature

Waste used	2. Biofuel produced	Key findings made	Reference
Rice husk	Syngas	Syngas calorific value of 4.46 mJ/m³ at 800°C was achieved.	Mukhtar et al., 2019
Corncob	Bioethanol	Corn cobs contain reducing sugar with a concentration of 0.167 mg/mL with a bioethanol concentration of 0.331 mg/L, and 76 cm³ of bioethanol was obtained.	Tambuwal et al., 2018
Neem seed oil	Biodiesel	A kinetic model for the prediction of biodiesel yield was developed.	Oyegoke & Ibraheem, 2021
Cassava peel	Bioethanol	4.85 kg of liquid biofuel was obtained in 1 hour 50 minutes when cassava peels are fermented with concentrated tetraoxosulfate (VI).	Adewumi & Akande, 2019
Sorghum bagasse	Bioethanol	Fuel-grade bioethanol was produced from 50,000 kg of sorghum bagasse.	Ajayi et al., 2020
Soybean straws	Biogas	0.19 m³ x 10⁹ BMP of biogas was produced based on 0.7 m³/kgVs at STP.	Akinbomi et al., 2014
Neem seed oil	Biodiesel	Impact of the time and temperature was established for the highest obtainable yield of biodiesel.	Ajadi et al., 2020
Groundnut peel and shell	Bioethanol	From 100 g of peels and shells of groundnut, 7.89% and 3.94% of ethanol were recovered, and the FT-IR spectral confirmed that the bio-liquid obtained from the samples was bioethanol.	Adejumobi & Ogunsuyi, 2020
Yam peels	Bioethanol	Pure bioethanol produced from cassava and yam peels reduces life cycle greenhouse gas emissions by 20% to 100%.	Adiotomre, 2015
Sugarcane bagasse	Bioethanol	Sugarcane bagasse is a good and sustainable feedstock for the production of bioethanol.	Abdulkareem et al., 2015
Coconut peels	Biochar	Because of its solid/fuel and carbon-source features, the feedstock (coconut peels) can be a viable source of energy.	Kabir et al., 2021
Plantain peels	Bio-oils and bio-char	Bio-chars made from plantain peel residues have a higher heating value of 22 MJ/kg and a pH of 10, making them potential renewable fuel sources in the country.	Ogunjobi & Labunmi, 2013
Sweet potato peels	Bioethanol	Fermentation of sweet potato peels increases the amount of bioethanol production (23.90 g/L).	Ebabhi et al., 2018

(*Continued*)

TABLE 14.2 (*Continued*)
Some Existing Biofuel Reports are Studied in the Literature

Waste used	2. Biofuel produced	Key findings made	Reference
Gliricidia sepium	Biodiesel	The feasibility of deploying the use of supercritical CO_2 and hexane to an optimized yield of oil extraction for biodiesel from seeds was established.	Macawile & Auresenia, 2022
Food wastes & human excreta	Biogas	The potential of utilizing the palm oil mill effluent for the production of biogas was established.	Ohimain & Izah, 2014
Molasses	Bioethanol	Process simulation was used to establish the feasibility of bioethanol and the minimum number of the stage (40) the distillation would require to attain the required purity.	Abemi et al., 2018
Waste cooking oil	Biodiesel	A long-time impact of deploying the use of biodiesel and bioethanol in Brazil vehicles was established.	Nogueira et al., 2015
Waste cooking oil and Bleached Palm oil	Biodiesel	Calcium oxide produced from waste eggshells was used as a catalyst to produce biodiesel from waste oil and expired palm oil.	Fasanya et al., 2022

shells, yam peels, sugarcane bagasse, coconut peels, plantain peels, sweet potatoes peels, and molasses. Biodiesel has also been produced from neem seed oil, jatropha oil, waste cooking oil, and *Gliricidia sepium*, while biochar has been obtained from waste of plantain peels and coconut peels.

14.5.1.4   SWOT Analysis of Agro-waste Transformation to be Biofuels

This approach to managing agro-waste would go a long way to not only improve our wastes management approach for better public health and environmental sustainability but also boost or strengthen our country's economy as well as job creation, and the opportunity for biofuels exportation from neighboring countries. However, one of the key challenges includes the poor waste management collection approach practice in most developing communities where waste disposal is done indiscriminately. Another is the economic viability of most of the processes which would rely on government incentives like tax-free policy for biofuel development and provision of subsidy for the biofuels processing equipment or products, despite its cheap and widely available feedstock (Table 14.3).

As a way to promote the deployment of the approach in our respective communities, the government can motivate the private entrepreneurs to invest in this approach via the use of opportunity—weakness where the opportunity offered by the approach can be used to design a better way of alleviating the weakness of deploying the method.

TABLE 14.3
SWOT Analysis for the Approach Deployment in Nigeria

Factor	3. Analysis outcome
Strength	Feedstock availability; cheap feedstock cost; no greenhouse gas emission.
Weakness	Poor waste collection approach; high cost of operation/production; economic viability is highly subjected to the presence of government incentives.
Opportunity	Export of bioethanol; job creation and promotion of economic growth; eco-friendliness promotion; better public health and environmental sustainability.
Threat	Changing government policies; unstable foreign exchange rating for local currency.

And the weakness—threat approach can be used to design a better way of addressing the weakness while eliminating any potential of threats reflecting the process of its deployment. One of the innovative weaknesses—threat strategy from the SWOT analysis outcomes—could suggest the proposition of legal policy that would promote the feasibility of establishing such a processing plant in Nigeria should be enacted as a law. It can be designed to charge the governor of the central bank of Nigeria and finance ministers with the responsibility of making incentives available for these biofuel investors as a possible way to sustain the viability of biofuels businesses in Nigeria.

14.5.2 BIOPOWER AND BIOENERGY GENERATION

14.5.2.1 Concept of Approach

Energy generated or sourced via the use of biomass (Tursi, 2019), bioresources, or green materials is known as bioenergy (Moustakas et al., 2021; Olujobi et al., 2022; Uddin et al., 2021), and is sometimes referred to as biopower (Levi-Oguike et al., 2022; Oyegoke & Jibril, 2016). It is also referred to as renewable energy that is derived from organic or agro-based materials which find application in the production of heat and electricity. It is capable of complementing or substituting other sources of energy generation like hydro, nuclear, solar, thermal, and much other power or energy (Olujobi et al., 2022; Rogers et al., 2017). It is a potential method for improving Nigeria's waste management approach.

14.5.2.2 Principle of Deploying Waste to Bioenergy Approach

The approach involves the transformation of the collected wastes into energy in the forms of heat and electricity. One of such existing technologies for the deployment of this approach employs the use of agro-wastes as the fuels used by the steam turbine power plant burners. The waste is used during the boiling of the water (highly pressurized via a pump) for the generation of steam that is used in turning steam turbines which in turn generates an electric current. The lower temperature exhaust steam then either passes to a lower pressure turbine for further electricity generation or is recycled. Other wastes to electricity technologies do not deploy the use of boiler or steam instead, and the agro-waste can be burnt as fuel in a combustor where flue gas (highly pressurized via a gas compressor) is generated and passed to the gas turbine. The gas is used to turn the turbine to generate electricity while releasing the resultant lower pressure exhaust gas; such technology is known as a gas turbine power plant. There also exists a hybrid form of this technology that tries to combine the use of the steam and gas turbine power plant which is the motive of maximizing the energy generation derivable from the agro-wastes (fuels), thereby improving the energy efficiency of the plant.

14.5.2.3 Report of Bioenergy Studies in the Literature

The literature review indicates that concerted work on this approach is still ongoing as researchers across the globe seek to identify the most sustainable way of generating affordable eco-friendly power. Research works are also studying the best ways of transforming all the agro-wastes widely generated, especially in the rural areas. Power generation has been one key threat to industrialization and has discouraged many investors from considering Nigeria for the establishment of their manufacturing plants due to the high cost of energy. The energy requirement demanded for the operation of the plant annually poses a threat to the economic viability of the investment in developing economies. To address some of these trending challenges, especially in Nigeria, the conversion of agro-wastes into energy (electricity and heat) would go a long way to help residential places and private businesses sustenance outside making our environment friendly as a result of the improved waste management approach that could be deployed. Some of the existing (recent) studies on bioenergy generation are presented in Table 14.4.

Reports in open literature indicate that some works have already established the technical viability of generation power via the use of these wastes (Ajewole et al., 2022; Moustakas et al., 2021; Ojolo et al., 2012). Going by the report of Oyegoke and Jibril (2016) and Levi-Oguike (2022), their reports have also proven that it is economically feasible for potential investors to consider. It is also important for further works to consider the exploration of other potentials that this approach of waste to energy technologies can offer for the development of the energy and waste management sector in Nigeria.

14.5.2.4 SWOT Analysis of Deploying Waste to Bioenergy Approach

The continuous struggles of several government administrations in Nigeria with the management of waste and provision of constant power supply for her residents have been a long-term subject of concern. A review of the deployment of this approach of converting agro-wastes into bioenergy for electricity and heat generation via the use of SWOT analysis indicates that this approach has several good qualities in form of strength which includes the ease of access to cheap fuels, absence of threat to food security, and promotion of better energy security as a result of the expansion that the approach would offer to the energy mix of Nigeria.

Among its weak features is the need for the importation of the equipment due to the absence of locally developed technology for its establishment in Nigeria. However, the government can partner with researchers from different institutes to work with them in the development of local technology for the generation of

TABLE 14.4
Some of the Existing Bioenergy Studies Report

Wastes used	4. Bioenergy generated	Key findings made	References
Biomass wastes	29.8 EJ (potential)	The continuous growth of Nigeria's bioenergy potential and the need to harness this huge resource for the rising energy demand in the country.	Ojolo et al., 2012
Sugarcane bagasse	5 MW	With the use of 50 tonnes of feed, a 5 MW per day was generated and was found to be viable economically.	Oyegoke & Jibril, 2016
Arable crops and agro-materials	775 GWh (electricity) 1.119 GWh (heat)	The maximum obtainable electricity and heat from the agrarian biomass of 715 km.	Moustakas et al., 2021
Cocoa-kolanut wastes	–	The feasibility of converting the wastes to power is continuously established, and agricultural areas where the wastes are generated can take advantage of this development.	Ajewole et al., 2022
Sugarcane bagasse	5 MW	A life cycle analysis indicates that the plant profitability increased correlatively as the operational efficiency of the plant changed from 85% to 91%	Levi-Oguike et al., 2022

TABLE 14.5
SWOT Analysis for the Waste to Bioenergy Approach in Nigeria

Factor	5. Analysis outcome
Strength	Cheap and readily available fuels (agro-wastes); new energy strategy for the country; promotion of energy security; not a threat to food security.
Weakness	Absence of local technology for its establishment; unequal access of an individual to government loan; poor government maintenance practice.
Opportunity	Diversification of national energy mix; gradual alleviation of power supply problem; creation of direct and indirect jobs; foreign direct investment.
Threat	High foreign exchange rate for importing equipment; political interest, monopolization of biopower production; the rise in fuel price and higher labor cost; potential of imposing levies, heavy tax and others on investors; changing government energy policies.

bioenergy that would commercialize. To better harness the full benefit of deploying this approach in Nigeria, the government has to design measures to address the potential threats (which are enumerated in Table 14.5) that could hinder the investment from flourishing and that could totally discourage investors from considering such enterprises.

14.5.3 Use of Agro-wastes in Adsorbents Productions

14.5.3.1 Concept of Approach

Adsorption has been employed as a separation technique for well over a century and is still actively in use today, especially in pollution control (Fang et al., 2020). The separation usually involves the separation of a liquid, gas, or solid unto another solid material called an adsorbent. Important characteristics of any adsorbent include its capacity, compatibility, selectivity as well as cost. Cost in particular has facilitated the exploration of unconventional materials for adsorbent synthesis. The utilization of agricultural waste as core components in adsorbent design has garnered momentum in the research community in the last few decades.

Agro-based waste materials are classified as renewable as they are readily available in large quantities at a minimal cost. Using them as adsorbents is a valid means of reducing waste generation. Adsorbents designed for a number of purposes have been successfully synthesized solely or partially comprised of biomass from agro-waste. Adsorbents for heavy metal removal such as chromium from water (Mouhamadou et al., 2022), removal of dyes (Adegoke & Bello, 2015; Subramaniam & Kumar Ponnusamy, 2015), crude oil sorption (Nwadiogbu et al., 2016) and among others have been successfully synthesized from agro-waste.

There are two general ways of getting value from agricultural waste (agro-wastes) as adsorbents. These include carbonization; extraction, and utilization of lignocellulosic content.

14.5.3.2 Principle & Report of Studies for the Use of Carbonization

Carbonization of agro-waste is a widely used method before it can be used as an adsorbent or catalyst. The means of carbonization, temperature, heating rate, particle size as well as morphology are pertinent factors that affect the density of active sites and other important properties (Khan et al., 2021). Calcination or combustion of agricultural biomass tends to heavily reduce the high carbon and oxygen contents present in the biomass (Basumatary et al., 2018), thus leaving behind carbonates and alkali metal oxides which possess active sites that promote transesterification reactions such as biodiesel synthesis. They tend to possess hydrophobic sites which minimize catalyst deactivation and facilitate adsorption of triglycerides and other long-chain organic molecules (Alfredo Quevedo-Amador et al., 2022).

Carbonization process can occur in a number of different ways each yielding different properties. Gasification (> 600°C), pyrolysis (350°C–1,000°C), torrefaction (200°C–320°C), and hydrothermal carbonization (HTC) (180°C–250°C) are well-known methods of actively carbonizing agricultural waste.

   a. **Biochar (Pyrolysis Char)**
   The pyrolysis (heating at high temperatures under limited flow of oxygen) of biomass often results in reasonable amount of solid char deposits often referred to as biochar. Biochar is often characterized by large surface area which is largely dependent on pyrolysis temperature. Pyrolysis temperatures typically range between 500°C and 700°C with higher surface areas and porosity being obtained at higher temperatures. This is as a result of

TABLE 14.6
Textural Properties of Some Biochar

Feedstock	6. Temp. (°C)	Pyrolysis type	Surface area (m²/g)	Pore vol. (cm³/g)	References
Rice husk	450–500	Fast pyrolysis	117.8	0.073	Liu et al., 2012
Date seeds	600	Slow	265.42	0.10	Ogungbenro et al., 2018
Oil palm fiber	387	Slow	352.5	0.1487 m³/g	Adelodun et al., 2020
Cassava waste	750	Slow	75.30	0.05	Luo et al., 2021
Rice husk	500	Slow	1226	0.59	Yaumi et al., 2018
Rice husk	500	Slow	29.18	0.058	Chen et al., 2019
Rice husk	600	Slow	10.995	0.0044	Kizito et al., 2015
Sugarcane bagasse	600	Slow	4.65	9.96 x 10⁻⁴	Ma et al., 2021
Sugarcane bagasse	600	Slow	388	–	Creamer et al., 2014
Sugarcane bagasse	750	Fast	622	0.38	Guo et al., 2020

increased decomposition of cellulose, lignin, and other organic matter at higher temperatures which encourages micropore formation (Tomczyk et al., 2020). Increased surface area and porosity generally portend an increase in adsorption capacity as well as improved catalytic activity. Table 14.6 highlights some of the differing textural properties of biochar. Slow pyrolysis is preferred as lower temperatures and prolonged residence time encourage repolymerization of the components of rice husk leading to improved biochar (Bushra & Remya, 2020).

The other major factor affecting the nature of biochar is the source material (Tang et al., 2013). Differences have been observed in the properties of biochar depending on the starting feedstock. These differences are equally highlighted in Table 14.6. Ageing of feed material as reported concerning sugarcane bagasse could also affect the adsorptive properties of synthesized biochar (Hass & Lima, 2018).

b. **Hydrothermal carbonization**

HTC is a thermochemical process wherein biomass is heated at temperatures between 180°C and 350°C at high pressures for periods ranging from 5 minutes to as much as 4 hours (Heidari et al., 2019). The process can also be looked at as a rapid process for coalification due to high pressures and is quite useful for tackling biomass with high moisture content due to the presence of water during carbonization.

A series of reactions occur during HTC: One of which is hydrolysis wherein ester and ether bonds contained in hemicellulose, cellulose, and lignin are broken down. Hemicellulose fragmentation occurs at $T > 180°C$, while cellulose and lignin fragment at $T > 200°C$ and $T > 220°C$, respectively. Dehydration and decarboxylation reactions are two equally important reactions that occur alongside hydrolysis and lead to the reduction of H/C and O/C ratios (Funke & Ziegler, 2010; Heidari et al., 2019).

c. **Gasification Char**

Gasification is a thermochemical process that involves the conversion of biomass to gas at temperatures between $500°C$ and $1,400°C$ in the presence of air, steam, O_2, CO_2, N_2, or a combination of these gases. Aside from syngas, liquid products in the form of tar and a solid product char are also formed. The method of gasification, as well as reaction conditions, greatly affect the properties of char formed. Differences have been seen in the nature of char formed using CO_2, air, and steam gasification with larger pores being formed at low temperatures using CO_2 while air gasification facilitated decomposition (Klinghoffer et al., 2012) and adsorption of naphthalene (Gradel et al., 2021).

d. **Torrefaction Char**

Similar to the other thermochemical processes listed above, torrefaction also involves heating biomass though temperatures are limited to between $200°C$ and $300°C$ under an inert atmosphere. Torrefied biomass is primarily used as a fuel source or a pretreatment step prior to pyrolysis or gasification (Niu et al., 2019; Zhang et al., 2016). Biochar from torrefaction has been applied as an adsorbent primarily in wastewater treatment for the removal of dyes such as methylene blue (Singh et al., 2021). 5.3.2.2 Char Activation.

It is also critical to note that activation of char is often required prior to its use as an adsorbent. Activation can either be thermal (physical) or by chemical means. The method of activation and agents of activation often have a remarkable impact on the physiochemical properties of the synthesized adsorbent. This is evident in the work of Guo et al. (2020) wherein air-activated biochar from sugarcane bagasse presented a surface area of $99 \, m^2/g$ while CO_2-activated biochar recorded a surface area of $622 \, m^2/g$.

Chemical activation can be conducted by impregnating the derived chars with strong alkalis such as NaOH or KOH, $ZnCl_2$, H_3PO_4, $CaCl_2$ and heating in an inert atmosphere at temperatures between $450°C$ and $900°C$ (Yaumi et al., 2018). Activation using potassium hydroxide leads to an increased presence of micropores within the char structure albeit with corresponding low carbon yield.

14.5.3.3 Principle and Report of Studies for the Non-char-based Adsorbents

Adsorbents can also be synthesized from agro-waste by extracting lignocellulosic content from them. Cellulose-based oleophilic adsorbents can be produced by first of all extracting cellulose using a suitable alkali such as sodium hydroxide (Fasanya et al., 2020). The obtained cellulose is further purified using bleaching agents such as sodium hypochlorite, sodium chlorite, or hydrogen peroxide to remove leftover lignin. For oil sorption, inducing hydrophobicity on the extracted fibers is then carried out by acetylation (Nwadiogbu et al., 2016), salinization (Do et al., 2020), or treatment with fatty acids such as stearic acid (Asadu et al., 2021).

Acetylation involves the introduction of the acetyl group (CH_3CO) onto cellulose fibers by reacting with acetic anhydride in the presence of a suitable catalyst such as N-bromosuccinimide. Depending on the degree of reaction, sufficient hydrophobicity is usually introduced, thereby leading to a higher degree of oil sorption when used as an oil absorbent.

Rice husk can be considered as a flexible resource that can be used as an adsorbent without chemical processing. Reports have shown its application as an adsorbent for the removal of contaminants such as free fatty acids, methylene blue, pyridine, and formaldehyde among others (Kenes et al., 2012). Also worthy of note are silica gel adsorbents which are produced from rice husk. The synthesis of silica gel proceeds via thermal treatment at temperatures within the range of 500° followed by sodium hydroxide treatment to extract the high silica content within the rice husk ash-forming sodium silicate. Lastly, precipitation of silica gel occurs by the addition of sulfuric acid (AbuKhadra et al., 2020).

14.5.3.4   SWOT Analysis of Transforming Agro-wastes into Adsorbent

A SWOT analysis of using agricultural waste for adsorbent synthesis is shown in Table 14.2. Key on the list of strengths is the large availability of agricultural waste which is usually discarded by burning in Nigeria (Oguntoke et al., 2019). Prior to the oil boom, agriculture was the mainstay of the Nigerian economy and still generates a substantial amount of agricultural waste, especially from rice and other grains (Obianefo et al., 2021). Added value can be placed on agricultural products by being able to effectively convert non-edible components of farm produce to adsorbents (Table 14.7).

A major weakness lies in initial cost of setting up facilities for the production of agro-waste adsorbents. There is still relatively poor access to credit facilities for agricultural activities (Osabohien et al., 2022) which in turn could affect the rate of raw materials available to adsorbent synthesis.

TABLE 14.7
SWOT Analysis of Agricultural Waste for Adsorbent Synthesis

Factor	7. Analysis outcome
Strength	Cheap and readily available raw materials due large amount of agricultural waste; multiple types of adsorbents for wide variety of purposes can be gotten from biochar; adsorbents from agricultural wastes tend to be biodegradable.
Weakness	Initial cost of setting up; desorption step may prove costly when used as an adsorbent.
Opportunity	Pollution control and waste water treatment; establishment of chemical process industries; generation of valuable foreign exchange by offering competitive prices on a global scale.
Threat	Creating awareness to convince investors; resistance by populace to change orientation with regards to discarding agricultural waste by burning; increasing cost of energy.

14.6 CONCLUSIONS & RECOMMENDATIONS

14.6.1 CONCLUSIONS

Improper disposal of agricultural waste is clear and presents danger. Present means of disposal tend to result in environmental pollution which can have detrimental effects on human life. Due to the biological nature of agro-waste, there is increased propensity for disease outbreak due to vectors that are attracted to decayed waste. There are potentially viable economic opportunities in harnessing the large quantity of agro-waste generated in developing countries such as Nigeria. Energy in terms of fuel (liquid, gaseous, or solid) or electricity is an efficient way of disposing or utilizing biomass. Another viable route to consider is conversion of agro-waste to useful materials and chemicals such as adsorbents. These are proven to be valuable in pollution control, water treatment, and also chemical process industries. Numerous opportunities exist in conversion of agricultural biomass to chemicals which should be critically looked into.

14.6.2 RECOMMENDATIONS

Conversion of agro-waste to useful forms is a viable means of wealth generation from waste. A worthwhile set to consider by developing nations especially those situated on the African continent would be mass re-orientation of their citizens to see the value in agricultural waste. Education of the dangers of burning waste also has to be conducted on a continuous basis and incorporated into basic school curriculum. Initial significant capital cost that may be required in terms of setting up process plants, tax breaks, and other such government incentives would be a welcome step in encouraging investors to turn their attention to agro-waste conversion to value products.

ACKNOWLEDGEMENT

The authors acknowledge the support of Pencil Team (PT) members and other affiliates that aid in driving this report to a completion stage.

REFERENCES

Abdulkareem, A. S., Afolabi, A. S., & Ogochukwu, M. U. (2015). Production and characterization of bioethanol from sugarcane bagasse as alternative energy sources. *Proceedings of the World Congress on Engineering at London, UK, II*, 1–5.

Abemi, A., Oyegoke, T., Dabai, F. N., & Jibril, B. Y. (2018). Technical and economic feasibility of transforming molasses into bioethanol in Nigeria. *National Engineering Conference at Zaria, Nigeria, 145*, 1–8. https://www.researchgate.net/publication/337223844_Technical_and_Economic_Feasibility_of_Transforming_Molasses_into_Bioethanol_in_Nigeria

AbuKhadra, M. R., Mohamed, A. S., El-Sherbeeny, A. M., & Elmeligy, M. A. (2020). Enhanced photocatalytic degradation of acephate pesticide over MCM-41/Co3O4 nanocomposite synthesized from rice husk silica gel and peach leaves. *Journal of Hazardous Materials, 389*, 122129. https://doi.org/10.1016/j.jhazmat.2020.122129.

Achi, H. A., Adeofun, C. O., Ufoegbune, G. C., Gbadebo, A. M., & Oyedepo, J. A. (2012). Disposal sites and transport route selection using geographic information systems and remote sensing in Abeokuta, Nigeria. *Global Journal of Human Social Science (B)*, *12*(12), 15–23. https://globaljournals.org/GJHSS_Volume12/3-Disposal-Sites-and-Transport-Route.pdf

Adegoke, K. A., & Bello, O. S. (2015). Dye sequestration using agricultural wastes as adsorbents. *Water Resources and Industry*, *12*, 8–24. https://doi.org/10.1016/j.wri.2015.09.002.

Adejumobi, I. B., & Ogunsuyi, H. O. (2020). Production of bioethanol from groundnut (Arachis Hypogaea l.) peels and shells. *International Journal of Engineering Applied Sciences and Technology*, *4*(11), 38–44. https://www.ijeast.com/papers/38-44,Tesma411,IJEAST.pdf

Adelodun, A. A., Adeniyi, A. G., Ighalo, J. O., Onifade, D. V., & Arowoyele, L. T. (2020). Thermochemical conversion of oil palm Fiber-LDPE hybrid waste into biochar. *Biofuels, Bioproducts and Biorefining*, *14*(6), 1313–1323. https://doi.org/10.1002/bbb.2130.

Adeoye, P. A., Adebayo, S. E., & Musa, J. J. (2011). Agricultural post-harvest waste generation and management for selected crops in Minna, Niger State, North Central Nigeria. *Journal of Applied Sciences in Environmental Sanitation*, 6(4), 427–435.

Adesogan, S. O. (2013). Wooden materials in building projects: Fitness for roof construction in southwestern Nigeria. *Journal of Civil Engineering Construction Technology*, 4(7), 217–223. https://doi.org/10.5897/JCECT2013.0278.

Adewumi, I. O., & Akande, M. A. (2019). Production of liquid bio-fuel from cassava peel using pilot scale plant. *International Journal of Energy and Environmental Research*, 7(2), 19–29.

Adeyemi, I. G., & Adeyemo, O. K. (2007). Waste management practices at the Bodija abattoir, Nigeria. *International Journal of Environmental Studies*, *64*(1), 71–82. https://doi.org/10.1080/00207230601124989.

Adiotomre, K. O. (2015). Production of bioethanol as an alternative source of fuel using cassava and yam peels as raw materials. *International Journal of Innovative Scientific & Engineering Technologies Research*, 3(2), 28–44.

Adogu, P. O. U., Uwakwe, K. A., Egenti, N. B., Okwuoha, A. P., & Nkwocha, I. B. (2015). Assessment of waste management practices among residents of Owerri Municipal Imo State Nigeria. *Journal of Environmental Protection*, 6, 446–456. https://doi.org/10.4236/jep.2015.65043.

Aisien, F. A., & Aisien, E. T. (2020). Biogas from cassava waste. *Detritus Multidisciplinary Journal for Waste Resource & Residues*, *10*(10), 108. https://doi.org/10.31025/2611-4135/2020.13910.

Ajadi, I. K., Oyegoke, T., Geoffrey, T. T., Fasanya, O., & Ojetunde, A. O. (2020). Biodiesel Production from Neem Seed Oil: Catalyst Synthesis, Effect of Time, and Temperature on Biodiesel Yield. *Proceedings of the Materials Science & Technology Society of Nigeria at Ife, Nigeria*, *19*, 189–194. https://www.researchgate.net/publication/356443109_Biodiesel_Production_from_Neem_Seed_Oil_Catalyst_Synthesis_Effect_of_Time_and_Temperature_on_Biodiesel_Yield#fullTextFileContent.

Ajayi, O. O., Onifade, K. R., Oyegoke, T., & Onadeji, A. (2020). Techno-economic assessment of transforming sorghum bagasse into bioethanol fuel in Nigeria : 1- Process modelling, simulation, and cost estimation. *Journal of Engineering Studies and Research*, 26(August), 154–164. https://doi.org/10.29081/jesr.v26i3.219.

Ajewole, T. O., Aworinde, A. K., Okedere, O. B., & Somefun, T. E. (2022). Agro-residues for clean electricity: In-lab trial of power generation from blended cocoa-kolanut wastes. *Heliyon*, 8, e09091. https://doi.org/10.1016/j.heliyon.2022.e09091.

Ajibade, L. T. (2007). Indigenous knowledge system of waste management in Nigeria. *Indian Journal of Traditional Knowledge*, 6(4), 642–647.

Akcali, I. D., Ince, A., & Guzel, E. (2007). Selected physical properties of peanuts. *International Journal of Food Properties*, 9(1), 25–37. https://doi.org/10.1080/10942910500471636.

Akhihiero, E. T., Ayodele, B. V., Alsaffar, M. A., Audu, T. O. K., & Aluyor, E. O. (2021). Kinetic studies of biodiesel production from jatropha curcas oil. *Journal of Engineering*, 27(4), 33–45. https://doi.org/10.31026/J.ENG.2021.04.03.

Akinbomi, J., Brandberg, T., Sanni, S. A., & Taherzadeh, M. J. (2014). Development and dissemination strategies for accelerating biogas production in Nigeria. *BioResources*, 9(3), 1–31.

Alemma-Ozioruva, A. (2017). Challenges of managing waste disposal in Nigeria. *The Guardian Nigeria News - Nigeria and World News — Saturday Magazine — The Guardian Nigeria News – Nigeria and World News.* https://guardian.ng/saturday-magazine/challenges-of-managing-waste-disposal-in-nigeria/.

Alfredo Quevedo-Amador, R., Elizabeth Reynel-Avila, H., Ileana Mendoza-Castillo, D., Badawi, M., & Bonilla-Petriciolet, A. (2022). Functionalized hydrochar-based catalysts for biodiesel production via oil transesterification: Optimum preparation conditions and performance assessment. *Fuel*, *312*, 122731. https://doi.org/10.1016/j.fuel.2021.122731.

Asadu, C. O., Anthony, E. C., Elijah, O. C., Ike, I. S., Onoghwarite, O. E., & Okwudili, U. E. (2021). Development of an adsorbent for the remediation of crude oil polluted water using stearic acid grafted coconut husk (Cocos nucifera) composite. *Applied Surface Science Advances*, 6, 100179. https://doi.org/10.1016/j.apsadv.2021.100179.

Babayemi, J., & Dauda, K. (2009). Evaluation of solid waste generation, categories and disposal options in developing countries: A case study of Nigeria. *Journal of Applied Sciences and Environmental Management*, 13(3), 83–88. https://doi.org/10.4314/jasem.v13i3.55370.

Basumatary, S., Nath, B., & Kalita, P. (2018). Application of agro-waste derived materials as heterogeneous base catalysts for biodiesel synthesis. *Journal of Renewable and Sustainable Energy*, 10(4), 043105. https://doi.org/10.1063/1.5043328.

Bjornlund, V., Bjornlund, H., & van Rooyen, A. F. (2020). Why agricultural production in sub-Saharan Africa remains low compared to the rest of the world – a historical perspective, *International Journal of Water Resources Development,* sup1, 1–34. https://doi.org/10.1080/07900627.2020.1739512.

Bushra, B., & Remya, N. (2020). Biochar from pyrolysis of rice husk biomass—characteristics, modification and environmental application. *Biomass Conversion and Biorefinery*. https://doi.org/10.1007/s13399-020-01092-3.

Chen, S., Qin, C., Wang, T., Chen, F., Li, X., Hou, H., & Zhou, M. (2019). Study on the adsorption of dyestuffs with different properties by sludge-rice husk biochar: Adsorption capacity, isotherm, kinetic, thermodynamics and mechanism. *Journal of Molecular Liquids*, 285, 62–74. https://doi.org/https://doi.org/10.1016/j.molliq.2019.04.035.

Chinwendu, A. O., Catherine, A.-O. B., Bassey, A. E., Okon, E. U., Chinwendu, A. O., Catherine, A.-O. B., Bassey, A. E., & Okon, E. U. (2019). The potential of biogas production from fruit wastes (Watermelon, Mango and Pawpaw). *World Journal of Advanced Research and Reviews*, 1(3), 052–065. https://doi.org/10.30574/WJARR.2019.1.3.0026.

Creamer, A. E., Gao, B., & Zhang, M. (2014). Carbon dioxide capture using biochar produced from sugarcane bagasse and hickory wood. *Chemical Engineering Journal*, 249, 174–179. https://doi.org/https://doi.org/10.1016/j.cej.2014.03.105.

Daffi, R. E., Chaimang, A. N., & Alfa, M. I. (2020). Environmental impact of open burning of municipal solid wastes dumps in parts of Jos Metropolis, Nigeria. *Journal of Engineering Research and Reports*, 12(3), 30–43. https://doi.org/10.9734/jerr/2020/v12i317083.

David, T. (2021). *Nigeria at a Glance.* https://www.fao.org/nigeria/fao-in-nigeria/nigeria-at-a-glance/en/.

Do, N. H. N., Luu, T. P., Thai, Q. B., Le, D. K., Chau, N. D. Q., Nguyen, S. T., Le, P. K., Phan-Thien, N., & Duong, H. M. (2020). Advanced fabrication and application of pineapple aerogels from agricultural waste. *Materials Technology*, 35(11–12), 807–814. https://doi.org/10.1080/10667857.2019.1688537.

Ebabhi, A. M., Adekunle, A. A., & Adeogun, O. O. (2019). Potential of some tuber peels in bioethanol production using Candida tropicalis. *Nigerian Journal of Basic and Applied Sciences, 26*(2), 17–22. https://doi.org/10.4314/njbas.v26i2.3.

Ebrayi, K. N., Pathak, H., Kalra, N., Bhatia, A., & Jain, N. (2007). Simulation of Nitrogen Dynamics in Soil using InfoCrop Model. *Environmental Monitoring and Assessment 131*(1), 451–465. https://doi.org/10.1007/S10661-006-9491-3.

Ekenta, C. M., Ajala, M. K., Akinola, M. O., & Oseni, Y. (2017). Abandoned Nigerian economic resources: The case of oil palm. *International Journal of Agricultural Extension and Rural Development Studies, 4*(2), 1–16. https://www.eajournals.org/wp-content/uploads/Abandoned-Nigerian-Economic-Resources-The-Case-of-Oil-Palm.pdf.

Elbasiouny, H., Elbanna, B. A., Al-Najoli, E., Alsherief, A., Negm, S., Abou El-Nour, E., Nofal, A., & Sharabash, S. (2020). Agricultural waste management for climate change mitigation: Some implications to Egypt. *Water*, 149–169. https://doi.org/10.1007/978-3-030-18350-9_8.

Eneji, C. V. O., Eneji, J. E. O., & Ngoka, V. N. (2016). Attitude towards waste management and disposal methods and the health status of cross river state, Nigeria. *SCIREA Journal of Agriculture, 1*(2), 231–247.

Ezechi, E. H., Nwabuko, C. G., Enyinnaya, O. C., & Babington, C. J. (2017). Municipal solid waste management in Aba, Nigeria: Challenges and prospects. *Environmental Engineering Research, 22*(3), 231–236. https://doi.org/10.4491/eer.2017.100.

Fang, D., Zhuang, X., Huang, L., Zhang, Q., Shen, Q., Jiang, L., Xu, X., & Ji, F. (2020). Developing the new kinetics model based on the adsorption process: From fitting to comparison and prediction. *Science of The Total Environment, 725*, 138490. https://doi.org/https://doi.org/10.1016/j.scitotenv.2020.138490.

FAO. (1997). Agriculture food and nutrition for Africa. In *A Resource Book for Teachers of Agriculture*. Publishing Management Group, FAO Information Division. https://www.fao.org/3/W0078E/w0078e05.htm.

FAOSTAT. (2022). *Statistical Database for Production*. Food and Agriculture Organization of United Nations. https://www.fao.org/faostat/en/#data.

Fasanya, O. O., Adesina, O. B., Okoduwa, U. J., Abdulkadir, J., Winful, E., Obidah, T. Y., Adamun, S. I., Audu, E. A., Myint, M. T. Z., Olabimtan, O. H., & Barminas, J. T. (2020). Characterization of Sansevieria liberica & Urena lobata fibers as potential sorbent materials for crude oil clean up. *Journal of Natural Fibers*, 1–16. https://doi.org/10.1080/15440478.2020.1788486.

Fasanya, O.O., Gbadamasi, S., Osigbesan, A.A., Ahmed, O.U., Isa, A.R., Ozogu, A.N., Hayatudeen, A., Yusuf, A.I. and Gano, Z.S. (2022), Effect of hydrothermal treatment on the properties of calcium oxide from eggshells used as a biodiesel catalyst. *Chemical Engineering Technology, 45*: 283–290. https://doi.org/10.1002/ceat.202100377.

Folorunsho, J. O., & Mamman, M. (2016). Assessment of open burning of agricultural waste in Biu local government area of Borno State, Nigeria. *Advances in Social Sciences Research Journal, 3*(6), 52–60. https://doi.org/10.14738/assrj.36.1801.

Funke, A., & Ziegler, F. (2010). Hydrothermal carbonization of biomass: A summary and discussion of chemical mechanisms for process engineering. *Biofuels, Bioproducts and Biorefining, 4*(2), 160–177. https://doi.org/10.1002/bbb.198.

Funmilayo Aransiola, E., Daramola, M. O., Ojumu, T. V., Aremu, O., Kolawole Layokun, S., & Solomon, B. O. (2012). Nigerian jatropha curcas oil seeds: Prospect for biodiesel production in Nigeria. *International Journal of Renewable Energy Research, 2*(2), 318–325.

Ganesan, R., Manigandan, S., Samuel, M. S., Shanmuganathan, R., Brindhadevi, K., Lan Chi, N. T., Duc, P. A., & Pugazhendhi, A. (2020). A review on prospective production of biofuel from microalgae. *Biotechnology Reports, 27*, e00509. https://doi.org/10.1016/J.BTRE.2020.E00509.

Gareth, A. (2009). Cash crops and freedom: Export agriculture and the decline of slavery in colonial West Africa on JSTOR. *International Review of Social History*, *54*(1), 1–37. https://www.jstor.org/stable/44583114.

Gradel, A., Wünning, J. A., Plessing, T., & Jess, A. (2021). Adsorption of naphthalene on activated wood charcoal derived from biomass gasification. *Chemical Engineering & Technology*, *44*(6), 972–979. https://doi.org/10.1002/ceat.201900632.

Gujba, H., Mulugetta, Y., & Azapagic, A. (2015). The household cooking sector in Nigeria: environmental and economic sustainability assessment. *Resources*, *4*, 412–433. https://doi.org/10.3390/resources4020412.

Guo, Y., Tan, C., Sun, J., Li, W., Zhang, J., & Zhao, C. (2020). Porous activated carbons derived from waste sugarcane bagasse for CO_2 adsorption. *Chemical Engineering Journal*, *381*, 122736. https://doi.org/10.1016/j.cej.2019.122736.

Hansen, C. L., & Cheong, D. Y. (2019). *Agricultural waste management in food processing*. In *Handbook of Farm, Dairy and Food Machinery Engineering*. Elsevier Inc. https://doi.org/10.1016/B978-0-12-814803-7.00026-9.

Heidari, M., Dutta, A., Acharya, B., & Mahmud, S. (2019). A review of the current knowledge and challenges of hydrothermal carbonization for biomass conversion. *Journal of the Energy Institute*, *92*(6), 1779–1799. https://doi.org/10.1016/j.joei.2018.12.003.

Idowu, A. P., Adagunodo, E. R., Esimai, O. A., & Olapade, T. C. (2012). Development of A Web based GIS Waste Disposal Management System for Nigeria. *International Journal of Information Engineering and Electronic Business*, *3*, 40–48. https://doi.org/10.5815/ijieeb.2012.03.06.

Insam, H., Franke-whittle, I. H., & Podmirseg, S. M. (2014). Agricultural waste management in Europe, with an emphasis on anaerobic digestion. *Journal of Integrated Field Sciences*, *11*, 13–17.

Ireti Olamide, O. (2015). Economic effects of changes in cash crops on farmers' welfare in Ikere LGA of Ekiti State. *International Journal of Innovative Social Sciences & Humanities Research*, *3*(1), 119–125. https://seahipaj.org/journals-ci/mar-2015/IJISSHR/full/IJISSHR-M-10-2015.pdf.

Isah, Y., Kabiru, H. D., Danlami, M. A., & Kolapo, S. F. (2019). Comparative analysis of bioethanol produced from cassava peels and sugarcane bagasse by hydrolysis using saccharomyces cerevisiae. *Journal of Chemical Society of Nigeria*, *44*(2), 233–238.

Itodo, I. N., Agyo, G. E., & Yusuf, P. (2007). Performance evaluation of a biogas stove for cooking in Nigeria. *Journal of Energy in Southern Africa*, *18*(4), 14–18.

Iwuagwu Ben, U., & Iwuagwu Ben, C. M. (2015). Local building materials: affordable strategy for housing the Urban poor in Nigeria. *Procedia Engineering*, *118*, 42–49. https://doi.org/10.1016/j.proeng.2015.08.402.

Japhet, J. A. (2020). The potential of wood and agricultural waste for pellet fuel development in Nigeria – A technical review. *International Journal of Engineering Applied Sciences and Technology*, *4*(11), 598–607.

Jatau, A. A. (2013). Knowledge, attitudes and practices associated with waste management in jos south metropolis, Plateau State. *Mediterranean Journal of Social Sciences*, *4*(5), 119–127. https://doi.org/10.5901/mjss.2013.v4n5p119.

Javier, M., Verónica, C., Antonio, G., & Diana, G. (2019). Evaluation of rice straw yield, fibre composition and collection under Mediterranean conditions. *Acta Technologica Agriculturae*, *2*(1), 43–47. https://doi.org/10.2478/ata-2019-0008.

Kabir, R., Anwar, S., Yusup, S., Sham, S., & Inayat, M. (2021). Exploring the potential of coconut shell biomass for charcoal production. *Ain Shams Engineering Journal*, *13*(1), 101499. https://doi.org/10.1016/j.asej.2021.05.013.

Karshima, S. N. (2016). Public health implications of poor municipal waste management in Nigeria. *Vom Journal of Veterinary Science*, *11*, 142–148. https://irepos.unijos.edu.ng/jspui/bitstream/123456789/2992/1/98-1479037386.pdf.

Kenes, K., Yerdos, O., Zulkhair, M., & Yerlan, D. (2012). Study on the effectiveness of thermally treated rice husks for petroleum adsorption. *Journal of Non-Crystalline Solids, 358*(22), 2964–2969. https://doi.org/10.1016/j.jnoncrysol.2012.07.017.

Kersten, I., Baumbach, G., Oluwole, A. F., & Obioh, I. B. (1998). Urban and rural fuelwood situation in the tropical rain-forest area of south-west Nigeria. *Energy, 23*(10), 887–898. https://www.cabdirect.org/cabdirect/abstract/19980616185.

Khan, H. M., Iqbal, T., Yasin, S., Ali, C. H., Abbas, M. M., Jamil, M. A., Hussain, A., M. Soudagar, M. E., & Rahman, M. M. (2021). Application of agricultural waste as heterogeneous catalysts for biodiesel production. *Catalysts, 11*(10), 1215. https://doi.org/10.3390/catal11101215.

Khotmanee, S., & Pinsopon, U. (2021). A study on biogas production potential in Thailand 2019. *7th International Conference on Engineering, Applied Sciences and Technology at Pattaya, Thailand, ICEAST 2021- Proceedings*, 269–272. https://doi.org/10.1109/ICEAST52143.2021.9426287.

Kizito, S., Wu, S., Kipkemoi Kirui, W., Lei, M., Lu, Q., Bah, H., & Dong, R. (2015). Evaluation of slow pyrolyzed wood and rice husks biochar for adsorption of ammonium nitrogen from piggery manure anaerobic digestate slurry. *Science of The Total Environment, 505*, 102–112. https://doi.org/10.1016/j.scitotenv.2014.09.096.

Klinghoffer, N. B., Castaldi, M. J., & Nzihou, A. (2012). Catalyst properties and catalytic performance of char from biomass gasification. *Industrial & Engineering Chemistry Research, 51*(40), 13113–13122. https://doi.org/10.1021/ie3014082.

Kuhe, A., Ibiang, F. A., & Igbong, D. I. (2013). Potential of low pressure agricultural waste briquettes : An alternative energy source for cooking in Nigeria. *Journal of Renewable and Sustainable Energy, 5*, 013109. https://doi.org/10.1063/1.4781048.

Levi-Oguike, J., Sandoval, D., & Ntagwirumugara, E. (2022). A comparative life cycle investment analysis for biopower diffusion in rural Nigeria. *Sustainability (Switzerland), 14*(3), 1423. https://doi.org/10.3390/SU14031423/S1.

Liu, P., Liu, W.-J., Jiang, H., Chen, J.-J., Li, W.-W., & Yu, H.-Q. (2012). Modification of biochar derived from fast pyrolysis of biomass and its application in removal of tetracycline from aqueous solution. *Bioresource Technology, 121*, 235–240. https://doi.org/10.1016/j.biortech.2012.06.085.

Luo, J., Li, X., Ge, C., Müller, K., Yu, H., Deng, H., Shaheen, S. M., Tsang, D. C. W., Bolan, N. S., Rinklebe, J., Ok, Y. S., Gao, B., & Wang, H. (2021). Preparation of ammonium-modified cassava waste-derived biochar and its evaluation for synergistic adsorption of ternary antibiotics from aqueous solution. *Journal of Environmental Management, 298*, 113530. https://doi.org/10.1016/j.jenvman.2021.113530.

Lutz, G., Amandla, O.-O., & Gillian, P. (2019). *Winning in African Agriculture*. https://www.mckinsey.com/industries/agriculture/our-insights/winning-in-africas-agricultural-market.

Ma, Y., Qi, Y., Yang, L., Wu, L., Li, P., Gao, F., Qi, X., & Zhang, Z. (2021). Adsorptive removal of imidacloprid by potassium hydroxide activated magnetic sugarcane bagasse biochar: Adsorption efficiency, mechanism and regeneration. *Journal of Cleaner Production, 292*, 126005. https://doi.org/10.1016/j.jclepro.2021.126005.

Macawile, M. C., & Auresenia, J. (2022). Utilization of supercritical carbon dioxide and co-solvent n-hexane to optimize oil extraction from gliricidia sepium seeds for biodiesel production. *Applied Science and Engineering Progress, 15*(1), 1–10. https://doi.org/10.14416/J.ASEP.2021.09.003.

Maton, S. M., Dabi, D. D., Dodo, J. D., & Nesla, R. A. (2016). Environmental hazards of continued solid waste generation and poor disposal in municipal areas of Nigeria. *Journal of Geography, Environment and Earth Science International, 6*(3), 1–10. https://doi.org/10.9734/jgeesi/2016/26469.

Molino, A., Larocca, V., Chianese, S., & Musmarra, D. (2018). Biofuels Production by Biomass Gasification: A Review. *Energies, 11*(4), 811. https://doi.org/10.3390/EN11040811.

Mouhamadou, S., Dalhatou, S., Dobe, N., Djakba, R., Fasanya, O. O., Bansod, N. D., Fita, G., Ngayam, C. H., Tejeogue, J. P. N., & Harouna, M. (2022). Linear and non-linear modelling of kinetics and equilibrium data for Cr(VI) adsorption by activated carbon prepared from Piliostigma reticulatum. *Chemistry Africa*. https://doi.org/10.1007/s42250-022-00324-5.

Moustakas, K., Sotiropoulos, D., & Vakalis, S. (2021). Evaluation of the biogas potential of agricultural biomass waste for energy applications in Greece: A case study of the western Greece region. *Waste Management and Research*, *39*(3), 438–447. https://doi.org/10.1177/0734242X20970607.

Muazu, A. G., & Alibaba, H. Z. (2017). The use of traditional building materials in modern methods of construction (a case study of northern Nigeria). *International Journal of Engineering Science Technology and Research, 2*(6), 30–40.

Mukhtar, B., Salisu, J., Muhammad, M. B., & Atta, A. (2019). Theoretical and experimental studies of rice husk gasification using air as gasifying agent in a downdraft gasifier. *Nigerian Research Journal of Engineering and Environmental Sciences*, *4*(2), 645–657.

Nagendran, R. (2011). Agricultural waste and pollution. In *Waste*. Elsevier Inc. https://doi.org/10.1016/B978-0-12-381475-3.10024-5.

NBS. (2022). *National Bureau of Statistics Database*. National Bureau of Statistics. https://www.nigerianstat.gov.ng/.

Ndidi, N., Nelson, O., Patricia, O., & A, J. S. (2009). Waste management in healthcare establishments within Jos Metropolis, Nigeria. *African Journal of Environmental Science and Technology*, *3*(12), 459–465.

Nguyen, T. L., Brian, R. O., & Thomas, R. P. (1991). *Cassava Root Silage for Crossbred Pigs Under Village, Livestock Research for Rural Development*. https://www.fao.org/ag/aga/agap/frg/lrrd/lrrd9/2/loc922.htm.

Nicolas, D. C., & Guido, P. (2011). *Market Competition in Export Cash Crops and Farm Income in Africa – ACET*. https://acetforafrica.org/publications/working-papers/market-competition-in-export-cash-crops-and-farm-income-in-africa/.

Nigam, P. S., & Singh, A. (2011). Production of liquid biofuels from renewable resources. *Progress in Energy and Combustion Science*, *37*(1), 52–68. https://doi.org/10.1016/J.PECS.2010.01.003.

Niu, Y., Lv, Y., Lei, Y., Liu, S., Liang, Y., Wang, D., & Hui, S. e. (2019). Biomass torrefaction: properties, applications, challenges, and economy. *Renewable and Sustainable Energy Reviews, 115,* 109395. https://doi.org/10.1016/j.rser.2019.109395.

Nkwachukwu, O. I., Chidi, N. I., & Charles, K. O. (2010). Issues of roadside disposal habit of municipal solid waste, environmental impacts and implementation of sound management practices in developing country "Nigeria." *International Journal of Environmental Science and Development*, *1*(5), 409–418. https://doi.org/10.7763/ijesd.2010.v1.79.

Nogueira, T., Cordeiro, D. de S., Muñoz, R. A. A., Fornaro, A., Miguel, A. H., & Andrade, M. de F. (2015). Bioethanol and biodiesel as vehicular fuels in Brazil — Assessment of atmospheric impacts from the long period of biofuels use. In *Biofuels - Status and Perspective*. InTech. https://doi.org/10.5772/60944.

Nwadiogbu, J. O., Ajiwe, V. I. E., & Okoye, P. A. C. (2016). Removal of crude oil from aqueous medium by sorption on hydrophobic corncobs: Equilibrium and kinetic studies. *Journal of Taibah University for Science, 10(1)*, 56–63. https://doi.org/10.1016/j.jtusci.2015.03.014.

Nwaokocha, C. N., & Giwa, S. O. (2016). Investigation of bio-waste as alternative fuel for cooking. *(CU-ICADI) International Conference on African Development Issues, at Ota-Nigeria*, 548–551.

Nwoko, C. I. A., Nkwoada, A. U., Ogu, H. U., Nwoko, C. I. A., Nkwoada, A. U., & Ogu, H. U. (2019). Optimization of biodiesel development from non-edible indigenous feedstocks in Nigeria. *The International Journal of Biotechnology, 8*(1), 84–92. https://doi.org/10.18488/JOURNAL.57.2019.81.84.92.

Obi, F. O., Ugwuishiwu, B. O., & Nwakaire, J. N. (2016). Agri wastes. *Nigerian Journal of Technology*, *35*(4), 957–964.

Obianefo, C. A., Ng'ombe, J. N., Mzyece, A., Masasi, B., Obiekwe, N. J., & Anumudu, O. O. (2021). Technical efficiency and technological gaps of rice production in anambra state, Nigeria. *Agriculture, 11*(12), 1240. https://doi.org/10.3390/agriculture11121240.

Odimegwu, O. (2014). *Special Report: Nigeria Wastes 40% Of Food, But Millions Of Citizens Are Dying Of Hunger.* https://africaprimenews.com/2018/01/20/development/special-report-nigeria-wastes-40-of-food-but-millions-of-citizens-are-dying-of-hunger/.

Ofori, S. N., Fobil, J. N., & Odia, O. J. (2018). Household biomass fuel use, blood pressure and carotid intima media thickness; a cross sectional study of rural dwelling women in. *Environmental Pollution*, *242*, 390–397. https://doi.org/10.1016/j.envpol.2018.06.102.

Ogundele, O. M., Opeagbe, &, Rapheal, M., & Abiodun, A. M. (2018). Effects of municipal waste disposal methods on community health in Ibadan - Nigeria. *Polytechnica 1*(1), 61–72. https://doi.org/10.1007/S41050-018-0008-Y.

Ogungbenro, A. E., Quang, D. V., Al-Ali, K. A., Vega, L. F., & Abu-Zahra, M. R. M. (2018). Physical synthesis and characterization of activated carbon from date seeds for CO2 capture. *Journal of Environmental Chemical Engineering*, *6*(4), 4245–4252. https://doi.org/10.1016/j.jece.2018.06.030.

Ogunjobi, J. K., & Labunmi, L. (2013). The potentials of cocoa pods and plantain peels as renewable sources in Nigeria. *International Journal of Green Energy*, 2014, 37–41. https://doi.org/10.1080/15435075.2013.848403.

Ogunrinola, I. O., & Adepegba, E. O. (2012). Health and economic implications of waste dumpsites in cities: the case of Lagos, Nigeria. *International Journal of Economics and Finance*, *4*(4), 239–251. https://doi.org/10.5539/ijef.v4n4p239.

Oguntoke, O., Emoruwa, F. O., & Taiwo, M. A. (2019). Assessment of air pollution and health hazard associated with sawmill and municipal waste burning in Abeokuta Metropolis, Nigeria. *Environmental Science and Pollution Research*, *26*(32), 32708–32722. https://doi.org/10.1007/s11356-019-04310-2.

Ohimain, E. I., & Izah, S. C. (2014). Potential of biogas production from palm oil mills' effluent in Nigeria. *Sky Journal of Soil Science and Environmental Management*, *3*(5), 50–58. http://www.skyjournals.org/SJSSEM.

Ojolo, S. J., Orisaleye, J. I., Ismail, S. O., & Abolarin, S. M. (2012). Technical Potential of Biomass Energy in Nigeria. *Ife Journal of Technology*, *21*(2), 60–65. https://researchprofiles.herts.ac.uk/portal/files/16270972/TECHNICAL_POTENTIAL_OF_BIOMASS_ENERGY_IN_NIGERIA.pdf.

Okonkwo, E. C., Okafor, K. I., & Akun, E. (2018). The economic viability of the utilisation of biogas as an alternative source of energy in rural parts of Nigeria. *International Journal of Global Energy Issues*, *41*(5/6), 205–225.

Okoro, E. E., Okafor, I. S., Igwilo, K. C., Orodu, K. B., & Mamudu, A. O. (2020). Sustainable biogas production from waste in potential states in Nigeria–alternative source of energy. *Journal of Contemporary African Studies*, *38*(4), 627–643. https://doi.org/10.1080/02589001.2020.1825650.

Okotie, S. (2018). The Nigerian economy before the discovery of crude oil. *The Political Ecology of Oil and Gas Activities in the Nigerian Aquatic Ecosystem*, 71–81. https://doi.org/10.1016/B978-0-12-809399-3.00005-7.

Olotuah, A. O., & Taiwo, A. A. (2013). Housing the urban poor in Nigeria through low-cost housing schemes. *International Journal of Physical and Human Geography*, *1*(3), 1–8.

Olujobi, O. J., Ufua, D. E., Olokundun, M., & Olujobi, O. M. (2022). Conversion of organic wastes to electricity in Nigeria: legal perspective on the challenges and prospects. *International Journal of Environmental Science and Technology*, *19*(2), 939–950. https://doi.org/10.1007/S13762-020-03059-3/TABLES/2.

Olukanni, D. O., & Akinyinka, M. O. (2012). Environment, health and wealth: Towards an analysis of municipal solid waste management in Ota, Ogun State, Nigeria. *Proceedings ICCEM at Rome, Italy*, 138–145.

Ononugbo, C. P., Avwiri, G. O., & Tutumeni, G. (2015). Estimation of indoor and outdoor effective doses from gamma dose rates of residential buildings in Emelogu Village in Rivers State, Nigeria. *International Research Journal of Pure and Applied Physics*, *3*(2), 18–27.

Onyegiri, I., & Ugochukwu, I. Ben. (2016). Traditional building materials as a sustainable resource and material for low-cost housing in Nigeria: Advantages, challenges and the way forward. *International Journal of Research in Chemical. Metallurgical and Civil Engineering*, *3*(2), 247–252.

Osabohien, R., Mordi, A., & Ogundipe, A. (2022). Access to credit and agricultural sector performance in Nigeria. *African Journal of Science, Technology, Innovation and Development*, *14*(1), 247–255. https://doi.org/10.1080/20421338.2020.1799537.

Owoyemi, J. M., Zakariya, H. O., & Elegbede, I. O. (2016). Sustainable wood waste management in Nigeria. *Environmental & Socio-Economic Studies*, *4*(3), 1–9. https://doi.org/10.1515/environ-2016-0012.

Oyegoke, T., & Ibraheem, K. A. (2021). Kinetic modeling of the biodiesel production process using neem seed oil: An alternative to petroleum-diesel. *European Journal of Chemistry*, *12*(3), 242–247. https://doi.org/10.5155/eurjchem.12.3.242-247.2085.

Oyegoke, T., & Jibril, B. Y. (2016). Design and feasibility study of a 5 MW bio-power plant in Nigeria. *International Journal of Renewable Energy Research*, *6*(4), 1496–1505. https://doi.org/10.20508/ijrer.v6i4.4755.g6953.

Oyegoke, T., Obadiah, E., Adah, F., Oguche, J. E., Timothy, G. T., Mantu, I. A., & Ado, A. D. (2022). Trends of progress in setting up biorefineries in developing countries: A review of bioethanol exploration in Nigeria. *Journal of Renewable Energy and Environment*, *9*(1), 37–52. https://doi.org/10.30501/JREE.2021.278037.1197.

Pandey, J. (2020). Biopolymers and their application in wastewater treatment. *Emerging Eco-friendly Green Technologies for Wastewater Treatment*, pp. 245–266. https://doi.org/10.1007/978-981-15-1390-9_11.

Popoola, A., Tawose, O., Abatan, S., Adeleye, B., Jiyah, F., & Majolagbe, N. (2015). Housing conditions and health of residents in Ibadan north local government area, Ibadan, Oyo State, Nigeria. *Journal of Environmental Sciences and Resources Management*, *7*(2), 59–80.

Rim-Rukeh, A. (2014). An assessment of the contribution of municipal solid waste dump sites fire to atmospheric pollution. *Open Journal of Air Pollution*, *3*, 53–60. https://doi.org/10.4236/ojap.2014.33006.

Robert, K. A. G., & Akinlawon, L. M. (2022). Africa - Agriculture. In *Britannica*. https://www.britannica.com/place/Africa/Agriculture.

Rogers, J. N., Stokes, B., Dunn, J., Cai, H., Wu, M., Haq, Z., & Baumes, H. (2017). An assessment of the potential products and economic and environmental impacts resulting from a billion ton bioeconomy. *Biofuels, Bioproducts and Biorefining*, *11*(1), 110–128. https://doi.org/10.1002/BBB.1728.

Sabiiti, E. N. (2011). Utilising agricultural waste to enhance food security and conserve the environment. *African Journal of Food, Agriculture, Nutrition and Development*, *11*(6), 1–9.

Sangodoyin, A. Y. (1993). Domestic waste disposal in southwest Nigeria. *Environmental Management and Health*, *4*(3), 20–23. https://doi.org/10.1108/09566169310043061.

Singh, S., Prajapati, A. K., Chakraborty, J. P., & Mondal, M. K. (2021). Adsorption potential of biochar obtained from pyrolysis of raw and torrefied Acacia nilotica towards removal of methylene blue dye from synthetic wastewater. *Biomass Conversion and Biorefinery*. https://doi.org/10.1007/s13399-021-01645-0.

Sridhar, M. K. C., & Hammed, T. B. (2014). Turning waste to wealth in Nigeria: An overview. *Journal of Human Ecology, 46*(2), 195–203. https://doi.org/10.1080/09709274.2014.11906720.

Subramaniam, R., & Kumar Ponnusamy, S. (2015). Novel adsorbent from agricultural waste (cashew NUT shell) for methylene blue dye removal: Optimization by response surface methodology. *Water Resources and Industry, 11*, 64–70. https://doi.org/10.1016/j.wri.2015.07.002.

Statista. (2022). *Nigeria: Agriculture Contribution to GDP 2019–2021*. https://www.statista.com/statistics/1193506/contribution-of-agriculture-to-gdp-in-nigeria/.

Stephen, O., Anthony, P., Ogbue, U. C., Victor, C., & Okedu, K. E. (2020). Off-grid electricity generation in Nigeria based on rice husk gasification technology. *Cleaner Engineering and Technology, June*. https://doi.org/10.1016/j.clet.2020.100009.

Taiwo, A. M. (2011). Composting as a sustainable waste management technique in developing countries. *Journal of Environmental Science and Technology, 4*(2), 93–102. https://doi.org/10.3923/jest.2011.93.102.

Tambuwal, A. D., Baki, A. S., & Bello, A. (2018). Bioethanol production from corn cobs wastes as biofuel. *Direct Research Journal of Biology and Biotechnology, 4*(2), 22–26.

Tang, J., Zhu, W., Kookana, R., & Katayama, A. (2013). Characteristics of biochar and its application in remediation of contaminated soil. *Journal of Bioscience and Bioengineering, 116*(6), 653–659. https://doi.org/10.1016/j.jbiosc.2013.05.035.

Tomczyk, A., Sokołowska, Z., & Boguta, P. (2020). Biochar physicochemical properties: pyrolysis temperature and feedstock kind effects. *Reviews in Environmental Science and Bio/Technology, 19*(1), 191–215. https://doi.org/10.1007/s11157-020-09523-3.

Tongshuwar, G. T., & Oyegoke, T. (2021). A brief survey of biomass hydrolysis as a vital process in Bio-refinery. *FUDMA Journal of Sciences, 5*(3), 407–412. https://doi.org/10.33003/FJS-2021-0503-724.

Tursi, A. (2019). A review on biomass: importance, chemistry, classification, and conversion. *Biofuel Research Journal, 6*(2), 962–979. https://doi.org/10.18331/BRJ2019.6.2.3.

Uddin, M. N., Siddiki, S. Y. A., Mofijur, M., Djavanroodi, F., Hazrat, M. A., Show, P. L., Ahmed, S. F., & Chu, Y. M. (2021). Prospects of bioenergy production from organic waste using anaerobic digestion technology: A mini review. *Frontiers in Energy Research, 9*, 33. https://doi.org/10.3389/FENRG.2021.627093/BIBTEX.

UNDP. (2022). *Sustainable Development Goals*. United Nations Development Programme. https://www.undp.org/sustainable-development-goals.

Xing, Y., Zheng, Z., Sun, Y., & Agha Alikhani, M. (2021). A review on machine learning application in biodiesel production studies. *International Journal of Chemical Engineering, 2021*. https://doi.org/10.1155/2021/2154258.

Yaumi, A. L., Bakar, M. Z. A., & Hameed, B. H. (2018). Melamine-nitrogenated mesoporous activated carbon derived from rice husk for carbon dioxide adsorption in fixed-bed. *Energy, 155*, 46–55. https://doi.org/10.1016/j.energy.2018.04.183.

Yevich, R., & Logan, J. A. (2003). An assessment of biofuel use and burning of agricultural waste in the developing world. *Global Biogeochemical Cycles, 17*(4), 1095. https://doi.org/10.1029/2002gb001952.

Yusuf, S., Faiz, M., Abd, L., & Muhaimin, A. (2020). Evaluation of hybrid briquettes from corncob and oil palm trunk bark in a domestic cooking application for rural communities in Nigeria. *Journal of Cleaner Production, 124745*. https://doi.org/10.1016/j.jclepro.2020.124745.

Zhang, S., Hu, B., Zhang, L., & Xiong, Y. (2016). Effects of torrefaction on yield and quality of pyrolysis char and its application on preparation of activated carbon. *Journal of Analytical and Applied Pyrolysis, 119*, 217–223. https://doi.org/10.1016/j.jaap.2016.03.002.

15 Experimental Investigation of MPFI and GDI Engines using Gasoline Fuel

Ufaith Qadiri

CONTENTS

15.1 Introduction 274
15.2 The Test Rig's Description 274
15.3 Control Panel with Full Instrumentation Set... 275
    15.3.1 Torque Measurement Load Cell 275
    15.3.2 Flow Sensors for the Measurement of Water Flow 275
    15.3.3 Arrangement for Flow Control 275
15.4 Water Cooling Arrangement for Pressure Sensor
    (Combustion Pressure) – For Any One Cylinder Only................................ 276
    15.4.1 Fuel Flow Measurement 276
    15.4.2 System for Measuring Air Consumption... 276
    15.4.3 Measurement of Speed 276
    15.4.4 Crank Angle Measurement... 276
    15.4.5 Measurement of Temperature... 276
    15.4.6 Calorimeter for Exhaust Gas 276
    15.4.7 Software.. 277
    15.4.8 Experimentation Scope.. 277
15.5 The Performance and Combustion Parameters Calculated from
    Each Experiment Performed on the Test Rig Are as per the
    Given Parameters Below.. 277
    15.5.1 Brake Power.. 277
    15.5.2 Indicated Power 277
    15.5.3 Friction Power... 278
    15.5.4 Brake-Specific Fuel Consumption... 278
    15.5.5 Indicated Specific Fuel Consumption.. 278
    15.5.6 Swept Volume 278
    15.5.7 Volumetric Efficiency 278
    15.5.8 Brake Thermal Efficiency... 279
    15.5.9 Indicated Thermal Efficiency 279
    15.5.10 Mechanical Efficiency... 279

DOI: 10.1201/9781003359784-15

        15.5.11 Brake Mean Effective Pressure 279
        15.5.12 Indicated Mean Effective Pressure..279
        15.5.13 Heat Input 279
        15.5.14 Heat Equivalent to BP...279
        15.5.15 Heat Carried Away by Engine Cooling Jacket Water 280
        15.5.16 Heat Carried Away by Exhaust Gas 280
        15.5.17 Unaccounted Heat..280
    15.6 Engine Test Rig Observation Report..281
        15.6.1   Operating Parameters..281
        15.6.2   Performance Parameters of Test Rig Using a 3-Cylinder SI Engine.....281
        15.6.3   Various Efficiencies of Test Rig Using a 3-Cylinder SI Engine 281
        15.6.4   Engine Test Rig Combustion Report of Test Rig
                 Using a 3-Cylinder SI Engine...282
    15.7 Results and Discussion 282
        15.7.1   Experimental Results for MPFI SI Engine Using
                 Conventional Gasoline Fuel under Constant Speed of
                 2,500 rpm and Variable Load 282
        15.7.2   Combustion Graphs 285
    15.8 Experiments Performed on MPFI Engine 287
        15.8.1   Description of the GDI Test Rig...289
    15.9 Conclusion 289
    References..290

15.1   INTRODUCTION

TECH-ED created Engine Analysis software to test the properties of exhaust and combustion of a specified engine test setup. Engine Analysis software is compatible with Windows 7 and higher and may be installed, uninstalled, and run. With Engine Analysis software, we can

- Setup and user interfaces using menus and commands
- Configuring according to the Engine Test Setup currently in use
- Measurement and analysis of the test setup
- Data logging and data acquisition
- Online computations of performance and combustion data
- Printing and creation of reports and graphs.

15.2   THE TEST RIG'S DESCRIPTION

The configuration comes with a universal coupling that connects a hydraulic dynamometer with a water-cooled multicylinder four-stroke GDI petrol engine and installed on an Ms channel-based symmetrically stabilized base structure. The system includes a box for air supply, gasoline. A tank, manometer, fuel measurement equipment, digital indicators, and transmitters for measuring various parameters are all included within a stand-alone fully powder-coated panel box. It also includes the

TABLE 15.1

Test Rig Specification of Engine

Model/Make	Omni-Maruti
Power maximum	At 5,000 rpm/27.6 kW
Torque maximum	At 3,000 rpm/6.1 Kg M
Range of working	At 2,500 rpm/10 horse power
Engine bore	68.5 mm
Length of stroke	72 mm
Engine CR	8.5:1
Displacement	796 cc
Option for start	Self-Start
Ignition method	SI
Diameter of orifice	20 mm

necessary sensors, such as combustion pressure and crank angle, which are measured by transmitters. For computerization, a signal conditioner and a signal converter connect all of these signals to a computer.

15.3 CONTROL PANEL WITH FULL INSTRUMENTATION SET

On the control panel, there are digital indicators. Temperature, water flow, and speed, as well as a signal conditioner for PV-P (interface unit), a burette, and a manometer, are connected as stand-by option. For altering different loads, on the same panel, a load controller is installed. Push-button switches with indicators are provided to control the speed and to start/stop the engine.

15.3.1 TORQUE MEASUREMENT LOAD CELL

The loading arm of the dynamometer is connected to a load cell. As the loading arm impacts the load cell, the load is sensed and read out in terms of torque (N-m) via the digital torque indicator.

15.3.2 FLOW SENSORS FOR THE MEASUREMENT OF WATER FLOW

The flow rate of water to the engine water jacket and the exhaust gas calorimeter is detected using a turbine-type water flow sensor. The readout will be indicated indicator (flow sensor capacity = 150 cc/sec).

15.3.3 ARRANGEMENT FOR FLOW CONTROL

Water flow for the engine jacket, the engine exhaust calorimeter, and the pressure sensor-cooling adopter is managed by stainless steel ball valves with precise control.

15.4 WATER COOLING ARRANGEMENT FOR PRESSURE SENSOR (COMBUSTION PRESSURE) – FOR ANY ONE CYLINDER ONLY

To detect the combustion chamber pressure under varying loads, a piezo-electric pressure sensor is mounted at a suitable location on the engine head. A water cooling adapter, which is positioned in the centre of the cooling adapter, cools the sensor. The sensor and the signal conditioner are connected via a low-noise connection, allowing communication between them (range: 0–5,000 pounds per square inch).

15.4.1 FUEL FLOW MEASUREMENT

A 4.5-L gasoline tank is installed atop a 10 kg cantilever style load cell, which aids in fuel rate measurement. As a backup option, a burette is attached in parallel, and the fuel loss in weight is translated to a flow rate in kilogrammes per hour, which is displayed on the digital fuel rate indicator.

15.4.2 SYSTEM FOR MEASURING AIR CONSUMPTION

The manometer is connected to an air tank situated below the control panel through a 20-mm aperture. A differential pressure transducer is also linked to the pressure tappings. The air-rate indicator displays the difference between the manometers, which is calculated and converted to real volume in m^3/h.

15.4.3 MEASUREMENT OF SPEED

The proximity sensors are positioned below the connection, with a speed and crank angle reference measurement. The crank angle sensor is linked to the signal conditioner, while the speed sensor is linked to the speed indicator.

15.4.4 CRANK ANGLE MEASUREMENT

A TDC encoder is provided with a suitable arrangement to measure crank angle.

15.4.5 MEASUREMENT OF TEMPERATURE

A temperature gauge with multiple-channel PT-100-type temperature sensors is used to sense low temperatures, while a K-type thermocouple is utilized to sense extreme temperatures of engine exhaust.

15.4.6 CALORIMETER FOR EXHAUST GAS

A pipe in pipe-style exhaust gas calorimeter is placed to calculate the amount of heat lost from the Ms channel frame. It is a thermocouple-equipped counter-flow heat exchanger. There are inlets and outlets for temperature measurement.

15.4.7 SOFTWARE

Engine Assessment Software is an engine performance-monitoring solution for Win Xp and higher systems. It can handle most engine diagnostic application demands such as data monitoring, reporting, and recording, which are all part of the process. Performance and combustion analysis are assessed by the software. At various working conditions, several graphs and reports are generated. During the commencement of a test run, the necessary signals are read, stored, and shown in a graph. During the test run file mode, stored data is displayed as a graph, and the graphs and findings can be printed. The data can be exported to be analysed further.

15.4.8 EXPERIMENTATION SCOPE

Engine performance research (manual mode)
   Engine performance research and combustion analysis (computerized mode)

15.5 THE PERFORMANCE AND COMBUSTION PARAMETERS CALCULATED FROM EACH EXPERIMENT PERFORMED ON THE TEST RIG ARE AS PER THE GIVEN PARAMETERS BELOW

15.5.1 BRAKE POWER

$$\mathbf{BP} = \frac{2\pi NT}{60,000} \text{ kiloWatt}$$

where

N = speed of the engine in rpm
T = Torque in N-m

15.5.2 INDICATED POWER

$$\mathbf{IP} = \frac{NW_{net}}{2 \times 60,000} \text{ kW}$$

where

N = RPM of the engine
W_{net} = Net work done in joules
$W_{net} = W_{indicated} + W_{pumping}$

15.5.3 Friction Power

$$FP = IP - BP \text{ kW}$$

15.5.4 Brake-Specific Fuel Consumption

$$BSFC = \frac{mfc}{BP} \text{ kg/kW h}$$

where mfc = Mass of fuel consumption from the indicator.

15.5.5 Indicated Specific Fuel Consumption

$$ISFC = \frac{mfc}{IP} \text{ kg/kW h}$$

15.5.6 Swept Volume

$$V_s = \frac{\pi d^2}{4} \times L \times \frac{N}{2} \times 60 \text{ m}^3 /h$$

where

d = diameter of bore = 68.5 mm
L = length of stroke = 72 mm
N = Speed of the engine in RPM.

15.5.7 Volumetric Efficiency

$$\eta_v = \frac{V_A}{V_S} \times 100\%$$

where

V_A = Actual volume from the indicator in m³/h
V_S = Swept volume in m³/h

15.5.8 BRAKE THERMAL EFFICIENCY

$$\eta_{\text{bth}} = \frac{\text{BP} \times 3,600 \times 100}{\text{mfc} \times \text{CV}}\%$$

where CV = calorific value for petrol = 43,500 KJ/kg.

15.5.9 INDICATED THERMAL EFFICIENCY

$$\eta_{\text{ith}} = \frac{\text{IP} \times 3,600 \times 100}{\text{mfc} \times \text{CV}}\%$$

15.5.10 MECHANICAL EFFICIENCY

$$\eta_{\text{mech}} = \frac{\text{BP} \times 100}{\text{IP}}\%$$

15.5.11 BRAKE MEAN EFFECTIVE PRESSURE

$$\text{BMEP} = \frac{\text{BP} \times 10^{-2} \times 60}{\dfrac{\Pi d^2}{4} \times L \times \dfrac{N}{2}} \text{ bar}$$

15.5.12 INDICATED MEAN EFFECTIVE PRESSURE

$$\text{IMEP} = \frac{\text{BP} \times 10^{-2} \times 60}{\dfrac{\Pi d^2}{4} \times L \times \dfrac{N}{2}} \text{ bar}$$

15.5.13 HEAT INPUT

$$\text{Heat input} = \frac{\text{mfc} \times \text{CV}}{3,600} \text{ kW}$$

15.5.14 HEAT EQUIVALENT TO BP

$$\text{HBP} = \text{BP kW}$$

$$= \frac{\text{BP} \times 100}{\text{Heat input}}\%$$

15.5.15 HEAT CARRIED AWAY BY ENGINE COOLING JACKET WATER

$$\mathbf{HJW} = \mathrm{mw} \times \mathrm{cpW} \times (T2 - T1)\ \mathrm{kW}$$

$$= \frac{\mathrm{HJW} \times 100}{\mathrm{Heat\ input}}\ \%$$

15.5.16 HEAT CARRIED AWAY BY EXHAUST GAS

$$\mathbf{HGas} = \frac{(T4 - T6)}{(T4 - T5)} \times \mathrm{mw} \times \mathrm{cpW} \times (T3 - T1)\ \mathrm{kW}$$

where

mw = mass of water = Vw × δw × 10⁻⁶
Vw = flow rate of water flow into calorimeter
Cp_W = 4.18 kJ/kg K

$$\mathrm{HGas} = \frac{\mathrm{HGas} \times 100}{\mathrm{Heat\ input}}\ \%$$

15.5.17 UNACCOUNTED HEAT

HUn = Heat input − (HBP + HJW + H Gas) kW

$$\mathbf{Hun} = \frac{\mathrm{HUn} \times 100}{\mathrm{Heat\ input}}$$

TABLE 15.2
Measurements on Test Rig

S. No.	Torque (Nm)	Speed (rpm)	Fuel Rate (kg/h)	Air Rate (m³/h)	Water Flow Eng. (cc/s)	Water Flow Cal. (cc/s)
1	0.34	2,042	0.77	7.9	77.2	41.1
2	4.07	2,046	0.96	11.3	76.4	49.8
3	4.07	2,046	0.96	11.3	76.4	49.8
4	7.86	2,014	1.2	13.8	75.9	51
5	11.54	2,034	1.44	16.2	76.4	50.4
6	16.57	1,956	1.73	18.7	76.7	50.9
7	20.25	1,973	1.87	21.4	76	51.4
8	23.42	1,989	2.16	24.8	76.9	51.7
9	27.76	1,939	2.35	26.6	95.4	66.9
10	31.14	2,038	3.02	30.9	95.6	65.5

15.6 ENGINE TEST RIG OBSERVATION REPORT

15.6.1 Operating Parameters

15.6.2 Performance Parameters of Test Rig Using a 3-Cylinder SI Engine

TABLE 15.3
Performance Parameters of Test Rig Using a 3-Cylinder SI Engine

S. No.	BP (kW)	IP (kW)	FP (kW)	BSFC (kg/kWh)	ISFC (kg/kWh)
1	0.07	4.27	4.2	10.56	0.18
2	0.87	6.37	5.49	1.1	0.15
3	0.87	6.37	5.49	1.1	0.15
4	1.66	7.67	6.01	0.72	0.16
5	2.46	9.44	6.98	0.59	0.15
6	3.39	10.3	6.91	0.51	0.17
7	4.18	11.89	7.7	0.45	0.16
8	4.88	13.14	8.27	0.44	0.16
9	5.64	14.43	8.8	0.42	0.16
10	6.65	16.28	9.63	0.46	0.19

15.6.3 Various Efficiencies of Test Rig Using a 3-Cylinder SI Engine

TABLE 15.4
Various Efficiencies of Test Rig Using a 3-Cylinder SI Engine

VE (%)	BThE (%)	IThE (%)	ME (%)
16.2	0.77	44.99	1.7
23.13	7.35	53.64	13.7
23.13	7.35	53.64	13.7
28.69	11.18	51.71	21.61
33.35	13.81	53.04	26.04
40.03	15.89	48.22	32.95
45.42	18.08	51.37	35.2
52.21	18.27	49.23	37.11
57.45	19.39	49.65	39.05
63.49	17.78	43.55	40.83

15.6.4 ENGINE TEST RIG COMBUSTION REPORT OF TEST RIG USING A 3-CYLINDER SI ENGINE

TABLE 15.5
Engine Test Rig Combustion Report of Test Rig Using a 3-Cylinder SI Engine

Sl. No.	Crank Angle	Cy. Pressure (bar)	Volume (cc)	Heat Release
1	0	0	34.02	0
2	1	0	34.04	−9.53
3	2	1.6	34.12	−0.02
4	3	1.6	34.24	0.56
5	4	1.5	34.42	−0.04
6	5	1.5	34.64	0.55
7	6	1.4	34.91	−0.06
8	7	1.4	35.24	0.55
9	8	1.3	35.61	−0.07
10	9	1.3	36.03	0.55
11	10	1.2	36.5	0.56
12	11	1.1	37.02	−0.73
13	12	1.2	37.59	0.57
14	13	1.1	38.2	−0.09
15	14	1.1	38.87	−0.1

15.7 RESULTS AND DISCUSSION

15.7.1 EXPERIMENTAL RESULTS FOR MPFI SI ENGINE USING CONVENTIONAL GASOLINE FUEL UNDER CONSTANT SPEED OF 2,500 RPM AND VARIABLE LOAD

As may be observed in Figure 15.1, the maximum power is indicated power, which is available at the piston of the engine. This is the highest power obtained for the multi-cylinder spark ignition engine. The increase in load also increases the power of the engine. The friction power obtained is slightly less than the indicated power, as there are some friction losses in multi-cylinder engine. These frictional losses are due to the power utilized for radiation, coolant and driving auxiliary devices such as a feed pump and a valve mechanism, and so on. Brake power available refers to the remaining power. At the end of crankshaft for useful work, the net power obtained for useful work is brake power, which in case of spark ignition engine is up to 30%. In addition, after increasing the load on engine, the mixture turns, more towards the richer side, and the supply of fuel in the engine also increases, which in turn increases the air supply. This increase in air/fuel supply increases the number of cycle's revolution per second, and the power obtained shows an increasing trend with an increase in load [1] (Figure 15.2).

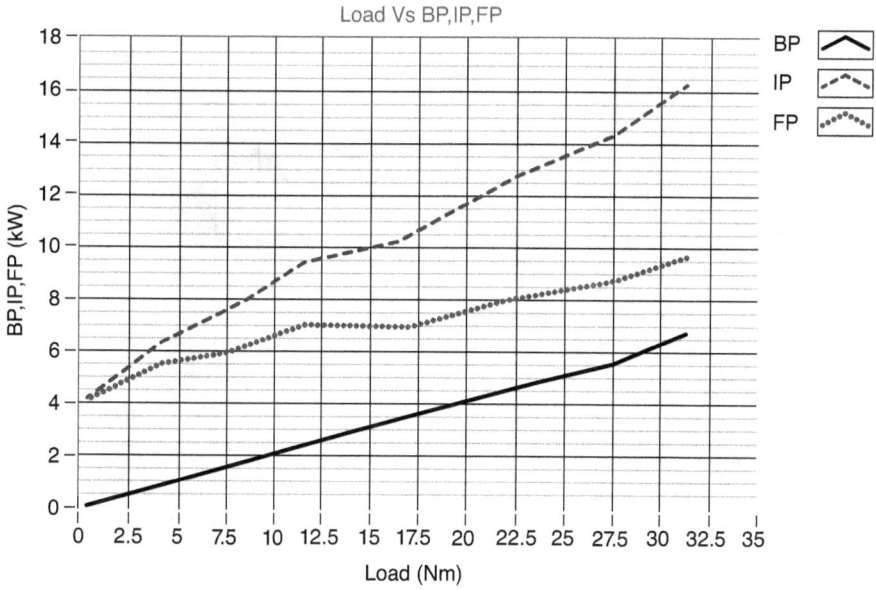

FIGURE 15.1   Variation of BP, IP, and FP with load.

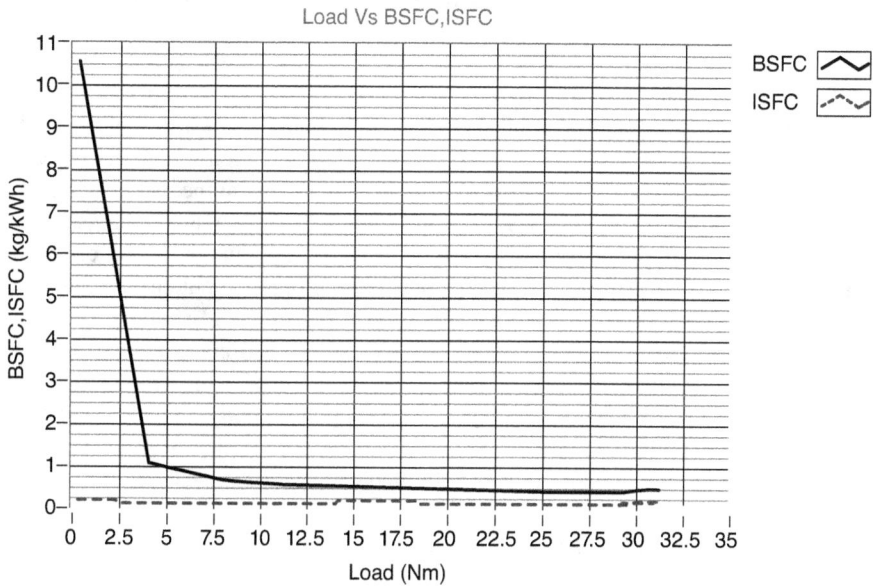

FIGURE 15.2   Variation of BSFC and ISFC with load.

Load Vs VE,BThE,IThE,ME

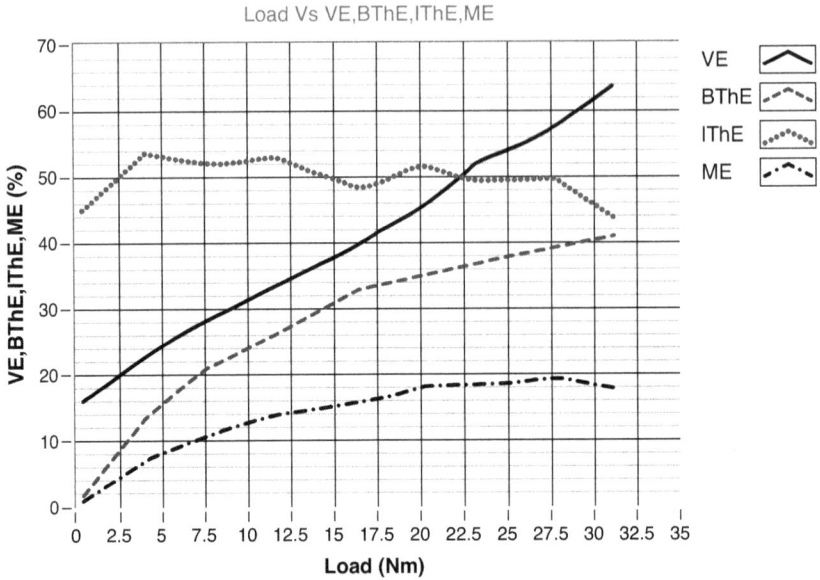

FIGURE 15.3   Variation of VE, BThe, IThE, and ME with load.

It is evident from the plot 2 that BSFC shows a decreasing trend with an increase in load. This happens because the mixture goes more towards the leaner side and combustion is being completed fully. When the combustion starts, the mixture is rich, so fuel consumed is more, until the mixture goes towards the stoichiometric air/ fuel ratio. The consumption of fuel continuously decreases with an increase in load as the mixture becomes lean and the combustion is being completed. This shows that the fuel is efficient and the multi-cylinder engine is performing as desired [2].

In Figure 15.3, it can be seen variations in efficiencies with varying loads. Volumetric efficiencies show the increasing trend with an increase in loads. As with an increase in loads, there is an increase in air supply to the engine, which improves the engine's volumetric efficiency. Again, mechanical efficiency also increases with an increase in loads. So, the engine in terms of efficiency proves to be effective. This gives the effectiveness of the engine in transforming its input energy to output energy. Brake thermal efficiency is also defined in Figure 15.3, which defines how well an engine converts heat from the fuel into mechanical energy, which can be clearly seen with an increase in load: the heat from the fuel also increases as the load increases and increases overall efficiency of the engine. Finally, the indicated thermal efficiency also increases with an increase in load and is highest at lower loads. This gives the concept of engine power generated within the engine in relation to heat provided in the form of fuel [3].

In Figure 15.4, there is a relation between HBP, HJW, H gas, Hun, and load. This plot gives the brief explanation of heat equivalent to brake power HBP that increases with an increase in load as engine produces a lot of heat at working conditions. This is obtained in terms of percentage. In this plot, there is also the heat emitted by gases

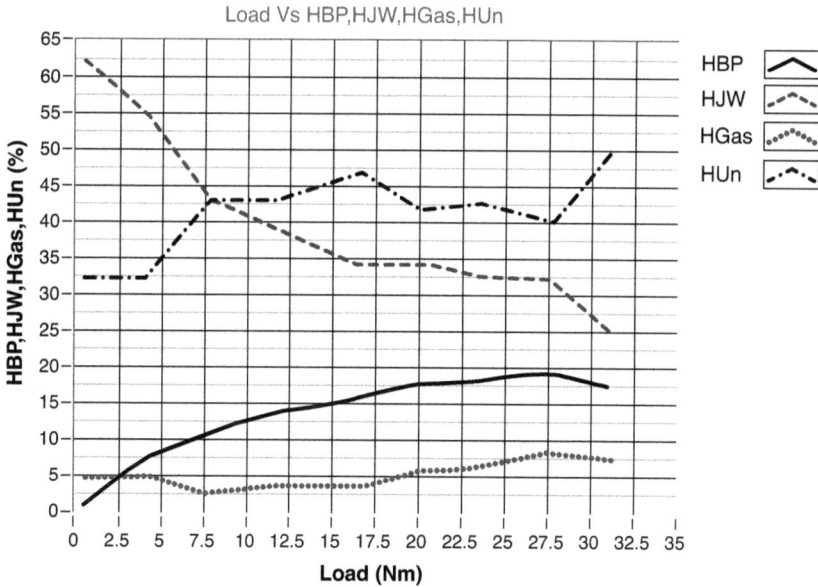

FIGURE 15.4   Variation of HBP, HJW, HGas, and HUn with load.

of exhaust with an increase in load, and the heat emitted by gases of exhaust also increases. Therefore, exhaust gases carry away heat produced from combustion of fuels. In addition, HJW is heat carried away by engine cooling jacket water, which clearly shows in the plot that with an increase in load, the cooling water also carries away heat from the engine with an increase in load. Therefore, the temperature of the engine also decreases, which helps in not overheating during running conditions and helps in reducing NOx emissions as the temperature is the main source for an increase in NOx emissions. Finally, HUn is unaccounted heat from the engine [4].

15.7.2 COMBUSTION GRAPHS

At 20 Nm Load

In Figure 15.5, there is a given explanation of pressure and crank angle, which increases at the start of compression and increases maximum up to the 40 bar, thus generating the pressure. This engine is producing pressure up to 40 bar. The maximum capacity of the engine is near about 10HP, which can produce torque up to 40 Nm. When the compression stroke starts, the pressure reaches up to maximum and increases the capability of engine to complete cycles [5].

In Figure 15.6, there is a variation of BMEP and IMEP with varying load conditions on engine. The graph clearly shows that IMEP is higher for gasoline engines than for the BMEP under varying load conditions. This increase in IMEP shows that the engine is producing enough pressure for the generation of the power, which is available at the engine. Similarly, the BMEP also shows the increasing trend of the

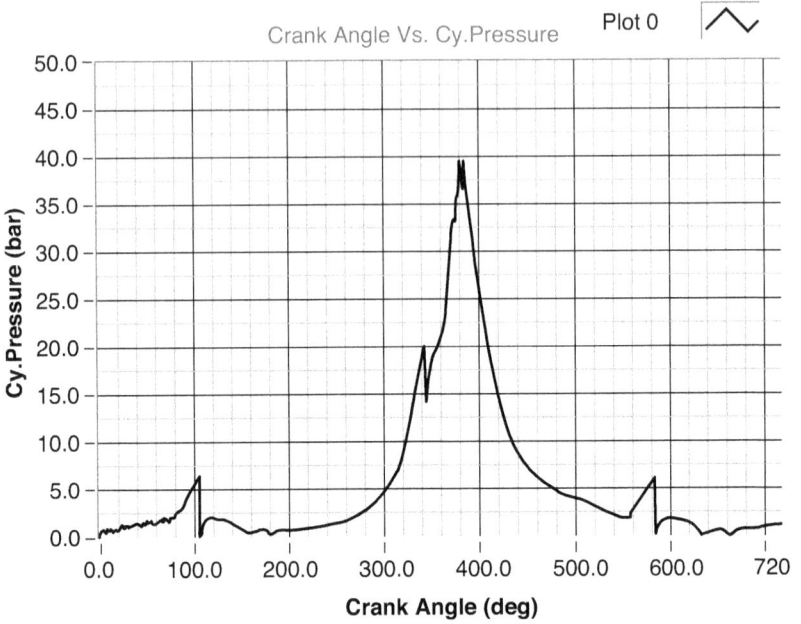

FIGURE 15.5   Changes in combustion pressure with varying crank angle.

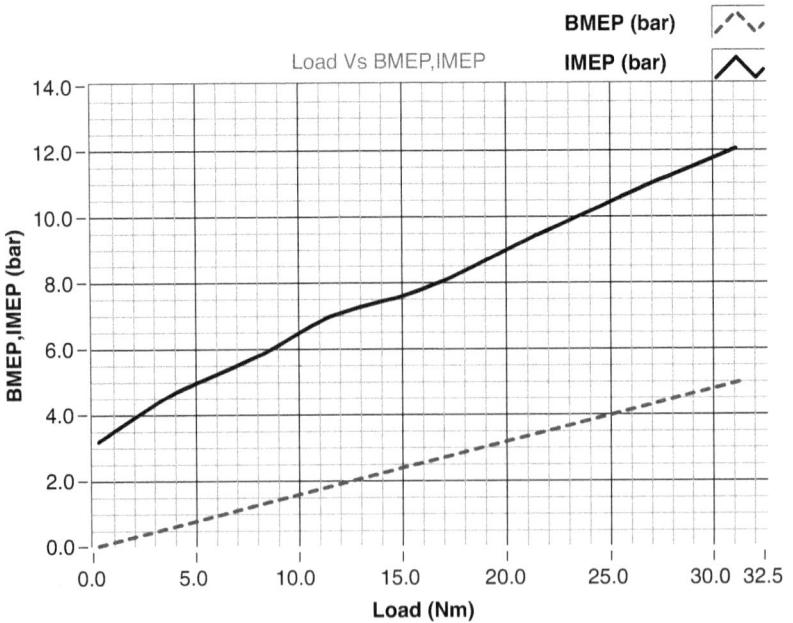

FIGURE 15.6   BMEP and IMEP variation with load.

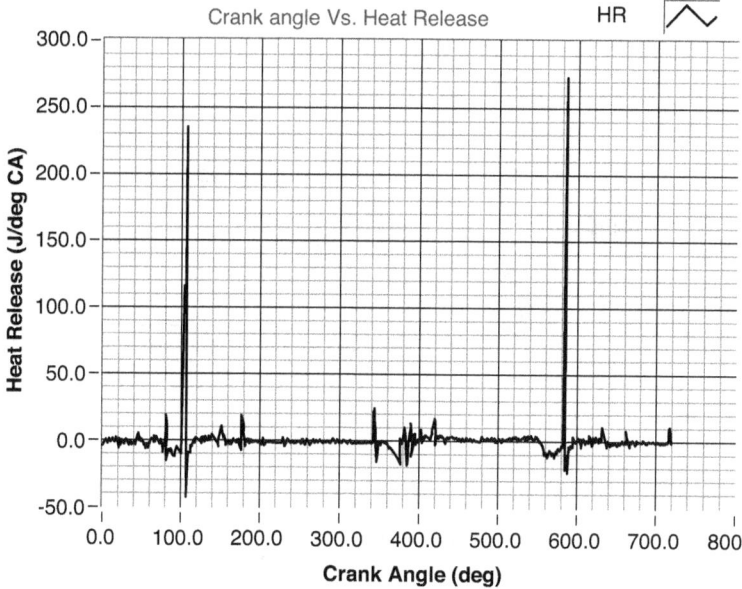

FIGURE 15.7   Heat release variation as a function of crank angle.

pressure for producing the power at the crankshaft, which is known as brake power. As there are some losses also because of friction and noise, so the BMEP is compara- tively less than the IMEP [6].

In Figure 15.7, there is the heat release from the engine with varying crank angle. The hear release rate from the engine shows that it increases at the start of combus- tion and then remains constant with varying crank angle. This heat released by the engine is in the form of conduction and radiation, which is released during running of engine, which produces lot of heat and is responsible for the loss of energy and frictional losses due to various parts of engine running during operation [7].

In Figure 15.8, there is the pressure volume diagram explained, which is normally known as Otto cycle for explaining spark ignition engines and various processes that take place during complete cycle from compression to exhaust. In Figure 15.8, four processes are taking place for completing a full power stroke. In PV diagram, heat addition takes place at constant volume, and then expansion takes place at adia- batic conditions, which actually gives the power stroke for SI engine. Finally, heat is rejected at constant volume again, which in actual is exhaust emission release [8].

15.8   EXPERIMENTS PERFORMED ON MPFI ENGINE

The micro-emulsion sample created in the IC Engine lab was utilized as a renewable fuel in an automobile to test its performance and exhaust on a multi-cylinder-based spark ignition engine (Figure 15.9). The emulsification fuel was made with a mixture

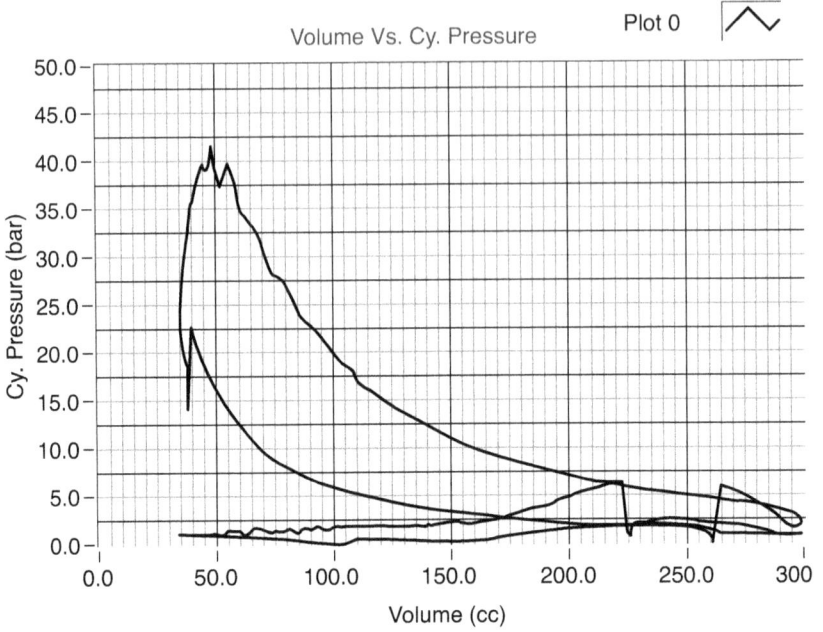

FIGURE 15.8   Variation in cylinder pressure as a function of volume.

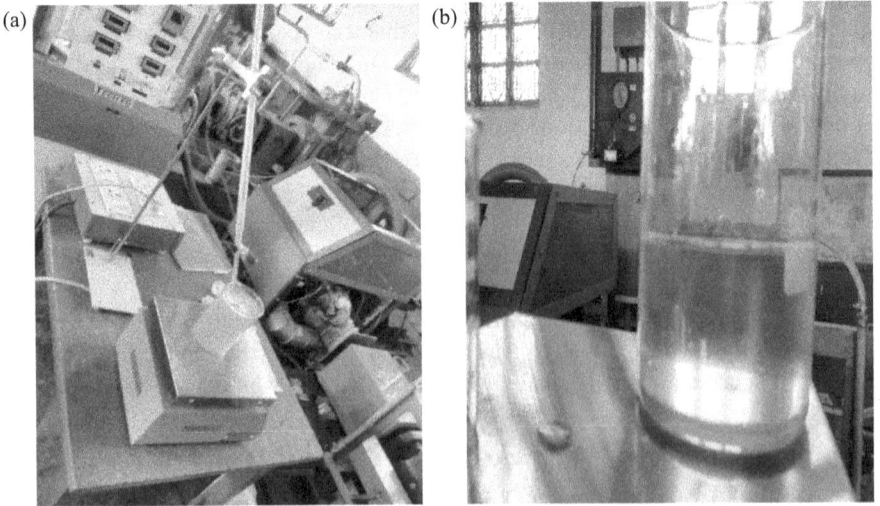

FIGURE 15.9   (a) MPFI engine test rig along with data acquisition system attached and (b) gasoline fuel.

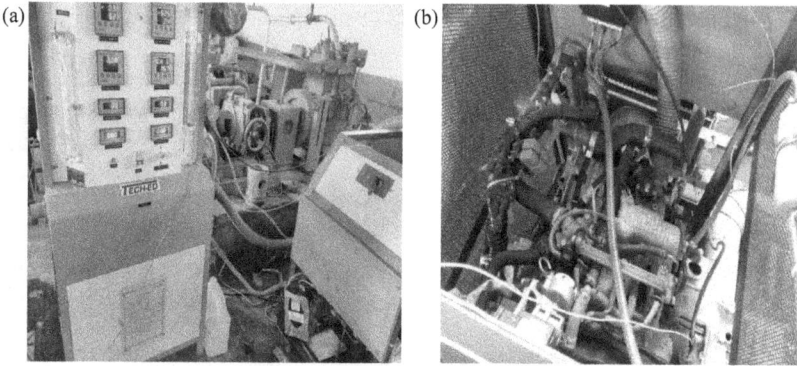

FIGURE 15.10 (a) The data is collected through an encoder. The PC is connected to a data acquisition system. (b) Test rig with GDI SI engine.

of 90% gasoline, 8% ethanol, and 2% water. The finished product was clear, translucent, and thermodynamically stable, isotropic liquid mixture of oil (conventional fuel), water, and co-surfactant ethanol. The magnetic stirrer was used for mixing all the fuels, including H_2O, which was added continuously into the blended mixture of the fuel [9].

15.8.1   DESCRIPTION OF THE GDI TEST RIG

Figure 15.10 shows the data acquisition system for the GDI SI engine-based test rig. This is linked to the computer, which shows performance and combustion data. Table 15.1 lists the characteristics of the test rig. The setup includes a petrol engine, three cylinders, four strokes, and water-cooled universal connection linked to a hydraulic dynamometer and mounted on an Ms channel-based centrally balanced foundation structure. In a fully powder-coated panel box that stands alone, the system incorporates a fuel tank, a manometer, and an air box measurement device; for measuring various parameters, digital indicators and transmitters are used. It also includes the necessary sensors and transmitters for detecting crank angle and combustion pressure. Signal conditioners and converters feed all of these signals into the computer [10].

15.9   CONCLUSION

- This work has been performed on multi-point fuel injection engine. The conclusion of this work is framed below.
- The primary goal of this research is to find out how the conventional engine works for various types of engine test rigs.
- It can be seen that all the engine types showed good performance and good efficiency.

- This work concludes that the highest power obtained is indicated power followed by brake power.
- In addition, the pressure produced in the combustion chamber reflects the range around 40 bar, which shows that the MPFI engine is working as per the specifications.

REFERENCES

[1] G.K. Prashant, D.B. Lata, P.C. Joshi, Investigation on the effect of ethanol blend on the combustion parameters of dual diesel engine, *Appl. Therm. Eng.* 99 (2016) 623–631.

[2] G. Khoobbakht, G. Najafi, M. Karimi, A. Akram, Optimization of operating factors and blended levels of diesel, biodiesel and ethanol fuels to minimize exhaust emissions of diesel engine using response surface methodology, *Appl. Therm. Eng.* 99 (2016) 1006–1017.

[3] A. Boretti, Advantages of converting diesel engines to run as dual fuel ethanol diesel, *Appl. Therm. Eng.* 47 (2012) 1–9.

[4] S. Padala, C. Woo, S. Kook, E.R. Hawkes, Ethanol utilization in a diesel engine using dual-fuelling technology, *Fuel* 109 (2013) 597–607.

[5] T.C.C. de Melo, G.B. Machado, C.R.P. Belchior, M.J. Colaço, J.E.M. Barros, E.J. de Oliveira, D.G. de Oliveira, Hydrous ethanol – gasoline blends – combustion and emissions investigations on a flex-fuel engine, *Fuel* 97 (2012) 796–804.

[6] M.A. Catagliola, M.V. Prati, S. Florio, P. Scorletti, D. Terna, P. Iodice, D. Buono, A. Senatore, Performances and emissions of a 4-stroke motorcycle fuelled with Ethanol/gasoline blends, *Fuel* 183 (2016) 470–477.

[7] G. Najafi, B. Ghobadian, A. Moosavian, T. Yusaf, R. Mamat, M. Kettner, W.H. Azmif, SVM and ANFIS for prediction of performance and exhaust emissions of a SI engine with gasoline-ethanol blended fuels, *Appl. Therm. Eng.* 95 (2016) 186–203.

[8] I. Schifter, R. Diaz, R. Rodriguez, J.P. Gomez, U. Gonzalez, Combustion and emissions behaviour for ethanol-gasoline blends in a single cylinder engine, *Fuel* 90 (2011) 3586–3592.

[9] X. Wang, Z. Chen, J. Ni, S. Liu, H. Zhou, The effects of hydrous ethanol gasoline on combustion and emission characteristics of a port injection gasoline engine, *Case Stud. Therm. Eng.* 6 (2015) 147–154.

[10] R.C. Costa, J.R. Sodre, Hydrous ethanol vs. gasoline-ethanol blend: engine performance and emissions, *Fuel* 89 (2004) 559–570.

16 Utilization of Waste Ceramic Tiles as Coarse Aggregates in Concrete

Priyanka Dhurvey, Harsh Panthi, Parth Verma, and Chandra Prakash Gaur

CONTENTS

16.1 Introduction 291
16.2 Materials and Methods 294
16.3 Results and Discussion 295
    16.3.1 Specific Gravity Test on Coarse Aggregates 295
    16.3.2 Water Absorption Test on Coarse Aggregates..................................296
    16.3.3 Aggregate Crushing Test 296
    16.3.4 Aggregate Abrasion Test...297
    16.3.5 Compressive Strength Test 297
    16.3.6 Economical Substantial 299
16.4 Conclusions...300
References..301

16.1 INTRODUCTION

Concrete is a versatile construction material. It consists of cement, fine aggregates, coarse aggregates, water, and admixtures, which vary according to the proportion and environmental conditions. Day-by-day rapid growth in construction areas results in the humongous need for concrete, and thus, it also contributes to the most critical problem on earth that is waste. Waste comes from construction, and demolition contributes the highest percentage of waste worldwide. Adding to it, the highest percentage of waste within this construction and demolition waste is due to ceramic products (Juan et al., 2012). A large number of ceramic tiles were produced in industries and consumed for various purposes like flooring, platform, roofing, and wall decoration in interiors and exteriors, which serves as protection from the weather as well. This large production and manufacturing also lead to the increased wastage that comes from it. Earlier researches and experimental studies show that these ceramic tiles have pozzolanic properties, resistant to aggressive chemicals as well as high strength. Therefore, many sustainable options are available for ceramic waste rather than conventional landfilling. These ceramic wastes from various sources can be accumulated and recycled for various purposes. In this way, we can reduce the need for raw

DOI: 10.1201/9781003359784-16

material, which can help us economically and also makes the efficient use of natural resources. Recycling waste ceramic tiles in concrete as coarse aggregate may forward us towards sustainable development. Topçu and Canbaz (2007) are the first who show their concern about the waste produced by the ceramic tiles and proposed the use of ceramic tiles waste in concrete as aggregate by quoting that tile waste is enough to use as an aggregate for concrete. They replace stone aggregate with crushed tiles aggregate only at three different variations, i.e., 0%, 50%, and 100%. They concluded that the water absorption ratio of crushed tiles aggregates produced with 4–16 mm increased by 200%. Unit weight of concrete formed by crushed tiles aggregate has decreased by 4% but the specific weight of crushed tiles aggregate is lower than that of ceramic tiles, whereas compressive strength and split tensile strength are decreased by 40% due to induced pores by crushed tiles aggregate. Wattanasiriwech et al. (2009) used waste ceramic mud from tiles production as the main component in paving blocks. They also determine the compressive strength values of paving blocks produced by waste mud and compared them with the nominal paving blocks as per Thailand Industrial Standards. X-ray diffraction analysis done by Wattanasiriwech shows that waste ceramic mud consists of five main minerals, which are albite, microcline, zircon, kaolinite, and quarts. They also concluded that water added during both the compaction and curing process plays an important role in achieving the true strength of paving blocks. As the water content increased, pore size became smaller and porosity was diminished. Pacheco-Torgal et al. (2010) have evaluated the properties of concrete and the durability of concrete containing ceramic waste experimentally. They formed concrete at a target compressive mean strength of 30 MPa with 20% of cement replaced by ceramic powder. They also made concrete with a mixture of natural sand and ceramic coarse aggregates as well as with ceramic sand and granite aggregates. In their experimental procedures, they analysed vacuum water absorption, capillary water absorption, water permeability, oxygen permeability, chloride diffusion test, and durability ageing test. Results that came from these tests by researchers concluded that the concrete with 20% replacement of cement with ceramic powder waste (made of ceramic bricks) has shown the highest compressive strength, i.e., it has higher pozzolanic reactivity at 20% composition. Vacuum water absorption test results show that concrete having ceramic bricks mixture has a higher absorption than control mixture up to 5%. However, the results of the oxygen permeability test reveal that concrete with ceramic bricks and concrete with sanitary ware have slightly higher permeability than the control mixture. Chloride test results with ceramic mix also show higher durability performance in concrete. Sekar et al. (2011) recommend the use of waste ceramic tiles as aggregate in concrete. They experimented and compared the properties of nominal concrete mix with the properties of concrete having tiles as aggregate and concrete having ceramic insulator scrap as coarse aggregate and also with concrete having glass as an aggregate. The mix proportion produced by them is such that it gives strength to M20 concrete. Results show that the compressive strength of concrete cubes made with 100% replacement of coarse aggregate is lower than that of nominal concrete cubes made with crushed stone as coarse aggregate. The compressive strength of concrete cubes with ceramic insulator scrap had 16% lesser strength than that of nominal concrete cubes, and the compressive strength of concrete cubes with crushed glass aggregate had 26.34% lesser strength than that of nominal concrete cubes. Tabak et al. (2012) show an environmentally friendly approach to recycling ceramic tiles waste and its possible use as aggregate in the production of concrete. In their study, they analysed the physical and mechanical properties of concrete having ceramic waste tiles as aggregate. Aggregates used in their experiments are crushed stones, stone dust, floor tile waste dust, and sea dust of a maximum size having 16.5 mm. The concrete cubes designed by the volumetric method had three different concrete mixes with a constant 0.54 ratio of water and cement. Experiment results show that the concrete with floor tiles waste aggregate compressive strength is higher than that of controlled mix concrete having nominal crushed stone aggregate. But concrete having both fine and coarse aggregates of tiles waste does not follow the European standards and has a lower strength value. Punit et al. (2014) mainly oriented towards the reduction of weight of concrete by using waste tiles as coarse aggregate in concrete. They conclude that using waste crushed ceramic tiles as coarse aggregate in concrete can reduce the weight of coarse aggregate in concrete by 50%, which in turn reduces the weight of concrete. They advised using waste ceramic tiles as coarse aggregate in concrete where the weight of concrete needs to be reduced. Elçi (2015) shows the use of crushed floor and wall tiles waste in concrete production as aggregates. He performed various chemical and physical tests on both wall tiles and floor tiles waste and compared it with limestone aggregates, which are generally used as coarse and fine aggregates in that region. He concluded that crushed tile aggregates are weaker than crushed stone aggregates, whereas the water absorption in crushed tiles aggregates is much higher than that of limestone aggregates due to their high porosity. He found similar values of crushed tiles aggregates and limestone aggregate in the Los Angeles crushing test and none of these aggregates were found alkali reactive. Aruna et al. (2015) used tile waste as aggregates while partially substituting OPC with fly ash. They recommended the use a concrete mix with 20% tiles and 40% fly ash since it is more cost effective to utilize tile waste instead of coarse aggregates and fly ash instead of OPC. Ch et al. (2015) investigated the acceptability of waste crushed tiles as a partial replacement of coarse and fine aggregate in concrete mixes, preparing a total of 9 mixes of M25 grade mixes. The compressive strength of all combinations is enhanced, with the highest compressive strength achieved for the mix containing 10% crushed tiles and 20% tile powder. Crushed waste ceramic tiles were utilized in concrete by Daniyal et al. (2015) with various percentage as an substitution for natural coarse aggregates. After examining the data, it was determined that the ideal value of waste ceramic tile to be around 30% replacement with a water/cement ratio of 0.5. The compressive and flexural strengths of ideal concrete were found to be 5.43% and 32.2% greater, respectively, than those of reference concrete. Singh and Singla (2015) did their studies in three different concrete mixes with crushed tiles of 20 mm size maximum as a replacement. The compressive strength was tested for different curing ages. They concluded that except for the M30 mix in other mixes, i.e., M20 and M25, no significant effects were observed for compressive strength up to 20% replacement. Then, with an increase in the proportion of tile aggregates, the strength started decreasing gradually. Chand and Ravi (2017) show a partial replacement of aggregate with waste crushed ceramic tiles in concrete for different grades of concrete M15, M20, and M25 with ceramic tiles powder and granite powder as fine aggregate. The strength of concrete increases by up to 30% percentage. Adekunle et al. (2017) utilize waste tiles as fine aggregate replacement in concrete. The control mix contains partial replacement with crushed tiles as fine in different proportion of 5%, 10%, 15%, and 20%. Also, another mix contains partial replacement of granite in place of crushed tiles in proportions of 25%, 50%, and 75%. They concluded that coarse aggregates can be substituted at 25% waste tiles, while fine aggregates can be substituted at 5% waste tiles. Sabău et al. (2021) investigate the effects of recycled aggregate concrete manufacturing on the environment and geochemistry with qualities appropriate for structural purposes. The findings of this paper suggest recycled aggregate concrete as an alternative to conventional concrete for achieving global sustainability standards in construction because it can have lower carbon emissions than conventional concrete while still having strength characteristics that are representative for structural members. Tavakoli et al. (2013) assess the properties of ceramic aggregate before grinding it and using it in concrete as a substitute for sand and coarse aggregates with a 0 to 100% substitution rate. The research showed that incorporating leftover ceramic tile typically improved the characteristics of concrete. Zimbili et al. (2014) research on the use of ceramic wastes as a partial substitute for fine aggregates or cement has not received as much attention as other fields. The purpose of this study was to investigate the potential benefits of partially substituting waste ceramic wall tile for cement and fine aggregates, the two components of concrete that are the most expensive to create. Miličević et al. (2015) demonstrate that making of precast concrete floor blocks with more than 50% replacement of natural aggregate is possible with recycled aggregates. They used crushed clay bricks and roof tiles in their experimental analysis. They study the effect on physico-mechanical properties of concrete with recycled and with natural aggregates. Rashid et al. (2017) studied about the cost benefits of utilizing the ceramic waste coarse aggregate (CWCA) in the concrete and also the environmental benefits as the use of CWCA reduces the CO_2 emission (CO_2e). Sabau et al. (2021) concluded that the CO_2e of concrete with natural aggregates is about 330 $kgCO_2e/m^3$. To the author's knowledge, not many researchers use waste tiles in pavement block as coarse aggregate in concrete for the strength of M30. So, the objectives of the present work is to study the various properties of waste ceramic tiles, to find the suitable composition of crushed tiles in concrete as coarse aggregate, and to determine the effects in compressive properties of concrete formed by using crushed tiles as coarse aggregates.

16.2 MATERIALS AND METHODS

This work was done to analyse the compatibility of waste tiles as coarse aggregate, nominal coarse aggregate, and waste tiles coarse aggregate, which are mixed in the proportion of 100:0, 90:10, 80:20, 70:30, 60:40, 50:50, 40:60, 30:70, and 20:80 by weight, respectively. These various proportions of concrete are mixed and tested for water absorption test, density, and compressive strength. The sizing of coarse waste tiles are done as per IS 2386-1(1963). The waste tiles as coarse aggregate are tested for various properties as follows: Los Angeles test, crushing strength test, impact test, water absorption, and specific gravity, as per IS 383 (1970), IS 2386-3(1963), IS

2386-4(1963) and then compared with nominal coarse aggregate. The mix design is done as per IS 10262 (2009). The compressive strength values are tested for concrete cubes after 28 days, respectively, while in this period, they are completely immersed in water till maturity period. The materials used in the present experimental work are OPC Grade 53 Cement, sand as fine aggregate, stone dust as fine aggregate, crushed stones as nominal coarse aggregate, and crushed ceramic tiles as the replacement of nominal coarse aggregate. The laboratory test consists of the following tests of aggregates:

1. Specific gravity test
2. Water absorption test of coarse aggregates
3. Determination of aggregate crushing value
4. Determination of aggregate abrasion value
5. Compressive strength test

16.3   RESULTS AND DISCUSSION

Various tests were performed on both types of concrete, i.e., concrete made with nominal coarse aggregates and concrete made with both crushed tiles aggregates and nominal aggregates. Also, some tests are individually run especially on both types of aggregates to get a research study about the properties of both types of aggregates and compare them so that we able to predict their future scope in various civil constructions.

16.3.1   SPECIFIC GRAVITY TEST ON COARSE AGGREGATES

As we change the composition of coarse aggregates in concrete cubes, it is very important to know the various properties of different types of coarse aggregates used in research. Specific gravity test is one of the major concerning criteria while selecting the aggregates for concrete structures since it contributes to the weight and density of the concrete. As per IS 2638-3, Specific gravity test results on both types of aggregates are given in Table 16.1.

Specific gravity of nominal coarse aggregates is 2.61, which is acceptable with reference to the theoretical value of aggregates, i.e., 2.56, whereas the specific gravity test results show that the crushed tiles aggregates have a specific gravity of 2.34. From these test results, it is clear that the aggregates formed by waste crushed tiles are lighter than the nominal coarse aggregates, which we are currently using. Lower

TABLE 16.1

Specific Gravity Test Results on both Types of Coarse Aggregates

Coarse Aggregates	Specific Gravity
Crushed stone aggregates	2.61
Crushed tiles aggregates	2.34

specific gravity results in lower weight of the concrete and thus structure. This lowered weight may reduce the transportation cost of those precast structures that are made with crushed tiles aggregates.

16.3.2 WATER ABSORPTION TEST ON COARSE AGGREGATES

Water absorption test on aggregates is very necessary while working with concrete because when we mix water in dry concrete mix, some percentage of water is also absorbed by the aggregates, which results in deficiency of free water available for cement hydration process. This deficiency causes error in the water content provided to the concrete and water content absorbed by the cement for hydration. Water content also factors the amount of water carried by the porous concrete. As per IS 2638-3, Water absorption test results are given in Table 16.2.

The water absorption of nominal coarse aggregates, i.e., crushed stone aggregates, is 0.4%. However, the absorption of crushed tiles aggregates is much more than that, i.e., 5.2%. Thus, water absorption test clearly states that the water requirement in concrete having crushed tiles aggregates is more than that of concrete having nominal coarse aggregates so that water/cement ratio will remain unaffected.

16.3.3 AGGREGATE CRUSHING TEST

Aggregates' crushing value is a relative term, which is used to determine the crushing resistance of aggregates under a gradual change in load application. This test is performed here to know the crushing value of nominal coarse aggregate and crushed tiles aggregates. Aggregate crushing test results are given in Table 16.3.

Test results of aggregate crushing value show that the crushing value of stone crushed aggregates is 21, whereas the crushing value of crushed tiles aggregates is 27. This result implies that the stone crushed aggregates have more crushing strength than crushed tiles aggregates, as only 21% of crushed stone aggregates passes through

TABLE 16.2
Test Results of Water Absorption Test on both Types of Coarse Aggregates

Coarse Aggregates	Water Absorption
Crushed stone aggregates	0.4
Crushed tiles aggregates	5.2

TABLE 16.3
Aggregates Crushing Value Test Results of both Types of Coarse Aggregates

Coarse Aggregates	Aggregate Crushing Value
Crushed stone aggregates	21%
Crushed tiles aggregates	27%

TABLE 16.4

Aggregate Abrasion Value Test Results of both Types of Coarse Aggregates

Coarse Aggregates	Aggregate Abrasion Value
Crushed stone aggregates	20%
Crushed tiles aggregates	24%

2.36-mm sieve. On the other hand, crushed tiles aggregates have lower crushing strength as 27% of fine aggregates were passed through 2.36-mm sieve. This increment in crushing value of crushed tiles aggregates with respect to the crushing value of crushed stone aggregates may be due to the reason that crushed stone aggregates are denser than crushed tile aggregates.

16.3.4 AGGREGATE ABRASION TEST

Aggregate abrasion value test confirms to IS: 2386 (part IV) – 1963, which is done on both types of coarse aggregates, i.e., crushed stone aggregates and crushed tiles aggregates. This test is performed on Los Angeles machine. Aggregate abrasion test results on both types of aggregates are given in Table 16.4.

The test results show that the abrasion value of crushed stone coarse aggregates is 20% and the abrasion value of crushed tiles coarse aggregates is 24%. This result indicates that stone aggregates are more resistive to abrasion as compared to the tile aggregates. The increased abrasion in tile aggregate as compared to the stone aggregates may be due to their angular structure, sharp edges, and flakiness.

16.3.5 COMPRESSIVE STRENGTH TEST

For compressive strength test, the total number of 27 cube specimens were casted to determine the compressive strength of concrete at a maturity period of 28 days, i.e., after curing of cube for 672 hours with different proportional mix of nominal and crushed waste tiles coarse aggregate. For each proportionate mix, three sample specimens were prepared and then cured for time period up to 28 days for all three cubes to find the compressive strength of 28 days, respectively. Concrete cube specimens are of size $150 \times 150 \times 150$ mm.

The amounts of various materials, i.e., cement, fine aggregate, and coarse aggregate, were defined by the concrete mix ratio of 1:1.692:3.053 by weight. And the amount of water is calculated by the water/cement ratio of 0.47. The compaction test results of mix proportion 100:0, 90:10, 80:20, 70:30, 60:40, 50:50, 40:60, 30:70, and 20:80 of nominal coarse aggregate and waste tiles coarse aggregate are shown in Figure 16.1.

The test results shown above indicate that the compressive strength of cubes was decreased as the amount of CTCA (crushed tile coarse aggregate) was increased in them. These cubes are of compressive strength M30 and their minimum average 28-day compressive strength was taken as 34.103 N/mm^2. This minimum average

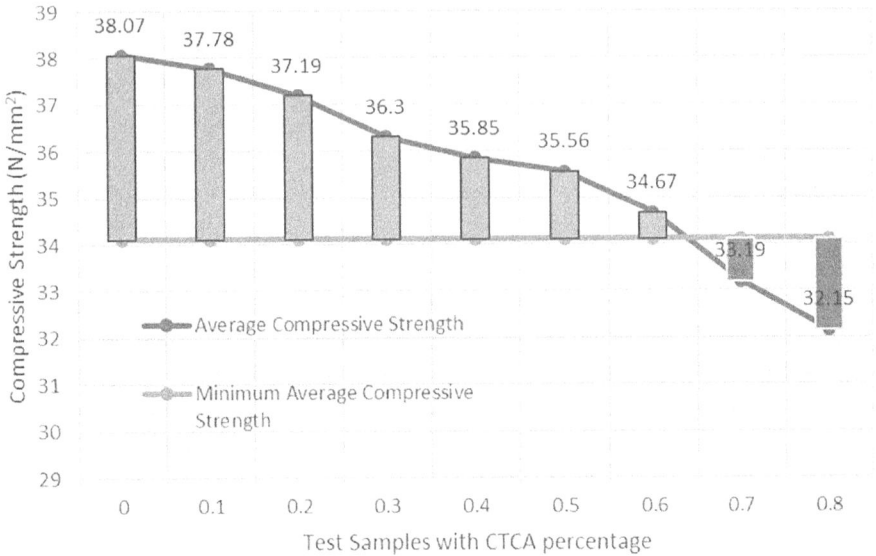

FIGURE 16.1 Average compressive strength of different mix proportions of CTCA cubes.

compressive strength is calculated by the formula given in clause 6.2.5.2 Table 16.3 of IS: 15658 (2006), which is for "Precast concrete blocks for paving", as our aim is on finding the maximum amount of waste crushed tiles that can be utilized in the construction of pavement blocks.

As per test results, the average strength of these cubes was over the minimum average mean compressive strength till 60% nominal coarse aggregates were replaced by crushed tiles aggregate as shown in Figure 16.1. But, in depth, we can see that one of test sample is failing at mix proportion of 40:60. The graph states that as we increase the amount of crushed tiles aggregates in the coarse aggregates form, this will reduce the compressive strength of cube but cube will not fail in terms of minimum average compressive strength up to 60% of CTCA in total aggregate weight in the cubes.

For the accuracy of the present work, the experimental results are compared with literature (Singh and Singla 2015), in which P. Singh has done the compressive strength test for three different concrete mixes, including M30 containing crushed tiles of different ratios having a maximum size of 20 mm as coarse aggregate with proportion of 0%, 10%, and 20%. For validation, mix proportion of 100:0, 90:10, and 80:20 is compared with the literature; the results show good arrangements with the literature as given in Table 16.5.

A regression-based model for estimating the compressive strength of concrete with different ratios of CTCA (0%, 10, 20, 30, 40, 50, 60, 70, and 80%). Therefore, based on experimental results, the equation for compressive strength vs percentage replacement of CTCA is determined by regression analysis of the best-fitted linear curve.

TABLE 16.5
Comparison of Results

Concrete Mix Ratio	Compressive Strength N/mm²		% Variation
	Literature (Singh and Singla 2015)	Present	
0%	38.73	38.07	1.70
10%	35.92	37.78	4.92
20%	34.96	37.19	5.99

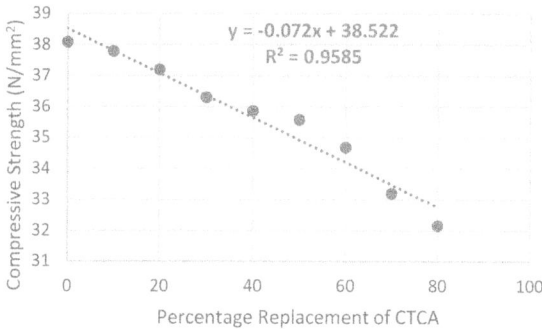

FIGURE 16.2 Compressive strength at 28 days as a function of the percentage replacement of CTCA.

In Figure 16.2, a relationship curve was stretched between compressive strength vs percentage replacement of CTCA. From Figure 16.2, analysis using regression to produce the following generalized compressive strength evaluation equation:

$$F_c = -0.072x + 38.522; R^2 = 0.9585 \qquad (16.1)$$

where F_c is the compressive strength of concrete, x is the percentage replacement of CTC, and R^2 is the coefficient of determination.

16.3.6 ECONOMICAL SUBSTANTIAL

The cost comparison was made in two different aspects: Firstly, cost of concrete, based on cost of raw materials, and secondly, on the basis of CO_2e taxes on the concrete. The cost benefits were more, if casting of concrete is done in bulk and also the percentage replacement of CTCA also reduced the cost of concrete, shown in Figure 16.3. From Table 16.6, it is very clear that the percentage replacement of CTCA provides about 2% and 4% cost benefits, with 50% and 100% replacements, respectively.

The carbon tax cost comparison shows about 17% cost benefits of using CTCA over natural aggregates as given in Table 16.7, and also, it will benefit the environment by reducing the global warming and waste ceramic tiles recycling problem.

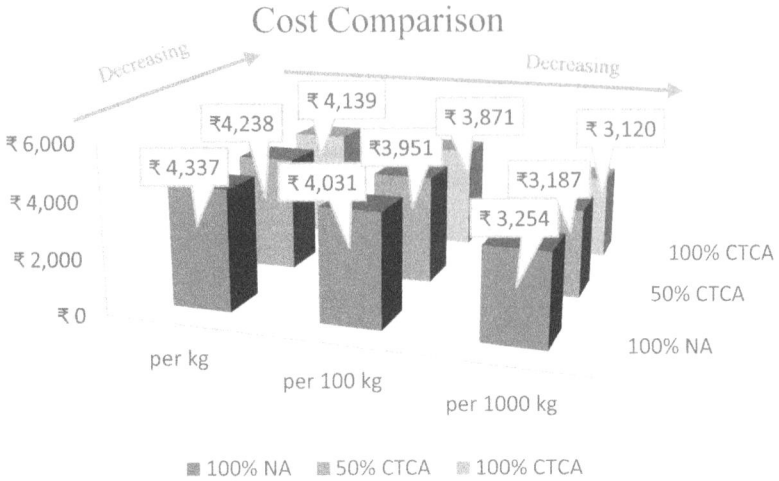

FIGURE 16.3 Cost comparison.

TABLE 16.6
Percentage Cost Benefits

Unit of Quantity	Cost (Rs.)			% Change on 50% CTCA	% Change on 100% CTCA
	100% NA	50% CTCA	100% CTCA		
per kg	₹ 4,337	₹ 4,238	₹ 4,139	2.3	4.6
per 100 kg	₹ 4,031	₹ 3,951	₹ 3,871	2.0	3.9
per 1000 kg	₹ 3,254	₹ 3,187	₹ 3,120	2.1	4.1

TABLE 16.7
Percentage Carbon Cost

Country	Tax in USD/ kgCO2e	Cost of NA in USD	Cost of CTCA in USD	% Change
Finland	0.075	24.75	20.55	16.97
Sweden	0.138	45.54	37.812	16.97
Norway	0.06	19.8	16.44	16.97
Denmark	0.028	9.24	7.672	16.97
EU	0.03	9.9	8.22	16.97

16.4 CONCLUSIONS

Based on the experimental analysis done on properties of crushed stone aggregates and crushed tiles aggregates, concrete having crushed stone aggregates as nominal coarse aggregates, and the mixture of crushed tiles and crushed stone as modified coarse aggregates, the following conclusions have been drawn.

- Specific gravity test results in our experiment show that crushed stone aggregates have 2.61 specific gravity and crushed tiles aggregates have 2.34 specific gravity. This test result indicates that crushed tiles aggregates are looser than crushed stone aggregates. This lower specific gravity in crushed tiles aggregates may be due to its porous structure.
- In water absorption test, results say that water-absorbing capacity of crushed tiles aggregates and crushed stone aggregates is 5.2 and 0.4, respectively. This test on both types of coarse aggregates shows a large difference in water-absorbing capacity between crushed stone aggregates and crushed tiles aggregates. This increase in water absorption is due to more porous structure of crushed tile aggregates compared to crushed stone aggregates.
- Aggregate crushing value test on crushed tiles aggregates and crushed stone aggregates shows a crushing value of 26% and 21%, respectively. Crushed tiles aggregates have 4% higher crushing value compared to crushed stone aggregates, which indicates that the crushing strength of tiles aggregates is lesser than that of stone aggregates. This reduction in strength of crushed tiles aggregates might be due to its angular shape and flakiness of coarse aggregates.
- Aggregate abrasion value test result gives an abrasion value of 20% for crushed stone aggregates and 24% for crushed tiles aggregates, which indicates more tear and wear of crushed tiles aggregates as compared to crushed stone aggregates. This higher value of abrasion in case of crushed tiles aggregates is due to the fact that coarse tiles aggregates are angular and flaky, whereas crushed stone aggregates are rounded to subangular in shape.
- Compressive strength test on cubes formed for design strength of M30 shows that cubes having 100% crushed stone aggregates have an average compressive strength of 38.07 N/mm^2. When the amount of crushed tiles aggregates was increased and replaced by crushed stone aggregates in concrete cubes, the strength of these cubes was decreased. Test result shows that the average compressive strength of concrete cubes having 60% of crushed tile aggregates as coarse aggregate is 34.67 N/mm^2, which is just above the target mean strength of M30 cubes. Individual test sample result shows that it is safer to use crushed tiles aggregates as coarse aggregate can replace up to 50% nominal coarse aggregates.
- The relationship between percentage replacement of CTCA and compressive was defined by linear regression with high coefficient of determination, $R^2 = 0.9585$.
- There is a 2% cost benefit in the use of 50% CTCA in concrete and about 17% carbon tax benefit.

REFERENCES

Adekunle, Adebola A., Kuye R. Abimbola, and Ayo O. Familusi. "Utilization of construction waste tiles as a replacement for fine aggregates in concrete." *Engineering, Technology & Applied Science Research* 7.5 (2017): 1930–1933.

Aruna, D., et al. "Studies on usage potential of broken tiles as part replacement to coarse aggregates in concretes." *International Journal of Research in Engineering and Technology* 4.7 (2015): 110–114.

Ch, Hemanth Kumar, et al. "Effect of waste ceramic tiles in partial replacement of coarse and fine aggregate of concrete." *International Advanced Research Journal of Science, Engineering and Technology* 2.6 (2015): 13–16.

Chand, G. and P. Ravi Kumar. "Partial replacement of aggregate with ceramic tile in concrete." Diss. Jawaharlal Nehru Technological University (2017).

Daniyal, Md, and Shakeel Ahmad. "Application of waste ceramic tile aggregates in concrete." *International Journal of Innovative Research in Science, Engineering and Technology* 4.12 (2015): 12808–12815.

Elçi, Hakan. "Utilisation of crushed floor and wall tile wastes as aggregate in concrete production." *Journal of Cleaner Production* 112 (2016): 742–752.

IS 383. Specification for Coarse and Fine Aggregates from Natural Sources for Concrete [CED 2: Cement and Concrete] (1970).

IS 2386-3. Methods of test for aggregates for concrete, Part 3: Specific gravity, density, voids, absorption and bulking [CED 2: Cement and Concrete] (1963).

IS 2386-4. Methods of test for aggregates for concrete, Part 4: Mechanical properties [CED 2: Cement and Concrete] (1963).

IS 10262. Guidelines for concrete mix design proportioning [CED 2: Cement and Concrete] (2009).

IS 2386-1. Methods of Test for Aggregates for Concrete, Part I: Particle Size and Shape [CED 2: Cement and Concrete] (1963).

Juan, Andrés, et al. "Re-use of ceramic wastes in construction." *Ceramic Materials* (2010): 197–214, ISBN: 978-953-307-145-9.

Miličević, Ivana, Dubravka Bjegović and Rafat Siddique, "Experimental research of concrete floor blocks with crushed bricks and tiles aggregate", Construction and Building Materials, 94 (2015) 775–783.

Pacheco-Torgal, Fernando, and Said Jalali. "RETRACTED ARTICLE: Compressive strength and durability properties of ceramic wastes based concrete." *Materials and Structures* 44.1 (2011): 155–167.

Punit Malik, J. Malhotra, A. Verma, P. Bhardwaj, A. Dhoundiyal, and N. Yadav, "Mix Design for Concrete with Crushed Ceramic Tiles as Coarse Aggregate." *International Journal of Civil Engineering Research* 5.2 (2014):151–154.

Rashid, Khuram, et al. "Experimental and analytical selection of sustainable recycled concrete with ceramic waste aggregate." *Construction and Building Materials* 154 (2017): 829–840.

Sabău, Marian, Dan V. Bompa, and Luis FO Silva. "Comparative carbon emission assessments of recycled and natural aggregate concrete: Environmental influence of cement content." *Geoscience Frontiers* 12.6 (2021): 101235.

Sekar, T., N. Ganesan, and N. V. N. Nampoothiri. "Studies on strength characteristics on utilization of waste materials as coarse aggregate in concrete." *International Journal of Engineering Science and Technology* 3.7 (2011): 5436–5440.

Singh, Parminder, and Rakesh Kumar Singla. "Utilization of waste ceramic tiles as coarse aggregate in concrete." *Journal of Multidisciplinary Engineering Science and Technology (JMEST)* 2(2015): 3294–3300.

Tabak, Y., et al. "Ceramic tile waste as a waste management solution for concrete." *3rd International Conference on Industrial and Hazardous Waste Management*, Chania, Crete, Greece (2012).

Tavakoli, D., A. Heidari, and M. Karimian. "Properties of concretes produced with waste ceramic tile aggregate." *Asian Journal of Civil Engineering* 14.3 (2013): 369–382.

Topcu, I. B., and M. E. H. M. E. T. Canbaz. "Utilization of crushed tile as aggregate in concrete." *Iranian Journal of Science & Technology, Transaction B, Engineering* 31(2007): 561–565.

Wattanasiriwech, D., A. Saiton, and S. Wattanasiriwech. "Paving blocks from ceramic tile production waste." *Journal of Cleaner Production* 17.18 (2009): 1663–1668.

Zimbili, O., W. Salim, and M. Ndambuki. "A review on the usage of ceramic wastes in concrete production." *International Journal of Civil, Environmental, Structural, Construction and Architectural Engineering* 8.1 (2014): 91–95.

17 E-Waste Generation, Flow, and Management in Eastern Region of Sri Lanka

A. K. Hasith Priyashantha,
N. Pratheesh, and P. Pretheeba

CONTENTS

17.1 Introduction 303
    17.1.1 E-Waste as a Global Crisis..304
    17.1.2 E-Waste as a Lucrative Business 305
    17.1.3 Addressing the Challenge in Sri Lankan Context 305
    17.1.4 Aim of the Study...306
17.2 Materials and Methods 306
    17.2.1 Study Setting...306
    17.2.2 Research Instrumentation 307
    17.2.3 Interview Schedule 307
    17.2.4 Data Collection Approach 307
    17.2.5 Sampling 307
    17.2.6 Data Analysis..307
17.3 Results and Discussion 309
    17.3.1 E-Waste Generation Scenario 309
    17.3.2 E-Waste Collection and Management... 312
    17.3.3 Role of Informal Collectors.. 314
    17.3.4 Formal E-Waste Collectors and Recyclers 316
17.4 Required Actions 316
17.5 Limitation of the Study... 318
17.6 Conclusions and Recommendations 318
References.. 319

17.1 INTRODUCTION

Along with the rising of the growing world population, it increases the generation of municipal solid waste. People believed that along with the advancement of technologies, there would be an answer to this problem, although it waved its hand at another threat—e-waste (Premalatha et al., 2014; Khan et al., 2022).

DOI: 10.1201/9781003359784-17

17.1.1 E-WASTE AS A GLOBAL CRISIS

Today, people all across the world are experiencing the rapid expansion of advances in science and technology; thus, the use of electrical and electronic equipment has become a necessity. In order to fulfill the demand, the production of electrical and electronic equipment has risen at a fast rate, and equally is being updated frequently. Consequently, it elevates the amount of electrical and electronic equipment that is outmoded, broken, or irreparable; thus, discarding by the owners becomes a frequent practice (Attia et al., 2021). Those end-of-life electronic devices or discarded by their consumers are generally called e-waste, and today, it represents 8% of all municipal solid waste; thus, the world is now being confronted with an ever-increasing e-waste challenge (Yang et al., 2021).

The data show that about 54 metric tons was generated worldwide in 2019. Though 82.6% (44.3 Mt) of the e-waste generated is uncertain, and just only 9.7 Mt was properly collected and recycled (Forti et al., 2020). It is estimated that around 4,000 tons of e-waste is exported worldwide every hour and 80% of this waste is exported to Asia (Yang et al., 2021). Matter of fact, Asia has outstripped all other regions in the world in terms of the quantity of e-waste generated. In 2019, about 24, 896 kt of e-waste was generated in Asia, which is about 46% of global e-waste generation. China (10,129 kt) becomes the world's largest e-waste producer in 2019, followed by the USA (6,918 kt), Japan (2,569 kt), India (3,230 kt), and Germany (1,607 kt). Despite the fact that developed countries are also producing e-waste at a higher rate, equally they have prominent waste collections and recycling mechanisms. However, the majority of developing countries are lacking such managerial practices (Forti et al., 2020).

Now the question arises, why there should be paid so much attention to e-waste? To answer this question, need to look at its composition. Of this, e-waste is a source of more than a thousand different compositions, including a number of toxic compounds; thus, it is recognized as hazardous waste (Needhidasan et al., 2014). Main toxic compounds in e-waste include heavy metals (e.g., arsenic, cadmium, lead, copper, mercury, zinc), polychlorinated biphenyls, and brominated flame retardants (Clarke et al., 2019; Dos Santos, 2021; Willner et al., 2021). The hazardous compound from e-waste may be released to the environment due to the irresponsible dumping/landfilling by communities, poor dismantling, incineration, and recycling practices (Rautela et al., 2021). Among those toxic compounds, the most attention has been given to the heavy metal as it can deliver a number of elevated health complications to humans, including cancer, muscular weakness, eye irritation, effect on the respiratory system, damage to the internal organisms like the heart, liver, kidney, and so on (Balali-Mood et al., 2021; Sarker et al., 2021; Witkowska et al., 2021). In addition to those, environmental contamination of such toxic compounds can change the microfauna community structure in the soil and natural water bodies (Needhidasan et al., 2014). As well, the concentration of heavy metals through the food chain and web will eventually donate the life threat to the consumers in top trophic levels (Dehghani et al., 2022; Nfor et al., 2022; Ruiz-Huerta et al., 2022). Therefore, it is obvious that immense attention should be given on e-waste in order to avoid the aforesaid negative consequences on the environment and the ecosystem, and also human well-being (Andeobu et al., 2021).

17.1.2   E-Waste as a Lucrative Business

It is well known that when e-waste is properly managed and recycled, there can be huge social and economic benefits that make it a profitable business (Adanu et al., 2020). However, aiming only the personal economical benefits and neglecting the other harmful effect due to the ill-handling, many warehouses have sprouted up, particularly in the developing countries, where most of them are illegal; thus, poor recycling and dismantling practices become common (Kêdoté et al., 2022). Coming to the point, what is the reason behind the e-waste becoming such an excellent business is due to the presence of precious metals. Gold, silver, and platinum are among the most valuable metals contained in the e-waste. In addition, copper and aluminum are the secondary metal that added value to the e-waste (Kaya, 2016; Fathima et al., 2022).

Compared to others, the extraction of gold becomes eye-catching practice due to the presence of a higher volume in e-waste. The gold content in electronic scrap materials is typically in the range of 10–10,000 g/t, which is substantially greater than the concentration in a natural gold ore (0.5–13.5 g/t) (Natarajan & Ting, 2015). Interestingly, about 280 g/t[1] can be extracted just only from central processing units (CPUs) alone. Therefore, e-waste can rightly be considered as a secondary source of gold and also a cheaper alternative source of gold (Natarajan & Ting, 2015; Ma et al., 2020; Ye et al., 2022).

It has been estimated that in New Delhi alone, about 25,000 of workers are engaged in the informal solid waste recycling activities, including collection, dismantling, and metal extraction. The majority of them are coming from economically stressed families thus working for a daily income. In Ghana, 6,300–9,600 workers are employed in the informal e-waste business, and another 1,21,000–2,01,600 people rely on them (Yang et al., 2020). It is crystal clear that as the e-waste generation becomes speed-up, the necessity of extra labor force is also mounting. In Serbia, for example, in 2013 about 14,000–18,000 tons of e-waste was formally recycled, employing nearly 1,500 formal labor force. In addition, as the acceleration of e-waste collection, new employees would be required, and approximately 2,000 new green jobs could be created by 2020 (Tackling informality in e-waste management, 2014). According to the above, it is obvious that the e-waste is important by creating opportunities for the employers, and also boosting gross domestic product (GDP), which is particularly important to the developing countries.

17.1.3   Addressing the Challenge in Sri Lankan Context

As aforesaid, South-Asian countries also produced a huge amount of e-waste. Sri Lanka is a small island among them, though e-waste generation is also at an alarming rate. In fact, the country is ranked fourth place in South Asia by means of the volume of e-waste produced. In 2019, the country produced 138 kt. For the last couple of years, the e-waste generation rate was about 5 kt in the island (Forti et al.,

2020). To address the growing e-waste generation scenario, Sri Lanka implemented a legal regime in 2008 by the Central Environmental Authority (CEA) (Electronic Waste Management in Sri Lanka, 2016). Among the South-Asian countries, except India, Sri Lanka is the only country, which has developed dedicated national-level legal implementation to restrict handling of e-waste (Forti et al., 2020). Sri Lanka also signed the Basel Convention on the Control of Transboundary Movements of Hazardous Wastes and their Disposal in 1992; hence, the country is also obliged to collect, transport, and dispose of e-waste properly, including aftercare of disposal sites (Ranasinghe, & Athapattu, 2020). It is consumable that legal implementation alone is not enough to overcome the e-waste challenge. By recognizing the urgent need of e-waste management, other than the strengthening regulations, many countries have taken steps to improve the e-waste recycling mechanism, including building new infrastructure, conducting pilot projects, setting up funds, raising consumer awareness, and so on (Zuo et al., 2020; Abalansa et al., 2021). In Sri Lanka, however, all these movements seem to be swept aside, which is the major drawback in the country to overcome this rising issue (Ranasinghe & Athapattu, 2020). On the other hand, studies on the fact that developing countries are lagging, particularly in Sri Lanka, have been few and far between.

17.1.4 Aim of the Study

Considering nine provinces in Sri Lanka, the Eastern Province holds an important position and is one of the most famous tourist destinations. Batticaloa is the most densely populated district among three districts in Eastern Province, which is located 215 km east of the commercial city of Colombo. According to the recent statistics, over 574, 836 of the population live in the area (Department of Census and Statistics, 2019). After about 26 years of conflict period, Batticaloa district started to emerge in all the aspects and also started to increase the livelihood of the local people along with invading the new technologies. Under these circumstances, the purpose of this exploratory study is to figure out the mainstream of generation, end-point of e-waste generated and management by the Batticaloa district, eastern region, Sri Lanka, through a qualitative research approach. To the best of the author's knowledge, this was the first study to date that investigated the e-waste scenario in the eastern region. Additionally, as a representative study, the findings further offer a primary understanding of how far the legal enforcement works out and required changes, not just for the studied region but for the entire country.

17.2 MATERIALS AND METHODS

17.2.1 Study Setting

The study was conducted in six Divisional Secretary's Division (DS divisions), namely, Eravur Town, Eravurpattu, Kattankudy, Koralaipattu, Koralaipattu West, and Manmunai North, out of the 14 randomly selected divisions of Batticaloa district, Sri Lanka.

17.2.2 RESEARCH INSTRUMENTATION

The study was carried out during July and August 2021 employing a qualitative research approach.

17.2.3 INTERVIEW SCHEDULE

In order to obtain information, the questionnaires were conceptualized as a manner of exploratory and also open-ended. The interview questionnaires were designed to obtain the primary information on awareness, handling practices, collecting/supplying, processing, and recycling of e-waste etc. First, the demographic data (e.g., gender, age, residence place, monthly average income, highest education qualification etc.) were collected from the respondents. Then, primary questions were asked from all the respondents, including; (i) Do you know about e-waste? (ii) If you know, how do you get to know? (iii) Do you know about the hazardous nature of e-waste? If so, please share what you know. (iv) Do you know the legal background behind e-waste handling/management? (v) Do you know any e-waste recycling area or dismantling area located in the Eastern Province/Sri Lanka?

Other applied key interview questions are indicated in Table 17.1.

17.2.4 DATA COLLECTION APPROACH

Data were gathered via in-depth interviews. The interviews were performed through face-to-face discussions or telecommunication with purposely selected respondents. The interviews lasted about 15–40 minutes, with an average of 20 minutes, and was audio-recorded with the informant's permission. During the interviews, respondents were allowed plenty of time for personal contemplation and asked follow-up questions. Furthermore, site-specific validations were carried out to acquire additional information.

17.2.5 SAMPLING

The primary data were obtained by interviewing 57 individuals, including household respondents ($n = 15$), government stakeholders ($n = 3$), mechanics from electronic repair shops ($n = 11$), solid waste collectors ($n = 5$), managers of electronic showrooms ($n = 4$), and other electronic items selling shops ($n = 5$). In addition, a few more interviews were conducted with individuals from out of Batticaloa district (Colombo), including warehouse ($n = 2$), formal e-waste collectors and recyclers ($n = 11$), and policymakers ($n = 1$). These respondents were identified as key role players in the e-waste management in the Batticaloa district.

17.2.6 DATA ANALYSIS

The collected data were then analyzed using the content analysis approach (NVivo 12 software package) where the data were classified, summarized, and tabulated.

TABLE 17.1
Key-Interview Questions

Respondents	Questions
Household	What electrical and electronic equipment do you dispose, most often?
	What are the reasons for such frequent disposal?
	How do you dispose the e-waste? Mention all ways of disposal.
	What are the problems you encountered with e-waste? Any human or environmental threat?
	Do you know anything about e-waste recyclers/recycling facilities in the Eastern Province or any?
Government stakeholder	What are the types of e-waste frequently disposed (also quantity) from your institute/department?
	How do you dispose it?
	Do you know anything about e-waste recyclers/recycling facilities in Eastern Province or any?
Electronic repair shops	What are the electronic equipment you are repairing?
	What are the most frequently received electronic items to repair?
	What are the main reasons for the breakdown of the equipment, which receive for the service?
	Mention the quantities you are repairing and returning back to the consumers (%). What are the reasons that you failed to repair those?
	Are you aware of e-waste? If so, what do you know? Environmental concern? Human health. Explain
	What are the most common e-wastes you get? (Quantity?)
	How do you dispose the e-waste?
	If storing, how long you are storing, what would ultimately you do?
	If buyers are coming to collect e-waste, what do they collect? (metal parts, plastic? etc) Are you giving freely or receiving money? If so, how much.
	Do you know anything about e-waste recyclers/recycling facilities in Eastern Province or any?
Local government solid waste collectors	Do you collect e-waste? If so, what are the items you are collecting frequently?
	How do you dispose it?
	Are you selling? If so, for where and how much?
Electronic showroom	Do you have a payback offer?
	If so, for which items you are giving this offer?
	How much do you collect?
	When you are giving this offer?
	What will you do with the received old electronic items?
	Do you have any specific e-waste storerooms?
Other multielectronic items selling shops	Do you receive any sold items as broken/not functioning?
	If so, what do you do?
	Do you give any warranty period for the selling items? Please specify.

(Continued)

TABLE 17.1 (*Continued*)
Key-Interview Questions

Respondents	Questions
Plastic and metal collectors	Are you collecting e-waste? If so, from where you are collecting?
	What are the items you are collecting? Please mention the quantity (per month) of your collection for each item
	What are the items you are collecting most frequently?
	How much are you paying for items at the buying point?
	To whom/where you are selling and at what price?
Warehouse	What are the e-waste items you are receiving?
	What are the items most frequently received?
	What steps are you taking to receive e-waste? (subjected to dismantling, recycling, or any)
Formal e-waste collectors and exporters, recyclers	Could you please explain the nature of your business?
	What are the e-waste items you are collecting?
	If limited items are collected, what is the reason?
	What are the items you are frequently collecting?
	Please mention the quantity of e-waste collected, if possible with certain e-waste items?
	From where you are collecting? (districts/province/island-wide, households/ business centers/government sector etc.)
	Do you collect e-waste from the eastern region, particularly from the Batticaloa district?
	What are the challenges you are facing and your suggestions to overcome those?
	Are you exporting e-waste? If so, what are the items (quantity) you are exporting and to where? Before export, do you follow any pre-processing techniques? If so, please explain
	If you recycle, what are the items you are recycling?
	Explain the process that you are carrying out.
	How do you dispose of the company waste?
	What are the safety precautions you follow?
Policymakers	Could you please explain about the current legal regime of e-waste, in Sri Lanka?
	Is the current legal enforcement at a satisfactory level, please explain?
	What are the challenges faced by proper e-waste handling and management?
	Could you please share your thoughts about the e-waste management practices, particularly about the eastern region?

17.3 RESULTS AND DISCUSSION

17.3.1 E-WASTE GENERATION SCENARIO

Today, as e-waste becomes a global concern, worldwide many studies have been carried out in order to address the scenario. The current study revealed that the requirement of public participation, in the Batticaloa district, is urgently needed to minimize e-waste generation and improve its management.

It was identified that computers, smartphones, fans, TV sets, and grinders/blenders are the most frequent e-waste items directed for the repairers. Thus, the source of most of the e-wastes is probably those items, except mobile phones. However, from household bulbs (CFL), mobile/smartphones plus their accessories and batteries (e.g., triple A batteries) are the most frequently disposed of items. Furthermore, according to the respondents (respondent numbers 17 and 51), government institutions like schools and other higher education centers such as universities frequently dispose of the toners most, next the computers. In addition, the most frequently highlighted terms through the interviews are shown in Figure 17.1.

According to the respondents, most of the e-waste generation comes from the Kattankudy, Koralaipattu Central, and Manmunai North DS divisions of the district. It was revealed that the majority of the e-waste generated by the district is coming from multiple electronic equipment repair centers, which is about 17 kg per month on average. All the repair shop respondents were given similar information, and no new information was obtained later on the interviews.

According to the respondents of electronic repairers, e-waste generation in the Batticaloa district has been exacerbated in recent years. The reasons for this include an increase in the purchasing rate of technologically advanced electronic products, the majority of which have a short lifespan and are more prone to damage.

A respondent from Kattankudy (city) DS divisions has shared his common practices:

"… we are repairing only the TVs and home theaters. Also, about a hundred TV sets per month are receiving us to repair. Most of the repairs coming to us were due to power failures. 15% of received items are not taken back by the customers, due to the higher cost to prepare especially the LED TV sets. Another 10%–15% could not repair, since no any parts are available in the market, as most are obsoleted"

(Respondent 1)

Despite receiving more work and money, a respondent who owned a repair shop in Batticaloa city remained in his good old days, as he used durable and quality electronic equipment.

In his own word:

"Yaaaaaa…I remember at an early age, we used a durable TV setup that functioned for more than 25 years. Still, the radio I am using is about 35–40 years old".

Similarly, a household participant also shared his experiences.

"…[…]…Last 16–18 years I brought three Television sets. Initially, I bought a CRT (cathode ray tube) television and used it for more than 13 years, without any issues. Then later on, I sold my television to my friend and bought a Flat LED (light-emitting diode) TV and just used it for only three and a half years. It was broken and for repair (replacing the panel), they requested about Rs. 30,000 (about 148 USD) and just only with a 6 months warranty. That is the price to buy a new TV, the same size. Then I buy a more advanced larger brand new LED TV for just about double the price with a 3 years warranty. Even I can't even sell my broken LED TV, since no one is ready to buy it. It's still stored in the storeroom in my home…"

(Respondent 2)

FIGURE 17.1 The most frequently used terms during this study. The image was created using NVivo 12 qualitative data analyzing package. The larger letters show the most frequently used terms by the respondents, while the smallest letters show the least.

Further discussion with respondent 11, explained about main reasons for excess e-waste generation and how e-waste helps to earn extra income.

"..this is our family business and we have been repairing all the electric items except TVs and phones for over 30 years. [..]..On average monthly, we repair more than 300 electronic items, where blenders, grinders, motors, speaker boxes, and fans are the most frequently received us to repair. Overuse of the equipment, poor maintenance or handling and low-quality products are the main reasons for breakdown. 90% of the items received are repaired by us".

(Respondent 3)

17.3.2   E-Waste Collection and Management

As many other developing countries in the world, Sri Lanka also does not own a proper systematic e-waste collection mechanism (Kazancoglu et al., 2020). This has been proven in the current study, as the informal sector plays a major role by collecting the majority of e-waste generated in the district. The informal collection was observed as a common and frequent practice, while formal collections were minimal in the district. A significant knowledge gap on proper e-waste handling was observed among the respondents. Most of the generated waste is collected by the localized e-waste shops in the district and ultimately sent to the informal e-waste collectors located outside of the Batticaloa district as shown in Figure 17.2.

Meanwhile, the electronics showrooms collect old electronic products via a buy-back offer, which is the solely targeted e-waste collection in the district, which is finally sent to the hand of the formal collectors. While moving to the government sector, institutional stores of e-waste or sold at auctions are common. From the auctions, e-waste is then demanded by the formal e-waste collectors, though the transfer

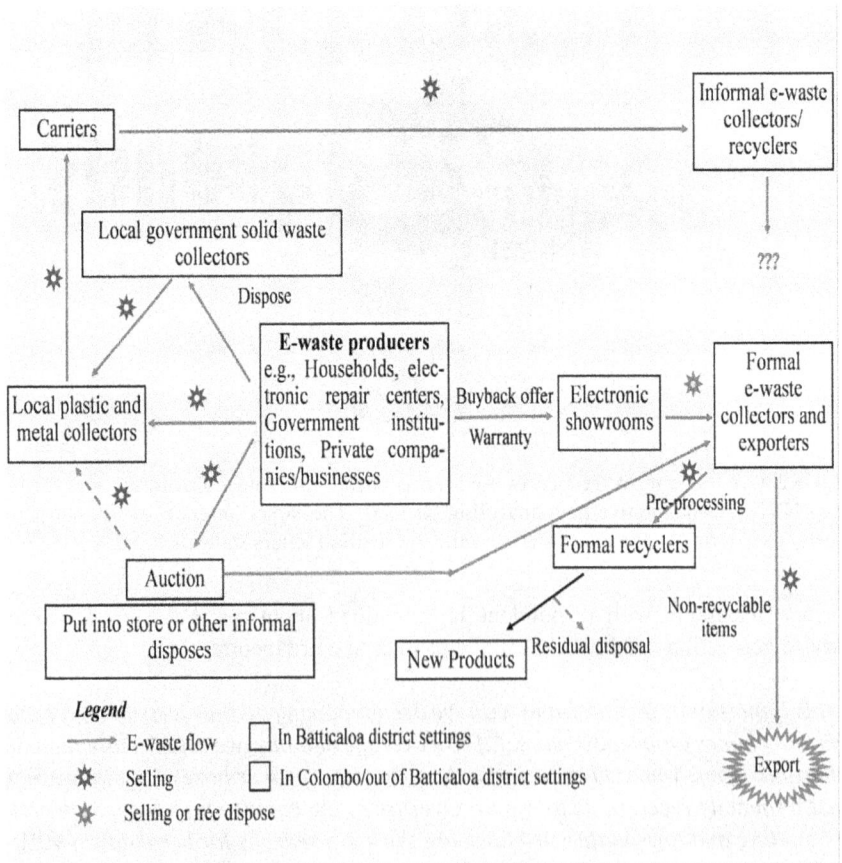

FIGURE 17.2   The generic flow and fate of e-waste Batticaloa district, Sri Lanka.

of all the e-waste to the formal collectors is doubtable. On the other hand, e-waste from recognized factories (e.g., textiles), other private companies, banks, and tourist hotels is also directed to formal collectors.

As aforementioned, knowledge on e-waste among the local communities is very limited and no householders could explain what effect informal e-waste handling could have. Thus, the lack of awareness is one of the major driving factors for the ill practices of e-waste handling in the district. By contrast, from the other 12 respondents, 3 household individuals from the rural area stated that no collectors visit to purchase their e-waste. Thus, in rural settings, it is common to store, bury, or burn e-waste. Moreover, turning again to the urban settings, throwing into the abundant open area is also practiced by the householders. Accounting overall, householders identified that e-waste is not going to be decayed soon, as it contains plastic or rubber coating: just think about the burning or storing.

According to the interviews of another two respondents (respondents 9 and 14), householders are daring enough to collect the parts from broken compact fluorescent lamp/CFL bulbs (which contain toxic heavy metal mercury) as soon as they get damaged with bare hands. The respondent 14 further stated:

".....aaaaaaaa.. No, I am not burning those broken CFL bulbs. Sometimes, I temporarily dispose them in the kitchen dustbin and later I damp those near the household fence to protect them from thieves".

Specifying the items, small unusable cell phone chargers and batteries are never seen as a danger and seem to be common to bury and dispose along with other solid wastes. While moving to the government institutions' prospect, the institutions are not allowed to dispose of the e-waste by any means rather than handing it over to formal collectors. However, in some cases, it is also allowed to withdraw unusable electronic equipment through the action, with proper legal documentation or recording. If there are no practices with those two, institutions need to store the e-waste as they need to respond to audit inquiries. As alluded to before, popular factories, other private companies, banks, and tourist hotels dispose of their complex e-waste to the formal collectors; the fact behind is to run their businesses as certified companies. Regrettably, e-waste generated by other local small business centers and hotels seems to be not directed to formal disposal as no formal collectors are looking to take the trash. Similarly, formal collectors consider minimally collecting e-waste from households too.

According to the interview conducted with four electronic showroom managers, among them, only one understood the troublesome nature of e-waste. However, all the respondents have only considered this as a part of their business strategy. Nevertheless, the practice is worthwhile considering the mitigating chances of local informal dumping of e-waste. According to the respondents, electronic repair shop owners/mechanics sometimes collect the interior electronic parts of the received electronic items from the buyback offer.

On the other hand, even if e-waste is collected at showrooms, it deems inadequate as collecting only restricted selected items. Further, respondents stated that the electronic collections are only for the specified items. Therefore, in this way, showrooms collected mainly the TV setups, and grinders/blenders/washing machines occasionally as those are rarely categorized under the buyback offer. In fact, buyback offers

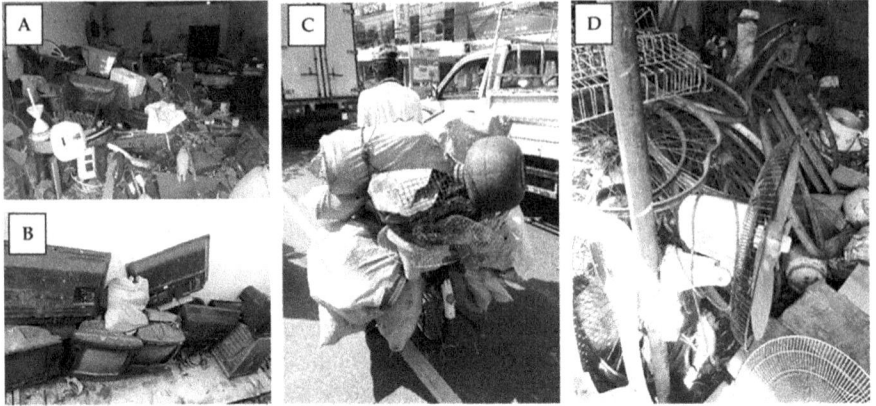

FIGURE 17.3 Some e-waste handling practices in Batticaloa district. A-Irreparable electronic equipment stored in a repair shop; B-Collected e-waste from a buyback offer stored in an electronic showroom's backyard; C-E-waste collected along with other solid waste by plastic and metal collector; D-Internal view of plastic and metal-collecting shop.

to encourage the local people to get rid of their old or short life spanned electronic items and own a new ones. Figure 17.3 shows the collected items via buyback offers and some other informal practices.

As further investigated, the collected products via buyback offer are being delivered to the company's warehouses in Colombo.

"Aaaa..yes...we are receiving the collected e-waste from our district branches and mainly we direct those to the authorized e-waste collectors. As a responsible company, we recently launched an e-waste collecting project (e-waste collecting project 2021, August 02nd to 6th) to collect e-waste from our company workers. There we collect many electronic items such as; personal computers and laptops, TVs, CRT and LCD monitors, telephones, mobile phones and other telecommunication items, all other electronic devices such as CD, VCD and digital cameras. However, we did not collect CFL and fluorescent bulbs, because of the handling difficulties"

(Respondent 4)

17.3.3 ROLE OF INFORMAL COLLECTORS

It is noteworthy to discuss the influence of plastic and metal collectors in managing district e-waste. In this regard, as aforementioned, plastic and metal collectors collect waste from households, and electronic repair shops, and are privileged to collect e-waste from the local government solid waste collectors. The majority of individual collectors use foot cycles to transport their collected e-waste. This has been a very common view in the district and collectors cycling daily throughout the selected area to search for waste items. Some waste collectors use buddy lorries to collect waste, including large e-waste items such as washing machines, and refrigerators. Unlike cycle collectors, those collectors use speakers to announce what items they are collecting and offer, thus taking more attention from the public.

Driven by the public toward the disposal of their e-waste to informal collectors is because of financial gifts.

A multiple electronic item repairer commented:

"......We are giving e-waste items to plastic and metal collectors. They are buying coppers (Rs. 600 or 3.01 USD per kg), iron (Rs. 50–60 or 0.25–0.30 USD per kg), Aluminum (Rs. 40 or 0.2 USD per kg), and plastic (Rs.20–30 or 0.1–0.15 USD per kg) from us. Per month, we are collecting around 10 kg of copper and 5 kg of iron and some amount of plastic and aluminum. By selling those, we are earning Rs. 8,000– 10,000 (about 40–50 USD) per month. The rest of the non-sellable parts are handed over to the local government garbage collecting trucks and I do not know about the e-waste or its issues actually, but I am not burning, burring, or damping elsewhere...."

(Respondent 5)

According to the computer repair shop respondents 17 and 21, printed circuit boards (PCB) of the broken equipment were sold at a rate of about Rs.300 (1.5 USD) per board. Compared to the multielectronic equipment repair shops, it was also identified that e-waste generated from the mobile repair shops is minor and negligible. The mobile repairers store all the wasted items to be used in future works, thus further reducing the amount of waste disposal.

What could be the role of local government solid waste collectors is interesting to mention. The local government solid waste collectors in the Batticaloa district collect all the biodegradable and non-biodegradable solid waste (including e-waste) from the urban area and bring the authorized dumping sites. The scenario behind the local government-damping sites involves collecting all the solid waste; sorting it into plastic, glass, degradable garbage, and polyethylene-like non-degradable items; and treating them accordingly. No evidence was found on the landfill of e-waste in damping areas nor manual dismantling practices.

Though, direct sales to the plastic and metal collectors were recorded. A municipal garbage collectors respondent:

"[..].actually we were instructed to collect only decomposable solid waste. Aaaaaaaa...but sir we also collect other non-degradable waste as you asked...until now, I actually don't know what e-waste was, but sir we are collecting trash including e-waste. We sell these items to local collectors. They come to our damping sites once or twice a week and give some money and take from us. [..] what to do sir.. what I earn with this job is absolutely not enough for the monthly expenses. I have a family and also school children. I don't know any other job I could do to earn extra, sir. By selling these collected items, we earn additional something rather than our salary".

(Respondent 6)

Individually gathered solid wastes by the local collectors are then sold at the plastic and metal-collecting shops located in town areas and are sorted at the collecting shops and then sold out to the buyers, without knowing the fact. Manual separations of plastic and metal from those collected electronic equipment are also recorded in the district.

17.3.4 Formal E-Waste Collectors and Recyclers

It is crystal clear that recycling is the best way to minimize greenhouse gas emissions, save energy and materials, reduce human health impacts, and meanwhile create employment opportunities (Maes and Preston-Whyte, 2022). Collectively in Sri Lanka, according to the Central Environmental Authority (2021), in Sri Lanka, about 200 Mt of crushed fluorescent lamps and powder was exported to Japan during 2019 and 2020. During this period, an equal amount of electronic scraps-A1180 (containing hazardous components such as accumulators, mercury switches, glass from cathode ray tubes, etc.) and printed circuit boards were exported to Korea. In addition 1,000 Mt of telecommunication battery scrap was exported to Europe. However, according to the formal e-waste collectors and recyclers (respondent numbers 48–53), most of such collections are gathered from Colombo based.

As noted earlier, formal e-waste collectors are limitedly collected, and the e-waste is generated from the Batticaloa district. The main challenge for the formal collectors is the distance as all the e-waste collectors and recyclers are concentrated in the Colombo district, and they need to travel a long distance to come to Batticaloa to collect the e-waste. Therefore, the formal e-waste collectors mostly gathered disposable electric items only in the nearby areas of their collecting centers. However, there are some formal collectors, who collect waste from throughout the island at request. As well, there are no dealers for them to collect the e-waste from the Batticaloa district and not accept such informal collecting owing to the government regulations.

In this regard, a participant from a formal e-waste collector and exporter commented:

"...we are one of the registered private companies to collect and export e-waste. About 20 laborers are working in our company both male (60%) and female (40%). Our role is to collect the e-waste from the generators and subject it to pre-processing. [...]Oh of course! We are following all the precautions to minimize the threats to the laborers as well as to the environment. We are dismantling the collected items (e.g., iron, plastic, motherboard, whirs), directing them to the local recycler for pre-processing, and directly exporting the non-recyclable items in Sri Lanka such as printed circuit boards and batteries to Korea and Europe. [...] We are not supposed to collect the e-waste from Colombo. Let's say, if we want to collect the e-waste from the Batticaloa district, it is too far away for us. From the e-waste we are collecting the limited item only, then there is higher transport cost to collect and we may face the losses"

(Respondent 7)

17.4 REQUIRED ACTIONS

There is ample evidence suggesting a number of management strategies for e-waste. Mundada et al. (2004) emphasize the influence of government on successful e-waste management. Furthermore, they have highlighted that rather than just implementing rules may not be sufficient, but also actively engaging with the public to direct them on proper e-waste management, and empowering formal collectors, exporters and recyclers are some mandating requirements. Extrapolating the e-waste scenario, it is logical that completing a few workouts may not be enough to achieve sustainable management. Of this, parallel management tactics need to be followed. Previous studies have shown that public awareness is one of the key factors in effective e-waste management (Ramzan et al., 2020; Parvez et al., 2021; Borthakur, 2022).

A comprehensive literature review conducted by Hasan (2004) showed that without active public participation, no legislation or enforcement could achieve the desired outcome of satisfactory waste management. Therefore, it is perceptible that the public needs to be conscious of the threats posed by e-waste and its best management practices. As stipulated before, the current study reviled that public awareness in the study area is extremely lacking and minimum. Nganji et al. (2010) suggested that web-based education could be quite compelling in this regard. The use of social media like YouTube and Facebook could play a vital role; equally according to the author's understanding, advertising in public places could also be much more effective when looking at the eastern region. In public transport, it is common to display the no smoke or health advice on not to smoke. Also recently, alongside COVID-19, it is also practiced the display of such advisable stickers in public places. Equally, if it can display such information about e-waste, the idea could be easily catchy to the public. In addition, such information could be displayed in the electrical and electronic equipment repairing and selling business centers that also could be a fruitful movement. Consequently, it is important to educate participants through awareness programs/workshops to increase social enticements and self-interest to support e-waste management. Since the district has several national higher education institutions and many schools, the steps can be easily initiated by involving these institutions.

A critical review conducted by Mmereki et al. (2015) noted that independent intervention by government, private sector, or other individuals might not be sufficient to deal with the impending crisis, though an integrated framework could do. Unfortunately, non-governmental organizations (NGOs), government institutions, and other community-based societies are making no efforts to address the e-waste challenge in the eastern region. Though, they have been involved in social activities in managing common solid waste. Obviously, if such practices initiate, indeed it will play a crucial role in underpinning environmental contamination of hazardous substances of e-waste.

In order to implement the systematic e-waste collection mechanism, several actions needed to be taken. Frequent conduct of e-waste drop-off events could enliven the public to dispose of their waste. It is also important to set up permanent e-waste bins in public places to maximize household participation. The establishment of city-based e-waste collection centers could be an ideal strategy, which also helps to collect household trash. According to respondents, all are willing to hand over their e-waste directly to such centers if established. To further improve the collection mechanism with strict rules and regulations, municipal garbage collectors can be used to collect e-waste, formally. However, in such implementation, continuous assessment is required to confirm that all the collections are being delivered only to the formal collectors. Besides that, the collectors should be adequately informed about the hazardous nature of the material and proper handling techniques. When delivering the message to those collectors, it is important to remember that they come from economically poor families with little or no formal education. As evidenced by this study, the provision of adequate funds from the government needs for effective collections and recycling. The government also needs to support formal collectors to utilize government-owned transportation to deliver the collected waste items from the eastern region to Colombo (or formal collecting centers), perhaps as free delivery or at a fair price. Also, this method confirms the accurate delivery of e-waste to formal collectors. It is worth noting that the required changes may not only apply to the eastern region but could also be implemented in the other districts and ultimately for the entire island.

17.5   LIMITATION OF THE STUDY

Notwithstanding the main findings of the current study, there are certain limitations. First, as the COVID-19 pandemic (3rd wave) spread across Sri Lanka more rapidly during the study period, one of the challenges the research group faced in designing this study was determining how best to interview people while maintaining social distance and adhering to other healthcare guidelines. Under these circumstances, the study aimed to get insight via a qualitative research approach, which minimized the quantity details of e-waste generation volume for the district. In future studies, a mixed method could be used to obtain in-depth information. Second, more priority given to conducting telecommunication, rather than face-to-face discussions, could be limited in extracting information.

17.6   CONCLUSIONS AND RECOMMENDATIONS

From the above, it can be concluded that more efforts are being made in many developed countries to quantify e-waste generation and identify e-waste flows, but efforts in developing countries are still sparse. Particularly in Sri Lanka, there is an extreme lack of studies in this aspect.

Despite the legality of e-waste handling, it is common to collect, sell, transport, and store the e-waste in the Batticaloa district, the eastern region of Sri Lanka informally. There is also evidence of illegal dismantling practices, where collected e-waste is manually separated into plastic and metal, and the rest of the waste is apparently disposed of unfavorably. Lack of awareness, the inadequacy of provisions, and the management exacerbated the problem. The families who are working to search for a bowl of rice, find solid waste collecting practices cycling throughout the day to find their resources. Similarly, many householders find them to sell their waste items to earn something. Therefore, in future proposed strategies if going to be accounted for, the challenge is there to drive the public to dispose of their waste for the formal collectors. It is perceived that gratification or rewarding is one of the ways to motivate all the parties to support formal e-waste management practices; thus, there may be a necessity.

Yet, there are no data available on the amount of e-waste produced, neither in Batticaloa district nor in the eastern region of the country; though, it was understood that the generation of e-waste becomes accelerated yearly, mainly due to the increase in consumption of electronic items that are having a shorter lifespan.

Therefore, it is consumable that attempts should be made to minimize the penetration of low-quality products and equally promote branded items. Moreover, the establishment of a proper e-waste disposal/collection mechanism for the district is urgently needed to avert the forthcoming crisis. Moreover, this is real time to think as a country, since the scenario could be similar for the many other districts as well. Of this, formally collected e-waste should be recycled in the country, at least from a minimum level, and the rest should be exported. The overall attempt of those will eventually lead to sustainable e-waste management in the country.

REFERENCES

Abalansa, S., El Mahrad, B., Icely, J., & Newton, A. 2021. Electronic waste, an environmental problem exported to developing countries: The GOOD, the BAD and the UGLY. *Sustainability*, *13*(9): 1–24. http://dx.doi.org/10.3390/su13095302.

Adanu, S. K., Gbedemah, S. F., & Attah, M. K. 2020. Challenges of adopting sustainable technologies in e-waste management at Agbogbloshie, Ghana. *Heliyon*, *6*(8): e04548. https://doi.org/10.1016/j.heliyon.2020.e04548.

Andeobu, L., Wibowo, S., & Grandhi, S. 2021. A Systematic review of e-waste generation and environmental management of Asia Pacific countries. *International Journal of Environmental Research and Public Health*, *18*(17): 1–18. https://doi.org/10.3390/ijerph18179051.

Attia, Y., Soori, P. K., & Ghaith, F. 2021. Analysis of households' e-waste awareness, disposal behavior, and estimation of potential waste mobile phones towards an effective e-waste management system in Dubai. *Toxics*, *9*(10): 236. https://doi.org/10.3390/toxics9100236.

Balali-Mood, M., Naseri, K., Tahergorabi, Z., Khazdair, M. R., & Sadeghi, M. 2021. Toxic mechanisms of five heavy metals: Mercury, lead, chromium, cadmium, and arsenic. *Frontiers in Pharmacology*, *12*: 1–19 https://doi.org/10.3389/fphar.2021.643972.

Borthakur A. 2022. Design, adoption and implementation of electronic waste policies in India. *Environmental Science and Pollution Research International*. Advance Online Publication. https://doi.org/10.1007/s11356-022-18836-5.

Central Environmental Authority. 2021. Licensed collectors of electronic waste management in Sri Lanka. http://www.cea.lk/web/en/index-php-option-com-content-view-article-layout-edit-id-983 (accessed March 07, 2022).

Clarke, C., Williams, I. D., & Turner, D. A. 2019. Evaluating the carbon footprint of WEEE management in the UK. *Resources, Conservation and Recycling*, *141*: 465–473 https://doi.org/10.1016/j.resconrec.2018.10.003.

Dehghani, A., Roohi Aminjan, A., & Dehghani, A. 2022. Trophic transfer, bioaccumulation, and health risk assessment of heavy metals in Aras River: Case study-Amphipoda- zander - human. *Environmental Science and Pollution Research International*. Advance Online Publication. https://doi.org/10.1007/s11356-021-18036-7.

Department of Census and Statistics. 2019. http://www.statistics.gov.lk/ (accessed March 07, 2022).

Dos Santos, K. L. 2021. The recycling of e-waste in the industrialised Global South: The case of Sao Paulo Macrometropolis, *International Journal of Urban Sustainable Development*, *13*: 1, 56–69. https://doi.org/10.1080/19463138.2020.1790373.

Electronic Waste Management in Sri Lanka. 2016. http://www.auditorgeneral.gov.lk/web/images/audit-reports/upload/2016/performance_2016/e_waste/Electronic-Waste-Management-in-Sri--Lanka----Performance-and-Environmental-Aiudit-Report_1-E.pdf (accessed February 27, 2022).

Fathima, A., Tang, J., Giannis, A., Ilankoon, I., & Chong, M. N. 2022. Catalysing electrowinning of copper from E-waste: A critical review. *Chemosphere*, *298*: 134340. Advance Online Publication. https://doi.org/10.1016/j.chemosphere.2022.134340.

Forti, V., Balde, C. P., Kuehr, R., & Bel, G. 2020. *The Global E-Waste Monitor 2020: Quantities, Flows and the Circular Economy Potential.* United Nations University (UNU)/United Nations Institute for Training and Research (UNITAR) – co-hosted SCYCLE Programme, International Telecommunication Union (ITU) & International Solid Waste Association (ISWA), Bonn/Geneva/Rotterdam.

Hasan, S. E. 2004. Public awareness is key to successful waste management. *Journal of Environmental Science and Health, Part A*, *39*(2): 483–492. https://doi.org/10.1081/ese-120027539.

Kaya, M. 2016. Recovery of metals and nonmetals from electronic waste by physical and chemical recycling processes. *Waste Management*, *57*: 64–90 https://doi.org/10.1016/j.wasman.2016.08.004.

Kazancoglu, Y., Ozbiltekin, M., Ozkan Ozen, Y. D., & Sagnak, M. 2020. A proposed sustainable and digital collection and classification center model to manage e-waste in emerging economies. *Journal of Enterprise Information Management*, *34*(1): 267–291. https://doi.org/10.1108/jeim-02-2020-0043.

Kêdoté, N. M., Sopoh, G. E., Tobada, S. B., et al. 2022. Perceived stress at work and associated factors among e-waste workers in French-speaking West Africa. *International Journal of Environmental Research and Public Health*, *19*(2): 851. https://doi.org/10.3390/ijerph19020851.

Khan, S., Anjum, R., Raza, S. T., Ahmed Bazai, N., & Ihtisham, M. 2022. Technologies for municipal solid waste management: Current status, challenges, and future perspectives. *Chemosphere*, *288*(Pt1), 132403. https://doi.org/10.1016/j.chemosphere.2021.132403.

Ma, T., Zhao, R., Li, Z., et al. 2020. Efficient gold recovery from e-waste via a chelate-containing porous aromatic framework. *ACS Applied Materials & Interfaces*, *12*(27): 30474–30482. https://doi.org/10.1021/acsami.0c08352.

Maes T., & Preston-Whyte F. 2022. E-waste it wisely: lessons from Africa. *SN Applied Sciences*, *4*(3):72. https://doi.org/10.1007/s42452-022-04962-9.

Mmereki, D., Li, B., & Li'ao, W. 2015. Waste electrical and electronic equipment management in Botswana: Prospects and challenges. *Journal of the Air & Waste Management Association*, *65*(1): 11–26. https://doi.org/10.1080/10962247.2014.892544.

Mundada, M. N., Kumar, S., & Shekdar, A. V. 2004. E-waste: a new challenge for waste management in India. *International Journal of Environmental Studies*, *61*(3): 265–279. https://doi.org/10.1080/0020723042000176060.

Natarajan, G., & Ting, Y. P. 2015. Gold biorecovery from e-waste: An improved strategy through spent medium leaching with pH modification. *Chemosphere*, *136*: 232–238. https://doi.org/10.1016/j.chemosphere.2015.05.046.

Needhidasan, S., Samuel, M., & Chidambaram, R. 2014. Electronic waste - an emerging threat to the environment of urban India. *Journal of Environmental Health Science & Engineering*, *12*(1): 1–9. https://doi.org/10.1186/2052-336X-12-36.

Nfor, B., Fai, P., Tamungang, S. A., Fobil, J. N., & Basu, N. 2022. Soil contamination and bioaccumulation of heavy metals by a tropical earthworm species (*Alma nilotica*) at informal e-waste recycling sites in Douala, Cameroon. *Environmental Toxicology and Chemistry*, *41*(2): 356–368. https://doi.org/10.1002/etc.5264.

Nganji, J. T., & Brayshaw, M. 2010. Is green IT an antidote to e-waste problems? *Innovation in Teaching and Learning in Information and Computer Sciences*, *9*(2): 1–9. https://doi.org/10.11120/ital.2010.09020006.

Parvez, S. M., Jahan, F., Brune, M. N., et al. 2021. Health consequences of exposure to e-waste: An updated systematic review. *The Lancet Planetary Health*, *5*(12): e905–e920. https://doi.org/10.1016/S2542-5196(21)00263-1.

Premalatha, M., Abbasi, T., & Abbasi, S. A. 2014. The generation, impact, and management of e-waste: State of the art. *Critical Reviews in Environmental Science and Technology*, *44*(14): 1577–1678.

Ramzan, S., Liu, C., Xu, Y., Munir, H., & Gupta, B. 2020. The adoption of online e-waste collection platform to improve environmental sustainability: An empirical study of Chinese millennials. *Management of Environmental Quality*. *32*(2):193–209. https://doi.org/10.1108/meq-02-2020-0028.

Ranasinghe, W. W., & Athapattu, B. C. L. 2020. Challenges in e-waste management in Sri Lanka. *Handbook of Electronic Waste Management*, 283–322. https://doi.org/10.1016/b978-0-12-817030-4.00011-5.

Rautela, R., Arya, S., Vishwakarma, S., Lee, J., Kim, K. H., & Kumar, S. 2021. E-waste management and its effects on the environment and human health. *The Science of the Total Environment*, *773*:145623. https://doi.org/10.1016/j.scitotenv.2021.145623.

Ruiz-Huerta, E. A., Armienta-Hernández, M. A., Dubrovsky, J. G., & Gómez-Bernal, J. M. 2022. Bioaccumulation of heavy metals and As in maize (Zea mays L) grown close to mine tailings strongly impacts plant development. *Ecotoxicology*, *31*(3): 447–467.

Sarker, A., Kim, J. E., Islam, A., et al. 2021. Heavy metals contamination and associated health risks in food webs-a review focuses on food safety and environmental sustainability in Bangladesh. *Environmental Science and Pollution Research International*, 1–16. Advance online publication. https://doi.org/10.1007/s11356-021-17153-7.

Tackling Informality in E-Waste Management: The Potential of Cooperative Enterprises. 2014. https://www.ilo.org/wcmsp5/groups/public/@ed_dialogue/@sector/documents/publication/wcms_315228.pdf (accessed February 05, 2022).

Willner, J., Fornalczyk, A., Jablonska-Czapla, M., Grygoyc, K., & Rachwal, M. 2021. Studies on the content of selected technology critical elements (Germanium, Tellurium and Thallium) in electronic waste. *Materials (Basel, Switzerland)*, *14*(13): 3722. https://doi.org/10.3390/ma14133722.

Witkowska, D., Słowik, J., & Chilicka, K. 2021. Heavy metals and human health: Possible exposure pathways and the competition for protein binding sites. *Molecules (Basel, Switzerland)*, *26*(19), 1–16. https://doi.org/10.3390/molecules26196060.

Yang, J., Bertram, J., Schettgen, T., et al. 2020. Arsenic burden in e-waste recycling workers - A cross-sectional study at the Agbogbloshie e-waste recycling site, Ghana. *Chemosphere*, *261*: 127712. https://doi.org/10.1016/j.chemosphere.2020.127712.

Yang, X. S., Zheng, X. X., Zhang, T. Y., Du, Y., & Long, F. 2021. Waste electrical and electronic fund policy: current status and evaluation of implementation in China. *International Journal of Environmental Research and Public Health*, *18*(24): 12945. https://doi.org/10.3390/ijerph182412945.

Ye, M., Li, H., Zhang, X., Zhang, H., Wang, G., & Zhang, Y. 2022. Simultaneous separation and recovery of gold and copper from electronic waste enabled by an asymmetric electrochemical system. *ACS Applied Materials & Interfaces*, *14*(7): 9544–9556. https://doi.org/10.1021/acsami.1c24822.

Zuo, L., Wang, C., & Sun, Q. 2020. Sustaining WEEE collection business in China: The case of online to offline (O2O) development strategies. *Waste Management*, *101*: 222–230. https://doi.org/10.1016/j.wasman.2019.10.008.

18 Sustainable Municipal Solid Waste Management through Membrane Science and Technology

R. K. Prajapati, Mohd. Ayub Ansari, and Haider Iqbal

CONTENTS

18.1 Introduction 323
    18.1.1 Management of MSWs 324
    18.1.2 Different Ages of Landfill 325
    18.1.3 Membrane Bioreactor (MBR) 326
18.2 Leachate Production and Characteristics 327
    18.2.1 Parameters of Leachates 328
18.3 Treatments of Landfill Leachate 329
    18.3.1 Treatment of Leachate with Domestic Sewage 329
    18.3.2 Recycling 329
    18.3.3 Biological Treatment 330
    18.3.4 Physicochemical Treatment 331
    18.3.5 Removal of Contaminant by Oxidation Using Common
            Oxidizing Agent 331
18.4 Separations of Leachate by Membrane Filtration 332
    18.4.1 Ultrafiltration (UF) 332
    18.4.2 Nanofiltration (NF) 333
    18.4.3 Treatment of Leachate by Using Reverse Osmosis (RO) 333
18.5 Conclusion 333
References 336

18.1 INTRODUCTION

The municipal solid waste (MSW) generated from household and industrial waste across the globe has created a major challenge to the environment, and their disposal requires a lot of money. MSWs that come from developing countries are mostly made up of organic wastes, while MSWs from developed countries are made up of a wide range of materials, like plastics and paper, as well as organic wastes. In countries

DOI: 10.1201/9781003359784-18

such as Thailand, China, Japan, and Brazil, more than half of the MSW fractions are organic food waste materials, plant residues, and animal derbies, are classified as organic fractions of municipal solid waste (OFMSWs). These OFMSWs certainly affect the environment during their haphazard decomposition. In turn, they pollute the different components of environment and change the climate by the release of greenhouse gases. MSWs lead to pollution into the natural environment, which is of two types: firstly, by the migration of leachates, which is rainwater or groundwater seepage that has soaked into the wastes and percolated through them, leading to groundwater contamination; and secondly, by the production of biogas by the fermentation and microbial decomposition of organic matter, causing air pollution (Calabrò et al., 2018; Abuabdou et al., 2020; Gao et al., 2015; Aftab et al., 2020; Jessica et al., 2022).

18.1.1 MANAGEMENT OF MSWS

Conveniently, the disposal of MSW and methods for its management have been done by the processes such as incineration, composting, and disposal by landfilling, which is relatively fast as well as cheap. More than 95% of MSWs generated worldwide have been disposed of through landfilling. Controlled landfilling is one of the most commonly used methods globally in the disposal of MSW, which produces leachate by the physicochemical degradation of waste in the presence of surrounding water in which it dissolves (Suresh et al., 2021; Ahn et al., 2002; Bakhshoodeh et al., 2020; Tara et al., 2021). Although landfills are done using highly sophisticated technologies to reduce the harmful impact of MSWs on the environment, the formation of leachate is still a major problem, which destroys the composition and fertility of the land as well as of surface and groundwater. The organic matter present in MSWs in the landfill produces landfill leachate (LFL) by different routes of degradation pathways represented in Table 18.1. Generally, LFL is a liquid formed when rain water enters into buried wastes at a landfill, and this liquid leaches out chemicals/constituents from the wastes.

TABLE 18.1
Balance of COD in the Organic Fraction in LFL

100% COD present in leachate due organic matter. This on hydrolysis form

Amino acid, sugars 66%		Fatty acids 34%
Amino acid, sugars by fermentation, acetogenesis		Fatty acid by Anaerobic oxidation
Fermentation Produces Acetogenesis Produces		Anaerobic oxidation Produces
Acetic Acid 46% **Acetotroph**	Intermediary Product: Propionate and butyrate	34% CO_2 and H_2 Hydrogenotroph
	By Methanogenesis	
70% CH_4,CO_2 **100% COD**		30% CH_4, CO_2

Pollutants present in MSW get dissolved into water with the action of rain, surface water, and microorganisms, which generate black- or brown-coloured LFL, and its smell is unpleasant and stingy. There is a high concentration of chlorinated organic matter, refractory organic materials, heavy and transition metal salts, suspended solids (SS), and nitrogenous-based inorganic compounds such as ammonia and nitrogen in LFL, which complicates water body quality. If these are not treated properly before disposal, LFL may pollute the soil environment and surrounding water bodies (Akgul et al., 2013; Giorgio et al., 2013; Ali et al., 2004).

The initial stage of landfilling that contains biodegradable organic matter is known as "young landfill", and it is the acidogenic phase. In the initial stage, LFL in the presence of high moisture or water content releases volatile fatty acids (VFAs) by fermentation, while the methanogenic phase occurs in a mature landfill in which the VFAs are changed into biogas methane and carbon dioxide (CO_2). Carbon dioxide, officially, has not been classified as an air pollutant, but its role in global warming, which arises due to cross-media metabolism, is an important issue. Carbon pollutants are not completely removed but cross the boundaries of the other environmental media during their removal and treatment process by metabolism, e.g., LFL generates landfill gases mainly CO_2, CO, CH_4, and more concentration of COD apart from nitrogenous, sulphur compound gases. During biochemical or other advanced treatment technologies, nearly half of the contaminants from LFL are changed into the sludge. When drying and incineration of sludge are carried out, carbon pollutants present in it are released back to the atmosphere in the forms of CO_2, CH_4, and HCs. Hence, continuous monitoring of the carbon metabolism across air-water, air-soil, and water-soil interfaces is important during different waste treatment processes. The percolation of leachate into underlying water bodies contaminates groundwater, threatening the health of the entire biomass, including humans. Hence, it is necessary to pre-treat the leachate by using the proper management and standard treatment strategies prior to its discharge into the sewer or ground and surface water (Jorge et al., 2021; Aftab et al., 2020; Samarasinghe et al., 2020; Abuabdou et al., 2020; Ahmad et al., 2020).

18.1.2 DIFFERENT AGES OF LANDFILL

By landfilling MSW over time, it causes the following significant negative impacts: epidemiological hazard; damage to the natural landscape; and contamination of the soil, underground and ground waters, and atmospheric air. The most important negative environmental impact of MSWs is the formation of LFL where MSWs are disposed of as landfill, which contains a variety of highly toxic chemical compounds and toxic metals leached from waste that are directly evolved in the process of biochemical decomposition of waste. The difficulties that arise in developing better leachate treatment technologies are because of the complexity of the compositions of leachates, which contain varieties of organic and inorganic substances, heavy metals, pathogenic microorganisms, etc. Thus, an integrated approach consisting of different treatment methods – mechanical, electrochemical, physicochemical, and biological – have been used for leachate removal. But landfilling, incineration, composting, and anaerobic digestion are the four important methods that are generally used in the treatment of MSWs. Obnoxious gases and wastewater present in leachate are treated simultaneously with the use of different technologies (Can 2019; Chen et al., 2020; Wang et al., 2020a; Li et al., 2009; Woldeyohans et al., 2014; Roy et al., 2021).

Historically, different technologies in combination of different methods of physicochemical as well as biological degradation have been applied globally in the removal of contaminants from leachate and are being used according to the local conditions of that country. Broadly, pre-treatment, main treatment, and advanced treatment processes are the steps in the treatment of leachate. Some insoluble inorganic salts, organic fractions, and heavy metals are removed by coagulation and adsorption in the first pre-treatment step, which subsequently reduces the treatment load in subsequent units. Biological treatment by adopting techniques such as up-flow anaerobic sludge blanket (UASB), sequencing batch reactor (SBR), and more effectively by membrane bioreactor (MBR) is done in the main treatment step, and during this step, most of the organic contaminants and nitrogenous compounds are eliminated from the leachate. Finally, in advanced treatment steps, advanced oxidation processes (AOPs) methods and high-pressure-driven membrane separation technologies are applied, which eliminate almost completely pollutants from the leachate. The various sustainable membrane separation technologies effectively remove the contaminants from leachate to the safe level. During AOPs, chemical oxidation and mineralization of refractory organic matter are achieved and also remove nearly 15%–25% of nitrogenous compounds as ammonia and gases, inorganic salts, and heavy metals from leachate (Wang et al., 2020b, 2013, 2014; Haifeng et al., 2021; Haizi et al., 2021).

18.1.3 Membrane Bioreactor (MBR)

MBR is a recently developed technology in which the treatment of leachate is carried out by membrane separation in combination with physicochemical method of separation. MBR can be effectively used in place of traditional secondary sedimentation tank separation. Here, leachate having large variations in quality and quantity of water is treated, the quality of effluent after treatment is quite stable, and the system has strong load impact resistance. It is effectively used in eliminating complex and recalcitrant compounds from domestic waste containing food and meat, industrial effluent from drugs, paper and pulp, leather, textiles, winery, and oil industries by retaining high biomass substances during the membrane separation step; and it has also been used in the treatment of MSWs, enabling its reuse. Conventional activated sludge (CAS) systems along with efficient membrane filtration processes have also been applied in MBR technology in which very fine particles (size of less than 0.1 µm) have been retained in its small pore and replace the gravity settling second stage in conventional methods used in wastewater treatment (Lucena et al., 2021; Hashemi et al., 2015; Srinivasaiah et al., 2021; Santos et al., 2010).

On the principle of working, MBR is made up of two units: a bioreactor where biological oxidation of pollutants is performed and a membrane separation module where the separation of contaminants from effluents is completed. Mechanically, it is applied in two alternative ways: (i) side-stream operation, where both the units work successively one after another, operating at constant pressure in elongated membrane modules and at variable permeating flux, and (ii) A method in which membrane is directly suspended within the bioreactor of submerged MBR technology, and separation of contaminant takes place at different transmembrane pressure with constant flow. The shapes of membranes used in MBR are vertical flat sheets and horizontal and elongated hollow fibres. The tubular form of membrane is less commonly used in submerged MBR, although it is preferred for side-stream operation. For example, by using the UASB-MBR process in the removal of contaminants from LFL, the concentration of COD and the ammonia nitrogen of contaminated water is 1,491–2,965 and 642–980 mg/L, which have decreased in effluent water by 588 and 323 mg/L, respectively (Gudeta et al., 2021; Woei et al., 2022; Caton et al., 2021).

In the last two decades, as compared to CAS systems, treatment of wastewater is mostly achieved by MBR technology in a 2–10 times smaller area, which produces two times less sludge, and the retention time of sludge is higher in MBR. It also shows higher removal efficiency of micropollutants and organic contaminants, and complete removal of dissolved and SS. Treatment of LFL by the application of MBR is much better as compared to the traditional activated sludge process, because it yields a lower amount of sludge with high hydraulic retention time, which facilitates the biological treatment by maintaining a high sludge concentration and, in such a way, produces minimum sludge. The following are the merits of the MBR process:

1. It has higher separation efficiency when used for mud water. The separation and filtration efficiency of suspended inorganic solids, polymer organic matter, and microbial particles is almost complete;
2. There is no loss of microbial flora, which also promote the microbial degradation of organic substances;
3. The larger duration of sludge residence time (SRT) in the reaction tank of MBR technology provides a favourable environment for the proliferation of bacteria, which enhances the degradation capability of microorganisms;
4. The MBR process produces minimum sludge output; and
5. In the MBR process, the effluent is stable by maintaining high microbial concentration and a good quantity of pollutant is removed (Mohammad et al., 2021; Tang et al., 2010; Hashisho et al., 2016; Yuri et al., 2021).

18.2 LEACHATE PRODUCTION AND CHARACTERISTICS

LFLs are liquid effluents produced by the degradation of MSWs when rainwater percolates through the solid waste or the moisture present in the waste. Its volume directly depends on precipitation, evaporation, transpiration, infiltration, and intrusion of surface as well as groundwater in the waste. The quantity of leachate produced can also be controlled by covering the landfill with the different protective techniques that control the percolation of water into it. Rainfall, apart from biological decomposition, groundwater inflow, and flow of surface water, is the main contributor of leachates generation. The water penetrates by percolation into the waste and dissolves the biodegrading waste by many physical and chemical processes. Water and other liquids that are present in the solid waste also increase the water content of leachate. Biochemical processes in wastes and water which has been percolated through wastes are responsible for the total quantity of leachates. Production of LFL is generally more when waste is less compactly packed, because compaction in waste certainly reduces the filtration rate. The characteristics of leachate change over time, the quantity of constituents applied, and the method applied in its treatment and leachate management (Zhang et al., 2020; Mavukkandy et al., 2016; Chen et al., 2020b; Alfaia et al., 2019; Luo et al., 2020).

18.2.1 PARAMETERS OF LEACHATES

The physical characteristics and quality of leachates depend on rainfall, snowfall, seasonal weather variation, age, types of waste and composition; however, the physical characteristics and chemical composition of leachates definitely vary as the age of landfill increases from young to old. The chemical oxygen demand (COD), ammonical nitrogen, and calcium in LFL are generally found to be in the range of 10^4–7.0×10^4 mg/L, 10^3–3.0×10^3 mg/NL, and 2.0×10^3–5.0×10^3 mg/L, respectively, and elimination of organic and nitrogen pollutants is the most tedious during the treatment of leachate.

Three types of leachates have been classified as per the age of landfilling (Table 18.2). At the beginning of the landfill, the amounts of the different chemical species are high, and after many years, when waste is stabilized, many chemicals are reduced continuously, without affecting the pH. The change in composition of leachate gradually changes with time due to changes in several physical characteristics as the landfill passes through stabilization phases, e.g., the COD in leachate decreases while the concentration of ammonia nitrogen increases. Ammonia is the major

TABLE 18.2
Typical Leachate Characteristics Constituents Inorganic Particles

Physicochemical Parameters	Age of Landfills		
	Young	Medium	Old
BOD (mg/L)	7.5×10^{-3}–1.7×10^{-2}	3.7×10^{-4}–1.1×10^{-3}	7×10^{-2}–2.6×10^{-2}
COD (mg/L)	1.0×10^{-2}–4.8×10^{-2}	1.2×10^{-3}–2.2×10^{-2}	6.7×10^{-4}–9×10^{-3}
$NH_{3+}N$ (mg/L)	4.0×10^{-3}–10^{-2}	3.0×10^{-5}–3.0×10^{-3}	1×10^{-5}–9×10^{-5}
Organic fractions (in %)	1.0–3VFAs (acetic, propionic, butyric acid)	5–40 VFA Humic & fulvic	50–100 VFA Humic & fulvic
pH	5.65–8.10	6.50–8.10	6.70–8.40
Parameters	Values in range	Parameters	Values in range
Hardness	4.0×10^2 to 2.0×10^3	Zn^{++}	0–1.35×10^3
TDS	0–4.23×10^3	Mercury	4.6×10^{-3} to 15×10^{-3}
Ca^{2+}	10^2–10^3	As^{+++}	0.10–0.44
Cl^-	20–2.5×10^3	Mn^{++}	0.06–1.4×10^2
Fe^{2+}	0.2–5.5×10^3	Pb^{++}	0–5
K^+	3–3.8×10^3	Mg^{++}	16.5–1.5×10^3
Na^+	0–7.7×10^3		16.5–1.5×10^3

contaminant in the leachate of old landfill sites. It has been estimated that young leachates mostly contain organic matter that is easily degraded by microbes, whereas ammonia-nitrogen compounds, heavy metals, chlorinated organic compounds, and inorganic salts are present in humic-type leachates. When leachate is stabilized, its composition changes and it mainly contains degradable carbon compounds with decreased COD and BOD but high concentrations of ammonia. Initially, landfill is in an acidic phase, where organic matter degrades easily into volatile acids. The mature landfill stage is the methanogenic phase, generating the organic fractions methane, humic, and fulvic acids (Argun et al., 2020; Chen et al., 2020c; Chang et al., 2018).

By and large, quality of landfill leachates greatly depends on the physical and chemical properties of the landfills, which are characterized by pH, requirements of biochemical oxygen demand (BOD), COD, ammonia nitrogen, salts, alkalinity, and amount of heavy metals. Generally, the chemical compounds present in leachate are heterogeneous and highly variable. Refractory compounds, humic and fulvic acids, dissolved organic compounds VFAs, aromatic hydrocarbons, phenols, pesticides, alkali metals (Na^+, K^+ ions), alkaline earth metals (Ca^{2+}, Mg^{2+} ions), iron, nickel, chromium, copper, zinc, manganese ions, chlorides, sulphates and bicarbonate ion, total nitrogen as ammonia-nitrogen, and heavy metals, e.g., Cd^{2+}, Pb^{2+}, are the most common chemicals present in LFL (Chang et al., 2019; Yin et al., 2021).

18.3 TREATMENTS OF LANDFILL LEACHATE

18.3.1 TREATMENT OF LEACHATE WITH DOMESTIC SEWAGE

Leachates are mostly disposed of by directly discharging through large drainage into the sewer system and ultimately flown away into the sea with a joint method of treatment at the sewage plant with domestic sewage. It is an easy and cost-effective method, despite the presence of organic inhibitory compounds in it that limits the complete removal of contaminants of low biodegradability and heavy metals and increases the effluent concentrations. This method of leachate treatment is used because nitrogen and phosphorus are not added at the plant. At a 10% concentration of leachate in the wastewater, the values of BOD reach 95%, whereas at the end of the day, half of the nitrogen has been removed, while the concentration of COD and NH_3-N_2 decreases with the increase in leachate/domestic wastewater ratio.

18.3.2 RECYCLING

It is one of the least expensive methods of waste treatment because recycling in a controlled reactor substantially reduces the percentage of moisture in the leachate system and distributes nutrients and enzymes uniformly between methanogens and solid/liquid juncture. Recirculation of leachate by recycling decreases methane production significantly, and optimum COD (63%–70%) is found when the volume of the recalculated leachate is reduced to 25% of initial values in an anaerobic pilot plant. The quality of leachate is also improved by recycling and treatments of leachate occur in a shorter period of time (De et al., 2019; Luo et al., 2020; Miao et al., 2019; Baderna et al., 2019; Sujetovie et al., 2019).

18.3.3 BIOLOGICAL TREATMENT

The biological treatments of leachates are cost-effective, easy to operate, show high treatment efficiency, and are being widely used around the world. This method involves the biological cycle of carbon, nitrogen, phosphorus, and other elements present in the living environment and decomposition of biological wastes with the synergetic action of microorganisms and micro-plant species present in the soil. Treatment of leachate by biological method has been mostly used for the elimination of nitrogen from LFL because it is simple to operate and is quite cheaper (Maia et al., 2015). But major drawbacks of biological treatment of leachates are a large floor area, its low treatment load, and inability to eliminate non-biodegradable plastics and refractory compounds from waste. Biological processes of purification are of two types: (i) aerobic and (ii) anaerobic. During aerobic processing, organic pollutants are oxidized with oxygen to remove the biodegradable organic pollutant by converting it into CO_2 and partially remove the ammonia nitrogen by nitrification. MBR uses aerobic oxidation method for the treatment of leachate (Zhang et al., 2019; Sadeghi et al., 2018; Silva et al., 2019; Lima et al., 2017; Mmereki et al., 2021; Sheng et al., 2020).

Treatment of MSWs by anaerobic digestion is a better choice in contrast to aerobic treatment because it produces nutrient-rich digestible substances and biogas that can be utilized in the production of heat and electricity (Figure 18.1). The organic fraction in MSWs (OFMSWs) containing high moisture content and its high biodegradability is the inherent quality suitable for anaerobic digestion as a prerequisite feedstock. Anaerobic treatment of OFMSW produces small monomers from hydrolysis of biopolymers (proteins, fats, carbohydrates, etc.), VFAs by acidogenesis, and acetic acid by acetogenesis, and methane by methanogenesis, respectively.

Anaerobic treatment consumes less energy, and produces mostly gases and very few solids; but it takes too much time; hence, the rate of reaction is slow. COD in anaerobic digestion decreases by up to 35% (Ye et al., 2018; Mahmut et al., 2021; Gell et al., 2011; Chen et al., 2019).

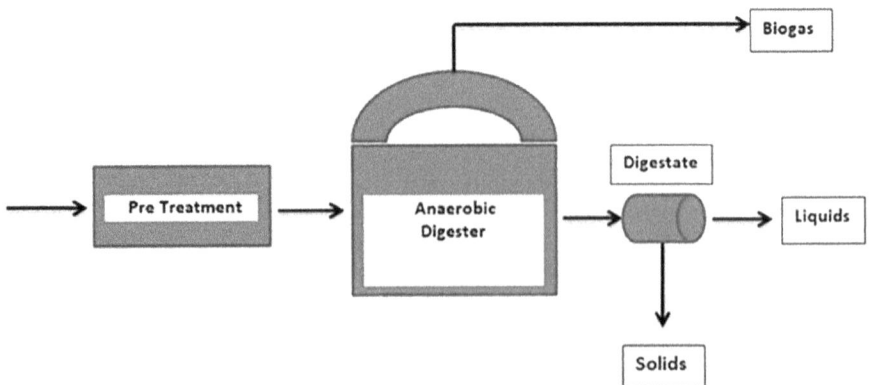

FIGURE 18.1 Diagrammatic representation of anaerobic digestion process.

18.3.4 Physicochemical Treatment

Physicochemical methods for the treatment of leachates have been applied along with the biological methods, which improve efficiency when the oxidation by biological process alone is unable to treat the waste containing heavy metals, humic, and fulvic acid. These are the most common physical and chemical methods used to treat MSWs. They include coagulation-flocculation, stripping of NH_3, chemical precipitations, removal of chemical ions by an ion exchanger, electrochemical treatment by membrane electro-dialysis, and the use of an electro-dialysis membrane to remove chemical ions. These techniques are mostly used in the removal of colloidal particles, ions, macromolecules, microorganisms, and fibres from the leachate.

18.3.5 Removal of Contaminant by Oxidation Using Common Oxidizing Agent

This method is used in the removal of soluble toxic organic compounds from the wastewater produced from LFL by using common oxidizing compounds, e.g., Cl_2, O_3, $KMnO_4$, and calcium hydrochloride, and by mineralizing recalcitrant organics in LFL, and has removed COD by around 20%–50% (Soomro et al., 2020; Yazici et al., 2019; Mondal et al., 2020; Cao et al., 2020). Recently, more effective advanced oxidation processes (AOPs) have been developed, which are an alternative to the conventional chemical oxidation treatment method, and the efficiency of chemical oxidation is increased by generating more hydroxyl radicals. In AOPs, oxidizing agent O_3 is used in combination with H_2O_2, with the use of radiation, e.g., ultraviolet (UV), or electron beam (EB), and transition metals as catalysts or photocatalysts, which have also been used in AOPs, which is shown in Table 18.3 (Yan et al., 2021; Mirko et al., 2020; Yu et al., 2018; Yan et al., 2021).

TABLE 18.3
List of Most Commonly Applied AOPs Systems in Chemical Oxidation

AOPs in Presence of Irradiation		AOPs in Absence of Irradiation	
In Homogeneous System			
1	Ozone/ultraviolet (UV)	1	Ozone /hydrogen peroxide
2	Hydrogen peroxide/ultraviolet (UV)	2	Ozone and hydroxyl radicals ions
3	With electron beam	3	Hydrogen peroxide and (Fenton's reagents, Fe^{+2})
4	With ultrasound (US)		
5	Hydrogen peroxide/ultrasound (US)		
6	With ultraviolet/ultrasound		
7	With hydrogen peroxide/ultraviolet (photo- Fenton's reagents)		
In heterogeneous systems			
8	With titanium dioxide/oxygen/ultraviolet	4	Electro-Fenton reagents
9	With titanium dioxide/hydrogen peroxide/ultraviolet		

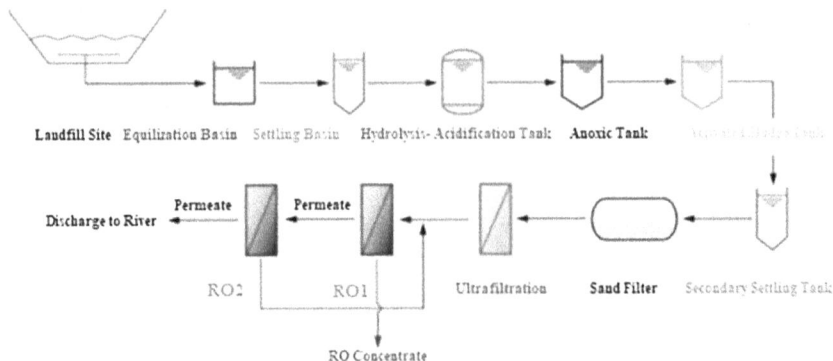

FIGURE 18.2 Sequential representation of the leachate treatment by membrane separation process.

18.4 SEPARATIONS OF LEACHATE BY MEMBRANE FILTRATION

Compared to biological treatment, membrane separation technologies, along with the AOPs, are the modern treatment methods, having stable performance for the most pollutants and used in the separation of pollutants from the leachate, especially for refractory organic compounds. By using the RO process after a rotating biological contactor, the removal rates of COD, BOD, and ammonia nitrogen concentration of LFL have improved to 99%. AOPs mainly remove refractory organic matter by degrading, mineralizing, and exhibiting a little bit of capability in removing organic particulates, especially ammonia nitrogen, total dissolved solids, and heavy metals from leachates. Therefore, in the treatment of leachate, microfiltration (MF), ultra-filtration (UF), nanofiltration (NF), and reverse osmosis (RO) membrane separation methods have been used to effectively remove contaminants. Flow diagram of treat-ment of LFL by the membrane separation process is depicted in Figure 18.2 (Ang et al., 2006; Keqiang et al., 2021; Xiaolin et al., 2021; Lei et al., 2022; Bai et al., 2021).

In MF, colloids and other suspended particles of size between 0.05 and 10 microns (i.e., fat) have been separated from leachates by using a cross-flow membrane operat-ing at a low-pressure process. MF is always used in combination with other methods of treatment, e.g., it is used in combination with chemical treatments, and in many cases, it is applied as a pre-treatment process along with other membrane separation processes (UF, NF, or RO).

But there is no significant reduction in COD by MF, and it is achieved up to 25%–35% (Tabet et al., 2002; Zhongsen et al., 2022; Mahar et al., 2016).

18.4.1 Ultrafiltration (UF)

This separation process is used to eliminate organic pollutants of larger molecular weight in a given leachate and hence provide information regarding the toxicity and recalcitrance in the leached fractions of LFL. Generally, UF is also applied as a pre-treatment method in RO (by inhibiting the fouling of RO membranes from high molecular weight pollutant particles present in leachate), as post-treatment method in biological oxidation of leachate and completely as membrane separation technology with bioreactors in MBR plants. But the removal of polluting substances is never satisfactory, and COD of treated effluent is found to be between 10% and 75% (Yu et al., 2022; Chen et al., 2021; Huiya et al., 2022).

18.4.2 NANOFILTRATION (NF)

This separation process is used to eliminate organic pollutants of larger molecular weight in a given leachate and hence provides information regarding the toxicity and recalcitrance of the leached fractions of LFL. Generally, UF is also applied as a pre-treatment method in RO (by inhibiting the fouling of RO membranes from high molecular weight pollutant particles present in leachate), as a post-treatment method in biological oxidation of leachate, and completely as a membrane separation technology with bioreactors in MBR plants. But the removal of polluting substances is never satisfactory, and the COD of treated effluent is found to be between 10% and 75% (Yu et al., 2022; Chen et al., 2021; Huiya et al., 2022).

18.4.3 TREATMENT OF LEACHATE BY USING REVERSE OSMOSIS (RO)

RO is energy-efficient separation method that requires high pressure and is mostly used in the purification of water where smaller contaminant molecules dissolved/suspended in solution are separated by concentrating solids pollutants in solution of LFL; and COD and heavy metals have been removed by more than 98% and 99%, respectively. Tubular, spiral-wounded RO membranes were initially operated at a pressure between 30 and 60 bar at ambient temperature for the purification of LFL with an 80% recovery rate. In spite of the high rejection performance in RO, two main difficulties arise: (i) the high concentration of foulants in LFL causes membrane fouling and (ii) the recovery of large volumes of semi-treated effluent. Fouling of membrane can be minimized by the tedious process of pre-treatment or cleaning of the membranes with chemicals, which certainly increases the operating cost and consequently decreases the efficiency of RO process (Fernandes et al., 2017; Bai et al., 2021; Chong et al., 2021). Various methods applied in the treatment of leachates are summarized in Table 18.4: young landfill ages (years 0–5), medium landfill (510 years), and old landfill (>10 years). SS suspended solid, TDS total dissolved solid (Figure 18.3).

18.5 CONCLUSION

Variations in concentration of pollutant in leachates with variation in time and place is the major challenge in the present era to use a universal separation technology of optimal and economical leachate treatment, which will reduce the harmful impact on the environment. The different method described here has its own merits and limitations when used in the treatment of LFLs. The better and suitable sustainable

TABLE 18.4
Summary of Different Methods Used in Leachate Treatments

Landfill Leachate Technique

Channelling	Biological Method			Physicochemical Method
Recycling	**Combined treatment with domestic sewerage.**	**Aerobic Oxidation**	**Anaerobic Oxidation**	This method involve by following processes.
• *Leachate quality improved.*	• *Remove (SS).*	• *Remove (SS) -Applicable effectively in all ages of LFL.*	• *Remove (SS)*	**1. Coagulation/flocculation**
• *Better in all age group of LFL*	• *Better in young, medium but poor in old age.*		• *Applicable effectively in all ages of LFL.*	• *Remove heavy metals, SS.*
Aerobic treatment is of two types			**It is carried out by**	• *poor in young, better in medium and old age of LFL*
A) Suspended growth of biomass is carried out by three methods			• *Suspended growth digester Anaerobic SBR.*	**2. Floatation**
• *Aerated lagoons*			• *p-flow anaerobic sludge blanket reactor (UASBR)*	**3. Chemical precipitation.**
• *Activated sludge process*				• *Removes Heavy metals and ammonical nitrogen.*
• *SBR.*				• *Better in young, medium but poor in old age of LFL.*
B) Attached growth of biomass				**4. Adsorption.**
• *oving bed biofilm reactor (MBBR)*				• *Removes Organic fractions.*
• *rickling filtration*				• *Poor in young, better in young and old age of LFL.*
B) Attached growth biomass system.				**5. Ammonia Stripping.**
It is carried out by				• *Remove ammoncal nitrogen.*
• *Anaerobic filter*				• *Effective in all age groups of LFL.*
• *Hybrid filters*				**6. Chemical oxidation.**
• *Fluidized bed filter*				• *Oxidizes organic fractions*
				• *Effective in all age groups of LFL.*
				7. Ion Exchange.
				• *Remove cations/anions, TDS.*
				• *Good in all age groups.*
				8. Electrochemical oxidation.

(Continued)

TABLE 18.4 (*Continued*)

Summary of Different Methods Used in Leachate Treatments

	Landfill Leachate Technique	
Channelling	Biological Method	Physicochemical Method
		9. Membrane **filtration** by-
		a. Microfiltration.
		• Remove for SS.
		• Applicable only in young age of LFL.
		b. Ultrafiltration.
		• Used for high molecular weight compound
		• Applicable only in young age of LFL.
		c. Nanofiltration.
		• Remove SO_4^{-2}, TDS
		• Applicable only in all ages of LFL.
		d. Reverse Osmosis.
		• Remove organic and inorganic salts, SS.
		• Applicable in all ages of LFL.

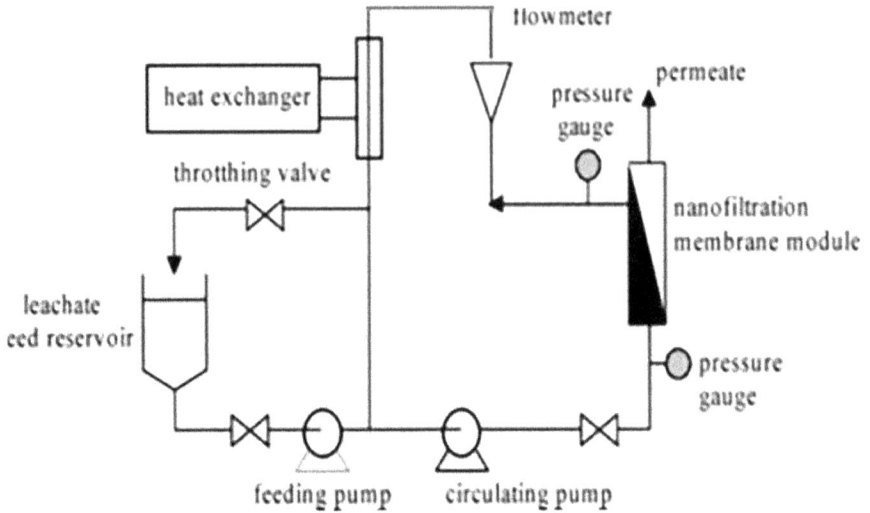

FIGURE 18.3 Schematic of the nanofiltration pilot plant.

treatment strategy adopted for treatments depends on certain major factors, which include the magnitudes and concentration of physicochemical parameters present in the leachates, e.g., COD, BOD, NH_3-nitrogen, and as well as age of leachate [98]. Different separation techniques with efficient results have been developed, by modification in conventional treatment methods combined with advanced sustainable membrane technology by which it is possible to achieve the better quality of treatment of MSW LFL. The water discharged after this treatment processes has been found to be standard as per the sanitary regulations and standards required for water discharged for sanitary and other domestic and industrial purposes by ensuring completely safety of the environment from leachate generated (Peng et al., 2017). Till late 1980s era, traditional biological treatments as well as simple physicochemical methods were mainly used in the separation and management of LFLs. Biological techniques in association of physical and chemical treatment methods are mostly used in the treatment of young leachate by removing COD, NH_3-nitrogen, organic refractory substances, and heavy metals. But hardening of discharge effluent progressively and ageing of leachates in the last three decades outdated the convention method of separation of leachates, and it has been replaced by the membrane separation technology. RO and NF provide the best method of separation and being more effective and easy to operate with rejection rates up to 98%–99% and reducing the concern of water pollution (Benedetti et al., 2020).

REFERENCES

Abuabdou, S. M. A., Ahmad, W., Aun, N. C., Bashir, M. J. K., 2020. A review of anaerobic membrane bioreactors (An MBR) for the treatment of highly contaminated landfill leachate and biogas production: Effectiveness, limitations and future perspectives. *J. Cleaner Prod.*, 255, 120215.

Aftab, B., Cho, J., Shin, H. S., Hur, J., 2020. Using EEM-PARAFAC to probe NF membrane fouling potential of stabilized landfill leachate pretreated by various options. *Waste Manage.* 102, 260–269.

Ahmad, T., Guria, C., Mandal, A., 2020. A review of oily wastewater treatment using ultrafiltration membrane: A parametric study to enhance the membrane performance. *J. Water Process Eng.*, 36, 101289.

Ahn, W. Y., Kang, M. S., Yim, S. K., Choi, K. H., 2002. Advanced landfill leachate treatment using an integrated membrane process. *Desalination*, 149(1–3), 109–114.

Akgul, D., Aktan, C. K., Yapsakli, K., Mertoglu, B., 2013. Treatment of landfill leachate using UASB-MBR-SHARON- anammox configuration. *Biodegradation,* 24(3), 399–412.

Alfaia, R.G.S.M., Nascimento, M.M.P., Bila, D.M., et al., 2019. Coagulation/ flocculation as a pretreatment of landfill leachate for minimizing fouling in membrane processes. *Desalin. Water Treat.*, 159, 53–59.

Ali, M. A. B., Rakib, M., Laborie, S., 2004. Coupling of bipolar membrane electro dialysis and ammonia stripping for direct treatment of wastewaters containing ammonium nitrate. *J. Membr. Sci.*, 244, 89–96. https://doi.org/10.1016/j.memsci.2004.07.007.

Ang, W. S., Lee, S. Y., Elimelech, M., 2006. Chemical and physical aspects of cleaning of organic-fouled reverse osmosis membranes. *J. Membr. Sci.*, 272, 198–210. https://doi.org/10.1016/j.memsci.2005.07.035

Argun, M. E., Akkus, M., Ates, H., 2020. Investigation of micropollutants removal from landfill leachate in a full-scale advanced treatment plant in Istanbul city, Turkey. *Sci. Total Environ.*, 748, 141423.

Baderna, D., Caloni, F., Benfenati, E., 2019. Investigating landfill leachate toxicity *in vitro*: A review of cell models and endpoints. *Environ. Int.*, 122, 21–30.

Bai, Z.T., Wang, Y. Q., Shan, M. J., et al., 2021. Study on anti-scaling of landfill leachate treated by evaporation method. *Water Sci. Technol.*, 84, 122–134.

Bakhshoodeh, R., Alavi, N., Oldham, C., Santos, R. M., Babaei, J., et al., 2020. Constructed wetlands for landfill leachate treatment: A review. *Ecol. Eng.*, 146, 105725.

Beekan, G., Feyessa, G., Fufa, F., et al., 2021. Household generated solid waste collection system management using arcgis: A case of Jimma town, Southwestern Ethiopia. *J. Solid Waste Technol. Manage.*, 47(2), 317–323.

Benedetti, F. M., De Angelis, M. G., Degli Esposti, M., 2020. Enhancing the separation performance of glassy PPO with the addition of a molecular sieve (ZIF-8): Gas transport at various temperatures. *Membr.*, 10, 56.

Calabrò, P. S., Gentili, E., Meoni, C., et al., 2018. Effect of the recirculation of a reverse osmosis concentrate on leachate generation: A case study in an Italian landfill. *Waste Manage.*, 76, 643–651.

Can, Y., 2019.Engineering application of pretreatment combined processes in treatment of leachate from waste-to-energy power plant (in Chinese). *Water Purif. Technol.*, 38, 102–107.

Cao, F., Liu, Q., Xiao, H. 2020. Experimental study of a humidification- dehumidification seawater desalination system combined with the chimney. *Int. J. Photoenergy,* 19, 704.

Caton, Patrick, Ernst, Howard, Flack, Karen, et al., 2021. Waste flow, recycling, and greenhouse gas emissions: A case study of the comparative environmental impact of recycling approaches on a college campus. *J. Solid Waste Technol. Manage.*, 47(3), 446–455.

Chang, H., Fu, Q., Zhong, N., 2019. Microalgal lipids production and nutrients recovery from landfill leachate using membrane photobioreactor. *Bioresoure Technol.*, 277, 18–26.

Chang, H., Quan, X., Zhong, N., 2018. High- efficiency nutrients reclamation from landfill leachate by microalgae Chlorella vulgaris in membrane photobioreactor for bio-lipid production. *Bioresour. Technol.*, 266, 374–381.

Chen, L., Chen, Z., Wang, Y., Mao, Y., Cai, Z., 2021. Effective treatment of leachate concentrate using membrane distillation coupled with electrochemical oxidation. *Sep. Purif. Technol.*, 267, 118679.

Chen, Q.B., Ren, H., Tian, Z., Sun, L., 2019. Conversion and pre-concentration of SWRO reject brine into high solubility liquid salts (HSLS) by using electrodialysis metathesis. *Sep. Purif. Technol.*, 213, 38, 587–598.

Chen, W., He, C., Gu, Z., et al., 2020a. Molecular-level insights into the transformation mechanism for refractory organics in landfill leachate when using a combined semi-aerobic aged refuse bio filter and chemical oxidation process. *Sci. Total Environ.*, 741, 140502.

Chen, W., He, C., Zhuo, X., et al., 2020b.Comprehensive evaluation of dissolved organic matter molecular transformation in municipal solid waste incineration leachate. *Chem. Eng. J.*, 400, 126003.

Chen, W., Zhuo, X., He, C., et al., 2020c. Molecular investigation into the transformation of dissolved organic matter in mature landfill leachate during treatment in a combined membrane bioreactor-reverse osmosis process. *J. Hazard. Mater.*, 397.

Chuah, C. Y., Binte, S. N., Anwar, M., 2021. Scaling-up defect-free asymmetric hollow fiber membranes to produce oxygen-enriched gas for integration into municipal solid waste gasification process. *J. Membr. Sci.*, 640, 119787.

De, A. R., Moraes Costa, A., de Almeida Oroski, F., 2019. Evaluation of coagulation–flocculation and nanofiltration processes in landfill leachate treatment. *J. Environ. Sci. Heal., Part A* 54, 1091–1098.

Fernandes, A., Labiadh, L., Ciriaco, L., 2017. Electro-Fenton oxidation of reverse osmosis concentrate from sanitary landfill leachate: Evaluation of operational parameters. *Chemosphere,* 184, 1223–1229.

Fong, S. Y., Lau, W. J., Tan, N. H. T., Chin, N., Chew, K. H., 2022. A case study of industrial MBR process for poultry slaughterhouse wastewater treatment. *J. Membr. Sci. Res.*, 8, (1).

Gao, J. L., Oloibiri, V., Chys, M., et al., 2015. The present status of landfill leachate treatment and its development trend from a technological point of view. *Rev. Environ. Sci. Bio-Technol.*, 14, 93–122.

Gell, K., Van, G. J., Cayuela, M. L., 2011. Residues of bioenergy pro duction chains as soil amendments: Immediate and temporal phytotoxicity. *J. Hazard. Mater.*, 186, 2017–2025.

Giorgio, B., Stefano, M., Franco, H. G., et al., 2021 Implementation of circular economy in the management of municipal solid waste in an Italian medium-sized city: A 30-years lasting history. *Waste Manage.*, 126 821–831.

Gu, H., Geng, H., Wang., D., Li, W., 2021. A new method for the treatment of kitchen waste: Converting it into agronomic sprayable mulch film. *Waste Manage.*, 126, 527–535.

Haizi, W., Xinming, P., Shibin, Z., 2021. Simulation analysis of implementation effects of construction and demolition waste disposal policies. *Waste Manage.*, 126, 684–693.

Hashemi, H., Hajizadeh, Y., Amin, M.M., et al., 2015. Macropollutants removal from compost leachate using membrane separation process. *Desalin. Water Treat.*, 10, 1080.

Hashisho, J., El-Fadel, M. 2016. Membrane bioreactor technology for leachate treatment at solid waste landfill. *Rev. Environ. Sci. Bio/Technol.*, 15, 441–463.

Huiya, W., Keqiang, D., 2022. Effect of self-made TiO2 nanoparticle size on the performance of the PVDF composite membrane in MBR for landfill leachate treatment. *J. Membr. Sci.*, 12, 216.

Jessica, M. H., Brenden, R. B., Rachel, B., 2022. Alternating current for the selective electrodeposition of cadmium, iron, and chromium method development for simulated industrial waste water. *Int. J. Water Wastewater Treat.*, 8(1), 1–5.

Jorge, M., Torrente, V., Maddalena, R., et al., 2021. Identification of inference fallacies in solid waste generation estimations of developing countries- A case-study in Panama. *Waste Manage.*, 126, 454–465.

Keqiang, W., Jianglin, L., Yiyou, L., 2021. Supervision behaviors of customs supervisors on solid waste import in Shanghai, People's Republic of China. *Waste Manage. Res.*, 40(4), 429–438.

Lei, C., Yufeng, Z., LeiNi, X., 2022. A novel loosely structured nanofiltration membrane bioreactor for wastewater treatment: Process performance and membrane fouling. *J. Membr. Sci.*, 644, 120128.

Li, G. D., Wang, W., Du, Q. Y., 2010. Applicability of nanofiltration for the advanced treatment of landfill leachate. *J. Appl. Polym. Sci.*, 116, 2343–2347.

Li, H., Zhou, S., Sun, Y., Feng, P., Li, J., 2009. Advanced treatment of landfill leachate by a new combination process in a full-scale plant. *J. Hazard Mater.*, 172, 408–415. https://doi.org/10.1016/j.jhazmat.2009.07.034.

Lima, L.S.M.S., De Almeida, R., Quintaes, B.R., et al., 2017. Analysis of quantification methodologies for humic substances in leachates from solid waste landfills. *Rev. Ambient. Água.*, 12, 87–98.

Lucena, L. G., de Carvalho, N. A., 2021. A systematic review on fenton optimization for leachate treatment via response surface methodology from 2005 to 2015. *J. Solid Waste Technol. Manage.*, 47(3), 409–416.

Luo, H., Zeng, Y., Cheng, Y., et al., 2020. Recent advances in municipal landfill leachate: a review focusing on its characteristics, treatment, and toxicity assessment. *Sci. Total Environ.*, 703, 135468.

Mahar, R.B., Sahito, A.R., Yue, D., etal., 2016. Modeling and simulation of landfill gas production from pretreated MSW landfill simulator. *Front. Environ. Sci. Eng.*, 10, 159–167.

Mahmut, A., Hamit, O., Erol, İ., 2021. Full scale sanitary landfill leachate treatment by MBR: Flat sheet vs. hollow fiber membrane. *J. Membr. Sci.*, 7(2), 118–124.

Maia, S.I., Restrepo, J. J. B., Castilhos, Junior, et al., 2015. Evaluation of the biological treatment of landfill leachate on a real scale in the Southern Region of Brazil. *Eng. Sanit. Ambient.*, 20, 665–675.

Mavukkandy, M.O., Bilad, M.R., Giwa, A., et al., 2016. Leaching of PVP from PVDF/PVP blend membranes, impacts on membrane structure and fouling in membrane bioreactor. *J. Mater. Sci.*, 51, 4328–4341.

Miao, L., Yang, G., Tao, T., et al., 2019. Recent advances in nitrogen removal from landfill leachate using biological treatments–A review. *J. Environ. Manag.*, 235, 178–185.

Mirko F., Francesca M., Enrico D, 2020. Progress of membrane engineering for water treatment. *J. Membr. Sci.*, 6, 3, 269–279.

Mmereki, D, Velempini, K, Mosime, S.L., 2021. Status of municipal solid waste management policy implementation in developing countries: Insights from Botswana. *J. Solid Waste Technol. Manage.*, 47(1), 46–55.

Mohammad, Arif, Goli, Venkata Siva Naga Sai, Chembukavu, Agnes Anto, et al., 2021. DecoMSW: A methodology to assess decomposition of municipal solid waste for initiation of landfill mining activities. *J Solid Waste Technol. Manage.*, 47(3), 465–481.

Mondal, P., Yadav, B.P., Siddiqui, N.A. 2020. Removal of lead from drinking water by bioadsorption technique: An eco-friendly approach. *Nat. Environ. Poll. Techn.*, 19(4), 1675–1682.

Peng, W., Pivato, A., 2017. Sustainable management of digestate from the organic fraction of municipal solid waste and food waste under the concepts of back to earth alternatives and circular economy. *Waste Biomass Valori*. E-Publication ahead of print, https://doi.org/10.1007/s12649-017-0071-2.

Ribera-Pi, J., Badia-Fabregat, M. E., et al., 2020. Decreasing environmental impact of landfill leachate treatment by MBR, RO and EDR hybrid treatment. *Environ. Technol.*, 42, 1–15.

Roy, C., Matteo, C., Leo, F. M., 2021. Municipal waste management: A complex network approach with an application to Italy. *Waste Manage.*, 126, 597–607.

Sadeghi, M., Fadaei, A., Tadrisi, M., et al., 2018. Performance evaluation of a biological landfill leachate treatment plant and effluent treatment by electrocoagulation. *Desalin. Water Treat.*, 115, 82–87.

Samarasinghe, S., Chuah, C.Y., Karahan, H. E., Sethunga, G., 2020. Enhanced O2/N2 separation of mixed-matrix membrane filled with pluronic compatibilized cobalt phthalocyanine particles. *Membr.* 10, 75.

Santos, A, Ma, W, Judd, S.J., 2010. Membrane bioreactors. Two decades of research and implementation. *Desalination*, 273(1), 148–154.

Sheng, B., Wang, D., Liu, X., 2020. Taxonomic and functional variations in the microbial community during the upgrade process of a full-scale landfill leachate treatment plant-from conventional to partial nitrification-denitrification. *Front. Environ. Sci. Eng.*, 14, 93. https://doi.org/10.1007/s11783-020-1272-7.

Silva, N.C.M., Moravia, W.G., Amaral, M.C.S., et al., 2019. Evaluation of fouling mechanisms in nanofiltration as a polishing step of yeast MBR-treated landfill leachate. *Environ. Technol.*, 40, 3611–3621.

Soomro, G.S., Qu, C., Ren, N., Meng, S., 2020. Efficient removal of refractory organics in landfill leachate concentrates by electrocoagulation in tandem with simultaneous electro-oxidation and in-situ peroxone. *Environ. Res.*, 183, 109249.

Srinivasaiah, R., Swamy, D.R., Krishna, A.S., 2021. Various models used in analysing municipal solid waste generation–A review. *J. Solid Waste Technol. Manage.*, 47(3), 569–578.

Sujetovien, G., Smilgaitis, P., Dagili, R., et al., 2019. Metal accumulation and physiological response of the lichens transplanted near a landfill in central Lithuania. *Waste Manage.*, 85, 60–65.

Suresh, A., Hill, G. T., Hoenig, E., Liu, C., 2021. Electrochemically mediated deionization: A review. *Mol. Syst., Des Eng.*, 6, 25–51.

Tabet, K., Moulin, P., Vilomet, J.D., Amberto, A., Charbit, F., 2002. Purification of landfill leachate with membrane processes: Preliminary studies for an industrial plant. *Separat. Sci. Technol.*, 37, 1041–1063.

Tang, S.J., Wang, Z.W., Wu, Z.C., etal., 2010. Role of dissolved organic matter (DOM) in membrane fouling of membrane bioreactors for municipal wastewater treatment. *J. Hazard. Mater.*, 178, 377–384.

Tara, R. Z., Frances, F., Vanessa, C., June, R., 2021. A systematic review on informal waste picking: Occupational hazards and health outcomes. *Waste Manage.*, 126, 291–308.

Wang, H., Cheng, Z., Sun, Z., et al., 2020a. Molecular insight into variations of dissolved organic matters in leachates along China's largest A/O-MBR-NF process to improve the removal efficiency. *Chemosphere*, 243,125354.

Wang, H., Ge, D., Cheng, Z., et al., 2020b. Improved understanding of dissolved organic matter transformation in concentrated leachate induced by hydroxyl radicals and reactive chlorine species. *J. Hazard. Mater.*, 387, 121702.

Wang, K., Wang, S., Zhu, R., et al., 2013. Advanced nitrogen removal from landfill leachate without addition of external carbon using a novel system coupling ASBR and modified SBR. *Bioresour. Technol.*, 134, 212–218.

Wang, H., Jiang, W., Ma, S., 2014. Design and operation of the treatment of refuse leachate (in Chinese). *Ind. Water Treat.*, 34, 87–89.

Woldeyohans, A. M., Worku, T., Kloos, H., et al., 2014. Treatment of leachate by recirculating through dumped solid waste in a sanitary landfill in Addis Ababa, Ethiopia. *Ecol. Eng.*, 73, 254–259.

Xiaolin, J., Kuiling, L., Baoqiang, W., et al., 2021. Membrane cleaning in membrane distillation of reverse osmosis concentrate generated in landfill leachate treatment. *Water Sci. Technol.*, 1, 244.

Yan, Z., Jiang, Y., Chen, X., et al., 2021. Evaluation of applying membrane distillation for landfill leachate treatment. *Desalination*, 520, 115358.

Yan, Z., Qu, F., Liang, H., et al., 2021. Effect of biopolymers and humic substances on gypsum scaling and membrane wetting during membrane distillation. *J. Membr. Sci.*, 617, 118638.

Yazici, G. S., 2019. Optimization of COD removal from leachate nanofiltration concentrate using H_2O_2/Fe^{2+}/heat–Activated persulfate oxidation processes. *Process Saf. Environ. Prot.*, 126, 7–17.

Ye, Y., Ngo, H.H., Guo, W., 2018. A critical review on ammonium recovery from wastewater for sustainable wastewater management. *Bioresour. Technol.*, 268, 749–758.

Yin, J., Roso, M., Boaretti, C., et al., 2021. PVDF-TiO2 core-shell fibrous membranes by microwave hydrothermal method: Preparation, characterization, and photocatalytic activity. *J. Environ. Chem. Eng.*, 9, 106250.

Yu, H., Du, C., Qu, F., et al., 2022. Efficient biostimulants for bacterial quorum quenching to control fouling in MBR. *Chemosphere,* 286, 131689.

Yu, W., Liu, T., Crawshaw, J., Liu, T., Graham, N., 2018. Ultrafiltration and nanofiltration membrane fouling by natural organic matter: mechanisms and mitigation by preozonation and pH. *Water Res.*, 139, 353–362.

Yuri, A. R. L., Victor, R. M., et al., 2021. A survey on experiences in leachate treatment: Common practices, differences worldwide and future perspectives. *J. Environ. Manage.*, 288, 112475.

Zhang, J., Xiao, K., Huang, X. 2020. Full-scale MBR applications for leachate treatment in China: Practical, technical, and economic features. *J. Hazard. Mater.*, 389, 122138.

Zhang, L., Lavagnolo, M. C., Bai, H., et al., 2019. Environmental and economic assessment of leachate concentrate treatment technologies using analytic hierarchy process. *Resour. Conserv. Recycl.*, 141, 474–480.

Zhongsen, Y., Zhenyu, L, Xiaolei, C., et al., 2022. Membrane distillation treatment of landfill leachate: Characteristics and mechanism of membrane fouling. *Separat. Purif. Technol.*, 289, 120787.

19 Study of Ultra-acoustic Behavior of Aspartic Acid in Water and Aqueous Potassium Sorbate

An Insight into Interactional Features

*Kshirabdhitanaya Dhala, Sulochana
Singh, and Malabika Talukdar*

CONTENTS

19.1 Rationale 343
19.2 Materials and Methods 345
    19.2.1 Materials 345
    19.2.2 Methods 345
19.3 Documentation and Discussion 345
    19.3.1 Density (ρ) and Ultrasonic Velocity (U) 346
    19.3.2 Isentropic Compressibiwlity (K_s) 346
    19.3.3 Apparent Molar Isentropic Compressibility $\left(K_{s,\phi}\right)$ and
            Isothermal Compressibility (K_T) 349
    19.3.4 Transfer Parameter of Limiting Apparent Molar Isentropic
            Compressibility $\left(\Delta_{tr}K_{s,\phi}^{0}\right)$ 351
    19.3.5 Pair and Triplet Interaction Coefficients 352
    19.3.6 Molar Free Length (L_f) and Molar Free Volume (V_f) 352
    19.3.7 Internal Pressure 355
    19.3.8 Relative Association (R_A) and Acoustic Impedance (Z) 356
19.4 Inferences 356
References 357

19.1 RATIONALE

Amino acids are bifunctional biomolecules having amino ($-NH_2$) and carboxylic ($-COOH$) groups attached to the same carbon atom of an aliphatic carbon backbone. This carbon chain is also connected with a side chain that is specific for each

DOI: 10.1201/9781003359784-19

amino acid. Peptides and proteins are macromolecules made up of long chains of amino acids. Each amino acid, whether synthesized metabolically (non-essential amino acids) or absorbed from amino acid-rich food (essential amino acids), has a specific function in our body starting from maintaining a healthy nervous system, production of different hormones, enzymes, and antibodies to promoting muscle growth.

Due to the presence of ionic sites, polar groups, H-bonding, and nonpolar hydrophobic parts in amino acids, a wide range of interesting physicochemical properties are expected for these biomolecules. Study of molecular interactions of amino acids in different solvent systems has opened a horizon of research in chemical, biochemical, pharmaceutical, medicinal, and many other fields of science. Electrostatic interactions, hydrogen bonding, ionic/hydrophilic-hydrophilic, and hydrophobic-hydrophobic interactions in solutions of amino acids are found to affect the functions of proteins. The presence of electrolytes as co-solute in aqueous amino acid solutions largely affects the structural properties, solubility, stability, and activity of proteins. As it is important to understand the effect of solvent on biomolecular behavior of amino acids, experimental investigations on volumetric, viscometric, and acoustic behavior of amino acids and in aqueous solutions of various inorganic and organic salts, sugars, ionic liquids, solutions of drug, and many other organic compounds have been made [1–10]. These physical parameters lead to derive many thermo-physical parameters helpful to understand the interactions among the different species present in the solutions. Harsh Kumar and his team reported volumetric and acoustic properties of amino acids in different mixed solvent systems and interpreted the molecular interactions occurring in those solutions by using co-sphere overlapping model [11,12]. Volumetric and compressibility behaviors of L-valine in aqueous polyethylene glycol solutions have been reported by Tahereh Moradian and his coworkers [13]. Rekha Gabba et al. have studied thermodynamic properties of amino acids in aqueous solutions of ionic liquids by volumetric and acoustic methods [14]. Density and acoustic studies of α-amino acids glycine, L-alanine, L-valine, and L-leucine in aqueous solutions of 1-butyl-3-propylimidazolium bromide were studied by Harpreert Kumar and coworkers [15]. We have chosen a non-essential amino acid with a hydrophilic side chain attached to another (–COOH) group, aspartic acid (Asp), as a substance of our research. Aqueous solutions of potassium sorbate (PS) were used as a medium of aspartic acid. PS is widely used as a food preservative. According to U. S. Food and Drug Administration review, PS is generally recognized as a safe preservative. Asp exists as zwitterions in the form of $-NH_3^+$ and COO^- in the solution. Another carboxyl group is present in the side chain of the Asp structure and is ionized at physiological pH (pH 7.4). Due to the acidic nature of the side chain (pK_a 3.9), it is highly hydrophilic and can form ionic bonds with almost any metal ion. Because of the presence of amino group and two carboxyl groups in Asp, a wide range of charge distribution takes place and aspartic acid participates in dipole interaction with water.

The sorbate anion has an unsaturated carbon chain, which may have hydrophobic interactions with the alkyl backbone of the Asp together with the interactions between ionic and polar groups of the molecules. This investigation is based on measurements of density and velocity of ultrasonic sound passing through the solutions with varying concentrations of Asp (0.01–0.05 mol/kg of solvent) and PS (0.5–2.5 mol/kg of water) at a temperature of 298.15 K and at standard atmospheric pressure (1.01×10^5 Pa).

19.2 MATERIALS AND METHODS

19.2.1 MATERIALS

Aspartic acid (CAS No. 56-84-8) was supplied by Loba Chemie Pvt Ltd, Mumbai, India. Potassium sorbate (CAS No. 24634-61-5) was procured from Sigma-Aldrich Chemicals Pvt Ltd., India. Both chemicals were used without further purification except drying over anhydrous $CaCl_2$ in a desiccator for over 24 hours.

19.2.2 METHODS

All solutions were prepared with deionized water (specific conductance $\sim 10^{-6}$/ (S/cm). 0.5, 1.0, 1.5, and 2.0 and 2.5 molal PS solutions were first prepared. Asp solutions of five different molality ($m = 0.01$–0.05) were prepared in water and (water + PS) solutions The uncertainty of the molality of the solutions of Asp and PS is $\pm 1 \times 10^{-3}$/(mol/kg).

A digital electronic balance of Citizen-Synchronics Electronics Pvt. Ltd with an accuracy of ±0.01/mg was used.

The density of the solvents (water and aqueous PS) and the solutions {(Asp + water) and (Asp + PS + water)} has been measured with the help of a Density analyser (DMA™ 5,000 M by Anton Paar). The uncertainty in density is ±0.12/(kg/m³).

Methods of solution preparation, density, and viscosity measurement of Asp in aqueous PS are described in detail at some other places [16].

Acoustic measurements were made at 298.15 K with an ultrasonic interferometer (Model M-05, by Mittal Enterprises, New Delhi) by measuring the wavelength (λ) of ultrasonic sound at a frequency (f) of 2 MHz.

Ultrasonic velocity (U) in different media was calculated by the following formula:

$$U = f \times \lambda \tag{19.1}$$

The ultrasonic velocity values are with ±0.5/(m/s) accuracy.

A thermostatically controlled water bath was used to maintain the experimental temperature at 298.15 with an accuracy of ±0.01/K.

19.3 DOCUMENTATION AND DISCUSSION

At lower concentration of aqueous PS, the dissociated salt particles, i.e., K^+ and $C_6H_7O_2^-$, occupy the interstitial spaces of water and restrict the solubility of the amino acid. However, solubility appreciably increases with increasing concentration of PS in the solution, which may be due to the increased electrostatic interaction between the zwitterionic centers of amino acid and the ions of the dissociated salt. Tightly packed clusters of ionic particles are easily trapped inside the solvation cages.

The expected interactions between the ionic and polar groups and hydrocarbon segments of Asp and PS are enlisted as follows:

- ions and polar groups of Asp and PS experience ion-hydrophilic and hydrophilic-hydrophilic interactions with polar water molecules
- ions and nonpolar hydrocarbon segments of Asp and PS experience ion-hydrophobic interactions with each other
- nonpolar segments of Asp and sorbate ions experience hydrophobic-hydrophobic interactions with each other

19.3.1 Density (ρ) and Ultrasonic Velocity (U)

Density (ρ) and ultrasonic velocity (U) values of solutions of Asp of different concentrations in water and (water + PS of different compositions) are documented in Table 19.1. The increased interactions between the ions of Asp and the ions of dissociated PS with polar water molecules in the solutions are validated by the rising order of density with concentration of the amino acid, as well as of PS. Due to strong electrostatic attraction between the components of the solutions, contraction of volume takes place, making the density of the solutions rise.

Ultrasonic sound wave when propagates through a solution, influences the structural behavior of the medium and the interactions between the specimens in the solution enhance with a concomitant rise in sound velocity with concentrations of aspartic acid and PS in solution.

19.3.2 Isentropic Compressibility (K_S)

Solvation layers are formed around solute particles in a solution. Similarly, hydration shells surrounding zwitterionic centers of Asp and K^+ and $C_6H_7O_2^-$ of dissociated PS form in the solutions. The water molecules in the hydration spheres are highly electrostricted by the charges on the ions and amino acid and are compressed to their maximum extent. The compressibility of the bulk water molecules is much higher than the water molecules in the vicinity of the hydration shells. Isentropic compressibility (K_s) of the solutions can be calculated by using the well-known Newton-Laplace's equation [17]:

$$K_s = 1/\left(\rho U^2\right) \tag{19.2}$$

The values thus obtained are presented in Table 19.1. As an obvious outcome, compressibility of the solution steadily decreases with concentration of Asp in aqueous medium. Compressibility further decreases in aqueous PS media with concentration of potassium PS as well. This phenomenon strongly supports the existence of ion-dipole interaction between solute and solvent. In Asp solution in aqueous PS media, ion-ion electrostatic attraction is also expected between the carboxylate ions and zwitterionic centers of Asp and the potassium and sorbate ions of dissociated PS. At higher concentrations of Asp and PS, the solvated particles approach

TABLE 19.1

Values of Concentration of Asp (m), Density ($c\rho$), Sound Velocity (U), Viscosity (η), Isentropic Compressibility (K_s), Apparent Molar Isentropic Compressibility ($K_{s,\phi}$), and Isothermal Compressibility K_T in Water and Aqueous of 0.5, 1.0, 1.5, 2.0 and 2.5 m PS Solutions at 298.15 K

m (mol/kg)	ρ (kg/m³)	U (m/s)	$\eta \times 10^3$ (kg/ms)	$K_s \times 10^{10}$ (m²/N)	$K_{s,\phi} \times 10^{13}$ (m⁵/mol N)	$K_T \times 10^{15}$ (m²/N)
			Water			
0.011	997.74	1,500.0	0.91	4.45	−9.82	6.06
0.019	998.53	1,508.0	0.93	4.40	−7.33	5.99
0.030	999.34	1,512.0	0.94	4.38	−5.71	5.95
0.039	1000.17	1,516.0	0.96	4.35	−4.90	5.91
0.051	1001.00	1,520.0	0.98	4.32	−4.41	5.87
			Water + 0.499 m PS			
0.009	1,018.39	1,549.2	1.05	4.09	−8.48	5.53
0.021	1,018.76	1,555.2	1.06	4.06	−5.67	5.48
0.032	1,019.18	1,562.0	1.07	4.02	−4.87	5.43
0.040	1,019.56	1,566.8	1.07	4.00	−4.20	5.40
0.051	1,019.92	1,568.4	1.08	3.99	−3.47	5.38
			Water + 1.013 m PS			
0.012	1,037.42	1,584.4	1.11	3.84	−7.11	5.16
0.019	1,037.81	1,589.2	1.13	3.82	−4.56	5.12
0.031	1,038.33	1,594.0	1.16	3.79	−3.74	5.09
0.041	1,038.71	1,598.8	1.18	3.77	−3.30	5.05
0.049	1,039.13	1,602.0	1.20	3.75	−2.90	5.03
			Water + 1.499 m PS			
0.009	1,053.65	1,634.0	1.21	3.55	−6.05	4.75
0.021	1,054.05	1,638.8	1.24	3.53	−3.92	4.72
0.032	1,054.44	1,644.0	1.27	3.51	−3.26	4.69
0.040	1,054.86	1,648.0	1.29	3.49	−2.80	4.66
0.051	1,055.33	1,651.2	1.33	3.48	−2.47	4.64
			Water + 2.011 m PS			
0.012	1,069.08	1,659.2	1.40	3.40	−4.10	4.52
0.021	1,069.52	1,664.4	1.44	3.38	−2.96	4.49
0.032	1,069.96	1,668.0	1.47	3.36	−2.38	4.46
0.042	1,070.44	1,670.0	1.50	3.35	−1.94	4.45
0.049	1,070.92	1,672.8	1.53	3.34	−1.73	4.43
			Water + 2.501 m PS			
0.009	1,082.06	1,670.8	1.51	3.31	−4.41	4.38
0.022	1,082.51	1,673.2	1.55	3.30	−2.57	4.37
0.030	1,083.00	1,676.0	1.58	3.29	−2.01	4.35
0.039	1,083.47	1,680.0	1.62	3.27	−1.84	4.33
0.051	1,083.96	1,684.0	1.65	3.25	−1.73	4.30

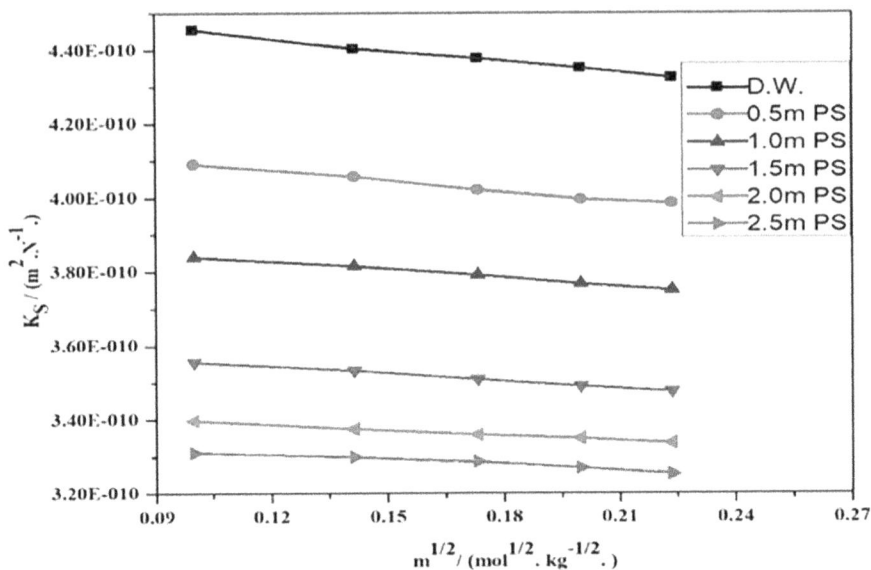

FIGURE 19.1   Plot of K_s versus $m^{1/2}$ of Asp solutions in water and aqueous PS (0.5–2.5 m) at 298.15 K.

so close to each other that overlapping of the solvation shells takes place [18]. At this condition, hydrophobic-hydrophobic interactions develop between the alkyl backbones of amino acid and sorbate ions. Therefore, variation of isentropic compressibility is a result of interactions of solute and co-solute with each other and with water. The linear relationship of K_s with molality (m) of Asp in water and (water + PS) media is expressed in form of equation (19.3) and is graphically presented in Figure 19.1.

$$K_s = K_s^0 + S_k \sqrt{m} \qquad (19.3)$$

Limiting isentropic compressibility $\left(K_s^0\right)$ is the intercept of the straight line and signifies the compressibility of the solution at infinite dilution, i.e., in the presence of infinitesimal amount of solute. Only solute-solvent interactions are expected, and the K_s value of the solution tends to merge with that of the solvent. However, in aqueous PS media in addition to the interaction of sorbate with water, there also exists sorbate-to-sorbate pairwise interaction. The slope of the straight line S_K is helpful to understand the solute-solute interaction. Solute-solute interaction is found to be much weaker compared to solute-solvent interaction for all the solutions under study as suggested by the values of K_s^0 and S_K displayed in Table 19.2. Overlapping of hydration shells and hydrophobic-hydrophobic interactions between Asp and PS are indicated by the negative values of S_K.

TABLE 19.2

Values of K_s^0, $K_{s,\phi}^0$, $S_{K,\phi}$, and $\Delta_{tr}K_{s,\phi}^0$ for Asp in Water and Aqueous Solutions of 0.5, 1.0, 1.5, 2.0, and 2.5 m PS at 298.15 K

Composition of aq. Potassium sorbate (mol/kg)	$K_s^0 \times 10^{10}$ (m²/N)	$S_k \times 10^{11}$ (m²/N mol$^{\frac{1}{2}}$ kg$^{\frac{1}{2}}$)	$K_{s,\phi}^0 \times 10^{13}$ (m⁵/mol N)	$S_{k,\phi} \times 10^{12}$ (m⁵/N mol$^{\frac{3}{2}}$ kg$^{\frac{1}{2}}$)	$\Delta_{tr}K_{s,\phi}^0 \times 10^{13}$ (m⁵/mol N)
0.000	4.56±0.006	−0.10±0.036	−13.86±0.083	4.43±0.048	
0.499	4.18±0.010	−9.01±0.060	−11.79±0.098	3.85±0.056	2.07
1.012	3.92±0.005	−7.44±0.032	−9.82±1.041	3.28±0.060	4.04
1.499	3.62±0.003	−6.53±0.022	−8.39±0.084	2.80±0.049	5.46
2.011	3.44±0.003	−4.83±0.018	−5.85±0.031	1.92±0.020	8.01
2.501	3.36±0.009	−4.61±0.057	−6.03±0.090	2.09±0.052	7.83

19.3.3 APPARENT MOLAR ISENTROPIC COMPRESSIBILITY ($K_{S,\phi}$) AND ISOTHERMAL COMPRESSIBILITY (K_T)

$K_{s,\phi}$ is the difference between the compressibility of water in the vicinity of the hydration shells and the bulk solution. This parameter can be computed by using the following equation:

$$K_{s,\phi} = \left(K_s \rho_0 - K_s^0 \rho\right)/m\rho\rho_0 + MK_s/\rho \qquad (19.4)$$

ρ and ρ_0 are the densities of solutions and solvent, respectively. M is the molar mass of Asp. Apparent molar compressibility provides information about the structural interaction in solution phase. There is a continuous rise in the values of apparent molar isentropic compressibility $\left(K_{s,\phi}\right)$ with concentration of Asp for all the systems. The negative values of $\left(K_{s,\phi}\right)$ signify the electrostrictive compression of the water molecules surrounding the solute particles [19,20]. With higher amount of solute particles in the solution, more water molecules move away from the bulk and enter the solvation shells. Due to the caging effect of the solvent, the solute ions are trapped at the core of the solvation spheres in tight clusters that offer electrostriction to water molecules in the close proximity of the solvation spheres. As the bulk is losing water molecules, its compressibility is decreasing and the gap between the compressibility of water molecules in these two environs narrows down. Hence, the increasing values of $\left(K_{s,\phi}\right)$ in water and all compositions of PS solutions, as apparent in Table 19.1, indicate strong interactions between the species present in the solutions. The variation of $K_{s,\phi}$ with concentration of Asp and PS is shown in Figure 19.2.

In the presence of infinitesimally small amount of solute, i.e., at infinite dilution, apparent molar isentropic compressibility, denoted by $K_{s,\phi}^0$, can be evaluated from the intercept of the straight line drawn by the following equation (19.5):

$$K_{s,\phi} = K_{s,\phi}^0 + S_{k,\phi}\sqrt{m} \qquad (19.5)$$

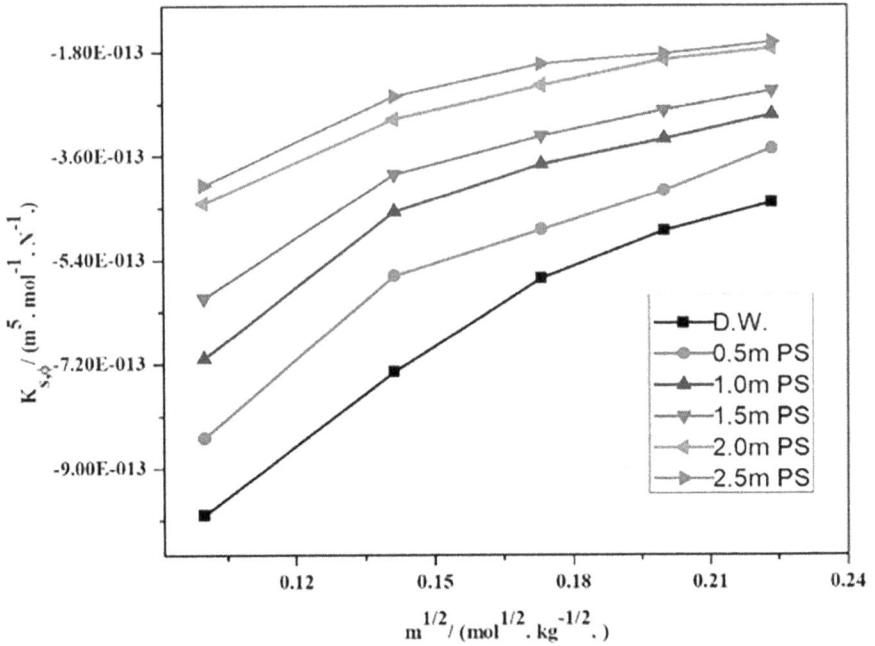

FIGURE 19.2   Plot of $K_{s,\phi}$ versus $m^{1/2}$ of Asp solutions in water and aqueous PS (0.5–2.5 m) at 298.15 K

$K^0_{s,\phi}$ values are of negative magnitude and are found to be less negative in solutions of Asp in (PS + water) than in water. At infinite dilution, the interactions in binary solutions are almost entirely between solute and solvent particles. Contrarily, for ternary solutions, e.g., (Asp + PS + water), interactions between different ionic/polar parts and nonpolar hydrocarbon segments of Asp and PS arise, which reduces the value of solute-solvent interactions. However, in the presence of more potassium and sorbate ions in solution, solute-solvent interactions start increasing again. Positive values of $S_{k,\phi}$ in water and aqueous PS media support the presence of solute-solute interactions even at infinite dilution.

Values of K^0_s, S_K, $K^0_{s,\phi}$, and $S_{k,\phi}$ along with their standard errors and $\Delta_{tr}K^0_{s,\phi}$ are given in Table 19.2.

Molar compressibility in isothermal condition (K_T) is another important parameter that further supports the existence of various interactions between the ionic specimens present in the solutions. Variation of K_T with composition of the solutions, displayed in Figure 19.3, resembles that of K_S. Values of K_T are calculated by equation (19.6) [21] and are documented in Table 19.2.

$$K_T = \left(17.1 \times 10^{-4}\right)\big/\left(T^{4/9}\rho^{4/3}u^2\right) \tag{19.6}$$

It is observed in Figure 19.3 that variation of K_T with concentrations of Asp and PS follows the same trend as that of K_s.

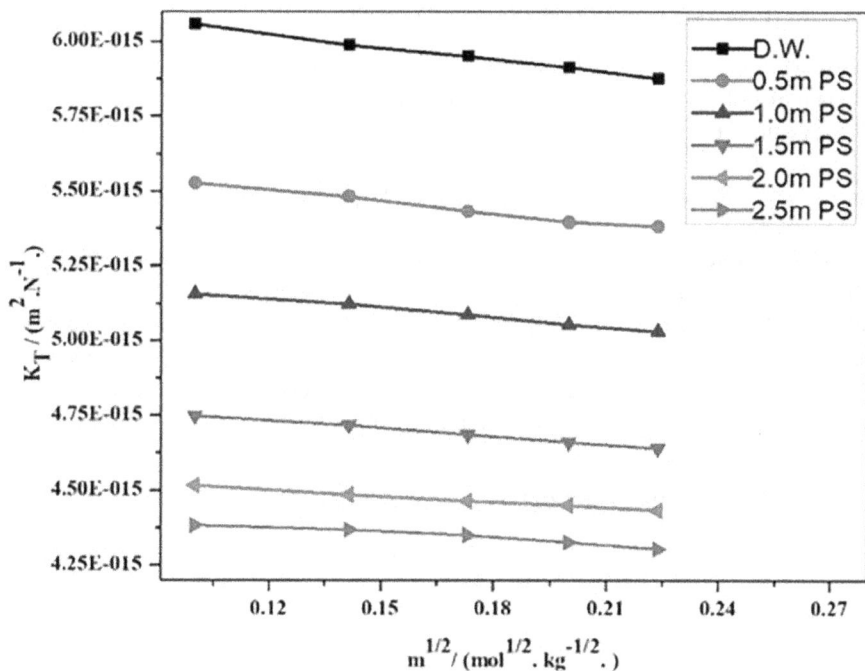

FIGURE 19.3   Plot of K_T versus $m^{1/2}$ of Asp solutions in water and aqueous PS (0.5–2.5 m) at 298.15 K.

19.3.4   TRANSFER PARAMETER OF LIMITING APPARENT MOLAR ISENTROPIC COMPRESSIBILITY $\left(\Delta_{tr}K^0_{s,\phi}\right)$

$\Delta_{tr}K^0_{s,\phi}$ bears a direct relationship with the molality of the co-solute (PS) according to the following equation:

$$\Delta_{tr}K^0_{s,\phi} = K^0_{s,\phi}\left(\text{in aqueous potassium sorbate}\right) - K^0_{s,\phi}\left(\text{in water}\right) \qquad (19.7)$$

There is a continuous increase in transfer values $(\Delta_{tr}K^0_{s,\phi})$ of Asp from water to aqueous PS solutions with molal concentrations of potassium sorbate (with an exception of 2.5 molal aqueous PS). It is evident from Table 19.2 that $\Delta_{tr}K^0_{s,\phi}$ is positive for all compositions of aqueous PS.

Positive transfer values indicate the dominance of ion-ion interactions between zwitterionic centers of amino acid and ions of dissociated PS. The increase in transfer values with concentration of PS must be due to the reduction in electrostriction of water molecules by Asp at higher concentration of PS and increase in compressibility of bulk water molecules gives rise to positive transfer values.

19.3.5 PAIR AND TRIPLET INTERACTION COEFFICIENTS

In order to analyze and calculate the interaction coefficients between two or more solute molecules, the following equation can be utilized and ion pair and triplet interactions can be estimated [22–24].

$$\Delta_{tr} K_{s,\phi}^0 = 2K_{AB} m_B + 3K_{ABB} m_B^2 + \ldots \tag{19.8}$$

A and B denote the amino acid and PS, respectively, and m_B is the molality of PS. Corresponding pair and triplet interaction coefficients K_{AB} and K_{ABB} are the intercept and slope of the straight line drawn from equation (19.8) and are found to be $2.20 \times 10^{-13} / (m^5 mol^{-2} N^{-1} kg)$ and $-1.35 \times 10^{-14} / (m^5 mol^{-3} N^{-1} kg^2)$, respectively. Positive value of K_{AB} and negative value of K_{ABB} suggest that ion pair interaction is predominant over ion triplet interaction in the investigated solutions.

19.3.6 MOLAR FREE LENGTH (L_f) AND MOLAR FREE VOLUME (V_f)

(L_f) and (V_f) are calculated by using the following set of equations (19.9–19.14):

$$L_f = kK_s^{1/2} \tag{19.9}$$

k is known as Jacobson constant and can be calculated as [25,26]:

$$k = (93.875 + 0.375\ T) \times 10^{-8} \tag{19.10}$$

$T = 298.15$ K is the experimental temperature for this investigation.

$$V_f = \left\{ \bar{M} U / (k_1 \times \eta) \right\}^{3/2} \tag{19.11}$$

$\bar{M}$ is the average molar mass of the solution and is calculated by using the following equation (19.12) [27]:

$$\bar{M} = \frac{m_1 n_1 + m_2 n_2 + m_3 n_3}{n_1 + n_2 + n_3} \tag{19.12}$$

where m_1 and n_1 are the molar mass and number of moles of Asp in $1\ m^3$ of solution, respectively. Similarly, m_2, m_3, and n_2, n_3 are the molar masses and numbers of moles of PS and water, respectively, in the same volume of solution.

k_1 is a temperature-independent constant term with the value of 4.28×10^9, and η, values of which are presented in Table 19.1, is the viscosity of the studied solutions.

Acoustic properties of a liquid mixture are controlled by the molecular properties of the components of the mixture. The simplest molecular property is the molar free length (L_f), that is, the distance between the surfaces of the molecules in a solution. Analysis of the variation of intermolecular free length with the concentrations of Asp strengthens the idea of formation of compact solvation spheres as a result of strong solute-solvent interactions. These tightly packed clusters approach each other at higher concentrations, and the distance between the surfaces of the solvated particles starts waning. Besides, with increasing concentration of potassium sorbate in

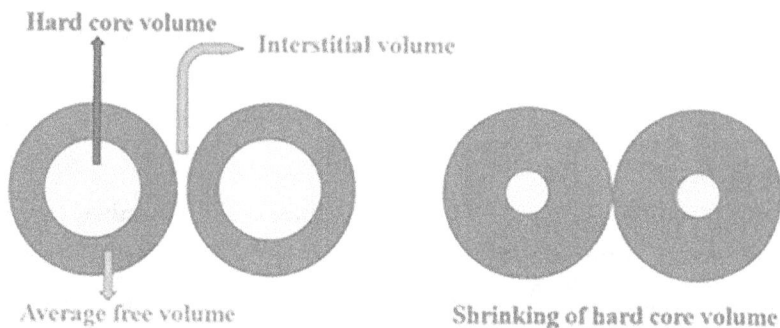

FIGURE 19.4 Schematic representation of shrinking of hard core spheres and increment of free volume.

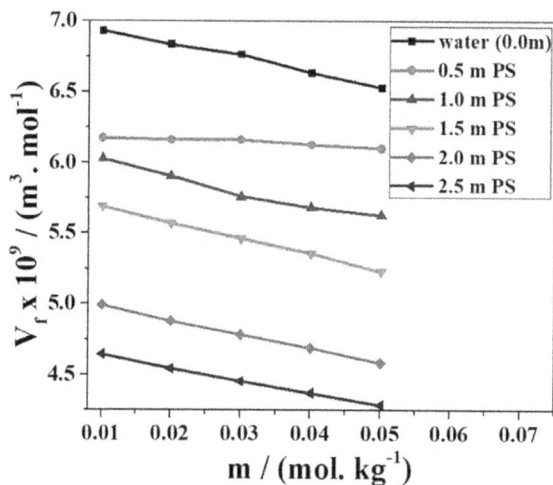

FIGURE 19.5 Plot of V_f against m of aspartic acid solutions in water and aqueous potassium sorbate (0.5–2.5 m) at 298.15 K.

solution, strong solute-co-solute interactions bring the particles even closer and L_f decreases with increasing concentration of potassium sorbate.

According to the average free volume model, solvation shells exist as hard core spheres formed due to strong attractive forces between solute and solvent molecules [28,29]. Each of these spherical particles is surrounded by an average free space (Figure 19.4).

Therefore, the total volume is the summation of hard core volume, the average free volume, and the interstitial volume. As the molecular interaction is expected to increase with increasing concentrations of solute and co-solute particles, the solvation spheres are increasingly shrinking in volume. These closely packed solvated particles fit well inside the interstitial spaces of water, and the unoccupied space that is the free volume shows a negative slope when plotted against concentration of aspartic acid solution in different molal composition of aqueous potassium sorbate as presented in Figure 19.5.

Diminishing trend of L_f with increasing concentrations of aspartic acid and potassium sorbate is given in Table 19.3.

TABLE 19.3
Concentration of Aspartic Acid (*m*), Internal Pressure (π_i), Relative ion Association (R_A), Acoustic Impedance (*Z*), Molar Free Volume (V_f), and Molar Free Length (L_f) in Water and Aqueous of 0.5, 1.0, 1.5, and 2 m Potassium Sorbate Media at 298.15 K

Composition of aq. Potassium Sorbate (mol/kg)	m (mol/kg)	$\pi_i \times 10^{12}$ N/m²	$R_A \times 10^{-1}$	$Z \times 10^{-5}$ (kg/m²s)	$V_f \times 10^9$ (m³/mol)	$L_f \times 10^{11}$ (m/mol)
0.000	0.011	5.100	9.999	14.966	6.931	4.341
	0.019	5.149	9.989	15.058	6.833	4.316
	0.030	5.174	9.988	15.110	6.763	4.303
	0.039	5.199	9.987	15.163	6.633	4.290
	0.051	5.224	9.987	15.215	6.528	4.277
0.499	0.009	5.470	9.992	15.777	6.175	4.160
	0.021	5.502	9.983	15.844	6.163	4.144
	0.032	5.540	9.972	15.919	6.161	4.125
	0.040	5.568	9.966	15.974	6.129	4.111
	0.051	5.579	9.966	15.996	6.101	4.106
	0.012	5.761	9.995	16.437	6.029	4.030
	0.019	5.789	9.988	16.493	5.904	4.018
1.013	0.031	5.819	9.983	16.551	5.759	4.004
	0.041	5.847	9.977	16.607	5.683	3.992
	0.049	5.867	9.974	16.647	5.628	3.983
1.499	0.009	6.128	9.997	17.217	5.692	3.878
	0.021	6.158	9.991	17.274	5.572	3.866
	0.032	6.189	9.984	17.335	5.463	3.853
	0.040	6.214	9.980	17.384	5.359	3.843
	0.051	6.235	9.978	17.426	5.232	3.834
2.011	0.012	6.363	9.992	17.738	4.993	3.791
	0.021	6.395	9.985	17.801	4.879	3.779
	0.032	6.419	9.982	17.847	4.782	3.769
	0.042	6.433	9.983	17.876	4.689	3.764
	0.049	6.452	9.982	17.914	4.585	3.757
2.501	0.009	6.507	9.994	18.079	4.646	3.742
	0.022	6.524	9.994	18.113	4.544	3.736
	0.030	6.544	9.993	18.151	4.455	3.729
	0.039	6.569	9.989	18.202	4.373	3.719
	0.051	6.596	9.986	18.254	4.289	3.709

19.3.7 Internal Pressure

Internal pressure (π_i) characterizes the change in the internal energy of the system during the process of small isothermal expansion and is described by equation (19.13) [30]:

$$\pi_i = (\alpha T / K_T) - P \tag{19.13}$$

Atmospheric pressure, P, is neglected in this equation. α is the isobaric thermal expansion coefficient and can be computed by the following equation [31]:

$$\alpha = (75.6 \times 10^{-3}) / (\rho^{1/3} U^{1/2} T^{1/9}) \tag{19.14}$$

In a liquid, internal pressure (π_i) develops due to the attractive and repulsive forces acting between the particles. This parameter measures the differences between expansive pressure, which is temperature dependent, and cohesive pressure arising due to temperature-independent attraction between molecules. Internal pressure holds the liquid molecules together and is more sensitive to attractive forces, whereas V_f is a function of repulsive forces. These two factors together describe the disorderness (entropy) of the solution. Internal pressure and its variation with the addition of solute in the solution are incredibly useful to understand the structural behavior of the solution, clustering phenomenon of solvated particles, and interactions between

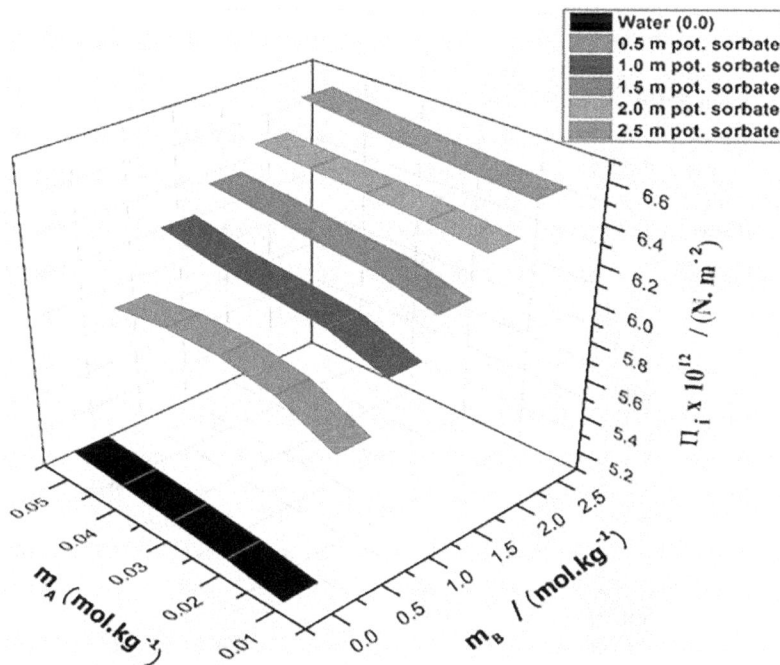

FIGURE 19.6 Plot of π_i against molality (m_A) of aspartic acid solutions in water and aqueous potassium sorbate ($m_B = 0.5 - 2.5$ m) at 298.15 K.

them. In Table 19.3 and Figure 19.6, we can see π_i increases steadily with the concentration of aspartic acid in water and aqueous sorbate media, suggesting strong solute-solvent interactions in all the investigated systems.

19.3.8 Relative Association (R_A) and Acoustic Impedance (Z)

Equations (19.15) and (19.16) are used to evaluate the relative association (R_A) and acoustic impedance (Z), respectively.

$$R_A = \left(\rho/\rho_o\right) \times \left(U_0/U\right)^{1/3} \qquad (19.15)$$

$$Z = \rho \times U \qquad (19.16)$$

Relative association is an outcome of strengthening or weakening of solvent structure on the addition of solute and solvation of solute particles by solvent [32]. We can see in Table 19.3 that R_A values diminish with molality of Asp. R_A values can be interpreted in light of ion-ion and ion-hydrophilic attractive forces between the polar groups $-NH_2$, $-COOH$, and $-COO^-$ on the side chain of Asp, K^+, and $C_6H_7O_2^-$ ions of PS and the polar water molecules and repulsive forces between the alkyl backbones of amino acid and sorbate ion. Association of solute and solvent molecules is more prominent at higher concentration, and an uphill variation of R_A with concentrations of PS is apparent in all solutions indicating strong solute-solvent interaction.

The longitudinal sound waves experience a resistance by the particles in a solution while traveling through it. This opposition to propagation of sound wave or acoustic impedance is a quantity depending on the molecular packing of the systems. Increment in Z values with concentration of Asp and PS, as evident in Table 19.3 and Figure 19.7, indicates the presence of strong interactions between ions and polar groups of Asp and PS with water molecules in all the solutions.

Existence of strong ion-solvent interactions in Asp solutions in aqueous PS is well supported by the volumetric and viscometric properties of the same system investigated by the authors and published elsewhere [16].

19.4 INFERENCES

A non-essential amino acid with a hydrophilic side chain, aspartic acid, is chosen for this experimental research. A potassium salt of sorbic acid was taken to serve as a co-solute. A number of solutions were made with different concentrations of aspartic acid in aqueous and aqueous sorbate solutions with different compositions. Density and acoustic velocity values for these solutions were experimentally measured. Acoustic parameters were computed from these data by using various mathematical relations. Density (ρ) and ultrasonic velocity (U) were found to produce regular increments with concentrations of solutions in all compositions of solvents. This fact indicates towards strong solute-solvent as well as solute-co-solute interactions present in all investigated systems. Compressibility parameters K_S, $K_{s,\phi}$, and K_T reinforce the presence of strong molecular interactions in the solutions. It establishes the fact that hydration of aspartic acid is higher in aqueous medium; however, K^+ and $C_6H_7O_2^-$ ions (produced from completely dissociated aqueous potassium sorbate) in solution

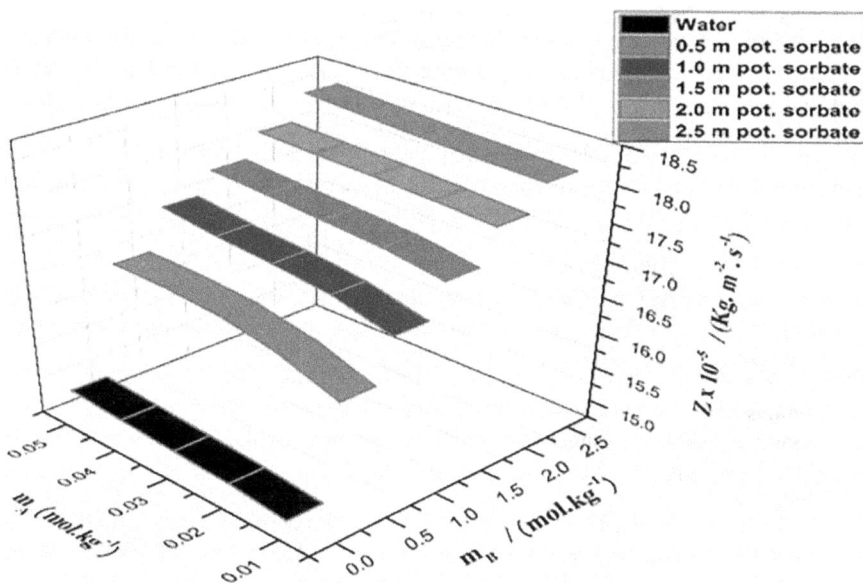

FIGURE 19.7   Plot of Z against molality (m_A) of aspartic acid solutions in water and aqueous potassium sorbate $(m_B = 0.5 - 2.5\,m)$ at 298.15 K.

offer strong attractive forces towards NH_3^+ and COO^- ions of aspartic acid, which reduces the electrostriction of water molecules to some extent. Compressibility values at infinite dilution, K_s^0 and $K_{s,\phi}^0$, diminish with concentration of sorbate in the systems, which reveal that solute-solvent interaction is brought down by the presence of interactions between the solute and co-solute particles. However, the negative values of S_k suggest that solute-solute interactions are lower than the solute-solvent interactions for all the systems. Positive $\Delta_{tr} K_{s,\phi}^0$ values reveal the predominance of ion-ion and ion-hydrophilic interactions between solute and co-solute particles. Study of other parameters like Z, π_i, and R_A implies the presence of hydrophobic interactions between the hydrocarbon backbones of aspartic acid and sorbate and hydrophobic solvation of these nonpolar parts of the molecules. Caging effect of self-associated water molecules encases the strongly bonded solute and co-solute particles, and the solvated particles are considered to be hard core spheres surrounded by water molecules. Variations of L_f and V_f with concentrations of solutions are in agreement with this interpretation.

REFERENCES

1. Arsule, A.D. Sawale, R.T. Kalyankar, T.M. Deosarkar, S.D. 2020. Thermodynamic behavior of systems containing amino acids in aqueous-lactose solutions. *J Soln Chem*, 49(1): 83–99. https://doi.org/10.1007/s10953-019-00945-4.
2. Khoshkbarchi, M.K. Vera, J.H. 1997. Effect of NaCl and KCl on the solubility of amino acids in aqueous solutions at 298.2 K: measurements and modeling. *Ind Eng Chem Res*, 36: 2445–2451. https://doi.org/10.1021/ie9606395.

3. Chauhan, S. Pathania, L. Sharma, K. Kumar, G. 2015. Volumetric, acoustical and viscometric behavior of glycine and DL-alanine in aqueous furosemide solutions at different temperatures. *J Mol Liq*, 212: 656–664. https://doi.org/10.1016/j.molliq.2015.09.042.

4. Shirvali, S. Iloukhani, H. Khanlarzadeh, K. 2019. Temperature dependent volumetric and acoustic properties of some amino acids in aqueous solutions of an antidepression drug. *J Mol Liq*, 295: 111651. https://doi.org/10.1016/j.molliq.2019.111651.

5. Kumar, D. Lomesh, S.K. Nathan, V. 2017. Molecular interaction studies of l-alanine and l-phenylalanine in water and in aqueous citric acid at different temperatures using volumetric, viscosity and ultrasonic methods. *J Mol Liq*, 247: 75–83. https://doi.org/10.1016/j.molliq.2017.08.057.

6. Riyazuddeen, Bansal, G.K. 2006. Intermolecular/interionic interactions in l-leucine-, l-asparagine- and glycylglycine-aqueous electrolyte systems. *Thermo Acta*, 445: 40–48. https://doi.org/10.1016/j.tca.2006.04.004.

7. Nain, A.K. Pal, R. Droliya, P. 2016. Study of (solute + solute) and (solute + solvent) interactions of homologous series of some α-amino acids in aqueous-streptomycin sulfate solutions at different temperatures by using physicochemical methods. *J Chem Thermo*, 95: 77–98. https://doi.org/10.1016/j.jct.2015.11.015.

8. Singh, S. Dhal, K. Talukdar, M. 2020. A comparative study on the effect of temperature and concentration on density, sound velocity and their derived properties for diclofenac potassium in aqueous urea media. *Biointerface Res Appl Chem*, 10(5): 6377–638815. https://doi.org/10.33263/BRIAC105.63776388.

9. Talukdar, M. Dhal, K. Dehury, S.K. 2021. Investigation on molecular interactions of multicharged electrolytes potassium pyrophosphate and potassium dichromate in aqueous d-sorbitol media at 298.15k by densiometric and acoustic methods. *Biointerface Res Appl Chem*, 11(1): 8075–8086. https://doi.org/10.33263/BRIAC111.80758086.

10. Sailaja, M. Sarangi, D. Mohapatra, P. Nanda, B.B. 2020. Ultrasonic studies on temperature dependence of ion solvent interactions of ionic liquid 1-butyl-2,3-dimethylimidazolium chloride in aqueous solutions of tetra-n-butylammonium bromide at T = (298.15 to 313.15) K. *Biointerface Res Appl Chem*, 10(2): 5259–5265. https://doi.org/10.33263/BRIAC102.259265.

11. Kumar, H. Kumar, V. Sharma, S. Katal, A. Alothman, A. A. 2021. Volumetric and acoustic properties of amino acids L-Leucine and L-Serine in aqueous solution of ammonium dihydrogen phosphate (ADP) at different temperatures and concentrations. *J Chem Thermo*, 155: 106350. https://doi.org/10.1016/j.jct.2020.106350.

12. Kumar, H. Singh, G. Kataria, R. Sharma, S.K. 2020. Volumetric, acoustic and infrared spectroscopic study of amino acids in aqueous solutions of pyrrolidinium based ionic liquid, 1–butyl–1–methyl pyrrolidinium bromide. *J Mol Liq*, 303: 112592. https://doi.org/10.1016/j.molliq.2020.112592.

13. Gaba, R. Pal, A. Kumar, H. Sharma, D. Navjot, 2017. Volumetric, acoustic and infrared spectroscopic study of amino acids in aqueous solutions of pyrrolidinium based ionic liquid, 1–butyl–1–methyl pyrrolidinium bromide. *J Mol Liq*, 242: 739–746. https://doi.org/10.1016/j.molliq.2020.112592.

14. Moradian, T. Iloukhani, H. Khanlarzadeh, K. 2018. Volumetric and compressibility behavior of l-valine in aqueous poly ethylene glycol solutions at T = (298.15, 308.15 and 318.15) K. *J Mol Liq*, 269: 869–887. https://doi.org/10.1016/j.molliq.2018.08.027.

15. Kaur, H. Thakur, R. C. Kumar, H. Katal, A. 2021. Effect of α–amino acids (glycine, l-alanine, l-valine and l-leucine) on volumetric and acoustic properties of aqueous 1-Butyl-3-propylimidazolium bromide at T = (288.15, 298.15, 308.15, 318.15) K. *J Chem Thermo*, 158: 106433. https://doi.org/10.1016/j.jct.2021.106433.

16. Dhal, K. Singh, S. Talukdar M. 2022. Elucidation of molecular interactions of aspartic acid with aqueous potassium sorbate and sodium benzoate: Volumetric, viscometric and FTIR spectroscopic investigation. *J Mol Liq*, 325:118659. https://doi.org/10.1016/j.molliq.2022.118659.

17. Rani, R. Kumar, A. Sharma, T. Sharma, T. Bamezai, R. K. 2019. Volumetric, acoustic and transport properties of ternary solutions of l-serine and l-arginine in aqueous solutions of thiamine hydrochloride at different temperatures. *J Chem Thermo*, 135: 260–277. https://doi.org/10.1016/j.jct.2019.03.039.

18. Rani, R. Kumar, A. Bamezai, R.K. 2017. Effect of glucose/lactose on the solution thermodynamics of thiamine hydrochloride in aqueous solutions at different temperatures. *J Mol Liq*, 240: 642–655. https://doi.org/10.1016/j.molliq.2017.05.127.

19. Ankita, Nain, A. K. 2020. Study on the interactions of drug isoniazid in aqueous-D-xylose/L-arabinose solutions at different temperatures using volumetric, acoustic and viscometric approaches. *J Mol Liq*, 298: 112086. https://doi.org/10.1016/j.molliq.2019.112086.

20. Devi, S. Kumar, M. Sawhney, N. Syal, U. Sharma, A.K. Sharma, M. 2021. Volumetric, acoustic and viscometric studies of L-histidine and L-serine in aqueous levofloxacin solutions at different temperatures and concentrations. *J Chem Thermo*, 154: 106321. https://doi.org/10.1016/j.jct.2020.106321.

21. Pandey, J. D. Sanguri, V. Yadav, M. K. Singh, A. 2008. Intermolecular free length and free volume of pure liquids at varying temperatures and pressures. *Ind J Chem*, 47A: 1020–1025. http://hdl.handle.net/123456789/2121.

22. Kozak, J. J. Knight, W.S. Kauzman, W. 1968. Solute-solute interactions in aqueous solutions. *J Chem Phys*, 48: 675–690.

23. Krishnan, C.V. Friedman, H.L. 1973. Enthalpies of alkyl sulfonates in water, heavy water, and water-alcohol mixtures and the interaction of water with methylene groups. *J Soln Chem*, 2: 37–51.

24. Franks, F. Pedley, M. Reid, D.S. 1976. Solute interactions in dilute aqueous solutions. Part 1.—Microcalorimetric study of the hydrophobic interaction. *J Chem Soc Faraday Trans*, 172: 359–367.

25. Pandey, J. D. Kumar, V. Saxena, M. C. 1979. Evaluation of Jacobson's constant and intermolecular free-length as a function of pressure and temperature for cryogenic liquids. *Ultrasonics*, 17(4): 153–158. https://doi.org/10.1016/0041-624X(79)90032-5.

26. Muley, G. G. Naik, A. B. Gambhire, A. B. 2014. Investigation on intermolecular interaction in supersaturation state of cadmium sulphate mixed zinc tris -thioureasulphate solutions. *J Mol Eng Mat*, 2(3): 1450002. https://doi.org/10.1142/S2251237314500026.

27. Meor, M.M.R. Affandi, M. Tripathy, M. Majeed, A.B.A. 2017. Solubility enhancement of simvastatin and atorvastatin by arginine: A solvodynamics study. *J Mol Liq*, 241: 359–366.

28. Yu, Y. Bejan, D. Krause-Rehberg, R. 2014. Free volume investigation of imidazolium ionic liquids from positron lifetime spectroscopy. *Fluid Phase Equil*, 363: 48–54. https://doi.org/10.1016/j.fluid.2013.11.011.

29. Saini, A. Prabhune, A. Mishra, A.P. Dey, R. 2021. Density, ultrasonic velocity, viscosity, refractive index and surface tension of aqueous choline chloride with electrolyte solutions. *J Mol Liq*, 323: 114593. https://doi.org/10.1016/j.molliq.2020.114593.

30. Dhondge, S.S. Ramesh, L. 2007. Isothermal compressibility and internal pressure studies of some non-electrolytes in aqueous solutions at low temperatures. *J Chem Therm*, 39(4): 667–673. https://doi.org/10.1016/j.jct.2006.08.007.

31. Sailaja, M. Sarangi, D. Dalai, B. Mohapatra, P. Nanda, B.B. 2020. Effect of presence of tetra alkyl ammonium bromides on solution behavior of 1-Butyl-2, 3-dimethylimidazolium chloride at different temperatures: Volumetric and acoustic studies. *Chem Data Coll*, 28: 100438. https://doi.org/10.1016/j.cdc.2020.100438.

32. Ganjare, P. J. Aswale, S. S. Aswale, S. R. 2016. Molecular interactions in solutions of sodium salt of 4-amino-2-hydroxy benzoic acid: An ultrasonic study. *World J Pharm Res*, 6(1): 564–573. https://doi.org/10.20959/wjpr20171-7554.

20 Proposal of an Improved Waste Collection System for Urban Environments

J. D. C. da Costa and R. D. S. G. Campilho

CONTENTS

20.1 Introduction 361
20.2 Design Considerations 364
    20.2.1 Objectives 364
    20.2.2 Methodology...365
    20.2.3 Applicable Regulations..365
    20.2.4 Safety Requisites 366
20.3 Results..367
    20.3.1 Overview of the Final Solution...367
    20.3.2 Main Structure...367
    20.3.3 Safety Device...369
    20.3.4 Top Bin...370
20.4 Structural Design...372
    20.4.1 Main Structure...372
    20.4.2 Safety Device...377
    20.4.3 Top Bin...380
20.5 Prototype Construction and Testing 380
20.6 Cost Analysis...384
20.7 Conclusions..385
References..385

20.1 INTRODUCTION

The integrated waste management is made possible by the implementation of waste recovery systems. The process of waste recovery needs a constant technological evolution of the existing infrastructures, aiming to answer to global and European standards that increasingly require a reduction in the deposition of waste in landfills and a progressive growth of the recycling levels per capita. The location, capacity, and aesthetics of the equipment inserted in urban collection points depend on factors such as accessibility and available logistics, the surrounding urban environment, and the main purpose (residential or industrial waste). To allow the insertion of waste equipment in places with limitations, such as the space available and the urban environment, buried equipment for the storage of waste appeared. This equipment has a

DOI: 10.1201/9781003359784-20

large storage capacity and, since they are completely buried underground, only permanent deposition top bins are visible. Thus, this equipment is aesthetically framed in the place since, usually, its floor is identical to the surroundings. Buried equipment with the function of urban solid waste (USW) storage has advantages and disadvantages, and to eliminate the disadvantages they entail, design companies need to comply with conditions imposed by international and/or European standards and specifications, with security obligations always at the forefront. Although the acquisition of these equipment is mostly carried out by subcontracted private companies, in the scope of public works, from the moment that design/fabrication is concluded by the contracting entity, the ownership and management of the equipment becomes responsibility of the city council in which the equipment is implemented. From the point of view of design companies, the large-scale production of these equipment allows the reduction of raw material and labor costs. For this purpose, the equipment of each manufacturer is usually based on standard modular structures, thus allowing the addition of parts to the structure up to reaching the desired load capacity and residue variety. In this way, the proposed solution can meet the customer's specifications with minor added design effort to the current portfolio. It is thus up to the customer to inform, depending on the previously defined project, the number of containers of each equipment. The various types of existing equipment can also contain common parts, thus reducing delivery times.

The mechanical design process of waste collection equipment is essential to propose efficient solutions to the clients, which satisfy the strength, stiffness, functionality, and safety requisites, while adding real-added value to the environment (Bi 2018). Mechanical design can be undertaken by analytical analyses (mechanics of materials-based) coupled with continuum mechanics criteria for failure assessment, Eurocodes or other specific standards, or numerical methods such as the FEM, either used with continuum mechanics criteria or fracture-mechanics-based models to promote structural failures. Analytical analyses are often limited to apply in real-life structures due to their complexity relating to the geometry, boundary conditions, and loadings, preventing the attainment of closed-form solutions to provide the stress/strain distributions for a given loading system (Okereke and Keates 2018). Continuum mechanics criteria are mostly used, which usually reduce a tri-axial and complex stress state to an equivalent uniaxial stress state (s_{eq}) defined from uniaxial tensile tests to assess yielding onset, i.e., considering $s_{eq} = s_y$, in which s_y is the yielding limit. These criteria can be based on particular stress components or the principal stresses (Budynas and Nisbett 2010). The maximum normal stress criterion is mostly employed to infer failure in brittle materials, when the largest normal principal stress reaches s_y (Dieter and Bacon 1986). The maximum shear stress or Tresca criterion predicts failure/yielding when the maximum shear stress (t_{max}) reaches the shear yield stress (t_y). In this case, t_y is not as straightforward, but it can be defined by the relationship between maximum shear and principal stresses (Flom 2013). The von Mises or maximum distortion energy criterion considers that failure takes place when the shear stress in the octahedral place attains a critical shear stress in the octahedral plane causing material yielding (Dowling 2012). However, in the most common form, this criterion is stipulated depending on principal stresses. Design by EN Eurocodes consists of complying with a series of ten European Standards

(EN 1990-EN 1999), developed by CEN (European Committee for Standardization) over the last 30 years, which provide verification formulae and criteria for the safety of structures, enabling a common method for the design of buildings and other civil engineering works, and construction products (CEN 2010). These are also the recommended reference for technical specifications to deliver by candidate companies to answer to public contracts. The Eurocodes cover the design of all types of structures and materials.

The FEM emerged from the limitation of analytical techniques in providing solutions for real-world applications, where the problems to be analyzed are much more complex (Haftka and Gürdal 2012). The FEM allows a numerical estimate of the integral or differential constitutive equations, which are established for the structure or for the elements of the structure individually, by using algebraic equations that return approximate solutions for the field variables, for a discrete number of points in the domain: the finite element nodes. Additionally, the pre-established laws of variation of the variables of the interior field of the finite elements are considered, given by the interpolation or shape functions (Bergan and Nygård 1984). The FEM formulation thus results in a system of equations to be solved at the expense of the original integral or differential functions. The FEM allows to obtain numerical solutions of engineering problems such as stress, temperature, fluid flow and aerodynamics, electromagnetic and electronic analysis, among many other cases (Schäfer 2006). Given the large potential of FEM modeling and possible application in the several domains of engineering, large companies resort to software existing in the market for this analysis. Software packages such as Solidworks®, Abaqus®, Ansys®, Inventor®, or Nastran™ provide integrated pre-processing, solving, and post-processing solutions. Added to the commercial versions, most of this software also has student versions with reduced functionality and costs. In FEM analysis, complex plasticity models can be used, providing a more realistic analysis of structures. Examples are the Drucker-Prager model for soils, concrete, and polymeric materials (Drucker and Prager 1952), the Mohr-Coulomb plasticity model for rock and soil mechanics, consisting in a linear failure envelope for the combinations of normal (σ) and shear (τ) stresses that would cause material's yielding (de Souza Neto et al. 2008, Labuz and Zang 2012), or the Johnson-Cook plasticity model, which is a variant of the Mises plasticity model with analytical forms of the hardening law and rate dependence, being applicable for high-strain-rate loading of different materials, including metals (Johnson and Cook 1985, Sobolev and Radchenko 2016). Complex fracture-mechanics-based models such as cohesive zone modeling (CZM) (Park and Paulino 2013) and extended finite element modeling (XFEM) (Moës et al. 1999) enable using the fracture toughness of materials to predict crack growth by using damage laws, which constitutes a step forward in numerical modeling and enable real material separation to take place, with corresponding load capacity assessment by measuring the reaction loads in the structure during displacement-control analyses.

The research work on waste collection equipment is highly limited, due to its specific nature, and usually associated with large companies producing this equipment. Aisa et al. (2006) developed, in partnership with Contenur, a 2-m^3 solid waste container. The study combined the use of several computer-aided engineering (CAE) tools (design, mechanical design, and rheological simulation). The main objective was the rapid manufacture of large-capacity plastic containers by injection molding, in order to compete with those made of welded sheet metal or through rotational molding, containing costly structural reinforcements. This study resulted, in a first phase, in more than 6,000 samples, in which no problems were detected in the injection, ejection, or in the life of the component. Laranjeira (2008) resorted to the FEM, using the Pro-Engineer software with the Pro-Mechanism module to design a container of a buried equipment (Citytainer-TNL), manufactured through rotational molding. To reduce the thickness, reinforcements and ribs were placed in strategic locations. With these changes, in addition to eliminating some deficiencies in the studied container, it was possible to reduce the container's mass by approximately 10% compared to the imposed requirements. Guimarães (2015) addressed the automation of a lifting platform for buried equipment (Bigtainer-TNL). A solution was proposed for oil leaks from the hydraulic system, resulting in a new synchronism solution for the hydraulic cylinders, where each one has its directional valve and a flow divider that makes the pressures in each cylinder independent, with an inclinometer to measure the relative difference between the positions of the cylinders. A significant cost reduction was also achieved with this solution.

This chapter proposes an improved equipment for underground waste disposal, with emphasis to safety devices in accordance with safety regulations, and automated mechanisms for platform opening and closing. The buried equipment is composed of three main components: the main structure, the liner, and the top bin, serving as user interface. Since the market-available equipment is typically overdesigned, the authors consider that these structures can highly benefit from design tools, leading to a higher competitiveness of companies. Thus, this chapter presents the equipment concept, followed by structural design using the FEM to perform successive iterations and improvements. The prototype fabrication is fabricated and tested, and a cost estimate is also presented.

20.2 DESIGN CONSIDERATIONS

20.2.1 OBJECTIVES

The main objective of this work is to design a complete solution for underground waste storage equipment, including the liner, the buried structure, and the interface with the depositor of waste (top bin), in partnership with Sitape, a Portuguese supplier company of industrial equipment, aiming to replace the existing overdesigned version, and also produce a new version that fully complies with the current safety regulations. The analysis and validation process of this equipment can be divided into the following features:

- Compliance check with the applicable legislation for the equipment and safety device;
- Design of an improved structure for the waste storage equipment, including a safety device that, in addition to its implementation in future equipment, is also possible to adapt to existing equipment;
- Design of the interface with the depositor of waste (top bin);

- Conducting a detailed study of the structure's behavior when subjected to the service loads and the load cases imposed by the applicable regulations, and respective prototype construction;
- Cost analysis, to provide an efficient solution for the equipment.

20.2.2 METHODOLOGY

The existing equipment, as well as a large part of the equipment on the market, does not meet the requirements imposed by the legislation regarding the safety device necessary to prevent people, animals, and objects from falling inside, during the container collection process. Moreover, it is over-sized, leading to unnecessary costs. Given the need for constant updating and continuous improvement, the following methodology was adopted for this work:

- Research and framework of existing products on the market designed for waste collection and storage;
- Investigation of European and national (Portuguese) legislation applicable for structural design, and published ordinances in order to overcome possible risks through specifically established requirements;
- Develop a solution to mitigate the lack of a safety device in the equipment. This solution must be possible to apply in currently operating equipment;
- Validate the equipment's structure, in case the implementation of the safety device implies changes at the structural level;
- Reduce, if possible, the cost of production of the equipment, by redesign of the structure;
- Design a new top bin compatible with all the company's equipment;
- Present the cost for the various items developed throughout the work and, if possible, propose a budget that includes the equipment with four top bins, equipped with all these adaptations.

20.2.3 APPLICABLE REGULATIONS

Directive (European Union or EU) 2018/850 on the landfill of waste sets the goal for member states for 2035 to reduce the total amount of municipal waste deposited in landfills to, at least, 10%. In the field of recycling, Directive (EU) 2018/852 establishes a common goal for member states to prepare to reuse and recycle 65% of packaging waste by 2025 (APA 2019). In order to comply with the aforementioned requirements, as well as to outline actions to be progressively implemented, a strategic plan was created in Portugal, which provides for the implementation of a set of actions that have proved to be fundamental in implementing the USW policy. This Strategic Plan is an instrument for the USW management for 2014–2020, which is called PERSU 2020 (Teixeira et al. 2014). PERSU 2020 contains measures that provide an increase in selective collection and recycling, driving a progressive reduction in direct landfilling, promoting an increase in the efficiency of infrastructure and USW management systems, with consequent rationalization, reduction, and sustainable recovery of costs (ERSAR 2016).

At national level, the design activity of metallic structures was, for a long period, regulated by the Regulation of Steel Structures for Buildings. However, this document, in addition to being technically outdated, has limitations in areas such as the plastic analysis of structures, bending, lateral bending, analysis and design of connections, among others (Simões 2005). Currently, according to the decree-law 211/86, for the design and dimensioning of structures, the Eurocodes created for this purpose must be complied with. The Eurocodes were conceived, at European level, with the objective of developing and standardizing calculation and design rules for different types of structures (Simões 2005). For the development of this work, the most relevant is Eurocode 3, divided into six parts, of which Part 1 (EN1993-1) stands out, which concerns general rules, and rules for buildings. This Part 1 is further subdivided into twelve subparts, the most relevant being the first (EN1993-1-1) for general rules, commonly referred to as EC3-1-1, and the eighth (EN1993-1-8) concerning to bonds, usually designated EC3-1-8.

Specific regulations for the type of structure under analysis can be divided into containers and structure. In the EU, dimensions, design, performance and testing criteria, health and safety, as well as requirements specifications for the manufacture of stationary waste containers, are governed by European standards EN840 and EN 12574. In these standards, the types of lifting devices and capacities are classified, as well as the test methods, and health and safety requirements (Rodrigues 2017). They contain terminology, dimensions, and requirements, as well as information on testing methods for containers. On the other hand, the structure of the buried equipment, with all its constituents, must comply with the requirements established by the following directives:

- Directive 2006/42/CE on machinery – Standards for marketing and putting into service of machines;
- Directive 2000/14/CE on the noise emission in the environment by equipment for use outdoors – Concerning the matter of noise emissions to the environment of equipment for use outdoors;
- Directive 2006/95/CE on the electrical equipment designed for use within certain voltage limits – Safety requirements of electrical equipment intended to be used in certain alternating current limits;
- Directive 2004/108/CE on electromagnetic compatibility – Regulates the electromagnetic compatibility of electrical and electronic equipment.

Apart from these directives, the most relevant technical standards are DIN ISO 2768 (General geometrical tolerances and technical drawings), DIN EN 10025 (Hot rolled products of structural steels), DIN EN ISO 13920 (General tolerances for welded constructions), UNE-EN ISO 12100 2004/A1:2010 (Safety of machinery – Basic concepts, general principles for design), EN ISO 14121-1:2007 (Safety of machinery – Risk assessment – Part 1: Principles), and ISO 13857:2008 (Safety of machinery – Safety distances to prevent hazard zones being reached by upper and lower limbs).

20.2.4 SAFETY REQUISITES

The equipment must comply with the requirements of standards EN 13071-1:2008 and EN 13071-2:2008, referring to this type of buried or semi-buried equipment. The EN 13071-2:2008 standard specifies the general characteristics of the equipment and its accessories, and establishes test methods and safety requirements. This standard specifies that the safety devices applied to this equipment must be resistant to all weather conditions, including humidity and dirt that may negatively affect its operation. For maintenance and cleaning, all parts of the buried or semi-buried system must be easily accessible. For the different parts that make up the buried system, the standard defines a series of parameters and requirements that must be met, namely:

- **Hole:** If the ditch is deeper than 500 mm, measured at any point, it must be equipped with safety devices, in order to prevent a possible accidental fall of pedestrians, thus avoiding injury by any mobile part of the system (CEN 2019). The safety device can be based on a platform or a barrier;
- **Pedestrian platform (or cover):** When the container is installed, the circulation of pedestrian traffic in the surface area of the equipment (usually pavement or other non-slip floor) should take place in safe conditions. All necessary precautions must be taken when defining the location of the system, so that the pedestrian platform is in a place with restricted access to vehicles. The platform must be subjected to a test whose objective is to verify its resistance to accidental overloads caused by vehicles or other heavy objects. After the test, no permanent deformation or failure capable of making impossible the use for which the platform was designed is allowed (CEN 2019).

20.3 RESULTS

20.3.1 OVERVIEW OF THE FINAL SOLUTION

The present work focuses on the Hook System (HS) buried equipment (as designated by the company) for USW disposal and collection. This equipment can contain from one (HS1) to four containers (HS4). When equipped with four containers, these are normally for depositing the three types of recyclable and unsorted USW. These residues are led to the container through the deposit top bin, where the users deposit the garbage. Figure 20.1 shows the three-dimensional (3D) design of the final HS4 system proposed in this work, which is subsequently divided into three subsystems: main structure (including the pedestrian platform in the figure and the hydraulic system support), safety device, and top bin. The predesign and selection process of best solution for each main subsystem are detailed in the next subsections (when applicable).

20.3.2 MAIN STRUCTURE

The equipment designed by Sitape is normally made up by combining different modules. In the case of the HS4, the structure is composed of two end modules and three intermediate modules. This HS equipment version is the most complete and commonly used, and it enables to extrapolate design results to equipment with smaller number of modules. Thus, this version was selected for analysis. Figure 20.2 shows the respective final design.

For the main structure, no design alternatives were equated, while the design process is discussed in subsequent Section 20.4.1. The main structure of the equipment is mainly composed of standard profiles, and profiles obtained from cold roll

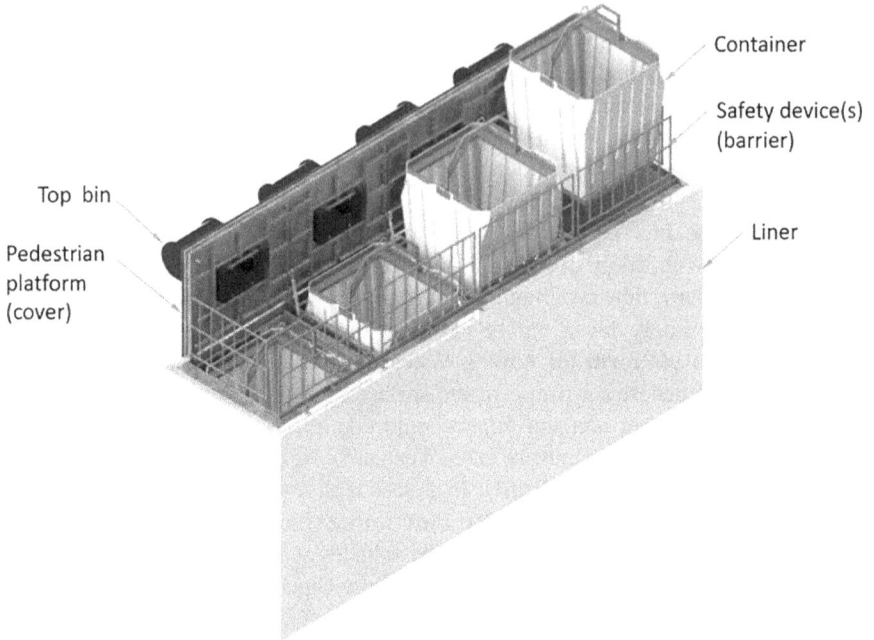

FIGURE 20.1 Overview and main subsets of the HS4 equipment.

forming, joined by fastened and mostly welded joints. Hydraulic cylinders fixed to the lower frame (indicated in yellow in Figure 20.2) promote the opening and closing movements to the cover, which hinges at one of the frame longitudinal edges. Design of this structure aimed to obtain a reduction in production costs and to make it compatible with the implementation of a safety device.

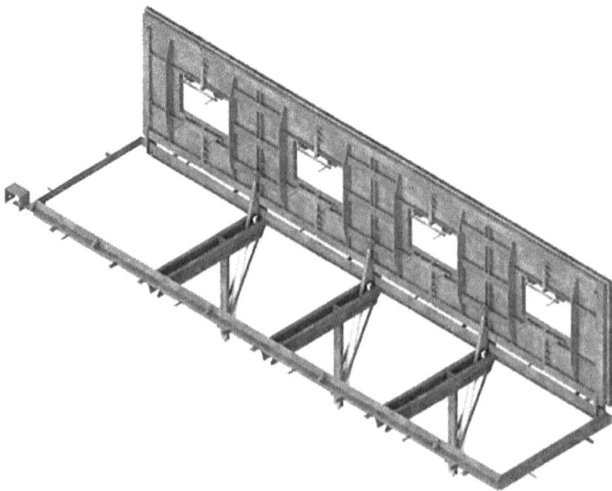

FIGURE 20.2 Opened HS4 model with hidden liner.

20.3.3 Safety Device

Currently, there are few devices that fully comply with applicable legislation in terms of safety requirements. These requirements are only fully fulfilled for buried structures with a single container, which are equipped with a safety platform. This solution involves the use of a trapdoor that automatically closes upon removal of the container at the time of waste unloading, where access to the ditch is totally prevented after the removal of the container. Another possible solution is a safety barrier that lifts concurrently with the cover opening to prevent access to the liner. However, this solution was not found either on the literature or as commercial product.

Aiming to implement a safety device transversal to the HS equipment and possibly adaptable to others, the design should comply with all applicable safety requirements. Due to the availability of more than one fundamental solution, and also complexity in the analysis, the selection matrix method (Ashby 2016) was used to choose the most appropriate mechanism to solve the problem (platform or security barrier, as legislated). Although the detailed process is not presented here, a criteria classification matrix was established, including the selected evaluation criteria by the design department of the company, and the respective score. The chosen criteria were the versatility of the solution, complexity and design time, ease of construction, ease of testing, useful space required, adaptability to current equipment, added difficulty to the waste collection process, production cost, and maintenance. For each criteria, the score ranges from 0 to 4, including the clear description of the requirements of a given solution to be classified within a certain interval. This procedure thus reduces ambiguities to the minimum, and enables to correctly frame the different solutions (Chang 2014). A criterion comparison matrix was then established to define the relative weights of each criterion in the overall performance index. With this purpose, the relative importance between each pair of criteria was defined, using variables of 1 (less importance than), 2 (as important as), and 3 (more important than), whose classification for each criterion and division by all criteria classifications gave the percentile weight of each criterion on the process. This data is then used as input in the selection/decision matrix shown in Table 20.1. In this table, A is the criterion score, established from the criteria classification matrix, and W is the weight (importance) of each criterion (extracted from the criterion comparison matrix). B is the weighted property, calculated as follows. If the criterion score should be as high as possible, then

$$B = \frac{\text{Criterion score for the device being analysed}}{\text{Highest criterion score between the two devices}} \times 100. \qquad (20.1)$$

On the other hand, if the criterion score should be as small as possible, then

$$B = \frac{\text{Small criterion score between the two devices}}{\text{Criterion score for the device being analysed}} \times 100. \qquad (20.2)$$

C is the performance index, given by $W_i \times B_i$ (i is the respective evaluation criterion). Finally, the global performance index of a given solution, which gives quantitative classification of each evaluation criterion and enables selecting the best one, is calculated by ΣC.

TABLE 20.1
Selection Matrix for the Safety System

Evaluation Criteria	W (weight)	Security Platform A	B	C	Safety Barrier A	B	C
Versatility of the solution	0.132	0	0	0	4	100	13.2
Complexity and design time	0.056	2	100	5.6	2	100	5.6
Ease of construction	0.111	2	100	11.1	2	100	11.1
Ease of testing	0.125	2	100	12.5	2	100	12.5
Useful space required	0.118	2	100	11.8	4	67	7.9
Adaptability to current equipment	0.132	2	50	6.6	4	100	13.2
Added difficulty to the waste collection process	0.139	2	100	13.9	4	50	7.0
Maintenance	0.083	3	100	8.3	3	100	8.3
Production cost	0.104	2	100	10.4	2	100	10.4
ΣC	1.000			80.2			89.2

The selection matrix results are depicted in Table 20.1, showing that the versatility of the security platform is graded as nil, due to the inability to be applied to more than one module. The results show that the device with the highest performance index is the security barrier with 89.2%, compared to the 80.2% obtained for the security platform. Based on this approach, the security barrier was chosen over the platform (Figure 20.1). The main advantages of this solution are the versatility ($C=0$ for the security platform since, for more than one module, when the platform is opened, access to the liner is only prevented at the position of the removed container) and adaptability to the existing equipment (due to an easier implementation and adaptation to different-sized structures). However, it performs worse on the required space (since it reduces the available container perimeter) and difficulty in waste collection (the required container elevation is higher to surpass the barrier).

20.3.4 TOP BIN

The top bins are a subsystem of the equipment that need, in addition to their aesthetic function, constant improvement in order to respond to the constant needs of the market. Figure 20.3 shows the final design of the top bin, including the following components: drum or lid for deposition; activation pedal, and commercial door.

The lid and pedal are usually interconnected by arm or lever mechanisms. The pedal will cause the drum or lid (Figure 20.3 (i) to open whenever the waste depositor presses it. In this way, in addition to the user being able to open the lid when both hands are occupied, no contact will take place with dirty surfaces. The commercial door, shown in Figure 20.3 (ii), is another optional feature, such as the foot pedal. This door is usually requested by local authorities in commercial areas, for merchants to dispose bulkier waste bags. In the top bin with lid, there is usually a system of counterweights or gas shock absorbers in order to slow down the closing speed of the lid, thus reducing the probability of the depositor getting trapped. With the same purpose, the top bin with drum normally has weights fixed to the sides of the drum.

Once the company intends to develop a new top bin solution to increase its product range in the market, the development of a new top bin was requested. For a more thoughtful analysis of the opening method under study, a strengths, weaknesses, opportunities and threats (SWOT) analysis was instead considered to additionally account for external effects (e.g., competition), in order to assist in selecting the most appropriate solution. In Table 20.2, a SWOT analysis of lid opening is performed, comparing this opening method with the drum method.

After discussing the two forms of opening, and considering the SWOT analysis results, the company decided to opt for the top bin with lid, giving particular preference to this mode of operation over the drum, due to the increased recent market demand for devices equipped with this system. Another major factor in the decision was the fact that there is, so far, no top bin with this mode of operation in the company's range of solutions. Some of the most emphasized attributes of this top bin include:

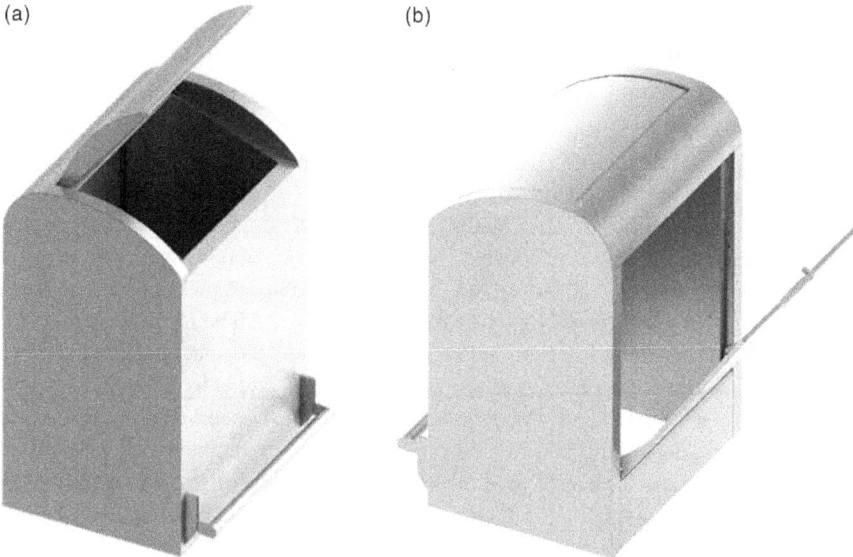

FIGURE 20.3 Front perspective (a) and rear perspective (b) of the "Cambridge" top bin with pedal.

TABLE 20.2
SWOT Analysis of the Top Bin Opening Method (Lid or Drum)

Strengths	Weaknesses	Opportunities	Threats
Deposition of larger volume bags;	Direct exposure to the waste; Minimum height required;	Few existing solutions on the market;	Unsuccessful affirmation of the new product in the
Lower manufacturing costs;	Heavier device; Less watertight system in terms of odors;	Recent increased market demand	market; More
Easier maintenance;	Easier access to waste pickers;	for devices with	technologically
More hygienic device.	Increased risk of damage caused by poor use.	this system.	advanced products.

- Framework in current urban furniture;
- **Geometry:** Absence of sharp edges, thereby reducing the possibility of injury to users or surroundings; and rounded lid contour to prevent residues from being deposited on it. This geometry facilitates the fall of the residue by gravity (applicable in cases of misuse);
- Top bin as versatile as possible, able to meet the maximum of legislation and costumer requirements;
- Light structure, not to overload the substructure, and also resistant to impact;
- Possibility to implement modular systems, consisting of the lid, lid/pedal, or lid/pedal/commercial door. The lid must have a handle to make manual open possible.

20.4 STRUCTURAL DESIGN

20.4.1 MAIN STRUCTURE

Since the structure is composed of modules, in the design phase, for simplicity, only a part of the structure was evaluated, consisting of an end module and an interior module, as shown in Figure 20.4. When the cover is closed, it is supported by the outer frame and intermediate crossbars that are part of it. In this position, a rubber element running the full frame perimeter seals the inside of the liner by the interaction promoted between the frame and cover. The cover opening, which hinges at one of the frame's longitudinal edges, is carried out through hydraulic cylinders. The Solidworks FEM software and respective simulation package were used. The model was properly simplified from its original design, including removal of non-structural components, using pin connections, and replacing several bolted connections by bonded contacts.

In the setup of Figure 20.4, the cylinder has the largest coverage area, leading to the critically loading condition, since it must support one of the intermediate modules of the structure, and also the end module (which, in terms of weight, is approximately half of an intermediate module). This analysis will also allow to extrapolate the results to the symmetric counterpart of the structure. By loading the model under different in-service working scenarios, the static FEM analyses made possible to predict displacements, strains, stresses (including von Mises equivalent), and safety factors, enabling structural optimization and improvements up to reaching the final structure.

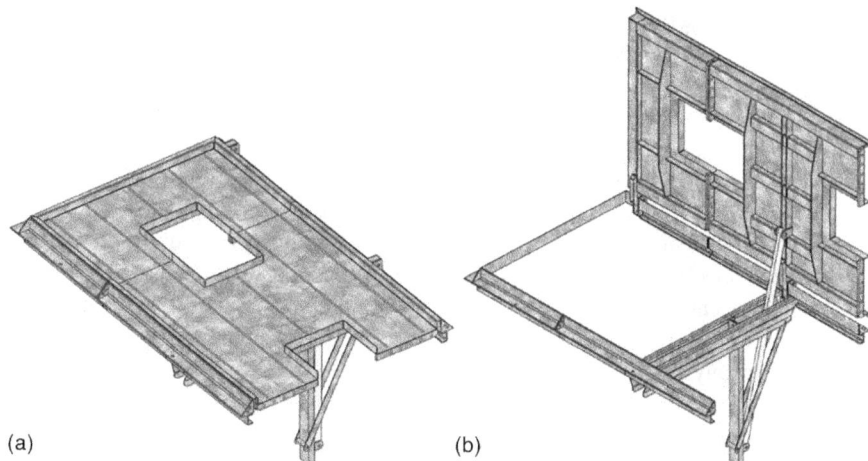

FIGURE 20.4 Simplified model perspective. (a) Closed cover. (b) Open cover.

Three loading cases were tested in the verification to the imposed loads: (i) closed cover, supported by the frame and crossbars, i.e., without cylinder action; (ii) cover opening onset, when it loses frame support and becomes supported by the cylinder; and (iii) fully opened cover. The applied loads are summarized in Table 20.3.

The imposed overload in the table is defined by legislation (standard EN 13071-2:2008) to emulate accidental overloads caused by vehicles or other heavy items, and it consists of applying and 500 kgf load in a surface with diameter of 200 mm at the worst possible location in the cover, but it only needs to be applied in case the cover remains closed. It was thus defined that, due to the waste collection workers' supervision that exists at the time of operating the equipment (who are instructed to perform all verification regarding the existing of obstructions to the cover operation), the overload is only applied to load case 1 (closed cover and resting cylinder). To perform the analyses, the materials were applied (consisting of either s235 or s275 structural steel), and the boundary conditions, loads, and connections were defined, which will be partially common to the three cases under study. The boundary conditions consisted of clamping the frame faces in contact with the soil and applying symmetry at the vertical symmetry plane. The loads were introduced in the form of gravity (structure self-weight), uniform pressure (floor, consisting of granite setts), remote load applied to the

TABLE 20.3
Loads Applied to the Structure

Load Type	Value
Metallic cover structure	248 kg (fixed value)
Floor	390 kg
Top bin	100 kg
Imposed overload	500 kgf (applied in a $\varnothing_{max} = 200$ mm)

center of mass (top bin), and concentrated force (overload). The connections are also essential in the pre-processing stage, since the geometrical relationships given in the model assembly no longer apply. Thus, a general bonded contact was enforced to all contacting surfaces, which results in rigid connections between elements. Furthermore, all cover/frame contacts were defined as no penetration, to ensure contact and perpendicular load transfer when entering in contact, by avoiding material overlap. All pins were modeled by the pin contact function, which emulates the real pin restrictions, i.e., concentricity between the holes it is connecting, and longitudinal restraint.

The mesh selection is a highly relevant part of the model preparation before running the analysis, to ensure converged and trustworthy results in view of the expected high stress concentrations due to sharp geometric changes (Faria et al. 2021). Thus, a mesh convergence analysis was carried out to choose the mesh parameters with the best ratio of accuracy/time spent in the study. This analysis was based on the resulting displacements (URES) and von Mises stresses measured in models with different mesh refinements. The load scenario 1 was selected, and the mesh defined as optimal was then applied to the other two scenarios. Five mesh refinements were tested, and Figure 20.5 shows the mesh for analysis 1 (lowest refinement), with emphasis to the reference points in which the measurements were performed. In this work, a curvature-based mesh was considered through all to avoid excessive refinement in regions with small stress gradients, by adjust the element size to the surrounding geometry, and enabling to save computational running time by coarsening the mesh at regions without excessive geometry distortions. Linear tetrahedrons were used thereof, as the standard elements used by the software, whose refinement depended on the results of the convergence check.

Through several analyses where stress and URES were plotted to verify convergence more easily, it was found that, from the third analysis (approximately 500,000 elements), the values converge. This effect was more evident in URES, whose results are depicted in Table 20.4. The relative variations between consecutive analyses (Δ_{URES1} and Δ_{URES2} for the respective points) for analysis 3 are in the order of 0.4% and 0.2%.

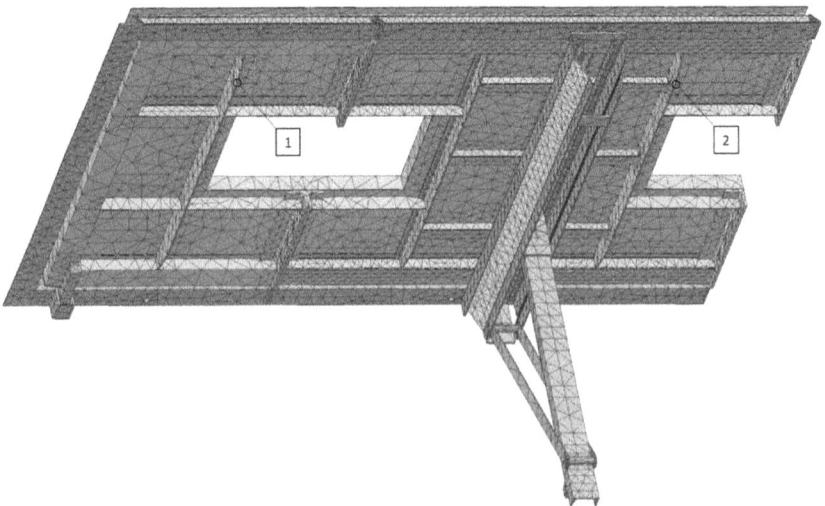

FIGURE 20.5 Location of the analyzed convergence points – example with analysis 1 mesh.

TABLE 20.4
URES Convergence Analysis Results

Analysis	URES 1	$\Delta_{URES\,1}$	URES 2	$\Delta_{URES\,2}$
1	4.303	–	2.502	–
2	4.382	1.84%	2.529	1.08%
3	4.398	0.37%	2.535	0.24%
4	4.394	−0.09%	2.534	−0.04%
5	4.492	−0.05%	2.534	0%

The decision to assume the parameters of analysis 3 was based mainly on the results obtained at the different points for the URES. On the other hand, the absence of unequivocal stress convergence is justified by the location selected for the measurement of stresses, i.e., at two vertexes of a reduced thickness rib, leading to potential miscalculations due to element distortions (Ataei and Mamaghani 2018). The final structure's model included minimum and maximum element sizes of 6 and 30 mm, respectively, in a total of 403,891 tetrahedral elements.

After the mesh convergence analysis, a strength analysis was performed to the three load scenarios. At this stage, it was also possible to define the critical scenario, in terms of displacements and stresses, leading to loading case 1 (closed cover with overload) as the one to focus on the design process. The initial run showed that was possible to lighten the structure, by clearly showing low stressed regions, whose redesign can bring weight advantages without compromising the structural safety. Moreover, input from the company design department was also included to optimize the structure. The main purposes of the optimization are the cost and weight reduction, the improved answer to the marked needs, and the possibility to build a modular structure. In brainstorming sessions with all involved actors, including the design department, production, and marketers, it was possible to highlight the main improvement points over the initial design, among which the following stand out:

- The structure is too robust compared to equipment from competitor companies;
- Some profiles can be removed or replaced by less robust profiles, accounting for the structural support of the concrete after equipment assembly;
- Connecting plates and fastening elements between modules can be lightened;
- The use of closed profiles should be reduced or eliminated, thus avoiding the further need to open holes, necessary for the penetration of the galvanizing liquid, as well as allowing the expansion of the material due to the temperature involved in the process;
- Explore the possibility of using standard hydraulic cylinders.

A model was built with these recommendations, which was then subjected to successive FEM analysis iterations until reaching a feasible solution. To perform this optimization, only loading case 1 was addressed, i.e., the critical one, although after reaching the final solution, the other cases were subjected to a final validation. All figures pertain to the final design. Figure 20.6 shows the von

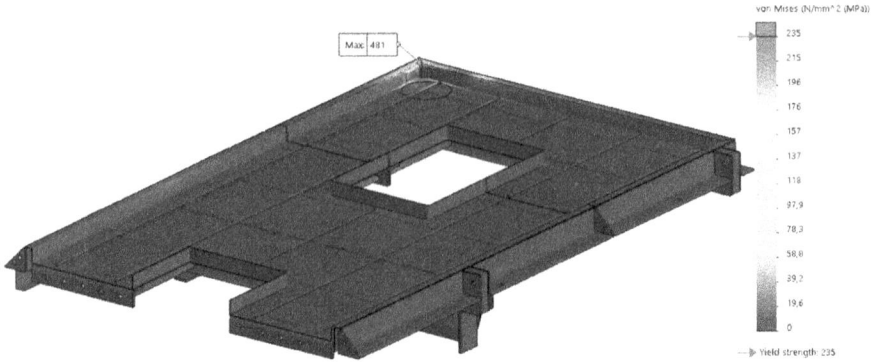

FIGURE 20.6 von Mises stresses in the structure for loading case 1.

Mises stresses for this loading case, emphasizing on the circular area for over-load application. For clarity, stresses are limited to the yield point of 235 MPa. The critically stressed region is clearly in the vicinity of the applied overload, as expected, while the remaining portions of the structure are markedly below the yielding point of the material. A peal value of 481 MPa is attained at the weld-ing junction between the longitudinal and transversal corrugated portions of the platform. However, this is a highly localized event, induced by sharp geometry changes and element distortion issues, which does not compromise structural safety. Moreover, the weld bead is executed with a stronger material than the structure to weld, locally allowing improved mechanical properties than in the structure in general. In the lower cover frame, including weld beads, the von Mises stresses do not exceed 100 MPa.

URES are plotted in Figure 20.7 for the same loading case (inside view). The maximum obtained URES was approximately 5.4 mm, close to the site of overload application. This value is not considered excessive, since it applies to a contact zone with the sidewalk where the equipment will be placed, and will be limited by the pavement. Additionally, the equipment's functionality is not compromised with this positional change to the original shape, thus validating the solution.

Table 20.5 compares von Mises stresses and URES between the initial and final designs, including the critical loading case (1) and the other 2, together with the max-imum stresses in the cylinder and hinge pins, used for design purposes. The reported pin stresses relate to shear stresses, which are under a double-shear load. Identical critical regions were identified for all three loading cases between the initial and final designs. Moreover, it was possible to keep or reduce stresses, and keep URES identi-cal, while reducing the overall structure's weight. Only initial values were considered for the pins, since the weight savings in the final analysis led to slightly smaller loads applied. Thus, design was accomplished for the initial loads applied to the pins.

Overall, a 15% mass reduction could be achieved between the initial and final designs. This reduction translates, for the complete equipment, into about 157 kg. The total value also adds to the mass of the components that have been removed from the simplified model, since these generically remain unchanged.

Max 5.43

URES (mm)

5.43
4.97
4.52
4.07
3.62
3.17
2.71
2.26
1.81
1.36
0.904
0.452
0

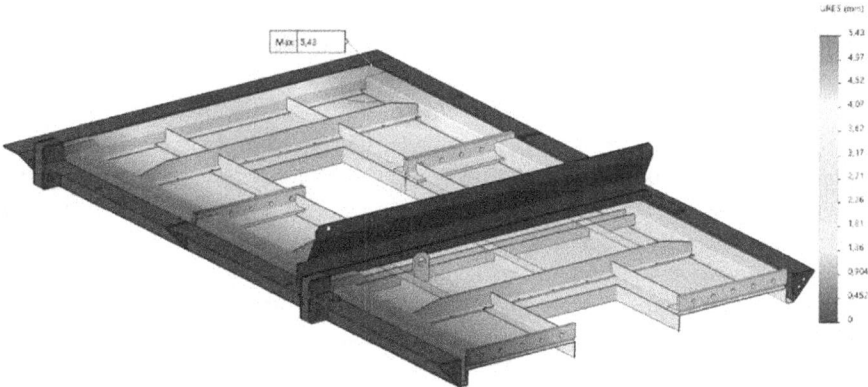

FIGURE 20.7 URES in the structure for loading case 1.

TABLE 20.5
Comparative Summary between the Initial and Final Structures

Analysis		Peak von Mises Stresses (MPa)	Peak URES (mm)
Closed cover with overload	Initial	578.0	5.4
(loading case 1)	Final	481.0	5.4
Cover opening onset	Initial	186.6	8.1
(loading case 2)	Final	183.0	9.3
Fully opened cover	Initial	31.6	0.3
(loading case 3)	Final	27.4	0.3
Hydraulic cylinder pins	Initial	31.4	–
Hinge pins	Initial	34.0	–

20.4.2 SAFETY DEVICE

After selecting the safety barrier as safety device for the equipment, it is first necessary to define that it should automatically deploy when the cover is opened, and retract to the safety position when the container is removed from the liner. According to standard EN 13071-2:2008, safety barriers must have a minimum height of 900 mm, from ground level to the top of the barrier, at all points around the equipment, without vertical interruptions exceeding 200 mm. If safety barriers consist of tubular profiles, they must have a maximum distance of 400 mm between horizontal elements (CEN 2019). The barrier shall be subjected to tests to verify its strength against static or impact loads while the container is not present. By the specifications of standard EN 13071-2:2008, the test procedure consists of the application of a horizontal load of 180 N, applied perpendicularly to the barrier surface, on a circular surface with maximum diameter of 200 mm. In the case of square or rectangular barriers, this load must be applied at their centroid. For circular barriers, the force of 180 N will be applied at 4 equidistant points along the barrier's perimeter, thus every 90°. The minimum test duration should be 1 minute. After the test, the barrier shall remain fully operational, and no permanent deformation or any other type of deformation that may hinder the use for which it was designed is permitted.

A simplified scheme of the barrier installed inside the concrete liner is presented in Figure 20.8, along with the respective nomenclature, for load calculation and supports' design. It was assumed that the safety device for each module should contain two linear guides for vertical motion, each one including a track rail and two slide units.

The slide units should be fixed to the concrete, while the rails are attached to the barrier. Actually, if the opposing solution was adopted, both slide units 1 and 2 would come out of the track rail, as the barrier would have to be slightly below ground level in order to allow the opening and closing of the cover of the buried equipment. The main variables are P (mass of the movable barrier assembly), M_i^x, M_i^y and M_i^z (bending moments over the three axes for slide unit i, with $i = 1$ and 2), R_i^y (reaction force in the direction y for slide unit i, with $i = 1$ and 2), R_z (reaction force required for barrier activation), T (total thickness of the safety barrier), and D (space required

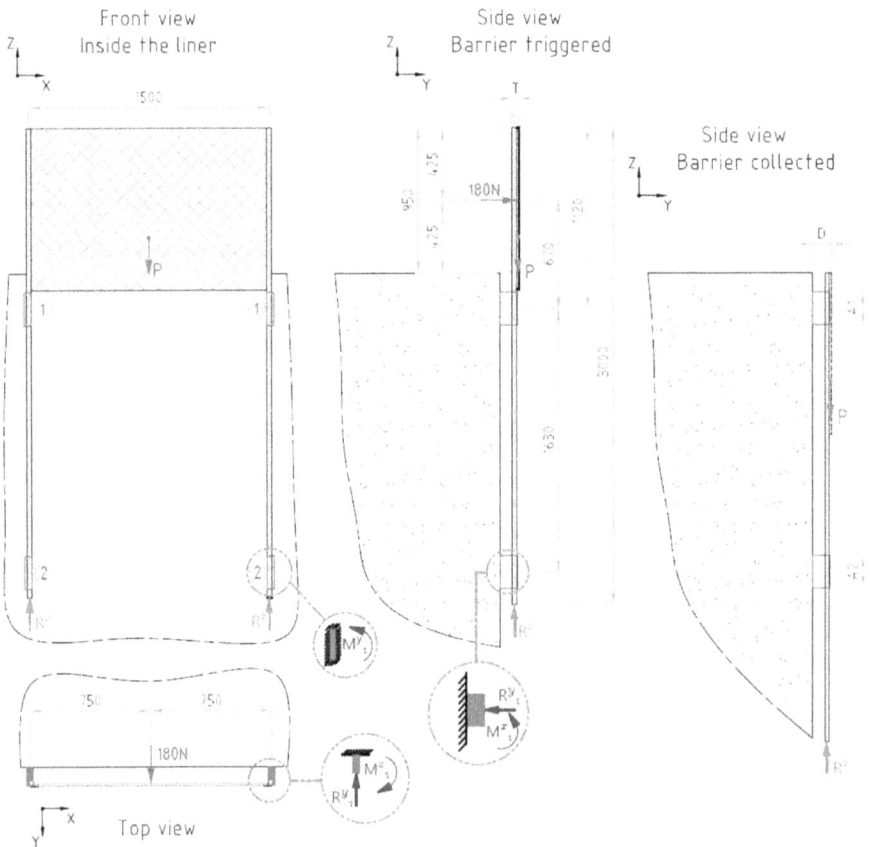

FIGURE 20.8 Simplified scheme for the predesign of the barrier and slide units.

for operation of the safety barrier). Basic strength of materials principles was used to calculate the reactions forces and moments, necessary to design the guides and actuator system.

After addressing several hypotheses of sliding mechanisms that can be applied to operate the barrier, it was established that sliding gate systems, usually applied in house garage gates, bring the most advantages. These systems, which are composed of the slide gate wheels and respective tracks, present a lower cost than precision solution, and enable a certain amount of backlash to the barrier system, both essential features for the application. Using two simplified models of the top and side barriers (where symmetry conditions were applied), in which the conditions imposed in the legislation were established, by mesh convergence analyses it was possible to optimize the geometry up to attaining the structural integrity of the barriers. The loading and boundary conditions for the side barrier are shown in Figure 20.9 as an example, including the symmetry conditions to model only haft barrier.

From the results presented in Figure 20.10, corresponding to the final design for the side barrier, it is visible that the highest von Mises stresses are located at the welding site between the tubular profile and the track. However, since the weld beads were not physically modeled, i.e., the different components were bonded together by their coincident faces as an approximation, it is considered that the peak values obtained at these junctions are overpredicted. Thus, the highest von Mises stress, of approximately 200 MPa, obtained at one of these junctions and at a very limited area, is regarded as acceptable. The maximum URES was 13.9 mm, which does not affect functionality and pertains to the barrier's elastic regime.

Thus, the design present in Figure 20.10, whose results can be qualitatively extrapolated to the top barrier, is considered validated. Actually, the studied barriers (side and top) structurally comply with the requirements imposed by the EN 13071-2 standard, thus eliminating the risk of falling into the liner. Moreover, for a given application, if it becomes more advantageous to apply another type of solution to ensure the maximum distances between horizontal and vertical elements, such as the application of a steel mesh, instead of the vertical elements applied in the simulation, this solution can be applied as long as it complies with the legislation in terms of response to the load imposed in the EN 13071-2 standard.

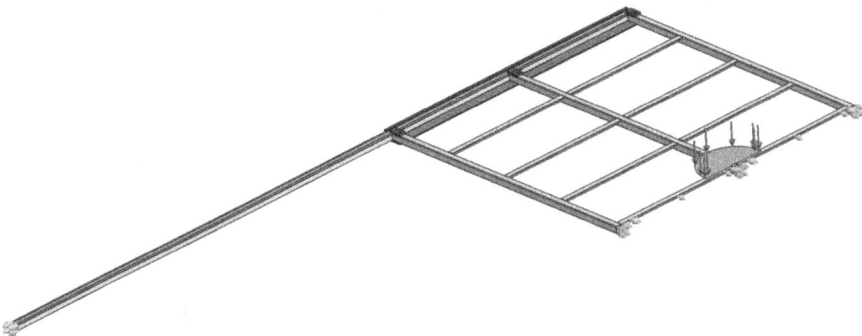

FIGURE 20.9 Application of the boundary conditions and loads in the simplified model.

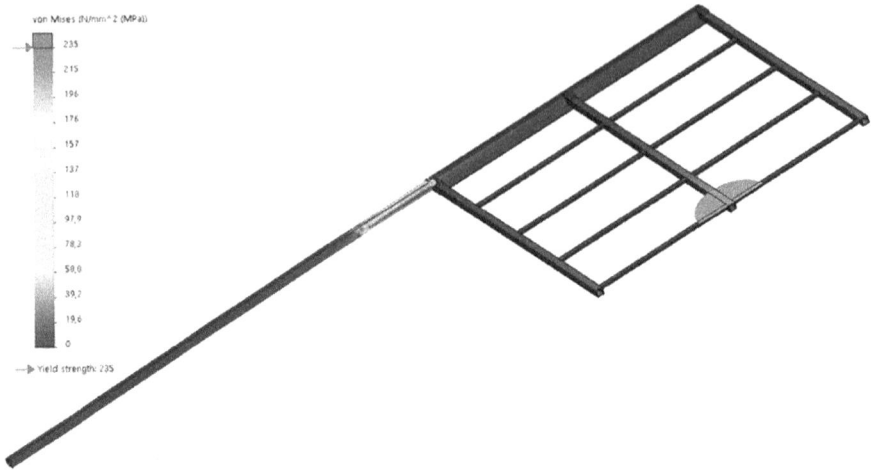

FIGURE 20.10 von Mises stress: frontal perspective of the symmetrical lateral barrier simulation.

20.4.3 Top Bin

The top bin design mainly consisted of the lid and commercial (back) door mechanisms, namely, the working principle and component stress verifications, since the overall bin structure design, namely, the plate thickness, is already stipulated by the client in the specifications, accounting for exceptional loads such as misuse, vandalism, and impacts. On the other hand, the mechanisms that will ensure functioning of the bin moving elements must provide correct functionality through the defined lifetime for the structure. For the lid drive mechanism, a lever system and a transmitting arm system were equated. A detailed analysis led to the conclusion that the transmitting arm would perform better, and thus, this solution was further developed. Figure 20.11 and Figure 20.12 present the working principle and free-body diagrams (FBD) serving as basis for the design process.

After the static analysis performed for the transmitting arm, it was concluded that, to actuate the lid, $P \approx 15$ kgf is required. However, P can be reduced by an upward force in C (e.g., a spring). The equilibrium equation is:

$$P = 0,25 \times 8,59M + 0,11C \qquad (20.3)$$

In addition to the lid, there is also the commercial door mechanism that allows the deposition of larger waste by authorized traders. This door is optional in the top bin. Nonetheless, Figure 20.12 presents the respective design principle and relevant diagrams. From the analysis, it follows that two gas dampers (G) are required, each loaded in compression with approximately 250 N, to ensure that the door will close after being externally opened. The closing function is promoted by a spring latch.

20.5 PROTOTYPE CONSTRUCTION AND TESTING

Prototype construction and testing (for functionality/structural safety check) were carried out. For both HS4 structure and top bin, a thoughtful check was initially

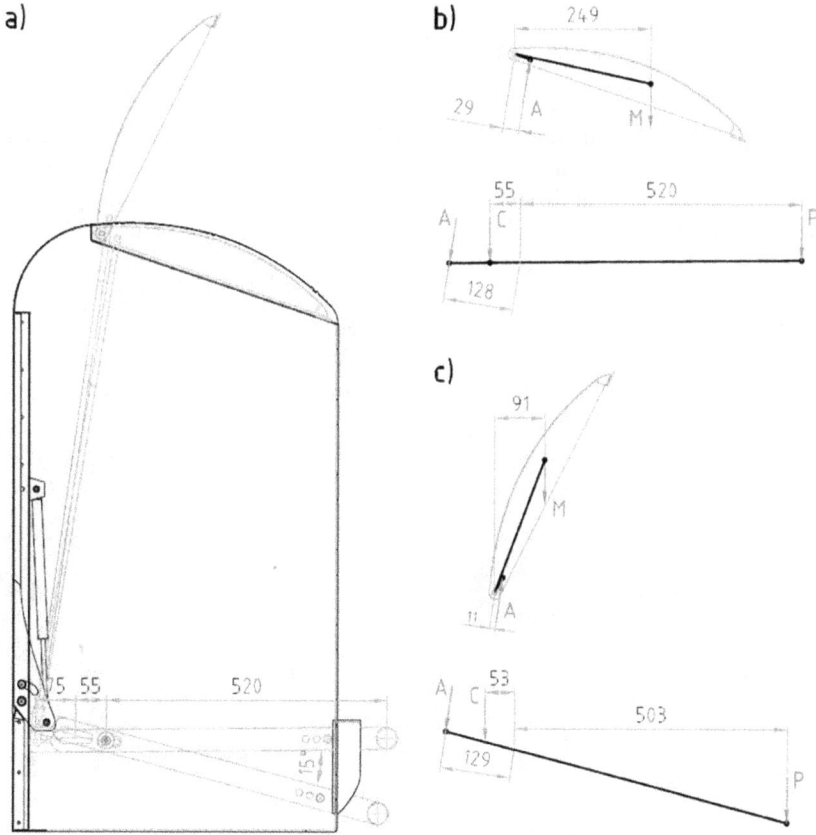

FIGURE 20.11 Top bin schematics (lid details): section view with opened/closed lid (a), FBD of the mechanism with closed lid (b) and FBD of the mechanism with opened lid (c).

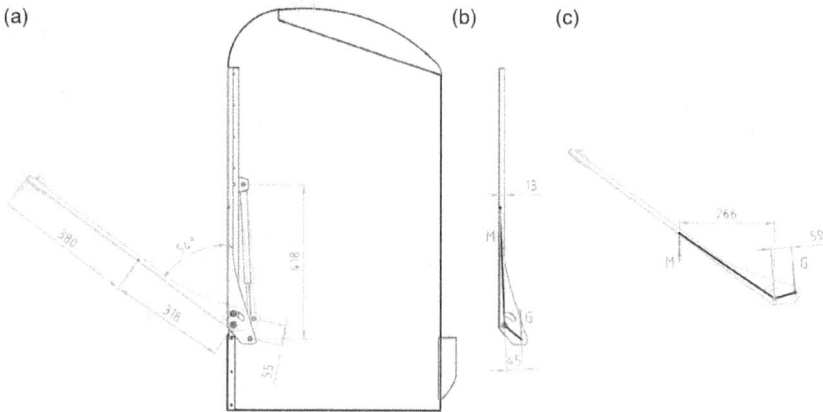

FIGURE 20.12 Top bin schematics (commercial door details): section view with opened/closed door (a), FBD of the mechanism with closed door (b) and FBD of the mechanism with opened door (c).

undertaken at the factory regarding the desired motion between components and correct functionality, and also replicating the in-service loads, including, for instance, the overload application to emulate the effect of vehicles or other heavy items over the platform. Then, the equipment/top bin components were carefully checked for damage and permanent deformations, leading to the absence of measurable issues. Thus, the designs were validated, and the equipment was subsequently deployed in the street for real future operation at different locations. This device is structurally consistent with the proposal developed in this paper, with the sole difference in the safety device consisting of applying perforated plate with constant pattern in the safety device using the same spacing between consecutive bars. This solution is more aesthetically appealing, but results in higher weight compared to the base solution (by approximately 20 kg). Figure 20.13 shows two examples of HS4 equipment implementation: in an inclined street (a) and opened base structure under assembly in flat ground

FIGURE 20.13   HS4 equipment in an inclined street (a) and opened base structure under assembly in flat ground (b).

(b). It should be mentioned that, for the equipment of Figure 20.13 (a), the top bins relate to a previous version (with drum).

Figure 20.14 shows a partial view of a container being loaded into the liner in a complete HS4 equipment including the safety device. The container loading and unloading operations took place smoothly, although the clearance between the container and liner can be enlarged for improved maneuverability regarding the process of collecting USW from the containers. As a result, the truck operators will be able to drive the container more easily at the time of collection and insertion in the liner.

Figure 20.15 presents an assembled top bin out of the factory (a) and an HS2 equipment with two top bins installed (b). With the proposed solution for the lid and commercial door, the functionality and operability were as planned, and the top bin was validated for mass use in HS equipment.

FIGURE 20.14 Partial view of a container being loaded into the liner in a complete HS4 equipment including the safety device (opened pedestrian platform to the right).

FIGURE 20.15 Assembled top bin (a) and HS2 equipment with two top bins installed (b).

20.6 COST ANALYSIS

The equipment cost analysis can be approximately divided into the cost of raw material, labor, and surface treatment costs. Table 20.6 presents an analysis for the HS4 equipment with a safety barrier (the values shown in the table were multiplied by an unknown factor due to company confidentiality issues). The obtained results show an approximate 6% reduction in the cost of the equipment after the study carried out throughout this paper, due to improvements in some aspects, such as redesign and changes to the current HS structure. In addition, the weight reduction was 15%. It should also be mentioned that the cost of the security barrier and its implementation represents about 13% of the cost of HS4 equipment, which represents an acceptable investment in view of fulfilling the legislation and ensuring operational safety. With the proposed changes, the possibility of the structure presenting damage from surface treatments also significantly decreases due to applying the cover plate after the galvanizing process.

Table 20.7 presents the estimated cost values (multiplied by the same unknown factor) for the Cambridge top bin, including the lid, commercial door and pedal mechanisms, and with its structure in AISI304 stainless steel.

Considering that each HS4 equipment has 4 bins, for an HS4 equipped with Cambridge top bins, the final cost value, considering the revised structure, will be approximately 9,450 €. This cost represents a significant unit reduction in the total equipment cost but, for a higher production quantity (series production), it is possible to further reduce the production costs of the equipment.

TABLE 20.6

Cost Analysis for the Current and Revised HS4 Structure Design

	Current Project Cost	Revised Project Cost
Raw material	1,919.44 €	1,827.54 €
Labor	1,099.10 €	964.10 €
Surface treatment	853.90 €	707.00 €
Assembly (including liner)	3,200.00 €	
Total cost	7,072.44 €	6,698.64 €
% cost reduction	5.93%	

TABLE 20.7

Cost Analysis for the Top Bin (4 Units)

	Cost
Raw material	455.50 €
Labor	232.25 €
Total cost	687.75 €
Total cost (4 units – HS4)	2,751.00 €

20.7   CONCLUSIONS

The waste storage process plays a fundamental role in the waste treatment cycle. With the existence of an increasingly restrictive legislation regarding the reuse of recycled materials and the reduction of waste deposited in landfills, the development and continuous improvement of collection equipment is a decisive factor to deal with regulation impositions enforced to avert the waste of recyclable resources. The design of suitable mechanical solutions requires an entire process of research and development, aiming to answer current market needs, taking costs into account, and avoiding overdesign, to produce competitive solutions within an ever-evolving market. With this study, it was possible to lower the cost of an HS4 equipment by about 6% compared to the current equipment, through the redesign of several components. A 15% reduction in weight was also accomplished, which culminates in a reduction in the cost of surface treatment. This improvement, in addition to lowering the cost (galvanizing operations), also facilitates assembly. The design of a safety device adaptable to existing and future equipment also assured full compliance with applicable standards at the level of the European community. Additionally, in response to the recent market demand for top bins with lid and pedal systems, a new device was designed (Cambridge top bin) that allows the insertion of higher volume waste through the lid, which has a significantly larger depositing area than the drum system that is currently the most used.

REFERENCES

Aisa, J., Javierre, C. and De la Serna, J. A. (2006). An example of simulation tools use for large injection moulds design: The CONTENUR™ 2400 l solid waste container. *Journal of Materials Processing Technology* 175(1): 15–19.

APA. (2019). *Persu 2020.* Zambujal, Portugal: Agência Portuguesa do Ambiente (in Portuguese).

Ashby, M. F. (2016). *Materials Selection in Mechanical Design.* Oxford, UK: Butterworth-Heinemann.

Ataei, H. and Mamaghani, M. (2018). Finite Element Analysis: Applications and Solved Problems Using Abaqus®. Scotts Valley, EUA: CreateSpace.

Bergan, P. and Nygård, M. (1984). Finite elements with increased freedom in choosing shape functions. *International Journal for Numerical Methods in Engineering* 20(4): 643–663.

Bi, Z. (2018). *Finite Element Analysis Applications.* Cambridge, USA: Academic Press.

Budynas, R. and Nisbett, K. (2010). *Shigley's Mechanical Engineering Design.* New York, USA: McGraw-Hill Education.

CEN. (2010). *Eurocode - Basis of Structural Design.* Brussels, Belgium: Comité Européen de Normalisation. EN 1990:2002+A1.

CEN. (2019). *Stationary Waste Containers up to 5 000 l, Top Lifted and Bottom Emptied - Part 1: General Requirements.* Brussels, Belgium: Comité Européen de Normalisation. EN 13071-1.

Chang, K.-H. (2014). *Design Theory and Methods Using CAD/CAE: The Computer Aided Engineering Design Series.* Cambridge, USA: Academic Press.

de Souza Neto, E. A., Perić, D. and Owen, D. R. J. (2008). *Computational Methods for plasticity, Theory and Applications.* Hoboken, USA: Wiley.

Dieter, G. E. and Bacon, D. J. (1986). *Mechanical Metallurgy.* New York: McGraw-Hill.

Dowling, N. E. (2012). *Mechanical Behavior of Materials: Engineering Methods for Deformation, Fracture, and Fatigue.* London, UK: Pearson.

Drucker, D. C. and Prager, W. (1952). Soil mechanics and plastic analysis or limit design. *Quarterly of Applied Mathematics* 10(2): 157–165.

ERSAR. (2016). Caraterização do setor de àguas e resíduos (in Portuguese). Retrieved October 5, 2019, from http://www.ersar.pt/pt/publicacoes/relatorio-anual-do-setor.

Faria, N., Campilho, R., Silva, F. and Ferreira, L. (2021). Concept and design of automated moving device for healthcare equipment. *FME Transactions* 49(3): 598–607.

Flom, Y. (2013). *Strength and Margins of Brazed Joints.* Advances in brazing. D. P. Sekulić. Sawston, UK: Woodhead Publishing.

Guimarães, J. D. O. (2015). *Comando de movimento de plataforma elevatória com sincronização de cilindros hidráulicos (in Portuguese).* M.Sc. Thesis, Faculty of Engineering of the University of Porto.

Haftka, R. T. and Gürdal, Z. (2012). *Elements of Structural Optimization.* Berlin, Germany: Springer.

Johnson, G. R. and Cook, W. H. (1985). Fracture characteristics of three metals subjected to various strains, strain rates, temperatures and pressures. *Engineering Fracture Mechanics* 21(1): 31–48.

Labuz, J. F. and Zang, A. (2012). Mohr–coulomb failure criterion. *Rock Mechanics and Rock Engineering* 45(6): 975–979.

Laranjeira, L. (2008). *Desenvolvimento de um sistema de recolha de residuos sólidos urbanos (in Portuguese).* M.Sc. Thesis, Faculty of Engineering of the University of Porto.

Moës, N., Dolbow, J. and Belytschko, T. (1999). A finite element method for crack growth without remeshing. *International Journal for Numerical Methods in Engineering* 46(1): 131–150.

Okereke, M. and Keates, S. (2018). *Finite Element Applications: A Practical Guide to the FEM process.* Berlin, Germany: Springer.

Park, K. and Paulino, G. H. (2013). Cohesive zone models: A critical review of traction-separation relationships across fracture surfaces. *Applied Mechanics Reviews* 64(6): 060802.

Rodrigues, S. S. M. (2017). *Classificação e benchmarking de sistemas de recolha de resíduos urbanos (in Portuguese).* Ph.D. Thesis, Nova University Lisbon.

Schäfer, M. (2006). *Computational Engineering: Introduction to Numerical Methods.* Berlin, Germany: Springer.

Simões, R. A. D. (2005). *Manual de dimensionamento de estruturas metálicas (in Portuguese).* Coimbra, Portugal: Associação Portuguesa de Construções Metálicas e Mistas (in Portuguese).

Sobolev, A. V. and Radchenko, M. V. (2016). Use of Johnson–Cook plasticity model for numerical simulations of the SNF shipping cask drop tests. *Nuclear Energy and Technology* 2(4): 272–276.

Teixeira, S., Monteiro, E., Silva, V. and Rouboa, A. (2014). Prospective application of municipal solid wastes for energy production in Portugal. *Energy Policy* 71(August): 159–168.

Index

Note: Bold page numbers refer to tables; *italic* page numbers refer to figures.

abrasion value test, aggregate 297, **297**
absorption test, vacuum water 292
acetylation 262
acid and enzymatic hydrolysis technology 132
acid hydrolysis 134
acoustic impedance 356
adsorbents
    performance of 60
    SWOT analysis of transforming agro-wastes
        into 262, **262**
advanced oxidation processes (AOPs) 61, 326,
        331, **331,** 332
aeration of wastewater 57
Africa
    medical waste management (MWM) in 122
    poverty, and human rights violations 148–151
aggregate abrasion test 297, **297**
aggregate abrasion value test 297
aggregate crushing test **296,** 296–297
agricultural soils, heavy metals in 227
agro-wastes
    in adsorbents productions 259–262
    conversion to biofuels 252–256, *253,* **254–255**
    for cooking, warming, and lighting 250
    disposal of 251
    generation in Nigeria 245, **247,** 248
    health and environmental implications of
        open burning of 248–249
    random dumping of 251–252
    transformation to biofuels 252–253,
        255–256, **256**
agro-wastes management in Nigeria 242–263
    agricultural production and contribution to
        national economy 244–245, *245, 246*
    methodology deployed in survey 243–244
air consumption, system for measurement 276
air pollutants, decontamination of 233–234,
        227–228
algal biomass 50
amino acids 343
    molecular interactions of 344
    non-essential 356
ammonia volatilization 52
anaerobic lagoons 53
anatomical waste 73
anoxic biodegradation 36
anoxic degradation 36
antimicrobial drugs 226
AOPs *see* advanced oxidation processes (AOPs)

apparent molar isentropic compressibility
        349–350
    transfer parameter of limiting 351
aquatic environment, pollution in 225–226
aquatic lentic ecosystems 225
Archimedes Principle Kit 110
aromatic hydrocarbon contamination 226
ash content (AC) 110
aspartic acid (Asp) 345, 346, **347, 354**
ASSOCHAM (Associated Chambers of
        Commerce and Industry of India) 6
atmospheric air 227
atomic bombing of Japan 146
autoclaving 83

bacterial biodegradation of dyes 232
Bamako Convention 153
Basel Convention 10, 152–153
batteries 169
benthic organisms, oil suffocates 48
benzene, toluene, ethylbenzene, and xylene
        (BTEX) 135–136
bioalcohols 253
biochar (pyrolysis char) 259–260
    gasification char 261
    textural properties of **260**
    torrefaction char 261
biodegradable waste 13
biodegradation 24, 35
    anoxic 36
    oxic 35
    oxo-biodegradations 34–35
bioenergy
    generation 256–258, **258**
    principle of deploying waste to 256–257
    studies report 257, **258**
biofuels
    agro-wastes conversion to 252–256, *253,*
        254–255
    agro-wastes transformation to 255–256, **256**
    studies report 253–255, **254–255**
    SWOT analysis for approach deployment in
        255–256, **256**
biogas 253
biological cycle of nutrients 24
biological degradation of microbes 36
biological oxygen demand (BOD) 53
biomass, algal 50

biomass cookstoves 96
    history of testing methods 96–101
    traditional 107
biomedical wastes (BMWs) 122
    COVID-19 122
Bio-Medical Wastes (Management and Handling)
        Rules 6
biomethane 253
biopower generation 256–258
bioreactor, membrane distillation 58
bioremediation 35
    integration of 40
    of pesticides 233
biosphere 145
bio-treatment 58
biowastes 110
    plant-based 110, *110*
    thermo-physical characteristics of 110–111, **111**
Black's Law 146
black water 49
BMEP *see* brake mean effective pressure
        (BMEP)
BMWs *see* biomedical wastes (BMWs)
brake mean effective pressure (BMEP) 279, 285
brake power 277
brake-specific fuel consumption (BSFC) 278, 284
brake thermal efficiency 279
breathing contaminated air 227–228
brown water 49
BSFC *see* brake-specific fuel consumption
        (BSFC)
burning of wastes 248–249
by-products
    of composting 36–37
    toxic 34

calcium nitrate (CN) 190, 191
calorimeter, for exhaust gas 276
camphor sulfonic acid (CSA) 134
    MCW with 134
carbonization
    hydrothermal 260
carbon nanomaterials 59–60
carbon nanotubes 59–60
cartridges, toner and ink 169
catalysts 61
catalytic degradation 35
catalytic pyrolysis
    value-added aromatics from wasted COVID-
        19 mask through 135–136
catalytic waste transforms 35
CCT *see* controlled cooking test (CCT)
cellulosic enzymatic hydrolysis 132
Central Environmental Authority (CEA) 306
ceramic tiles 291
ceramic waste coarse aggregate (CWCA) 294
chamber method 100
CHD *see* chemical disinfection (CHD)

chemical attack 194
chemical disinfectants 131
chemical disinfection (CHD) 83–86, 130–131
chemical oxidation, applied AOPs systems in **331**
chemical waste 73
chemolysis 34, 40
Chlorella vulgaris 51
circuit boards 169
class F fly ash 206–210
Clean Air Act 147
CNN 164, 180
coarse aggregates
    gravity test on **295,** 295–296
    water absorption test on 296, **296**
CO emission *117,* 117–118
cohesive zone modeling (CZM) 363
combustion pressure 276
complex fracture-mechanics-based models 363
compost at home
    concept generation 18, *19*
    design development 19, *20–22,* 23
    design intervention with the stakeholders
        23, *23*
    field study 18, *18*
    problems 18
    screening of concepts 18, **19**
composting
    by-products of 36
    of plastics 36–37
compressive strength 191, 216, *217–218*
    correlation between 194
    test 213–214, 297
    for waste replace concrete 214
computer-aided engineering (CAE) tools 363
concrete 205, 291
concrete cubes, electrical resistivity (ER) of 191
concrete test, durability of 216, **216**
constructed wetlands (CWs) 51–52, *52*
contaminants, plants for removal of 54, **55**
contamination, aromatic hydrocarbon 226
controlled cooking test (CCT) 96, 101
control panel, with full instrumentation set 275
conventional activated sludge (CAS) systems 326
conventional gasoline fuel, MPFI SI engine using
        282–285, *283, 284*
conventional remediation techniques 234
conventional techniques 29
cooking 95
    agro-wastes for 250
cooking practice, firewood in 108
cookstoves 95, 109, *109*
    biomass 96
    evaluation parameters 112–114
    performance of 96, 112
    sampling setup 112
    testing protocol 112
    thermal behaviors of 112
    traditional biomass 107

copper slag 206, 208, **209, 210**
coral reefs 48
COVID-19 70–71, 122, 125
   biomedical wastes (BMWs) 122
   infected person 74
   medical waste (MW) 130–131
   medical waste management (MWM)
      77–82, 128
   medical waste transportation 125
   new waste management provisions
      during **80**
COVID-19 masks
   conversion of COVID-19-infected masks 137
   generation of value-added aromatics from
      wasted 135–136
   production of butane from the monthly
      production of 136–137
cow dung powder 60
crack angle, measurement 276
crude oil-based hydrocarbons 233
crushed tile coarse aggregate (CTCA) 297–299
crushing test, aggregate **296,** 296–297
CWCA *see* ceramic waste coarse aggregate
   (CWCA)
CWs *see* constructed wetlands (CWs)
CZM *see* cohesive zone modeling (CZM)

decontamination
   of air pollutants 233–234
   of dyes 232
   of heavy metals 231–232
   of pesticides 232–233
   of petroleum hydrocarbons 233
degradation 34
   anoxic 36
   catalytic 35
   mechanochemical 35
   ozone 35
   photocatalytic 59
   photo-oxidative 34
   of plastics, microbes **38–39**
   thermal 35
depolymerization 35
DFC *see* dry fuel consumed (DFC)
dioxins 129
disinfection
   chemical disinfection (CHD) 83–86
   microwave 83
   ozone 61, *62*
   technology, chemical 130–131
disposable material culture 13
disposal of untreated waste 47
domestic cooking practice 107
drinking water, source of 226
drugs, antimicrobial 226
dry fuel consumed (DFC) 113
dry plastic 36
dumping wastes 190

dyes
   bacterial biodegradation of 232
   decontamination of 232

eco-friendly adsorption technology, wastewater
      treatment 60–61
ecosystems, aquatic lentic 225
electrical resistivity (ER) 190, 192–194, *193,* 194,
      196–197
   of concrete cubes 191
electronic components and effect on health **2**
emission
   CO *117,* 117–118
   mercury 129
   parameters 99–100
   testing 99–100, *100*
energy recovery efficiencies (EREs) 133
Engine Analysis software 274
Engine Assessment Software 277
environmental pollution, by pesticides 232–233
Environmental Protection Act 1986 6
environment and economic analysis 199
environment issues, human health and 9
Environment Protection Agency (US) 122
enzymatic hydrolysis of cellulose 132
enzymes 36
enzymolysis 132
EPR *see* extended producer responsibility (EPR)
EREs *see* energy recovery efficiencies (EREs)
European Union's policy 7
e-waste
   composition 165
   defined 164
   dumping 10
   economy in unorganized sector 172
   examples of **165**
   as global crisis 304–305
   image recognition 180–181
   impact on society 4–5
   inventory of 164–165
   items 2–3
   as lucrative business 305
   movement of 4
   primary e-waste generators 168
   recycling 168–169
   solution for manual e-waste retrieval 181
   sources 5
   toxic metals in **166**
   types 165
e-waste disposal 168–171, **169**
   landfill disposal 169
   microscopic piping strategy 170–171
   physical disposal processes 168–169
   technologies for disposal processes
      169–171, *170*
e-waste generation 9, 164
   in India 166–167, *167*
   stakeholder involvement in 167–168, *168*

e-waste management 1–2
  measurement in other countries 6–8
  measurement within India 5–6
  purposes of global legislation 8
  recycling 9
  social awareness 9
  Uniform Law 8–9
  using deep learning object classifier 179–180
e-waste management and handling, India rules
    for 172–179
  applicability 172–173
  collection centers' responsibilities 176
  consumer's/bulk consumer's responsibilities
    177–178
  dealers' responsibilities 176–177
  dismantler's responsibilities 178
  manufacturer responsibilities 174
  producer's responsibilities 174–176
  recycler's responsibilities 178–179
  refurbisher's responsibilities 177
  responsibilities 174–179
  rules 173–174, **174**
  state government responsibilities 179
exhaust gas, calorimeter for 276
extended finite element modeling (XFEM) 363
extended producer responsibility (EPR) 6, 8

FAOSTAT 244
Federal Environmental Protection Agency
    (FEPA, 1988) Decree 155
FEM 362, 363, 372
Fenton reagent 61
Fenton technology 61–62
fertiliser, topsoil 3
firewood, in cooking practice 108
flagships 29
flexural strength 220, *221–222*
  test 214, **215**
flexural test, for waste replacement concrete **215**
floating treatment wetland (FTW) 56–57, *57*
flood water 49
flow control, arrangement for 275
flow sensors, for measurement of water flow 275
fly ash
  class F 206–208
  debris 206
food waste 137
fracture-mechanics-based models, complex 363
free radical technique 29
fresh concrete, test on 213
frictional losses 282
FTW *see* floating treatment wetland (FTW)
fuel flow measurement 276
fumes, non-condensable 33

gasification *33,* 33–34
  char 261
GDI test rig 289, *289*

global e-waste creation 171
globalisation, effect of 4
global legal regimes 151–153
global positioning system (GPS) 182
global waste trade sector 171–172
GPS *see* global positioning system (GPS)
granite waste powder 190
gravity test, on coarse aggregates **295,** 295–296
gray water 49
ground dwelling plants 56
groundwater 225

Harmful Waste Act 147, 154
Hazardous and Solid Waste Amendment 7
hazardous liquid waste 49
hazardous waste 146–147
  control of hazardous waste in Nigeria
    154–157
  illegal trade in 148–149
  management of 74
health and environmental implications
  agro-wastes for cooking, warming, and
    lighting 250–251
  of open burning of the agro-wastes 248–249
  of randomly dumping wastes 251–252
  using traditional (undesigned) landfills
    249–250
healthcare waste 70, 124
health effects, microplastics and 28
health, electronic components and effect on **2**
heavy metals
  in agricultural soils 227
  decontamination of 231–232
hemodialysis, medical waste (MW) generated
    from 123
HF constructed wetlands (CWs), VF CWs with 52
high-temperature incineration technology
    128–129
high-temperature pyrolysis technology 129
homogeneous photocatalysis 63
Hook System (HS) buried equipment 367
hospital waste 70
HS4 equipment 367, *368, 382, 383, 383*
human health, and environment issues 9
hydrocarbon
  contamination, aromatic 226
  crude oil-based 233
hydrocracking *32,* 32–33
hydrogenation 32
hydrolysis
  acid 134
  acid and enzymatic 132
  cellulosic enzymatic 132
hydroponic plants, roots of 56
hydrothermal carbonization (HTC) 260
hydroxide in water 50
hydroxyl radical (OH) 61
hyperaccumulators 54

image-matching algorithms 183–185
IMEP *see* indicated mean effective pressure
(IMEP)
IMS 109, *109,* 116
thermal efficiency (TE) of 114–115
incineration
primary by-products of 30
technology, high-temperature 128–129
India
e-waste generation in 166–167, *167*
rules for e-waste management and handling
172–179
indicated mean effective pressure (IMEP)
279, 285
indicated specific fuel consumption (ISFC) 278
indoor air 227
industrial development 28
industrial emissions directive (waste incineration 8
infectious liquid medical waste (MW) 130
infectious waste 73
insects, attracted to kitchen 16, *16*
integrated waste management 361
integration technologies 37
interaction coefficients, pair and triplet 352
internal pressure 355–356
International E-Waste Day 8
International Union for Conservation of Nature
reports 28
isentropic compressibility 346–348
apparent molar 349–351
ISFC *see* indicated specific fuel consumption
(ISFC)
isothermal compressibility 349–350

Khian Sea waste disposal incident 3
kitchen, insects attracted to 16, *16*
kitchen performance test (KPT) 96, 101
kitchen waste *16,* 17, 24
disposal of 16, *16,* 17, *17*
reusing 23–24
Kota stone dust 190
KPT *see* kitchen performance test (KPT)

lagoons, anaerobic 53
landfill
different ages of 325–326
dumping of agro-wastes on traditional
249–250
e-waste disposal 169
technology, sanitary 127–128
landfill leachate (LFL) 324, 327
balance of COD in organic fraction in **324**
production of 327–328
technique **335–336**
LDCs 150
leachates
anaerobic digestion process *330*
biological treatments of 330

characteristics constituents inorganic
particles **328**
with domestic sewage, treatment 328
nanofiltration (NF) 333, *336*
parameters of 328–329
physicochemical treatment 331
production and characteristics 327–328
recycling 329
removal of contaminant by oxidation using
common oxidizing agent 331
separations of leachate by membrane filtration
332–333
treatment **334–335**
treatment by membrane separation
process *332*
treatment of leachate by using reverse
osmosis (RO) 333
ultrafiltration (UF) 332–333
LFL *see* landfill leachate (LFL)
lighting, agro-wastes for 250
liquefaction 33
liquid medical waste 71
liquid waste
hazardous 49
trade 49
treatment 49–50
types 49
load cell, torque measurement 275
L-valine, volumetric and compressibility
behaviors of 344

macro-algae 50
Management and Handling Rule 5
marine water 226
MBR *see* membrane bioreactor (MBR)
MC *see* moisture content (MC)
MCW *see* medical cotton waste (MCW)
mechanochemical degradation 35
medical cotton waste (MCW) 131–132
with camphor sulfonic acid (CSA) 134
medical waste (MW) 70
categories 71, **72**
classification 71, 73–74, 122–124
composition 71
functional progression of 124
generated from hemodialysis 123
generation in some countries 78–79
handling 124–125
infectious liquid 130
liquid 71
organic substances o 129
physicochemical composition of 71
solid 73–74
sorting 124
sources 70–71
systematic collection of 123
types of 70–71
medical waste disposal (MWD) 122, 124, 127, 128

medical waste generation (MWG) 124–125
medical waste generation rates (MWGRs)
     122–123, 138
medical waste management (MWM) 75–82, 123,
     126, 127, 133
  in Africa 122
  collection 76–77
  in context of pandemic COVID-19 77–82
  for COVID-19 limits 128
  and disposal 79–82, **80**
  landfill 77
  model 125
  packaging 75–76, **76**
  storage 77
  transportation 77
  treatment 77
  waste sorting and conditioning 75–76
medical waste treatment (MWT) 82, 125–126, *127*
  acid and enzymatic hydrolysis technology 132
  chemical disinfection (CHD) 83–86
  chemical disinfection technology 130–131
  energy, fuels, and materials as value added
        products generated by 134–135
  high-temperature incineration technology
        128–129
  high-temperature pyrolysis technology 129
  incineration 83
  medium-temperature microwave
        technology 130
  microwave disinfection 83
  plasma technology 131
  pressure steam sterilization technology 130
  reverse polymerization (RP) 83
  sanitary landfill technology 127–128
  steam/autoclave disinfection (STD) 83
  torrefaction technology 131–132
  transportation to MWT locations 125, *126*
medium-temperature microwave technology 130
membrane bioreactor (MBR) 58, 326–327
membrane configuration factor 58
membrane distillation bioreactor (MDBR),
        wastewater treatment 58, *59*
MENA countries, treatment process of medical
        wastes 82
  chemical disinfection (CHD) 83–86
  incineration 83
  microwave disinfection 83
  reverse polymerization (RP) 83
  steam/autoclave disinfection (STD) 83
mercury 169
  emissions 129
metal nanoparticles, zero-valent 60
metal oxide nanoparticles 59
micro-algae 50, 232
microalgal remediation 50–51
microbes 36
  biological degradation of 36

  for degradation of plastics **38–39**
microbial cells 36
microbial compositing, of plastic materials 36, *39*
microbial degradation, of plastics 35, *36*
microbial extracellular enzymes 36
microbial remediation mechanism 228, 231, *231*
microfiltration (MF) 332
microfluidics 170–171
microorganisms 50
  removal of pollutants by 228, **229–230**
microplastics and health effects 28
microstructural analysis 197, *198,* **199,** *199*
microwave disinfection 83
microwave technology, medium-temperature 130
mix proportioning, methodology and 210, **212,**
        212, 213
moisture content (MC) 110–111
molar free length 352–354
molar free volume 352–354
MPFI SI engine 287–289, *288*
  using conventional gasoline fuel 282–285
MST 83
MSW *see* municipal solid waste (MSW)
municipal solid waste (MSW) 323–324
  management of 324–325
  treatment of 330
Municipal Solid Wastes (Management and
        Handling) Rules 6
MWD *see* medical waste disposal (MWD)
MWG *see* medical waste generation (MWG)
MWGRs *see* medical waste generation rates
        (MWGRs)
MWT *see* medical waste treatment (MWT)
MWT technologies (MWTTs) 133, 137–138
MWTTs *see* MWT technologies (MWTTs)

nano-bioremediation 40
nanofiltration (NF) 333, *336*
nanomaterials, carbon 59–60
nanoparticles
  metal oxide 59
  silver 60
  zero-valent metal 60
  ZnO 59
nanotechnology, wastewater treatment 58–60
nanotubes, carbon 59–60
nano zero-valent Zn powders 60
National Environmental Protection Management
        of Solid and Hazardous Waste S.I. 15
        155–156
National Waste Action Plan 7
National Waste Policy (2009) Australia 7
natural catastrophes 47
NESREA Act 2007 156–157
Nigeria
  agro-wastes management in 242–263
  control of hazardous waste in 154–157

SWOT analysis for waste to bioenergy
    approach in 257–258, **258**
SWOT analysis of agro-waste transformation
    to be biofuels 255–256, **256**
non-biodegradable waste 13
non-char-based adsorbents, principle and report
    of studies for 261–262
non-condensable fumes 33
non-essential amino acid 356
non-hazardous health care waste 74
nutrients, biological cycle of 24

OFMSWs *see* organic fractions of municipal
    solid waste (OFMSWs)
oil suffocates benthic organisms 48
organic fractions of municipal solid waste
    (OFMSWs) 324, 330
Organisation for Economic Cooperation and
    Development (OECD) 146
oxic biodegradation 35
oxidation, wastewater treatment 61–63
oxo-biodegradations 34–35
ozonation 61
ozone
    degradation 35
    disinfection 61, *62*
    filtration/disinfection 61

palm shells 110, *110*, 118
Park, C. 137
pathological waste 73
PCB *see* polychlorinated biphenyl (PCB)
peanut shells 110, *110*
performance evaluation 200
pesticides
    bioremediation of 233
    decontamination of 232–233
    environmental pollution by 232–233
petroleum hydrocarbons, decontamination of 233
pharmaceutical waste 74
photocatalysis 63
    homogeneous 63
photocatalytic degradation 59
photo-oxidative degradation 34
photovoltaic (PV) values 52
phycoremediation, wastewater treatment 50–53
physical recycling methods 30
phytoaccumulation 54
phytodegradation 55
phytoextraction 54
phytoremediation, wastewater treatment 53–57
phytostabilization 55
phytovolatilization 54–55
plant-based biowaste 110, *110*
plants
    ground dwelling 56
    for removal of contaminants 54, **55**

plasma technology, medical waste treatment
    (MWT) 131
plastic materials, microbial compositing of 36, *39*
plastics
    composting of 36–37
    microbial degradation of 35, *36*
    polymers 30
    recycling methods for 30
    thermoplastics 29
    thermosetting plastics 29–30
    as versatile waste 28
    waste management 28
    waste, pyrolyzed 139
3-ply face masks (3PFM), thermogravimetry
    (TGA) with 136
pollutants
    in air 227–228
    removal of 228
    surfacewater 225
polluter-pays principle 8
pollution
    in aquatic environment 225–226
    in terrestrial environment 226–227
polychlorinated biphenyl (PCB) 4
polyethene, breakdown and components 36, *37*
polymerization 29
polymers
    plastic 30
    synthetic 34
pond system, waste 53, *54*
PPE 81
prediction model 199
pressure sensor, water cooling arrangement for
    276–277
pressure steam sterilization technology 130
primary e-waste generators 168
primary recycling 30
Priyadarsini, M. 37
Producer Responsibility Organisation (PRO) 6
Pro-Engineer software 364
Pro-Mechanism module 364
PS 346
pyrolysis 30–31, *31*
    high-temperature 129
pyrolyzed plastic waste 139
pyrolyzer–Rankine cycle 133, 135

radioactive waste 73
rare-earth materials, recovering 171
R-CNN 164, 180
RCRA *see* Resource Conservation and Recovery
    Act (RCRA)
recycling 30–31
    e-waste 168–169
    e-waste management 9
    waste ceramic tiles 292
relative association 356

Resource Conservation and Recovery Act
        (RCRA) 3, 7, 147
reverse osmosis (RO), leachates treatment of
        leachate by using 333
reverse polymerization (RP) 83
rhizofiltration 56, 56
rice husk 262
Rio and Johannesburg Conferences 151–152
RP see reverse polymerization (RP)

sanitary landfill technology 127–128
SARS-CoV-2 see severe acute respiratory
        syndrome coronavirus 2
        (SARS-CoV-2)
SBR see sequencing batch reactor (SBR)
Scheirs, J. 31
sea grass 48
secondary recycling 30
sequencing batch reactor (SBR) 326
severe acute respiratory syndrome coronavirus 2
        (SARS-CoV-2) 80, 122, 130
sewage 49
        water 49
SFC see specific fuel consumption (SFC)
sharps waste 73
silver nanoparticles 60
simulated kitchen method 100
sludge residence time (SRT) 327
social awareness, e-waste management 9
soils, heavy metals in agricultural 227
solid medical waste 73–74
Solid Waste Disposal Act (SWDA) 6
sorbate anion 344
Special Criminal Provision Act 1988 154–155
specific fuel consumption (SFC) 98, 99, 112,
        114–115
split tensile strength 192, 218, **219–220**
split tensile test 214, **215**
Sri Lanka 303–319
        addressing challenge in Sri Lankan context
                305–306
        data collection approach 307
        e-waste as global crisis 304–305
        e-waste as lucrative business 305
        e-waste collection and management 312–314
        e-waste generation scenario study 306–311,
                308–309, *311*
        formal e-waste collectors and recyclers 316
        informal collectors role 314–315
        required actions 316–318
stakeholder involvement, in e-waste generation
        167–168, *168*
steam/autoclave disinfection (STD) 83
steel slag 206
        and E waste 209–210, **210**
sterilization technology, pressure steam 130
Stockholm Conference 151

stone dust, Kota 190
stone waste powder (SWP) 191–192
        effects of 190
strengths, weaknesses, opportunities and threats
        (SWOT) analysis 371
strength test, flexural 214, **215**
support vector machine (SVM) classifiers 180
surfacewater, pollutants 225
sustainable liquid waste treatment 48
sustainable technologies, for wastewater
        treatment 50–63
Swachh Bharat Mission 29
SWDA see Solid Waste Disposal Act (SWDA)
SWOT analysis; see also strengths, weaknesses,
        opportunities and threats (SWOT)
        analysis
        of agro-waste transformation to be biofuels
                255–256, **256**
        for approach deployment in biofuels 255–256,
                256
        of transforming agro-wastes into adsorbent
                262, **262**
        for waste to bioenergy approach in Nigeria
                257–258, **258**
SWP see stone waste powder (SWP)
synthetic polymers 34

tannery waste 206–209, **209**
TECH-ED 274
Technical/Second Industry Revolution 3
temperature measurement 276
tensile strength 192
tensile test, split 214, **215**
terrestrial environment, pollution in 226–227
test rig 274–275, **275**
        brake mean effective pressure (BMEP) 279
        brake power 277
        brake-specific fuel consumption 278
        brake thermal efficiency 279
        combustion graphs 285–287, *286, 287*
        efficiencies of **281**
        engine test rig combustion report of **282**
        friction power 278
        GDI 289, *289*
        heat carried away by engine cooling jacket
                water 280
        heat carried away by exhaust gas 280
        heat equivalent to BP 279
        heat input 279
        indicated mean effective pressure (IMEP) 279
        indicated power 277
        indicated specific fuel consumption 278
        indicated thermal efficiency 279
        measurements on **280**
        mechanic al efficiency 279 performance
                parameters of **281**
        swept volume 278

unaccounted heat 280
    volumetric efficiency 278
Texaco methods 33
thermal behaviors of cookstoves 112
thermal degradations 35
thermal efficiency (TE)
    of TMS and IMS *114,* 114–115, *115*
    of top-lead-up-draft 108
thermogravimetry (TGA)
    with 3-ply face masks (3PFM) 136
thermolysis 30, 40
thermophiles 58
thermophilic bio-digestion conditions 134
thermoplastics 29
thermosetting plastics 29–30
titanium dioxide 59
TMS 109, *109,* 116
    CO emission in 117
    thermal efficiency (TE) of 114–115
toner and ink cartridges 169
top bins 370–372, *371, 372,* 380, *381*
topsoil fertiliser 3
torque measurement load cell 275
torrefaction char 261
torrefaction technology 134
    medical waste treatment (MWT) 131–132
toxic by-products 34
toxicity, economic opportunity with 171–172
toxic metals
    in e-waste, adverse effects on humans **166**
toxic sludges 226
toxic substances 147
toxic waste 73, 146, 150
    global regimes combating transfrontier
        shipment of 151–157
    transshipment of 153
trade liquid waste 49
traditional biomass cookstoves 107
traditional devices, advantage of 170–171
traditional landfill
    dumping of agro-wastes on 249–250
    health and environmental implications using
        249–250
triethanolamine (TEA) 190, 191

ultrafiltration (UF) 332–333
UN Convention on the Law of the Sea 1982 152
Uniform Law 8–9
United States Environmental Protection Agency 6
upflow anaerobic sludge blanket (UASB) 326
urban solid waste (USW) storage 362
URES 376, *377*
UV radiation 62–63

vacuum water absorption test 292
value-added aromatics 135–136
versatile waste, plastic as 28

VFAs *see* volatile fatty acids (VFAs)
VF constructed wetlands (CWs), with HF CWs 52
volatile fatty acids (VFAs) 325
volatile matter content (VMC) 110–111
volatilization, ammonia 52
Volunteer in Technical Assistance (VITA) 96,
        97, 101

warming, agro-wastes for 250
waste; *see also* e-waste
    ceramic tiles, recycling 292
    collection equipment, mechanical design
        process 362
    disposal 47
    disposal incident, Khian Sea 3
    materials 206
    pond system *54*
    powder, granite 190
    shipment regulation 8
    toxic 146
    transforms, catalytic 35
waste collection system 361–364
    cost analysis 384, **384**
    overview of final solution 367
    prototype construction and testing 380–383,
        383
    safety device 369–370, **370**
    structure 367–368, *368*
    top bin 370–372
waste collection system, design considerations
    applicable regulations 365–366
    methodology 365
    objectives 364–365
    safety requisites 366–367
waste collection system, structural design
    safety device 377–379, *378, 379*
    structure 372–376, *373,* **374,** *374,* **375,** *377*
    top bin 380, *381*
waste framework directive (2018) 7
waste generation 145
    by 2050, estimation 14, *14*
    quantity and types 14, *14*
waste management 13
    integrated 361
waste materials in concrete
    cement 207, **207**
    class F fly ash 208
    coarse aggregate 207, **208**
    copper slag 208, **209, 210**
    fine aggregates 207, **207**
    steel slag and E waste 209–210
    tannery waste 208, **209**
    water 208, **208**
waste replacement concrete
    compression strength for 214
    flexural test for **215**
waste stabilization ponds (WSPs) 52–53

waste treatment
  liquid 49–50
  sustainable liquid 48
wastewater
  aeration of 57
  evacuation of 48
  remediation method of 48
wastewater treatment, sustainable technologies
        for 50–63
  eco-friendly adsorption technology 60–61
  membrane distillation bioreactor 58, *59*
  nanotechnology 58–60
  oxidation 61–63
  phycoremediation 50–53
  phytoremediation 53–57
water
  hydroxide in 50
  properties of 208, **208**
water absorption test
  on coarse aggregates 296, **296**
  vacuum 292
water boiling test (WBT) 96
  emission testing and emission parameters
        99–100, *100*
  evolution of **98**

experimental setup for 113
field-based assessment methods 101
hood method of 112
laboratory-based assessment method 97
limitations of 101
phases and performance parameters of
        97–99, *99*
protocol 108
water cooling arrangement, for pressure sensor
        276–277
water flow, flow sensors for measurement of 275
water remediation techniques 48
WBT *see* water boiling test (WBT)
WEEE *166*
wet concrete, test on **213**
World Health Organization (WHO) 122
WSPs *see* waste stabilization ponds (WSPs)
WSSD 152

XFEM *see* extended finite element modeling
        (XFEM)

zeolites 135
zero-valent metal nanoparticles 60
ZnO nanoparticles 59

For Product Safety Concerns and Information please contact our EU
representative GPSR@taylorandfrancis.com
Taylor & Francis Verlag GmbH, Kaufingerstraße 24, 80331 München, Germany

www.ingramcontent.com/pod-product-compliance
Lightning Source LLC
Chambersburg PA
CBHW060749220326
41598CB00022B/2375